Advanced Surface Coating Techniques for Modern Industrial Applications

Supriyo Roy
Haldia Institute of Technology, India

Goutam Kumar Bose
Haldia Institute of Technology, India

IGI Global
PUBLISHER of TIMELY KNOWLEDGE

A volume in the Advances in Chemical and Materials Engineering (ACME) Book Series

Published in the United States of America by
IGI Global
Engineering Science Reference (an imprint of IGI Global)
701 E. Chocolate Avenue
Hershey PA, USA 17033
Tel: 717-533-8845
Fax: 717-533-8661
E-mail: cust@igi-global.com
Web site: http://www.igi-global.com

Library of Congress Cataloging-in-Publication Data

Names: Roy, Supriyo, editor. | Bose, Goutam Kumar, 1970- editor.
Title: Advanced surface coating techniques for modern industrial applications / Supriyo Roy and Goutam Kumar Bose, editors.
Description: Hershey, PA : Engineering Science Reference, an imprint of IGI Global, [2021] | Includes bibliographical references and index. | Summary: "This book contains research on trends in surface coating materials, deposition techniques, properties of coated materials, and their applications in engineering and industrial fields"-- Provided by publisher.
Identifiers: LCCN 2020014005 (print) | LCCN 2020014006 (ebook) | ISBN 9781799848707 (hardcover) | ISBN 9781799857266 (paperback) | ISBN 9781799848714 (ebook)
Subjects: LCSH: Coating processes.
Classification: LCC TP165.C57 A384 2021 (print) | LCC TP165.C57 (ebook) | DDC 667/.9--dc23
LC record available at https://lccn.loc.gov/2020014005
LC ebook record available at https://lccn.loc.gov/2020014006

This book is published in the IGI Global book series Advances in Chemical and Materials Engineering (ACME) (ISSN: 2327-5448; eISSN: 2327-5456)

British Cataloguing in Publication Data
A Cataloguing in Publication record for this book is available from the British Library.

For electronic access to this publication, please contact: eresources@igi-global.com.

Advances in Chemical and Materials Engineering (ACME) Book Series

J. Paulo Davim
University of Aveiro, Portugal

ISSN:2327-5448
EISSN:2327-5456

Mission

The cross disciplinary approach of chemical and materials engineering is rapidly growing as it applies to the study of educational, scientific and industrial research activities by solving complex chemical problems using computational techniques and statistical methods.

The **Advances in Chemical and Materials Engineering (ACME) Book Series** provides research on the recent advances throughout computational and statistical methods of analysis and modeling. This series brings together collaboration between chemists, engineers, statisticians, and computer scientists and offers a wealth of knowledge and useful tools to academics, practitioners, and professionals through high quality publications.

Coverage

- Wear of Materials
- Multifuncional and Smart Materials
- Computational methods
- Electrochemical and Corrosion
- Materials to Renewable Energies
- Chemical Engineering
- Heat Treatments
- Thermo-Chemical Treatments
- Ductility and Crack-Resistance
- Fatigue and Creep

IGI Global is currently accepting manuscripts for publication within this series. To submit a proposal for a volume in this series, please contact our Acquisition Editors at Acquisitions@igi-global.com or visit: http://www.igi-global.com/publish/.

The Advances in Chemical and Materials Engineering (ACME) Book Series (ISSN 2327-5448) is published by IGI Global, 701 E. Chocolate Avenue, Hershey, PA 17033-1240, USA, www.igi-global.com. This series is composed of titles available for purchase individually; each title is edited to be contextually exclusive from any other title within the series. For pricing and ordering information please visit http://www.igi-global.com/book-series/advances-chemical-materials-engineering/73687. Postmaster: Send all address changes to above address.

Table of Contents

Detailed Table of Contents

The ever-growing interest in nanocoating and its enthralling protective properties makes it a very capable candidate for next generation protecting systems. The future of these special nanocoating markets will be expanding in different industries such as marine, building, and defense. The main purpose of coatings involves the use of thin films (nanoscale dimensions) that are applied to the surface of materials, which improve the material functionalities. Some of the improved functionalities include anti-corrosion, easy-to-clean (anti-graffiti), anti-icing, anti-fogging, anti-fouling, etc. Some of the common techniques used for nanocoating are chemical vapor phase deposition, physical vapor phase deposition, Sol-gel methods, electro-spark deposition, electrochemical deposition, and laser beam surface treatment. Commercial application of nanocoating nanotechnology includes self-cleaning coatings, depolluting coatings, ultraviolet (UV) light protective coatings, anticorrosion coatings, thermal resistance, anti-fouling coatings, and anti-graffiti coatings.

The goal of the chapter is to address the fundamental theory of thermal spraying and its modern industrial applications, in particular, those involving flame spray, HVOF, plasma spray, and cold spray processes. During the last 30 years, thousands of manuscripts and various book chapters have been published in the field of thermal spray, displaying the evolution of thermally sprayed coatings in many industrial applications. Thermal spray coatings are currently interesting for different modern applications including prosthesis, thermal barriers, electrochemical catalysis, electrochemical energy conversion devices, biofouling, and self-repairing surfaces. The chapter will explain the fundamental principles of the aforementioned thermal spraying processes and discuss the effect of different controlling parameters on the final properties of the produced coatings. This chapter will also explore current and future industrial applications of thermal spray coatings.

Thermal barrier coatings protect the substrate from thermal diffusion, oxidation, phase transformations, elastic deformation, plastic deformation, creep deformation, thermal expansion, thermal radiation. It allows parts and components of gas turbines to withstand high temperature upto1650 °C. Cylic oxidation behavior of alumina incorporated, lanthanum titanium aluminum oxide (LaTi2Al9O19), and yttria stabilized zirconia (YSZ), that is LTA/YSZ top ceramic layer coating, was investigated. Two coating combinations, L 100 having top LTA layer thickness of 100 μm and L 150 having top layer of LTA having thickness 150 μm, were tested for thermal cycles at the temperature of 1100°C. The performances of these coatings were compared with conventional YSZ coatings. Microstructure studies, EDX, and XRD analysis demonstrated the formation of mainly LTA, LaAlO3, Al2TiO5, Al2O3, and TiO2 at 1100°C in both coatings. But in L 150 coating, the rate of oxidation was found slower than L 100 coating. Annealed L 150A and L 100A specimens show cyclic oxidation life of 272th and 250th cycles, respectively.

One of the areas of research that continue to attract researchers worldwide is the development of thermal barrier coatings (TBCs) especially associated with the design of new ceramic topcoats with low thermal conductivity and a high coefficient of thermal expansion. The purpose of this chapter is to present the advances that have been achieved regarding ceramic topcoats in the last decades, making a historical journey that culminates with the contributions of this decade. The introduction of new crystalline structures and chemical compositions have opened the door to the real possibilities of replacing yttria-stabilized zirconia (YSZ) to ensure the optimal thermomechanical-chemical properties required by TBCs. Future research directions associated with this topic are also provided.

Distinct methods of depositing the coatings are available according to the intended area of application with a sole objective to protect the surface of structural component like boiler, boiler tubes, and heat exchangers from any mechanical or chemical damage. The main benefit is to minimize the manufacturing cost of a new component and also its fabrication. Thermal spraying is the commonly used technique to tailor the surface properties with a cost-effective approach. The different techniques of thermal spraying such as plasma, HVOF, cold spraying, etc. were investigated and introduced. Although each of these

This chapter attempts to make a review of electroless metal deposition over various non-conducting substrates like for its application in the field of medical research, electrical and electronics units, household aesthetics, automobile and textile industries. Electroless coating of metals over conducting substrates have been developed, critically reviewed, and proven its worth by showing excellent desired properties over the years. This review aims to discuss the techniques that have been applied by the researchers to overcome the difficulties of coating on these materials, their influence in their physical and mechanical properties, and their prospects of use in the industries. With the discussion of the underlying coating fundamentals and its historical backgrounds, the emphasis was put into the coating deposition with sensitizations and activations of various substrates, electroless baths, and the characteristically changed properties of the materials observed in the analysis.

Chapter 9
Tribological and Micro-Structural Characterization of Ni-Cu-P-W Coatings.................................. 209
Bal Mukund Mishra, Haldia Institute of Technology, India
Supriyo Roy, Haldia Institute of Technology, India
Goutam Kumar Bose, Haldia Institute of Technology, India

Ni-Cu-P-W coating was deposited by electroless method on mild steel substrate to study the crystallization and tribological behavior at different annealing temperatures. Energy dispersive x-ray (EDX) analysis, scanning electron microscopy (SEM), x-ray diffraction (XRD), and differential scanning calorimeter (DSC) were used to study the composition, surface morphology, phase behavior, and thermal behavior of the coating, respectively. Tribological study was conducted using Pin-on-Disc tribotester. EDX analysis confirms the presence of Ni, Cu, P, and W in the deposit. SEM image shows the surface is dense, smooth, and without any observable nodule. Some of the samples were heat treated to 300°C, 500°C, and 700°C for 1 hour to observe the crystallographic change by XRD. One sharp crystalline peak of Ni (111) is present in all condition, but the intensity increases rapidly with the heat treatment temperature. The phase transition temperature of this quaternary coating analyzed by DSC was 431.8°C.

Chapter 10
Role of Alloying Elements on Powder Metallurgy Steels and Spectroscopic Applications on Them 226
Sefa Celik, Istanbul University, Turkey
Nuray Bekoz Ullen, Istanbul University-Cerrahpasa, Turkey
Sevim Akyuz, Istanbul Kultur University, Turkey
Aysen E. Ozel, Istanbul University, Turkey

The powder metallurgy (PM) technique is suitable for mass production and is a well-established process for the low production costs of net-shaped products close to long series. The properties of the iron-based PM parts, such as strength, hardness, magnetic properties, impact, wear, and corrosion resistance, can be improved with adding various alloying elements. The desired surface performance can be achieved with various surface coating technologies. Recently, various coating techniques have been developed as discussed in the chapter. The alloying elements have a significant effect on the coating quality of the final product. Surface coating can be analyzed by examining the surface-coated powder metallurgy (PM) using infrared and Raman spectroscopy technique. This contribution focuses on the role of alloying elements on properties and coating technologies of powder metallurgy steels and Fourier transform infrared (FT-IR) and Raman spectroscopic applications on them.

In this chapter, an overview of welding as a technology for surface engineering is explored. According to literature, all types of welding techniques are appropriate for coating applications. However, as a result of process characteristics, some welding processes stand out. The most used welding techniques in the metal coating are arc welding (MIG, TIG, and PAW) and oxyacetylene welding. In the coating of metals using welding techniques, the coatings produced usually have a thickness that ranges between 1 and 6 millimeters. Applications of surface coating have been studied extensively. Such applications include aeronautic industry, sports, transport industries, petroleum and chemical industries, mining, food, and in the electronic industry. Plasma MIG welding is an advanced plasma process that combines the advantages of both MIG and plasma welding. Applications of plasma MIG welding in the surface coating of metals are expected to be explored extensively in the future.

Surface engineering includes augmentation of intrinsic properties of the boundary of the component, which isolates the continuum from surroundings known as the surface. Two main purposes of surface engineering encapsulate primarily the hardness of the surface for enhanced wear resistance and also to poise up with inter-surface frictional behavior. Today, there are many different surface engineering techniques available: starting from vacuum to atmospheric pressure, wet to dry, simple to sophisticated, and low-cost to high-cost to obtain the required purposeful distinctiveness of material. Most methods used today are dry and thus environmentally sound. This chapter describes various types of coatings over materials to get an overall idea of the technique keeping prime focus on graduate and undergraduate students.

Preface

Recent advancement of material science has revolutionized the application of surface coatings in almost every field of engineering. It is nevertheless to say that, surface coating or surface modification has shown a new path to materials engineers, mechanical engineers, tribologists, chemical & corrosion researchers, bio-medical engineers, bio-material scientists, aerospace researchers and many more. In many cases the material of main component is unable to sustain long life or protect itself from adverse operating environment. As example, bearing of gas turbine undergoes high temperature tribo-corrosion damage; artificial bio-implant material undergoes rigorous friction and wear within a typical corrosive environment in presence of synovial fluid; space ships or rockets are launched from marine (corrosive) environment and during passing through atmosphere, enormous heat generated in surface due to huge friction with air and there are many more examples. In different cases different material properties like, anti-friction and wear, anti-corrosive, thermal resistive, super hydrophobic etc. are required as per the operating conditions. It is quite difficult to match the bulk structural property and the surface property of any component by a single material. If those bulk components are made of such materials, possessing those properties, the cost will be very high. In those cases, practical solution is surface coating, which serves as a protective barrier to the bulk material from the adverse environment. During last decade, with enormous efforts, researchers and scientists have developed suitable materials to overcome those unfavourable operating conditions. Not only that, they have used advanced deposition techniques to enhance the adhesion and surface texturing of the coatings. Thus all these developments are needed to be explored through a common broad platform to the scientists and researchers associated with it for their easy reference.

The objective of this book, *Advanced Surface Coating Techniques for Modern Industrial Applications*, is to publish most of the advanced concurrent surface deposition techniques, their properties and probable fields of applications under one cover. This book could be a quick guide for researchers and scientists working in this field. Selection of suitable material for any component, which will be working in a specified environment and operating condition, is the first challenge to any design engineer. In most cases, the working load and working environment impose a demand of dissimilar material properties and that creates a difficult situation for material selection. Surface coating can solve this difficulty with ease. As environment effects mostly on surface, thus a protective coating can be provided over the surface of the load bearing material. By virtue of new technologies in the surface deposition techniques, both the metal and non-metal can be used either as substrate or as coating materials. This book covers a wide range of surface deposition techniques and coating materials, which are being used in concurrent industrial applications. Hopefully, this book will be able to put a positive impact not only to the related researchers and scientists but graduate students at higher education also.

Thus, to provide recent technological information in surface coating used in various industrial applications, this book encompasses the details of scientific and technological findings along with review of coating materials, coating technologies and coating applications. To satisfy the interest of a broad audience, this book includes twelve chapters contributed by respective subject experts. Brief descriptions about the chapters are as follows:

Chapter 1: This chapter describes details of nano coating technologies. Emphasise has been given on various nano-materials, coating technologies and properties of coated surface. Application of these coatings in various industries, especially in medical and textile industries has been explained. Special focus has been given to various types of environment friendly coatings.

Chapter 2: This chapter deals with detailed overview of thermal spray coatings. Being one of the most popular coating technologies of modern time, thermal spray coatings gained immense popularity in various engineering field. This chapter encompasses the description of coating procedures and importance of operating parameters, which govern the characteristics of various thermal spray coatings.

Chapter 3: This chapter is focused on the characterization of a thermal barrier coating deposited by thermal spray technique. Cyclic oxidation of combined LTA/YSZ and alumina coatings has been investigated and reported in terms of micro-structural and phase transition behavior.

Chapter 4: This chapter deals with advances in low thermal conductivity materials for thermal barrier coatings. Hopefully, a well defined history, why to use low thermal conductivity and solutions and recommendations for various industrial problems related to this coating will be helpful to the readers.

Chapter 5: This chapter briefly elaborates the progressive advancements made in the field of thermal spray coatings to combat hot corrosion. Along with corrosion, effect of composite coatings on micro-structural characteristics and quality of the surface developed has been discussed.

Chapter 6: A systematic review of cold gas dynamic spraying has been introduced in this chapter. This covers the methodology, the equipment, process parameters, material system, the metallurgical bonding aspects, advantage and disadvantages and finally the present status of this process.

Chapter 7: This chapter is focused towards the functionally graded materials developed by electroless coating and powder metallurgy. Emphasis is given to the fundamentals of functionally graded materials and their importance in material science, electroless coating process, powder metallurgy and most importantly, fabrication and characterization of functionally graded metal and ceramic powders by electroless coating method.

Chapter 8: This chapter deals with the general overview of electroless deposition of metals over non-conductive surfaces. This includes electroless nickel, copper, gold, silver coatings on different non-conductive surfaces like, ceramic powders, polymers, textiles etcetera. However, the characterization and properties of those coated surface has also been discussed.

Chapter 9: This chapter reports the experimental findings of electroless quaternary coatings. Detailed investigations on micro-structural, phase behavior, thermal behavior and tribological characteristics have been presented.

Chapter 10: This chapter explains the correlation between alloying elements and surface coating of powder metallurgy steel products. Emphasize has been given on the variety of alloying and surface coating techniques in powder metallurgy steel products and FTIR and Raman Spectroscopic Applications on them.

Chapter 11: This chapter presents an overview of surface engineering of materials through weld-based technologies.

Chapter 12: This chapter deals with various deposition methods by conventional and non-conventional coating techniques. A classification of different types of coatings used for industrial purpose is also proposed. Brief descriptions of nano-coatings and polymer coatings have been described. Finally, recent progress in coating technologies is reported.

Thus this book includes different surface coating/modification methods (thermal spray, cold gas spray, electroless deposition), coating materials (nano materials, role of alloying elements) and their applications (thermal barrier, corrosion resistance, friction and wear resistance, coating compatible for boiler tube applications) under single cover from the subject experts of related fields as a easy reference to the readers.

Chapter 1
Coating Materials:
Nano-Materials

Mintu Maan Dutta
Gauhati University, India

Mridusmita Goswami
Gauhati University, India

ABSTRACT

The ever-growing interest in nanocoating and its enthralling protective properties makes it a very capable candidate for next generation protecting systems. The future of these special nanocoating markets will be expanding in different industries such as marine, building, and defense. The main purpose of coatings involves the use of thin films (nanoscale dimensions) that are applied to the surface of materials, which improve the material functionalities. Some of the improved functionalities include anti-corrosion, easy-to-clean (anti-graffiti), anti-icing, anti-fogging, anti-fouling, etc. Some of the common techniques used for nanocoating are chemical vapor phase deposition, physical vapor phase deposition, Sol-gel methods, electro-spark deposition, electrochemical deposition, and laser beam surface treatment. Commercial application of nanocoating nanotechnology includes self-cleaning coatings, depolluting coatings, ultra-violet (UV) light protective coatings, anticorrosion coatings, thermal resistance, anti-fouling coatings, and anti-graffiti coatings.

INTRODUCTION

Nanotechnology is an emerging branch of science and technology that deals with the study of matter at dimensions roughly in the range of 1-100 nm (1 nm = 10^{-9} m). (Joshi & Adak, 2019) Materials in nanometer scale possess specific physical, chemical and biological properties that differ in fundamental and valuable ways from properties of bulk matter. (Kim, 2010) Potential application of nanotechnology includes its use in catalysis, scientific tools, textile industries, electronics, biomedical application, agriculture, environmental remediation etc. (Joshi & Adak, 2019) Recently, reports have been found which explains the use of nanotechnology in solving various problems that are persisting in various develop-

DOI: 10.4018/978-1-7998-4870-7.ch001

ing countries. One of the important industrial applications includes the use of nanostructured coating where nanotechnology has been extensively used is the coating. (Achanta *et al.*, 2011) It is therefore now possible to use nanomaterials and nanostructures in the coating industry for its application in various fields. (Dubbert *et al.*, 2014)

Nano-Coating

"A coating is defined as a coherent layer formed from a single or multiple application of a coating material to a substrate" (DIN EN ISO 4618; 2.52). According to the existing standard (DIN EN ISO 4618; 2.53) "a coating material is a material in liquid, paste or powder form which, when applied, forms a protective and decorative coating." (Dubbert et al., 2014)

Coating materials mainly consists of following four types of ingredients (Dubbert et al., 2014):

a) **Binders:** They help in forming a coherent film during drying and hardening.
b) **Pigments and extenders:** Pigments are usually used as colourant and mainly comprises of insoluble color particles. While extenders are used to modify the physical properties.
c) **Solvents:** Solvents are single liquids or mixture of liquids that dissolve other substances to form solutions without reacting with the substance to be coated.
d) **Additives:** Very small quantities of additives when added to the coating material considerably modify its properties. Some of the properties include its flow behaviour, surface tension, gloss, structure, UV and weather resistance. (Dubbert *et al.*, 2014)

The main purpose of coatings involves the protection and decoration of materials, and its use has been widened with increasing social and industrial development. (Makhlouf, 2011)

There are thousands of coating systems, ranging from simple systems to multilayers systems and complicated instruments. Most of these systems have an adverse effect on the environment. To combat with this problem continuous research and development work has been undergoing in coatings science and surface technology. The main driving forces are:

Increase in industry supplies at relatively low cost for high performance coatings;

To reduce the hazardous waste produced during coating processes (such as hexavalent chromate and volatile organic compounds (VOC)) which results in environmental pollution. (Makhlouf, 2011)

Nanocoating are one-phase solids structures that are applied onto the surface which dimension is lesser than 100 nm, thereby adding a specific property to the surface. These coatings are made up of layers or by combination of particles thinner than 100 nm, which enhances the surface properties or improves the material functionalities. Properties of nanocoating are greatly dependent on the constituent nanostructures. Small particle size of nanoparticles offers more surface area per unit mass that improves its application with low production costs. Enhanced surface properties encompass improved mechanical properties, wetting properties, thermal and chemical properties, electronic and magnetic properties, biological properties and optical properties. Improved properties may considerably alter the reactivity and capacity that is highly desirable depending upon the applications. (van Lente and van Til, 2008)

Some of the applications of nanocoatings include its use in medicine industry, electronics, food packaging, soft material industries (such as polymers, wood, textiles, leather, etc.), automotive industries, solar cells etc. (van Lente and van Til, 2008)

For nanocoating, the material is usually composed of at least two immiscible phases that are separated from one another by interface region. Material consists of a major component referred to as matrix in which fillers are dispersed. (Nguyen-Tri *et al.*, 2018) Classifications of nanocoating are done based on the type of nanostructured fillers and type of matrix where filler nanostructures are dispersed. (Nguyen-Tri *et al.*, 2018)

Some of the commonly used nanomaterials used in coatings are: Carbon black, Oxides (TiO_2, Fe_2O_3, Fe_3O_4, SiO_2,Cr_2O_3) (on mica flakes or SiO_2 spheres, with metal pigments), ZnO, Tungsten oxide (WO_3) (electrochromic), Polymer gel, specific organic-inorganic hybrid polymers, Oxide (synthetic amorphous silica), SiO_2, Al_2O_3), Zinc or aluminium coated with nano-TiO_2, nanoclay (like hydrotalcite $Mg_4Al_2(OH)_{12}CO_3xH_2O$). Its properties and uses are listed in table: 1 below: (Nanocoating, 2014; Dubbert *et al.*, 2014; Mathiazhagan and Joseph, 2011)

Table 1. Nanomaterials used in coating and their properties

Nanocoating materials	Properties and Uses
Titanium dioxide (TiO_2)	Possess photocatalytic properties: It is found in paints, coatings and inks; anti-corrosive; self-cleaning effect (with Anatase crystalline TiO_2); anti-microbial
Silicon dioxide (SiO_2)	Used in self-cleaning wall paints. Thermal resistant and fire resistant Scratch resistant
Iron oxide	UV protection, anti-corrosive
Zinc oxide (ZnO)	UV protection and anti-microbial
Silver (Ag)	Anti-microbial effect. Used as one of the important constituent in wall paintings in hospital and food processing industries.
Polymer	Monolayer adhesive films
Carbon (Graphene and carbon-nanotube)	Anti-oxidation and anti-corrosive (Bao *et al.*, 2019)
Copper oxide	Anti-microbial (Mishra *et al.*, 2013)
Indium/Antimony Tin Oxide	IR-absorption (Mishra *et al.*, 2013)

Nano-coating methods:

Several techniques are employed for the application of Nano-coating to the substrates. (Makhlouf, 2011) Few of the techniques includes chemical vapor phase deposition, physical vapor phase deposition, Sol-gel methods, electro-spark deposition, electrochemical deposition, laser beam surface treatment and plasma spray coating. (Khanna, 2008) (van Lente and van Til, 2008) Each of these methods has their own advantages and drawbacks.

Chemical Vapor Deposition (CVD)

This is one of the common techniques used to produce thin films in semiconductor industries. This method is used to coat most of the metallic and ceramic materials. (Makhlouf, 2011) In CVD process, the wafer (substrate to be coated) is exposed to one or more precursor gases. Commonly used precursor gases includes inert gases such as He, H_2 or N_2 are introduced into the reaction chamber and fragmented.

UV, laser or infrared light supplies the energy required during this process. (van Lente and van Til, 2008) This method also uses photo and plasma-assisted means to enhance the chemical reaction rates of the precursors. (P. O'Brien, 2001; Tavares *et al.*, 2008) The precursor gases passes over the surface of the heated wafer that reacts or decomposes on the substrate surface to produce the desired deposit thereby depositing a solid material from gaseous precursors. (Makhlouf, 2011) Commonly, volatile by-products are produced, which are removed along with unreacted precursor gas through the reaction chamber. Altering various experimental conditions such as substrate material, substrate temperature, composition of the reaction gas mixture, total pressure gas flows, etc., can give rise to material with wide range of physical, tribological, and chemical properties. (Carlsson and Martin, 2011)

There are multitude of CVD process, which includes atmospheric pressure CVD (APCVD), ultra-high vaccum CVD (UHVCVD), metal-organic CVD (MOCVD), low pressure CVD (LPCVD), laser CVD (LCVD), chemical beam epitaxy, microwave plasma-assisted CVD (MPCVD) and plasma-enhanced CVD (PECVD) etc. (Makhlouf, 2011; Carlsson and Martin, 2011)

Uses of CVD: (Carlsson and Martin, 2011)

a) Used in many thin film applications such as corrosion-resistant coatings, heat-resistant coatings and epitaxial layers for microelectronics.
b) Synthesis of high-temperature materials (such as tungsten, ceramics, etc.)
c) Synthesis of solar cells, high-temperature fiber composites and particles of well-defined sizes.
d) Synthesis of high Tc superconductors and carbon nanotubes.

Advantages of CVD: (Creighton and Ho, 2001)

a) CVD films are generally quite conformal.
b) Wide variety of materials can be deposited with very high purity.
c) High deposition rates
d) CVD often does not require high vacuum as required by PVD processes.

Disadvantages of CVD: (Creighton and Ho, 2001)

Precursors of CVD process need to be volatile at near-room temperatures that restrict the use of number of elements in the periodic table.

a) High toxicity [$Ni(CO)_4$], explosive (B_2H_6), or corrosive nature ($SiCl_4$) of CVD precursors.
b) Hazardous byproducts (such as CO, H_2 or HF) are produced from CVD reactions.
c) Films are deposited at high temperature.

Physical Vapor Deposition (PVD)

The technique is similar to CVD, only difference lies with the precursors (i.e. the material to be deposited on the substrate). In PVD, precursors are introduced to the reaction chamber in solid form, whereas in CVD, the precursors are introduced to the reaction chamber in gaseous form. (Makhlouf, 2011) The Solid coating material is thermally evaporated by heat or by bombardment with ions (sputtering). Thermal evaporation technique relies on vaporization of material by heating the material using suitable methods in vaccum. Whereas sputtering technique is a plasma assisted technique which requires the presence of a reactive gas such as argon, which reacts with the metal vapor that assists in depositing a thin layer onto the substrate surface. (Yousaf *et al.*, 2015) The figure below represents the flow diagram of PVD process where a solid material goes from a condensed phase to a vapor phase and then back to a thin film condensed phase.

Figure 1. Flow diagram of PVD process (ref. Wikipedia)

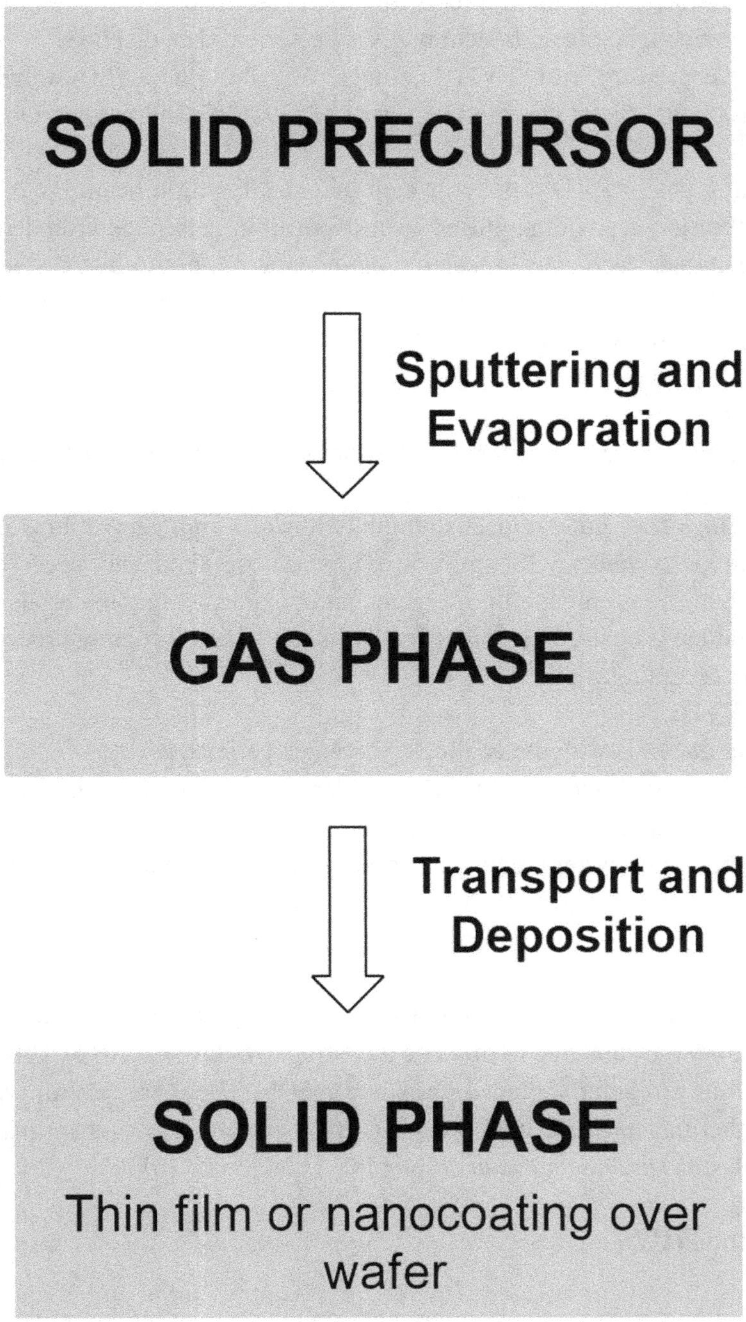

Uses of PVD:

a) Synthesis of semiconductor devices such as thin film solar panels, (Selvakumar and Barshiliaa, 2012)
b) Aluminized PET film for food packaging and balloons, (Hanlon, 1998)
c) Titanium nitride coated cutting tools for metalworking etc.
d) Optical coatings, i.e., in architecture, TV screens and data displays.
e) Automotive industry. (ref: PVD Coatings. Website. <https://www.pvd-coatings.co.uk/applications/>; 2014 [last accessed February 11, 2014); (van Lente and van Til, 2008)

There are several types of PVD process, which includes Electron beam PVD (EBPVD), cathodic arc deposition, evaporation deposition, pulsed laser deposition, pulsed electron deposition etc. Various characterization techniques such as calo tester, nanoindentation, pin to disc tester, scratch tester have been used to measure the physical properties of coating i.e. thickness, hardness, wear and friction test and coating adhesion test respectively.

Advantages of PVD:

a) PVD coatings are harder, as a result they are more corrosion resistant than coatings applied by the electroplating process.
b) Most coatings have good impact strength, withstand high temperature and scratch resistant. They have high durability that protective topcoats are almost not necessary.
c) This technique is promising for inorganic and organic coating materials.
d) This technique is environmentally friendly than traditional coating processes such as electroplating and painting.

Dis-advantages of PVD:

a) This technique fails with the coating of complex geometries.
b) Some PVD technologies operates at considerably high temperatures and vacuums, so it requires special attention.
c) High temperatures system requires a cooling water system to dissipate large heat loads.

Spray Coating

Spray coating is a technique in which molten or solid powder particles are accelerated in a supersonic gas jet and sprayed onto the substrate to produce a coating. (Makhlouf, 2011) This technique has been widely used for coating irregularly shaped glass and metals. There are several type of spray coating techniques such as thermal spraying, cold spraying, High-velocity oxy-fuel spraying (HVOF), Plasma spraying, vacuum plasma spraying and warm spraying. (Makhlouf, 2011)

Cold Spraying method (CS):

In the cold spraying technique, particles are accelerated to very high speed by supersonic gas jet (carrier gas), which is forced through a nozzle. Upon interaction with the substrate, particles undergo plastic deformation and bind to the surface thereby forming a thin film. This method is usually conducted at lower temperature i.e. at temperature lower than the melting points of the sprayed material to avoid oxidation and decomposition as well as phase transition. The coating obtained have high strength (>280 MPa), strong adhesion (>70 MPa), low porosity (<1%) and low oxygen concentration. Several factors that determines the quality of cold-sprayed coatings are as follows: (Makhlouf, 2011, Kuroda *et al.*, 2008)

a) Gas type: Such as Air, He, N_2
b) Gas pressure
c) Particle size
d) Gas temperature
e) Standoff distance: Distance between cold spray nozzle and the substrate.

There are two types of cold spray process: (M. Faizan-Ur-Rab*et al.*, 2015; Eric Irissou *et al.*, 2008)

Table 2. Two types of cold spray process and specifications

Conditions	High-pressure cold spraying (HPCS)	In low-pressure cold spraying (LPCS)
Gas	Nitrogen or Helium	Compressed gas
Pressure	> 1.5 MPa	0.5–1.0 MPa
Flow rate	> 2 m^3/min	0.5-2 m^3/min
heating power	18 Kw	3-5 kW
Application	Spraying pure metal powders within the range of 5–50 μm	Spraying a mechanical mixture of metal and ceramic powders

Uses of cold spraying:

a) Used for deposition of ceramic materials on metals, notably titanium dioxide for photocatalytic effects.
b) Used to repair machine parts in a matter of minutes. For example; Metal (nickel alloys) particles travel in a mixture of nitrogen and helium gas and progressively it get stack up on the damaged part to repair the damaged surface. A robot controls movement of the sprayer.

Advantages of cold spraying: (Makhlouf, 2011)

a) As cold spraying is a cold process, physical and chemical properties of the particles are retained. Thus, probability of using soft metals such as copper or aluminum as well as elements with high melting points such as tungsten, titanium and tungsten carbide and cobalt becomes possible.
b) Nitrogen or helium can be used as an inert carrier gas instead of oxygen. Thus, this method become useful for those materials that are sensitive to oxygen and gets oxidized. Example: aluminum, copper, titanium, and carbide composites (such as tungsten carbide).
c) High thermal and electrical conductivity of coatings.

Disadvantages of cold spraying: (Makhlouf, 2011)

a) Low deposition efficiency,
b) This method requires very fine powder to acquire higher velocities, which makes it industrially unattractive.
c) High cost of helium.

Thermal Spraying

In thermal spraying technique, melted or heated materials are sprayed onto the substrate to be coated. The precursor can be in powder or wire form. These particles are melted by either electrical (plasma or arc) or chemical means (combustion flame) and accelerated to high velocities. Combustion or electrical arc discharge is usually used as the source of energy for thermal spraying. This method can produce thick coating ranging from 20 μm to several mm over a large area at high deposition rate as compared to other coating processes.

Quality of coating as a result of thermal spraying is assessed by measuring bond strength, porosity, oxide content, macro and micro-hardness and surface roughness. Thus, the coating quality increases with increasing particle velocities.

Uses of Thermal Spraying: (Makhlouf, 2011)

a) Used to produce coatings on structural materials, which provides protection against high temperatures, corrosion, erosion and wear control.
b) When precursors are sprayed on substrates of various shapes and removed, freestanding parts in the form of plates, tubes, shells, etc. can be produced.
c) Used to create porous structures, suitable for bone ingrowth, as a coating for medical implants.

Advantages of Thermal Spraying: (Makhlouf, 2011)

a) This method has the ability to provide coatings ranging from 20 μm to a few mm thick for substrates with large surface areas.
b) Feeding powders of different coating materials such as ceramics, plastics and composites or pure metal can be sprayed over the substrate surface.

Dis-advantages of Thermal Spraying:

a) Thermal spray application is not compatible with complex geometries or surface blocked by other bodies.

Sol-Gel Method

The sol-gel process is a wet-chemical process in which monomer are converted into colloidal solution (or sol) that gradually forms an integrated network (gel) comprising both liquid phase and solid phase. Metal alkoxides and metal chlorides are generally used as precursors, which on hydrolysis and poly-condensation reactions transforms to colloids. The structure may vary from distinct colloidal particles to continuous chain-like polymer networks. Sol-gel method is suitable for the synthesis of high quality films with its thickness ranging from10 Å (10^{-10} m) to few micrometers (10^{-6} m). This approach has control over the product's chemical composition. Introduction of dopants, such as organic dyes and rare-earth elements in the sol considerably ends with the uniformly dispersed final product.

Application of Sol-gel method: (Makhlouf, 2011)

This method is used for the fabrication of both glassy and ceramic materials. It has varied application, whichincludes its use in

a) Optics, as antireflective coatings for glass lenses.

b) Electronics,
c) Thin film and fibers
d) (Bio)sensors,
e) Medicine (e.g., controlled drug release),
f) Reactive material and separation (e.g., chromatography) technology,
g) Protective coating

Advantages of Sol-gel method:

a) This method is used to cast suitable container with the desired shape.
b) This method is cost-effective and low-temperature technique that allows the fine control of the product's chemical composition.

Dis-advantages of Sol-gel method: (Makhlouf, 2011; Nguyen-Tri *et al.*, 2018)

a) Coating done using this method has poor adhesion performance,
b) Crack formation while drying in the film which are thicker than a critical value,
c) Absence of high-performance surface treatment based on environmentally acceptable salts.

Plasma Spray Coatings

Plasma spray coatings is a thermal process which make use of plasma, a relatively inert spraying medium (usually argon, combination of N_2 and H_2, combination of Ar and H_2 or He) powder (metals, ceramics etc.) and high particle velocities to produce a high quality coating onto a substrate. Plasma refers to the heating of gas at considerably high temperature so that the gas ionizes and becomes electrically conductive.

In this process, the coating material is introduced as a powder into a plasma beam. Plasma jet melts the material and scatter material particles across the substrate with minimum waste. The particles solidify rapidly with exceptional bond strength thereby producing a coating. This process can be used for spraying of almost any metallic or ceramic on to a large range of materials.

Uses of Plasma spray coatings:

a) Fretting wear: Spray of tungsten carbide/cobalt are used.
b) High temperature protection, thermal barrier coatings.
c) Wear resistance: Spray of chrome oxide ceramic are used.
d) Anti-galling: Spray of molybdenum alloys are used.
e) Erosion/abrasion resistance, corrosion protection.

Advantages of Plasma spray coatings:

a) Maximizes the lifespan of the sunstrate i.e. the become resistance to wear, corrosion, abrasion, heat.
b) This coating process is suitable for variety of materials such as any type of metal, glass, ceramic, plastics.
c) Coating materials such as metals, ceramics, alloys, cermets and refractory materials can be individually selected and introduced into plasma beam for spraying depending on the substrate. (ref: https://www.irsltd.co.uk/what-is-plasma-spraying)

Disadvantages of Plasma spray coatings: (https://www.irsltd.co.uk/what-is-plasma-spraying/; https://degradation2.wordpress.com/disadvantages-of-plasma-spray-coating-2/)

a) Plasma spraying equipment is expensive to buy, use and maintain.

b) Parts of plasma spraying equipment such as internal components of plasma gun requires regular replacements.

c) As like other thermal spraying process this process is not suitable for its use to coat internal bores of small diameters

Advantages of using coating on the surface of nanomaterials:

There are some unique properties of the nanocoatings, which make them highly applicable in the field of coating. Nanocoating is a surface engineering process by deposition of atomic or molecular films less than 100 nm thin. Hence, the thickness of nanocoating films obtained from these processes is quite different from bulk coating methods such as dipping and roll-over-knife etc. There are various techniques to achieve the atomistic or molecular deposition on the surface of a given substrate namely physical vapor deposition (PVD), chemical vapor deposition (CVD), electroplating, laser vaporization, plasma enhanced chemical vapor deposition (PECVD), etc.

There are several advantages of using coating on the surface of the nanomaterials. Some of those are scratch and fire and thermal resistance; controlling the nanomaterials from UV, IR or other radiations; hydrophobic or oil repellent nature of the coating materials and have anti- fouling, anti-corrosion and anti-microbial properties.

1. **Scratch resistance**:

Many external factors and mechanical loads may have a negative effect on the surfaces of substances, which can eventually destroy the surface structure. Scratches and cracks are harmful to the aesthetics and functionality of a product and degrade the value of the same. The anti-scratch coating neutralizes these problems by producing a surface with higher scratch resistance than the substrate.

Researchers nowadays have accepted the challenge of improving the scratch resistance of a coating without adversely affecting its other properties. If a greater number of cross-links are incorporated in the coating's binder, scratch resistance can be achieved. However highly cross linked films become hard and have poor impact resistance due to less flexibility. A less-cross linked film is naturally soft and will show better performance with regard to the properties like ant-fingerprint and impact resistance but it will definitely have less scratch and abrasion resistance. Hence, in order to obtain best possible scratch resistance, the correct combination of hardness as well as flexibility is required.

Silica is naturally very strong, hard and resistant against chemicals and variation. All these characteristics allow the nano-SiO_2 particles in coatings to improve various properties such as modulus, stiffens, toughness etc. resulting in a very good scratch resistance property.

Recent progress in the field of nanotechnology plays a very vital role in the development of scratch-resistant coatings. Gläsel *et al.* have shown the use of siloxane encapsulated SiO_2 nanoparticles to develop scratch resistant coatings (Gläsel *et al.* 2000). Conradi and coworkers separately put a 0.5 kg load on the surface of pure epoxy and 130 nm SiO_2/epoxy coating, which showed the indent in the epoxy coating were significantly large compared to those in the SiO_2/epoxy coating, and the hardness was found to be increased by almost 40% (Conradi *et al.*, 2013). Other researchers (Thomas and Stoks 2001) have also reported coatings with good abrasion and scratch resistant properties.

2. **Anticorrosion coatings**

Corrosion generally refers to metals, even if nonmetallic substrates such as plastics, concrete or wood also depreciate in the environment. Due to corrosion, huge industrial losses with a depletion of our natural resources always occur. The coatings offer anti- corrosion properties following several mechanisms available. They are namely sacrificial means, barrier effect and inhibition.

Corrosion of metallic substrates is protected by nanocoating systems via anodically active metal effect or galvanic effect, barrier effect, inhibitive effect etc. (Müler *et al.*, 2017).

One of the examples of galvanic effects is the Zn based nanocoating protective mechanism, in which it sacrificially corrodes by the oxygen and moisture present in the coating surface and forms a corrosion product such as ZnO. Shen and his coworkers reported that corrosion resistance properties of 316L stainless steel could be improved significantly by covering its surface with uniform TiO_2 nanoparticles films, the corrosion resistance increasing over three orders of magnitudes (Shen *et al.*, 2005). Graphene and carbon nanotubes are also used in coating for anti-corrosion application. Yu *et al.* introduced modified graphene oxide on the metal surface, which showed better corrosion resistance than the bare metal (Yu *et al.*, 2016).

Polymeric coatings are also applied to surface of metallic substrates to provide a barrier against corrosive species. Pigments having lamellar or plate-like shapes are introduced to polymeric coatings to increases the length of the diffusion paths for the corrosive species. The pigments in the coating are parallel to the surface, and they are highly compatible with the matrix resin because of which they can provide a good barrier effect. Montmorillonite clay which is layered clay can be introduced into organic resin systems to increase the barrier effect towards oxygen and water molecules, in that way enhancing the anticorrosive property of the coating

Aluminium oxide is generally accountable for the resistance of metallic aluminium to weathering. Metallic aluminium reacts very quickly with atmospheric oxygen, and a thin layer of aluminium oxide forms on the exposed aluminium surface (Zaman 1997). Formation of this passivation layer protects the metal from further oxidation. A number of alloys, such as aluminium bronzes, exploit this property by including a proportion of aluminium in the alloy to enhance corrosion resistance. The aluminium oxide generated by anodizing is typically amorphous, but discharge assisted oxidation processes such as plasma electrolytic oxidation result in a significant proportion of crystalline aluminium oxide in the coating, enhancing its hardness. (Chinglenthoiba *et al.*, 2017)

Chen and his coworkers used as-grown graphene for the protection of the surface of metallic growth substrate of Cu or Cu/Ni alloy away from air oxidation (Chen *et al.*, 2011). The covered graphene resulted in a passivation of metal underneath, which offered an inserted block between the reactants and the protected metal. The graphene-based nano-coating was very effective as it extended the materials thermal stability and corrosion-resistance. Yu and coworkers introduced metronidazole modified graphene oxide on the metal surface to obtain better corrosion resistance than the bare metal (Yu *et al.*, 2016). Kwon *et al.* also used graphene and metal-chloride-doped graphene in which the evaporation of chlorine was almost blocked, and was able to maintain the doped state even after thermal annealing (Kwon and Kim, 2014).

Conducting polymers have been used as host matrices in various composite films due to their electrochemical properties and their mixed ionic/electronic conductivity properties (Rout *et al.*, 2003). Depending upon the application, the morphology, conductivity and different physical properties of the conducting polymers can be modified by incorporating organic or inorganic particles into them. Yeh *et al.* reported the synthesis of nanocomposites with polypyrrole and montmorillonite clay, which showed better corrosion protection compared to bare polypyrrole (Yeh *et al.*, 2003). In other reports, it was found that polypyrrole nanocomposites with oxides, especially with Fe_3O_4 have prospects for use in corrosion protection of iron (Gracia *et al.*, 2002).

3. Self-cleaning coatings: super hydrophobicity property of nanocoatings

Leaves of the lotus plant *(Nelumbo nucifera)* are an enthralling example of nanostructure in nature. A dual texture in nano- and micro-scale is responsible for making the waxy, hydrophobic surface of the lotus leave superhydrophobic (i.e. the contact angle with water is greater than 150°). This super hydrophobicity makes water droplets roll on the surface, thereby cleaning dirt particles that might be collected. This phenomenon helps in keeping the leaves of lotus flowers clean themselves. This is called **lotus effect**. It is inspired by this property of lotus leaves; researchers applied it to the surface of substrates so that the surfaces could acquire the self-cleaning property.

Due to their excellent water-repellent and self-cleaning properties, superhydrophobic surfaces have gained immense attention in fundamental research as well as practical applications. Self-cleaning is a situation when water drops of a set volume/mass can go through sliding, rolling, or both when the self-cleaning substrate is twisted beyond the critical angle. The water droplets slide down the surface leaving no trail because of the significant decrease in surface tension between water droplets and self-cleaning surface. Under this circumstance, dust particles are easily washed off when water droplets roll/slide down the self-cleaning surface. Self-cleaning surfaces are generally manufactured in one of the following ways: through the creation of hierarchical nanostructures on the hydrophobic surface or the chemical modification of the nanostructured surface with molecules having low surface free energies. Usually self-cleaning coatings utilize complicated deposition methods, such as chemical vapor deposition, layer-by-layer assembly and micro patterning, which may not be suited for a deposition to large surfaces.

Introduction of nano-TiO_2 into the coating systems can lead the nanocoating with enhancement in protection and can even endow it with self-cleaning properties (Bao *et al.* 2019). TiO_2 exposed under UV exhibits photo catalytically induced super hydrophilicity that converts the hydrophobic character of surface to hydrophilic and forms the uniform water film, which prevents the adhesion of inorganic or organic components on its surface, with retained cleanliness. The TiO_2 thin films may be deployed on the surface of various substrates, such as glass, ceramics, metals, textiles, cement, bricks or fibers to provide layer that exhibits self-sterilization and self-cleaning properties, when it is exposed to the light.

Goffredo and coworkers (Goffredo and Munafo, 2015) had applied titanium dioxide (TiO_2) for the conservation of architectural, monumental or archaeological stone surfaces. In their study, TiO_2 nanoparticles were dispersed in an aqueous colloidal suspension and applied directly on a light-colored limestone (travertine), by spray coating in order to obtain a nanometric film on stone samples. In another study, Ferrari and his coworkers evaluated the environmental sustainability of the nano-TiO_2-based coatings over the surfaces of architectural stones. In order to appraise the balance between the potential risks due to the utilization of engineered nanomaterials in the life cycle phases and the benefits of the self-cleaning and de-polluting properties of the Titania coating, they had adopted a safe behaviour in all life cycle steps (Ferrari *et al.* 2015).

Table 3 below represents some of the superhydrophobic textiles and its evaluations that have shown promising results in recent years.

Table 3. Types of textiles, its synthesis and evaluation of properties

Type of textiles	Synthesis method	Evaluation	Ref.
ZnO nanorod array film on cotton fabrics	Wet chemical route; hydrophobic modification with *n*-dodecyltrimethoxy silane (DTMS)	WCA=161° for 8μL water droplet; CAH=9° for 8μL water droplet	Xu and Cai, 2008
Superhydrophobic textiles using fluoro-free organosilanes applied to various textiles (e.g., polyester, cotton and wool)	Dip-coating method using (R-polymer) of alkyltriethoxylsilane ($RSi(OEt)_3$) and $RSi(OEt)_3$-modified silica nanoparticles (R-silica)	Excellent mechanical (e.g., abrasion, laundering and scratching), chemical (e.g., acid, base and organic solvents) and environmental (e.g., UV irradiation and outdoor) stability	Wu *et al.*, 2013
11 textile fabrics made from natural and man-made fibers	One-step gas phase coating with polymethylsilsesquioxane nanofilaments	WSA; textile friction analyzer (TFA)	Zimmermann *et al.*, 2008
Superhydrophobic cotton fabric	Electrostatic layer-by-layer assembly of polyelectrolyte/silica nanoparticle followed by fluoroalkylsilane treatment	static contact angle range between 151° and 157°; contact angle hysteresis for 1-3 layer > 45°; for 5-7 layer < 10°.	Zhao *et al.*, 2010
Superhydrophobic cotton textiles	Ag NPs with Hexadecyltri methoxysilane (HDTMS)	High antibacterial activity against the gram-negative bacteria, Escherichia coli	Xue *et al.*, 2012
Superhydrophobic surfaces on cotton textiles	Amino- and epoxy-functionalized silica nanoparticles on epoxy-functionalized cotton textiles followed by hydrophobization with stearic acid, 1H,1H,2H,2H-perfluorodecyltrichlorosilane	WCA=170° for a 5μL droplet; water roll-off angle < 5°	Xue *et al.*, 2009
Superhydrophobic cotton fabrics	Sol-gel coating method of TiO_2 followed by hydrophobization with stearic acid, 1*H*, 1*H*, 2*H*, 2*H*-perfluorodecyltrichlorosilane	UV-shielding properties; water static CAs range from 151.5° to 163.5° for a 5 μl droplet	Xue *et al.*, 2008
Superhydrophobic behaviour of cellulose nanofiber-coated microfibers	Electrospinning method, fibers are plasma modified with trifluoromethane	WCA >150°	Thorvaldsson *et al.*, 2012
Surface modifications of polymer surfaces such as poly (ethylene terephthalate) (PET), polyetherimide (PEI), polyetheretherketone (PEEK), aramides, or polyethylene	UV-light treatment; Fluorinated Carboxylates with Zirconium (IV) acetylacetonate Coupling Agents	WCA up to 160°	Textor *et al.*, 2004
Surface treatment of cotton fabrics	Ultra sonication, Dip-coating method silica nanoparticles, mixed silane (n-octyltrimethoxysilane (OTMS), n-decyltriethoxy silane (DTES), bis(triethoxysilyl) ethane (BTEOSE), tetramethoxysilane (TMOS), and tetraethoxysilane (TEOS))	Durability test, crocking and laundering tests. When fabrics was treated with Aerosil® 90-containing OTMS and BTEOSE solutions highest contact angle 139.1° was obtained.	Roe and Zhang, 2009
Functionalization of cotton	Dip-coating method; Treated with carbon nanotubes (CNTs)	Flame retardancy, improved UV-blocking and properties, WAC > 150°.	Liu *et al.*, 2008
Fabrication on cotton substrates	Dipping-drying-curing method; fabricated with pristine carbon nanotubes (CNTs) and surface modified carbon nanotubes (PBA-g-CNTs) PBA: poly(butylacrylate)	WCA > 150°.	Liu *et al.*, 2007
Superamphiphilic Janus Fabric	Electrospinning method using polyacrylonitrile (PAN)	WCA = 151.2°.	Lim *et al.*, 2010
Chemical change to fabric surface such as silk and cotton	Plasma surface-treatment in hexafluoropropene (C_3F_6) atmosphere	WCA = 119.4±2.08 for silk and from 08 to 124.2±0.98 for cotton.	Li and Jinjin, 2007
Surface modification of silk fabric	Magnetron sputter coating method was used to deposit polytetrafluoroethylene (PTFE)	WCA = 152 (±3)° CA hysteresis = 5°	Huang *et al.*, 2007
Modifications of thermoactive membrane textile materials	Plasma deposition method using hexamethyldisiloxane (HMDSO)	WCA = 145°	Twardowski *et al.*, 2012
Impart functionality to cotton/ polyester blend	Plasma deposition method where atmospheric glow plasma was used. Fluorocarbon based monomers such as 1,1,2,2-tetrahydro perfluorodecyl acrylate (THPFDA) and 1,1,2,2-tetrahydroperfluorododecyl acrylate (THPFDDA) were deposited, thereafter second plasma exposure with antimicrobial agent, diallyldimethylammonium chloride (DADMAC), a quaternary ammonium salt.	Anti-microbial activity	Davis *et al.*, 2011
Modifications of glass slides	Chemical vapor deposition method by using silicone Nano filaments		Artus *et al.*, 2006
Modifications of cotton fabric surfaces	Microwave plasma treatment using oleic acid as hydrophobic agent	WCA > 150°	Cabrales and Abidi 2012

4. UV radiation protecting coatings

Harmful UV radiations can have a negative effect on any substrate and its light sensitive ingredients and lead to permanent damage. UV degradation is a significant problem for automobile and aircraft industries. Protection of wood substrates from UV degradation is also very important.

UV barrier layers can effectively protect the surface against UV light. This is achieved by reducing the UV transmission to a minimum through a reflective or absorbent layer in the UV range. Coatings always play a significant role in controlling the adverse effects of electromagnetic radiation (UV, Visible, IR, other) on many surfaces. UV protection categories of textiles based on UPF range and rating is shown in table 4 below. (Akgun, 2010)

Table 4. Ultraviolet protection factor (UPF) range and rating

Protection category	UPF range	UPF Rating
Excellent	40-50, 50+	40, 45, 50, 50+
Very good	25-39	25, 30 35
Good	15-24	15, 20

The metal oxide nanoparticles for example nano-TiO_2, nano-ZnO and nano-ceria (CeO_2) etc. can efficiently block UV radiations resulting a better and sustainable performance compared to other UV absorbers (Duan *et al.* 2011, Yang *et al.* 2004, Kathirvelu *et al.* 2009). Their UV blocking property comes from (i) high reflective index causing higher reflection and/or scattering and (ii) absorption of UV radiation because of their semi-conductive nature.

Nano-ZnO used in the coating can also improve the UV resistance. When alkyd-based water borne coating was exposed under UV-radiation for a period of 110 h, there were several cracks on the surface, whereas for the ZnO-based nanocoating in the same condition; no cracks were observed (Dhoke *et al.*, 2009). Nano-ZnO particles of particle size in between 20 and 40 nm, can scatter UV radiation between 200 and 400 nm, thereby providing excellent UV blocking property to the coating system without interacting with the UV radiation. The size of the nanoparticles plays an important role in enhancing the UV resistance.

The mechanism of TiO_2-based nanocoating for the protection from UV radiation is that electron and hole are generated in the coating once nano-TiO_2 was exposed to UV-irradiation. In this process, the electrons were transferred to the metal substrate making its electrode potential more negative than its corrosion potential, meanwhile, TiO_2 is not consumed.

5. Bio textile applications:

When we apply nano-coatings to textile or bio textile items, like clothing or furniture, the surface will be protected against contamination and staining, which will definitely extend the life of the product, where it is applied. Nano-coatings are nowadays immensely used in bio textile applications, which include bio filtration materials, medical outfits, medical fabrics, textile implants, textile substrates for cell growth, and many other textile products for medical systems (Chu 2002, Anand *et al.* 2005, Su *et al.* 2004).

Nano-coatings are formed on textile surfaces using multiple ways (e.g. by thin films, nanocapsules, and nanoparticles). Usually they have dimensions of different range varying from several tens of nanometers up to hundreds of nanometers. Carbon nanofibers (Lynam *et al.* 2007, Yang *et al.* 2007) possessing high mechanical strength, lightweight, high electrical conductivity, thermal and chemical resistance, may also be used in textile composites and coatings for biomedical applications. Surface properties such as liquid repellence, stain resistance, antimicrobial activity, odor control and delivery of biologically active agents may also be added to bio textiles by the application of nano-coatings.

Plasma coatings are particularly appropriate for textile-made biomaterials to improve their biocompatibility or improve (or reduce) biological activity on the surface. Different plasma technologies may be used to apply metallic nanoparticles onto textile surfaces. They include plasma polymerization of organometallics, chemical vapor deposition and plasma sputtering (physical vapor deposition or PVD) which is a low-pressure process that allows the direct application of metallic nanoparticle coatings onto fabrics. Plasma surface treatments have multiple medical applications. Plasma activation confers hydrophilic properties to woven or nonwoven textiles, allowing them to be used as blood filters or filter-

Table 5. Types of textiles, its synthesis and evaluation of properties

Type of textiles	Synthesis method	Evaluation	Ref.
Ultraviolet properties of cotton fabrics dyed with natural colorants.	Dyed with natural plant colorants madder (*Rubia tinctorum*) and indigo (*Indigofera tinctoria*) and the natural colorant of insect origin cochineal (*Dactylopius coccus*).	Ultraviolet protection factor (UPF)	Sarkar, 2004
Improvement of the UPF values of cotton fabrics	UV-absorber based on the oxalanilide.	Ultraviolet protection factor (UPF)	Riva *et al.*, 2006
Modifications of linen fabrics	Mordanting and dyeing Using Henna and copper sulfate.	Thermal stability, antioxidant activity, antibacterial action and UV protection.	Yadav *et al.*, 2019
Modifications of pure cotton textiles	Sol-gel method using zinc oxide nanosol. Nanosol is prepared using TEOS, ethanol, hydrofluoric acid, deionized water and zinc acetate.	Ultraviolet protection factor (UPF) >50+	Vihodceva and Kukle, 2013
Modifications of cotton fabric	Pad-dry, pad-dry-hydrothermal and pad-dry-solvothermal using TiO_2 sol	Ultraviolet protection factor (UPF)	Mishra *et al.*, 2019
Functionalization of Polyester Fabric	Atmospheric pressure glow discharge technique has been use to load silver chloride and titanium dioxide	antibacterial activity, UV-protection	El-Zairy and Morgan, 2015
Modifications of woven fabrics such as cotton/polyester, nylon	Different metal including copper and aluminium yarns are added into the fabric structure.	UV-protection	Ghane and Ghorbani, 2016
Modifications of cotton fabrics	In-situ chemical polymerization method was deployed for the synthesis of ppy-ZnO and ppy-ZnO-CNT composites. ppy: polypyrrole	Flame retardant; UV-protection	Yazhinia and Prabu, 2015
Modifications of cotton fabrics	Natural dyeing method using 3-chloro-2-hydroxy propyl trimethylammonium chloride, followed by dyeing with Henna extract	Colour strength, fastness properties, UV-blocking and tensile strength	Alebeida *et al.*, 2015

ing membranes for various applications. Plasma coatings are particularly considered for textile-based biomaterials, as biomaterials may involve plasma polymerization to deposit coatings, thus making their surfaces more or less bioactive. Medical applications could include sterile fabrics for use in bandages, implants, blood bags or surgical garments. Plasma-coated, pre-sterilized textiles could result in energy savings, because the fabric would not have to be sterilized with high temperature and pressure before use.

Table 5 below represents some of the different textiles that have been modified and shown promising results in recent years.

6. Thermal resistance and fire resistance property

In our day today life we need thermal resistant coating in many of our household or other utensils such as nonstick cookware, boilers and barbecues etc. Although fluorine and silicon-based products are usually used in nanocoatings to achieve a high thermal resistance, fluorinated coatings are not considered suitable for high-temperature applications because they were degraded above ~300°C and they produce toxic byproducts. Hence, nowadays silicon containing coatings are in maximum demand as they offer better thermal resistance due to the high energy required to cleave silicon bonds compared to carbon bonds in analogous molecules (Mathiazhagan and Joseph 2011). Silicon based coatings can resist the temperature upto 1000 °C. Some other examples of thermal resistant coatings are also available which includes the combination of titanium esters and aluminium flakes incorporated into binders and the composite can resist the temperature up to 400 °C (Gangotri and Chaware 2004).

The use of flammable materials in different household or other stuffs is increasing day-by-day resulting in a significantly increased threat to the safety of human life and to the cost of damage repairing resulted from fire. In 2013, the Unites states had 1.24 million fire cases where 3240 civilian lost their lives, 15,925 were injured and also there was a huge property lost (according to the report by National Fire Protection Association) (Karter 2014). Therefore, there is a huge requirement of fire resistant coatings to overcome this type of problems. Although it is not possible to get protection by these coatings for a very long period, but it can certainly delay the spread of fire allowing enough time to take the required safety measures.

Halogenated molecules comprise the most varied class of fire retardants. Amongst them, the most efficient materials, which find their frequent use, are those containing chlorine or bromine. These are very successfully used for many engineering plastics and epoxy resins (Litzenburger 2000). Owing to the environmental risks associated with the use of halogenated fire resistant, phosphorus containing materials nowadays find their immense application in this field (Schartel 2010). Inorganic hydroxides, commonly containing aluminum or magnesium have been successfully used as fire resistant as they release water at higher temperature (Bras *et al.* 2005).

7. Anti-graffiti coating

Anti-graffiti is a property of the nano coating which does not allow graffiti, stickers, paints or any type of markers to bond to the dry film (Dewan 2009). This is a permanent coating. Anti-graffiti coatings can be applied by following two ways. In the first classification, coatings are applied to a surface and after the graffiti is applied, it is removed from the surface. Thus, surface under the coating will remain clean. In another way a permanent coating is applied which prevents the graffiti from sticking to the surface.

Stable nanocoating always offers a protective surface with which spray paint cannot form bond. After the surface has been damaged, the paint can be removed by using a simple solvent (toluene) and some manual labor. The underlying surface and the protective coating will remain undamaged.

There are some important properties of anti-graffiti coatings as follows:

a) Sufficient adherence without damage to surface
b) Hydrophobicity
c) Environmentally friendly composition
d) Resistance to UV radiation
e) Good Cleaning Efficiency

8. **Antifouling coatings**

Antifouling coatings are very important in marine industries. It can effectively help in maintaining the speed and energy efficiency of ships by preventing fouling. Fouling is a natural process caused by marine organisms such as microbial slimes, microalgae and barnacles etc. It is a worldwide problem in marine system where unwanted growth or deposition of such organisms takes place. Because of this process, fuel consumption of the ships as well as the cost of maintenance of the different equipment's increases.

Two types of coatings biocidal and nonbiocidal coatings are used for avoiding fouling. In a traditional approach, fouling has been controlled by antifouling paint with biocides. One of the main biocide effectively used for antifouling was tributyl tin (TBT) which does not find its application nowadays as their widespread use has caused severe pollution in the marine ecosystem (Chen *et al.* 2008). Recently, micro-encapsulation technology is used for the production of a wide number of antifouling products (Windsor 2005). For the nonbiocidal approach of antifouling coatings, polymers that have lower surface energy are used in order to avoid the adhesion of marine organisms and silicone elastomers find their extensive use for this purpose.

9. **Anti-mist coatings:**

We can get rid of foggy windows and lenses with the help of nanocoatings (Mishra *et al.* 2013). Researchers have been trying to develop anti-fog technology for many years. Anti-fog sprays are available that help to reduce fogging on the inside of car windows, but the process must be repeated in regular interval to get effective result. Glass containing titanium dioxide also shows potential for decreased fogging, but the technique only works in the existence of ultraviolet (UV) light. Recently, a typical polymer coating made of silica nanoparticles have been reported, which can create surfaces that never fog. These coatings are transparent and hence can be applied to eyeglasses, camera lenses, bathroom mirrors etc. When thousands of tiny water droplets condense on glass or other surfaces, the surfaces become translucent or foggy. This frequently happens when a cold surface suddenly comes into contact with a warm, moist air. The nanocoating prevents this process through its super-hydrophilic nature. The nanoparticles in the coating effectively attract the water droplets resulting the droplets to flatten and merge into a uniform, transparent sheet rather than forming numerous individual light-scattering spheres. "Basically, this nanocoating powers the water that strike the surfaces to build up a sustained sheeting effect, and that prevents fogging".

10. Antibacterial coatings

Microorganisms such as bacteria, fungi or viruses represent potential threats for many applications like biosensors, biomedical implants and devices, food packaging, industrial as well as different marine equipment's. In order to deal with this problem, the researchers synthesize surface coating with coating materials having antibacterial properties. There are different types of commercially available organic or inorganic biocides which demonstrate a wide variety of biocidal and biostatic mechanisms. Organic biocides contain polymers, tertiary alkyl amines and different organic acids (Sauvet *et al.*, 2003). Some examples of inorganic biocides are silver, zinc oxide (ZnO), copper oxide (CuO), TiO_2, and selenium etc. (Wagener, 2005). Ivanovaa and Philipchenko (Ivanovaa 2012) build up a effortless method to design superhydrophobic anti-bacteria chitosan-based coatings. This anti-bacterial functionality of coating is supported by using chitosan-based nanoparticles. Biocides containing microcapsules have also been developed in order to increase the endurance and effectiveness of antibacterial coatings (Xu, 2003).

Heinonen *et al.* presented a superhydrophobic silver-containing coating, which was prepared by following a sol-gel technique (Heinonen *et al.*, 2014). In the process, silver nanoparticles were reduced on the surface of the prepared γ- alumina layer by Tollens process then the composite coating was further modified to obtain a low surface energy.

11. Transparency and anti-reflection

Transparent nanocoating has found its extensive applications mainly in windows, eyeglasses, camera lenses and solar cell system, etc. (Rahmawan *et al.*, 2013) Transparency will be reduced with surface roughness increases because the light scattering (Yu *et al.*, 2014). Si/SiO_2 based nanocoating shows excellent optical property and is easy to control coating thickness and roughness. Owing to this advantage, Si/SiO_2 based material received a lot of attention in this research field. Xu and coworkers (Xu *et al.*, 2010) used a liquid polysiloxane (PSO) for the successful fabrication of highly transparent and durable superhydrophobic hybrid nanoporous coatings through a simple solidification-induced phase separation method. Wang *et al.* reported the production of highly transparent, and stable organic-inorganic composite nanocoating by following a well-designed sol-gel dip-coating method (Wang *et al.*, 2014). In another report, Her and coworkers prepared a superhydrophobic coating on poly (methyl methacrylate) (PMMA) using a dry etching with CF_4 plasma and a subsequent hydrolysis process with water immersion (Her *et al.*, 2013).

Reflection of light on the surface generally causes harms like energy loss and imaging interference. (Xi *et al.*, 2007) Hence, anti-reflection is extremely important to our daily life. Anti-reflective micro-nanostructures that can lead to a successful refractive index gradient between air and the substrate must have a period smaller than the irradiated light wavelength. Based on these theories, a number of considerable micro-nanostructures have been designed (Xu and He, 2012). Along with transparency and self-cleaning, the application of transparent nanocoating on solar cells has also found its useful application as it could improve the optical absorption as well as remove the pollutants from the surface (Zhu *et al.*, 2010).

12. Application of Nanocoating technology in industries:

Use of nanocoating technology can help to solve common current industrial problems.

a) Medical Applications:

A rapid increase in the use of nanocoatings towards various medical applications such as orthopedic and dental implants has been observed. This has improved the ***osseointegration***. *Osseointegration* refers to the structural and functional connection between living bone and the surface of artificial implant. (Mavrogenis *et al.*, 2009; Yousaf *et al.*, 2015) *Osseointegration* has certainly improved the area of medical bone and joint replacement, dental implants techniques. In addition to this emphasis has been done towards coating of cells and stents.

b) Textile Industry:

With increase in customers' demands for durable, multi-functional and smart textiles nanotechnology plays an important role in textile industry. Some of the nanotechnology enhanced textile products includes nanofiber, polymer nanocomposites (PNs), nanostructured surfaces. These products make the textiles more functional and smart. (Joshi *et al.*, 2008, Kim, 2010) Improved functional properties in textiles includes UV protection, water, oil, stain repellency, antibacterial, antimicrobial, self-cleaning, wrinkle-resistance, flame retardancy, improved dye ability etc. without comprising the comfort properties. Figure 2 represents the different nanotechnology enhanced textile products and their applications. Coating technique such as sol-gel, spray coating, self-assembly, dip coating, plasma polymerization

Figure 2. Nanotechnology enhanced textile products and its application (Joshi and Adak, 2018)

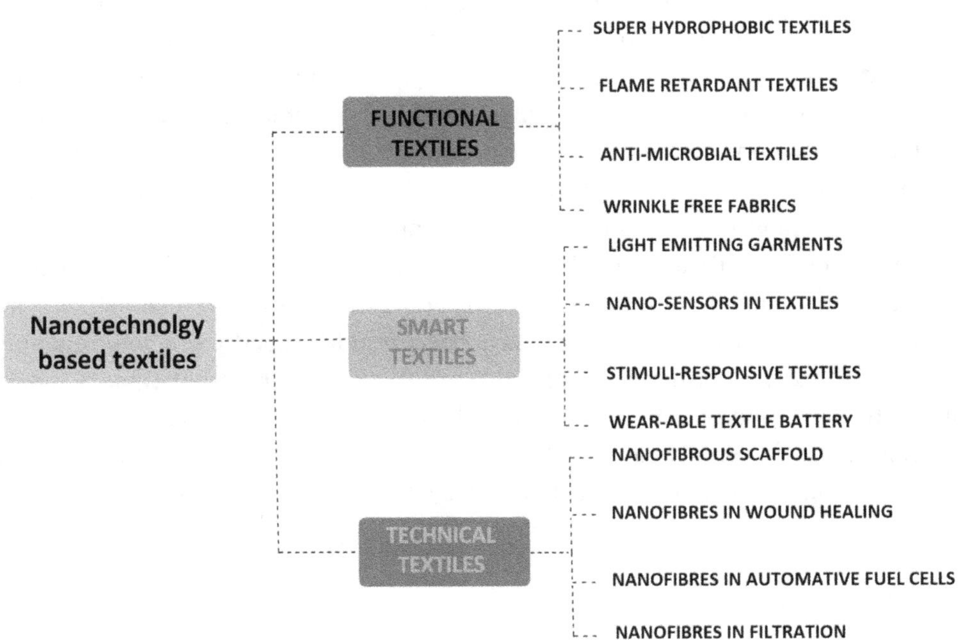

etc. have been used in textile industries for the synthesis of high performance textiles. (Joshi and Bhattacharyya, 2011; Gashti *et al.*, 2016)

Functional Textiles:

Functional textiles are textiles, which are incorporated with chemicals to improve various functions such as temperature control, humidity control etc. upon manufacturing so as to withstand various condition based on its application area. Rayon and polyester fibers are mostly used fibers in the functional textiles. (Functional Textiles Market) Demand for these functional textiles in markets has increased because of its increased use in various sportswear (such as cycling, ski sports, swimming etc.) Further, the functional textiles are broadly classified into four categories based on its type, functions and applications (Figure 3) and attracts the market globally. (Functional Textiles Market)

Water and oil repellant textiles:

Busy schedule, working environment, lifestyle has changed the clothing pattern of the people. People prefer using cloths that are hygienic, easy to use and wash. Usually a material having low surface energy is used in a textile material to impart water and oil absorbency. There are two types of water/oil repellent chemicals namely fluorocarbon based and non-fluorocarbon (e.g., silicon) that are used in commercially available textiles. (Joshi; Joshi and Adak, 2018) Fluorocarbon based chemicals are costly. This imparts both water and oil absorbency properties in the textiles whereas non-fluorocarbon or silicon-based chemicals can impart only water repellency property to the fabric. Presently, nanocoatings or nanofinishing are able to fulfill most of shortcomings of conventional finishes with water and oil repellent textiles also referred to as 'superhydrophobic' textiles.

The term superhydrophobic refers to the surface that exhibits excellent water repellency properties with lower surface energy (< 10dynes/cm) water contact angle (WCA) > 150° and contact angle hysteresis (CAH) < 10°. This is also referred to as self-cleaning effect or Lotus effect. (Joshi and Adak, 2018; Park *et al.*, 2015) Lotus-leaves are known to be one of the best example exhibiting natural superhydrophobic surfaces. This surface can effectively roll out impurities, such as mud, with water and make the surface clean. The phenomenon is due to presence of nano-level hydrophobic wax crystals on the top of lotus-leaf surfaces that makes it not- wettable. (Joshi and Adak, 2018; Abdulhussein *et al.*, 2016) Lotus effect is not only limited to lotus leaf but it is also shown by various plants and insects such as rice leaves, fish scales, butterfly wings, mosquito eyes and shark skin. (Ahmad and Kan, 2016) Lotus effect is considered and based on that several modification has been done on the surface of textile fabric by nanofinishing or nanocoating. Literature search reveals that various materials such as zirconia, nano-silica, ZnO nanorods or surface modified carbon nanotubes (CNTs) have been used for superhydrophobic coatings. (Joshi *et al.*, 2012; Das and De, 2015; Vasiljevic´ *et al.*, 2013; Ates and Unalan, 2012) Table 3 and 5 above represents some of the superhydrophobic textiles and UV protecting textiles and its evaluations that have shown promising results in recent years. Table 6 below represents textiles that have been modified and it has shown promising results towards anti-microbial activities in recent years.

Figure 3. Classification of Global functional textile market (Functional Textiles Market)

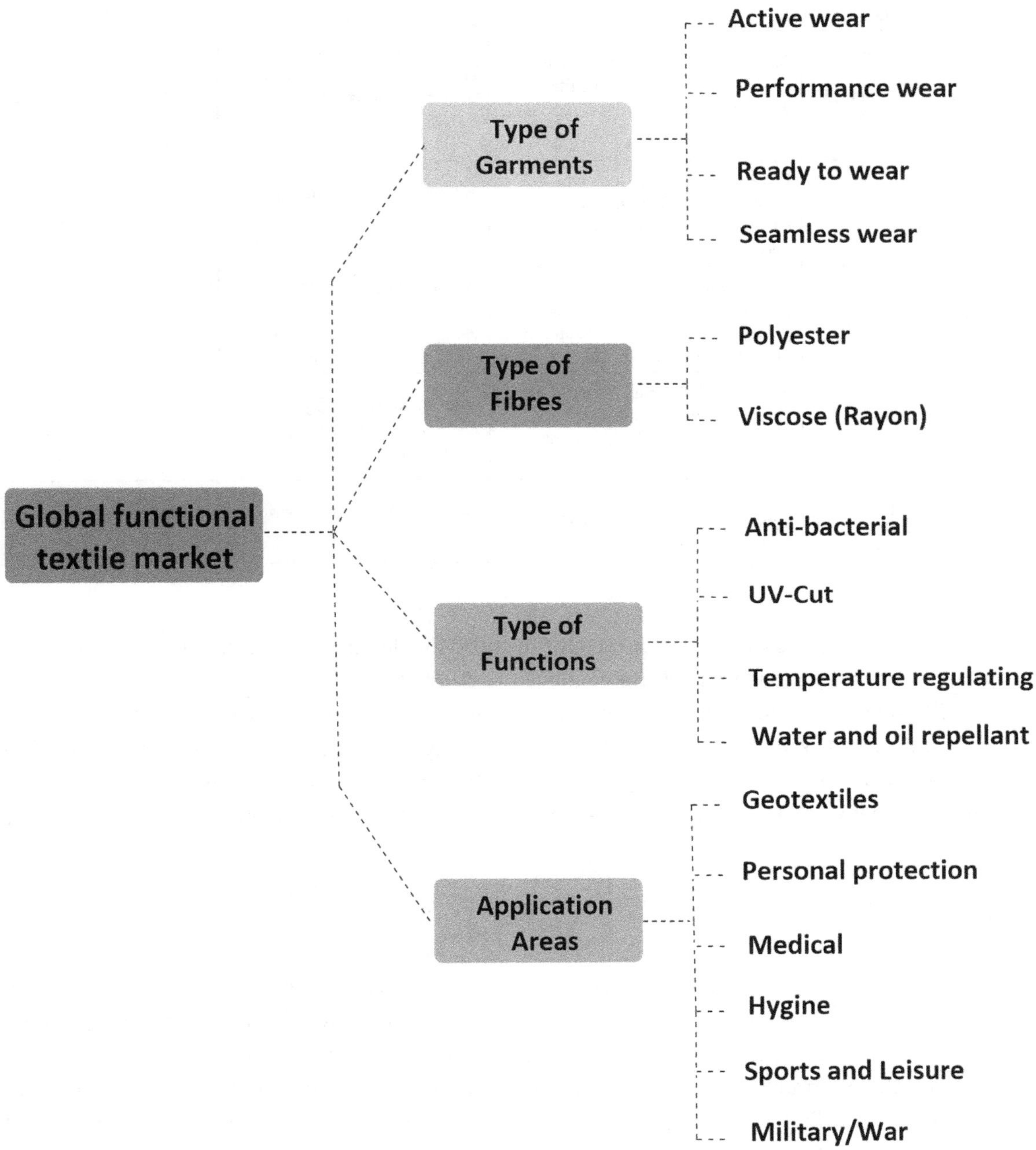

Environmental Benefits of Nanocoatings

Use of nanocoating systems may reduce environmental pollution even further (Steinfeldt *et al.* 2004). Nanocoatings usually fabricate thinner coating layers, which certainly saves raw materials. Moreover, the lighter-weight products can bring environmental benefits in the use stage also, particularly in the transportation area. Another potential environmental benefit could be the minimization or substitution of solvents and toxic compounds used in paints and lacquers (e.g. chromium compounds).

Table 6. Types of textiles, its synthesis and evaluation of properties

Type of textile	Synthesis method	Evaluation	Ref.
Modifications of natural materials such as cotton, wool, and silk and also synthetic materials such as PET fabrics	Spraying, the pad dry cure method; nanomaterials used include titanium dioxide, zinc oxide, silver oxide, and silver nanocolloids	Anti-microbial activity against wide range of bacteria and fungi.	Gokarneshan *et al.*, 2012
Modifications of woven cotton fabric	Pad-dry-plasma-cure method; 5,5-dimethylhydantoin (DMH) was coated via atmospheric pressure plasma treatment (APP)	Antibacterial properties against Staphylococcus aureus (*S. aureus*)	Zhou *et al.*, 2017
Modifications of cotton fabric	Modifications via cationization process and carboxymethylation process with different concentration of both Neem and Tulsi Extract	Anti-microbial properties, tensile strength, elongation, roughness, water absorbency (wettability)	El-Shafei *et al.*, 2017
Modifications of cotton textiles	Sol-gel method; treating cotton textile with water glass as silica source and then with silver nitrate solution	Anti-microbial activity towards *Escherichia coli* (E. coli), a gram-negative bacterium.	Xing *et al.*, 2007
Modifications of cotton- and cellulose containing fabrics	Treating with dimethylol dimethylhydantoin (DMDMH), followed by chlorine bleaching	Anti-microbial activity, wrinkle resistance	Sun *et al.*, 2001

Self-cleaning or easy-to-clean surfaces reduce the need for cleaning. Especially in industrial cleaning self-cleaning or easy-to-clean surfaces may reduce the consumption of energy and cleaning agents and extend the life span of coated objects. UV resistant coatings are based on inorganic-organic binder. They are able to dry under UV light within seconds, thus reducing the energy required all through the drying process. In addition, UV resistant coatings usually contain no or less volatile solvents.

Technically, many environmental benefits are to be expected from nanomaterial-containing coatings.

CONCLUSION

The ever-growing interest in nanocoating and its enthralling protective properties makes it very capable candidates for next generation protecting system. Nanocoating has proved itself as a valuable technology in different fields including medical, textile industries and almost every dimension of our life. It promises to unleash the enormous potential in the field of coatings. The future of these special nanocoating markets will be expanding in different industries such as marine, building and defense. A huge number of environmental benefits are also to be expected from nanocoatings, for example, thinner coating layers, minimization or substitution of solvents and toxic compounds used in paints etc. In addition, self-cleaning or "easy-to-clean" surfaces may reduce the need for cleaning, decrease the consumption of energy and cleaning agents and lengthen the life span of the coated objects. Nanotechnology developed coatings may conserve many of our natural resources and improve the overall performance of different substances. Every technology has some limitations. The effect of this technology on human health is yet to be studied in detail.

REFERENCES

Abdulhussein, A. T., Kannarpady, G. K., Wright, A. B., Ghosh, A., & Biris, A. S. (2016). Current trend in fabrication of complex morphologically tunable superhydrophobic nano scale surfaces. *Applied Surface Science*, *384*, 311–332. doi:10.1016/j.apsusc.2016.04.186

Achanta, S., Drees, D., & Celis, J.-P. (2011). *Nanocoatings for tribological applications*. Woodhead., doi:10.1533/9780857094902.2.355

Ahmad, I., & Kan, C. (2016). A Review on Development and Applications of Bio-Inspired Superhydrophobic Textiles. *Materials (Basel)*, *9*(11), 1–34. doi:10.3390/ma9110892 PMID:28774012

Akhtar-uz Zaman, I. M. (1997). *Development of nanocoating on cutting tools*. Academic Press.

Alebeida, O. K., Taoa, Z., & Seedahmed, A. I. (2015). New Approach for Dyeing and UV Protection Properties of Cotton Fabric Using Natural Dye Extracted from Henna Leaves. *Fibres & Textiles in Eastern Europe*, *23*(5), 60–65. doi:10.5604/12303666.1161758

Anand, S. C., Miraftab, M., Rajendran, S., & Kennedy, J. F. (2005). *Medical Textiles and Biomaterials for Healthcare*. Woodhead.

Artus, G. R. J., Jung, S., Zimmermann, J., Gautschi, H.-P., Marquardt, K., & Seeger, S. (2006). Silicone Nanofilaments and Their Application as Superhydrophobic Coatings. *Advanced Materials*, *18*(20), 2758–2762. doi:10.1002/adma.200502030

Ates, E. S., & Unalan, H. E. (2012). Zinc oxide nanowire enhanced multifunctional coatings for cotton fabrics. *Thin Solid Films*, *520*(14), 4658–4661. doi:10.1016/j.tsf.2011.10.073

Bao, W., Deng, Z., Zhang, S., Ji, Z., & Zhang, H. (2019) Next-generation composite coating system: Nanocoating. *Frontiers in Materials, 6*.

Cabrales, L., & Abidi, N. (2012). Microwave plasma induced grafting of oleic acid on cotton fabric surfaces. *Applied Surface Science*, *258*(10), 4636–4641. doi:10.1016/j.apsusc.2011.12.130

Carlsson, J-O., & Martin, P. M. (2010). *Chemical Vapor Deposition*. Elsevier. DOI: . doi:10.1016/B978-0-8155-2031-3.00007-7

Chen, M. L., Qu, Y. Y., Yang, L., & Gao, H. (2008). Structures and antifouling properties of low surface energy non-toxic antifouling coatings modified by nano-SiO_2 powder. *Science in China. Series B, Chemistry*, *51*(9), 848–852. doi:10.100711426-008-0069-5

Chen, S., Brown, L., Levendorf, M., Cai, W., Ju, S. Y., Edgeworth, J., Li, X., Magnuson, C. W., Velamakanni, A., Piner, R. D., Kang, J., Park, J., & Ruoff, R. S. (2011). Oxidation Resistance of Graphene-Coated Cu and Cu/Ni Alloy. *ACS Nano*, *5*(2), 1321–1327. doi:10.1021/nn103028d PMID:21275384

Chinglenthoiba, C., Ramkumar, K., Shanmugaraja, T., & Sharma, S. (2017). Study on Nanotechnology, Nanocoating and Nanomaterial. *IJCAM,* 17-25.

Chu, C. C. (2002). *Textile-based biomaterials for surgical applications*. CRC.

CoatingsP. V. D. (n.d.). https://www.pvd-coatings.co.uk/applications/

Conradi, M., Kocijan, A., Kek-Merl, D., Zorko, M., & Verpoest, I. (2014). Mechanical and anticorrosion properties of nanosilica-filled epoxy-resin composite coatings. *Applied Surface Science*, *292*, 423–437. doi:10.1016/j.apsusc.2013.11.155

Creighton, J. R., & Ho, P. (2001). *Introduction to chemical vapor deposition (CVD)*. ASM International.

Das, I., & De, G. (2015). Zirconia based superhydrophobic coatings on cotton fabrics exhibiting excellent durability for versatile use. *Scientific Reports*, *5*(1), 18503. doi:10.1038rep18503 PMID:26678754

Davis, R., El-Shafei, A., & Hauser, P. (2011). Use of atmospheric pressure plasma to confer durable water repellent functionality and antimicrobial functionality on cotton/polyester blend. *Surface and Coatings Technology*, *205*(20), 4791–4797. doi:10.1016/j.surfcoat.2011.04.035

Dewan, P. (2009). Nanotech enabled anti-graffiti clear coating system. *Bombay Technologists*, *59*, 79–83.

Duan, W., Xie, A., Shen, Y., Wang, X., Wang, F., Zhang, Y., & Li, J. (2011). Fabrication of Superhydrophobic Cotton Fabrics with UV Protection Based on CeO_2 Particles. *Industrial & Engineering Chemistry Research*, *50*(8), 4441–4445. doi:10.1021/ie101924v

Dubbert, W., Schwirn, K., Völker, D., & Apel, P. (2014). *Use of nanomaterials in coatings*. Federal Environment Agency.

El-Shafei, A., El-Bisi, M. K., Zaghloul, S., & Refai, R. (2017). Herbal Textile Finishes-Natural Antibacterial Finishes for Cotton Fabric. *Egyptian Journal of Chemistry*, *60*(2), 161–180.

El-Zairy, E. M. R., & Morgan, N. N. (2015). Functionalization of Polyester Fabric with UV-Protection and Antibacterial Property by Means of Atmospheric Glow Discharge. *International Journal of Textile Science*, *4*(5), 113–117.

Faizan-Ur-Rab, M., Zahiri, S. H., Masood, S. H., Phan, T. D., Jahedi, M., & Nagaraja, R. (2016). Application of a holistic 3D model to estimate state of cold spray titanium particles. *Materials & Design*, *89*, 1227–1241. doi:10.1016/j.matdes.2015.10.075

Ferrari, A. M., Pini, M., Neri, P., & Bondioli, F. (2015). Nano-TiO_2 Coatings for Limestone: Which Sustainability for Cultural Heritage. *Coatings*, *5*(3), 232–245. doi:10.3390/coatings5030232

Gangotri, D. L., & Chaware, A. D. (2004). *Paint India*. Academic Press.

Garcia, B., Lamzoudi, A., Pillier, F., Le, H. N. T., & Deslouis, C. (2002). Oxide/Polypyrrole Composite Films for Corrosion Protection of Iron. *Journal of the Electrochemical Society*, *149*(12), B560. doi:10.1149/1.1517581

Gashti, M. P., Pakdel, E., & Alimohammadi, F. (2016). *Nanotechnology-based coating techniques for smart textiles*. Academic Press.

Ghane, M., & Ghorbani, E. (2016). Investigation into the UV-protection of woven fabrics composed of metallic weft yarns. *AUTEX Research Journal*, *16*(3), 154–159. doi:10.1515/aut-2015-0021

Gläsel, H. J., Bauer, F., Ernst, H., Findeisen, M., Hartmann, E., Langguth, H., Mehnert, R., & Schubert, R. (2000). Preparation of scratch and abrasion resistant polymeric nanocomposites by monomer grafting onto nanoparticles, Characterization of radiation-cured polymeric nanocomposites. *Macromolecular Chemistry and Physics*, *201*(18), 2765–2770. doi:10.1002/1521-3935(20001201)201:18<2765::AID-MACP2765>3.0.CO;2-9

Goffredo, G. B., & Munafò, P. (2015). Preservation of Historical Stone Surfaces by TiO_2 Nanocoatings. *Coatings*, *5*(2), 222–231. doi:10.3390/coatings5020222

Gokarneshan, N., Gopalakrishnan, P. P., & Jeyanthi, B. (2012). Influence of Nanofinishes on the Antimicrobial Properties of Fabrics. *ISRN Nanomaterials*, *2012*, 193836. Advance online publication. doi:10.5402/2012/193836

Hanlon, J. F., Kelsey, R. J., & Forcinio, H. (1998). *Handbook of Package Engineering*. Technomic. doi:10.1201/b18045

Heinonen, S., Huttunen-Saarivirta, E., Nikkanen, J. P., Raulio, M., Priha, O., Laakso, J., Storgards, E., & Levänen, E. (2014). Antibacterial properties and chemical stability of superhydrophobic silver-containing surface produced by sol-gel route. *Colloids and Surfaces. A, Physicochemical and Engineering Aspects*, *453*, 149–161. doi:10.1016/j.colsurfa.2014.04.037

Her, E. K., Ko, T. J., Shin, B., Roh, H., Dai, W., Seong, W. K., Kim, H. Y., Lee, K. R., Oh, K. H., & Moon, M. W. (2013). Superhydrophobic Transparent Surface of Nanostructured Poly(Methyl Methacrylate) Enhanced by a Hydrolysis Reaction. *Plasma Processes and Polymers*, *10*(5), 481–488. doi:10.1002/ppap.201200131

Huang, F. L., Wei, Q. F., Xu, W. Z., & Li, Q. (2007). Preparation and characterizations of PTFE gradient nanostructure on silk fabric. *Surface Review and Letters*, *14*(4), 547–551. doi:10.1142/S0218625X07009827

Irissou, E., Legoux, J.-G., Ryabinin, A. N., Jodoin, B., & Moreau, C. (2008). Review on cold spray process and technology: Part I- Intellectual Property. *J. Therm. Spray Technol.*, *17*(4), 495–516. doi:10.100711666-008-9203-3

Ivanovaa, N. A., & Philipchenkob, A. B. (2012). Superhydrophobic chitosan-based coatings for textile processing. *Applied Surface Science*, *263*, 783–787. doi:10.1016/j.apsusc.2012.09.173

Joshi, M., & Adak, B. (2018). *Advances in Nanotechnology Based Functional, Smart and Intelligent Textiles: A Review Comprehensive*. Elsevier. doi:10.1016/B978-0-12-803581-8.10471-0

Joshi, M., & Adak, B. (2018). *Nanotechnology-based textiles: A solution for the emerging automotive sector*. Walter de Gruyter GmbH.

Joshi, M., & Bhattacharyya, A. (2011). Nanotechnology-a new route to high-performance functional textiles. *Textile Progress*, *43*(3), 155–233. doi:10.1080/00405167.2011.570027

Joshi, M., Bhattacharyya, A., Agarwal, N., & Parmar, S. (2012). Nanostructured coatings for super hydrophobic textiles. *Bulletin of Materials Science*, *35*(6), 933–938. doi:10.100712034-012-0391-6

Joshi, M., Bhattacharyya, A., & Ali, S. W. (2008) Characterization techniques for nanotechnology applications in textiles, *Indian Journal of Fibre and Textile Research, 33*(3), 304-317.

Karter, M. J. (2014). *Fire loss in the United States during 2013*. National Fire Protection Association.

Kathirvelu, S., D'souza, L., & Dhurai, B. (2009). UV protection finishing of textiles using ZnO nanoparticles. *Indian Journal of Fibre and Textile Research*, *34*(3), 267–273.

Khanna, A. S. (2008). Nanotechnology in High Performance Paint Coatings. *Asian J. Exp. Sci.*, *21*(2), 25–32.

Kim, Y. K. (2010). *Nanotechnology-based advanced coatings and functional finishes for textiles*. Woodhead. doi:10.1533/9781845697785.2.251

Kuroda, S., Kawakita, J., Watanabe, M., & Katanoda, H. (2008). Warm spraying- A novel coating process based on high-velocity impact of solid particles. *Sci. Technol. Adv. Mater.*, *9*(033002), 1-17. DOI: 1088/1468-6996/9/3/033002

Kwon, K. C., & Kim, S. Y. (2014). Extended thermal stability in metal-chloride doped graphene using graphene over layers. *Chemical Engineering Journal*, *244*, 355–363. doi:10.1016/j.cej.2014.01.100

Le Bras, M., Wilkie, C. A., & Bourbigot, S. (2005). *Fire retardancy of polymers: new applications of mineral fillers*. Academic Press.

Li, S., & Jinjin, D. (2007). Improvement of hydrophobic properties of silk and cotton by hexafluoropropene plasma treatment. *Applied Surface Science*, *253*(11), 5051–5055. doi:10.1016/j.apsusc.2006.11.027

Lim, H. S., Park, S. H., Koo, S. H., Kwark, Y.-J., Thomas, E. L., Jeong, Y., & Cho, J. H. (2010). Super amphiphilic Janus Fabric. *Langmuir*, *26*(24), 19159–19162. doi:10.1021/la103829c PMID:21073162

Litzenburger, A. (2000). Criteria for, and examples of optimal choice of flame-retardants. *Polymers & Polymer Composites*, *8*, 581–592.

Liu, Y., Tang, J., Wang, R., Lu, H., Li, L., Kong, Y., Qia, K., & Xin, J. H. (2007). Artificial lotus leaf structures from assembling carbon nanotubes and their applications in hydrophobic textiles. *Journal of Materials Chemistry*, *17*(11), 1071–1078. doi:10.1039/B613914K

Liu, Y., Wang, X., Qia, K., & Xin, J. H. (2008). Functionalization of cotton with carbon nanotubes. *Journal of Materials Chemistry*, *18*(29), 3454–3460. doi:10.1039/b801849a

Lynam, C., Moulton, S. E., & Wallace, G. G. (2007). Carbon-Nanotube Biofibers. *Advanced Materials*, *19*(9), 1244–1248. doi:10.1002/adma.200601986

Makhlouf, A. S. H. (2011). *Current and advanced coating technologies for industrial applications*. Woodhead., doi:10.1533/9780857094902.1.3

Mathiazhagan, A., & Joseph, R. (2011). Nanotechnology-A New Prospective in Organic Coating –Review. *International Journal of Chemical Engineering and Applications*, *2*(4), 225–237. doi:10.7763/IJCEA.2011.V2.108

Mathiazhagan, A., & Joseph, R. (2011). Nanotechnology-a new prospective in organic coatings-review. *International Journal of Chemical Engineering and Applications*, *2*(4), 225–237. doi:10.7763/IJCEA.2011.V2.108

Mavrogenis, A. F., Dimitriou, R., Parvizi, J., & Babis, G. C. (2009). Biology of implant osseointegration. *Journal of Musculoskeletal & Neuronal Interactions*, *9*(2), 61–71. PMID:19516081

Mishra, A., Mohapatra, H. S., & Butola, B. S. (2019). Imparting protection against UV radiations In situ coatings of Titanium dioxide on textiles. *International Letters of Chemistry. Physics and Astronomy*, *82*, 14–20.

Mishra, R. S., Tiwari, D., & Singh, K. K. (2013). Nanocoating: A novel drift in nanotechnology. *International Journal of Engineering Innovation & Research*, *2*(2), 128–131.

Müler, K., Bugnicourt, E., Latorre, M., Jorda, M., Sanz, Y. E., Lagaron, J. M., & … . (2017). Review on the Processing and Properties of Polymer Nanocomposites and Nanocoatings and Their Applications in the Packaging, Automotive and Solar Energy Fields. *Nanomaterials (Basel, Switzerland)*, *7*(74), 74. doi:10.3390/nano7040074 PMID:28362331

Nguyen-Tri, P., Nguyen, T. A., Carriere, P., & Xuan, C. N. (2018). Nanocomposite Coatings: Preparation, Characterization, Properties, and Applications. *International Journal of Corrosion*, *2018*, 1–19. doi:10.1155/2018/4749501

O'Brien, P. (2001). Encyclopedia of Materials: Science and Technology. Academic Press.

Park, S., Kim, J., & Park, C. H. (2015). Superhydrophobic Textiles: Review of Theoretical Definitions, Fabrication and Functional Evaluation. *Journal of Engineered Fibers and Fabrics*, *10*(4), 1–18. doi:10.1177/155892501501000401

Physical Vapor Deposition, P. V. D. (n.d.). http://www.sigmaaldrich

Rahmawan, Y., Xu, L. B., & Yang, S. (2013). Self-assembly of nanostructures towards transparent, superhydrophobic surfaces. *Journal of Materials Chemistry. A, Materials for Energy and Sustainability*, *1*(9), 2955–2969. doi:10.1039/C2TA00288D

Riva, A., Algaba, I. M., & Pepio, M. (2006). Action of a finishing product in the improvement of the ultraviolet protection provided by cotton fabrics. Modelisation of the effect. *Cellulose (London, England)*, *13*(6), 697–704. doi:10.100710570-006-9085-9

Roe, B., & Zhang, X. (2009). Durable Hydrophobic Textile Fabric Finishing Using Silica Nanoparticles and Mixed Silanes. *Textile Research Journal*, *79*(12), 1115–1122. doi:10.1177/0040517508100184

Rout, T. K., Jha, G., Singh, A. K., Bandyopadhyay, N., & Mohanty, O. N. (2003). Development of conducting polyaniline coating: A novel approach to superior corrosion resistance. *Surface and Coatings Technology*, *167*(1), 16–24. doi:10.1016/S0257-8972(02)00862-9

Sarkar, A. K. (2004). An evaluation of UV protection imparted by cotton fabrics dyed with natural colorants. *BMC Dermatology*, *4*(15), 1–8. doi:10.1186/1471-5945-4-15 PMID:15509304

Sauvet, G., Dupond, S., Kazmierski, K., & Chojnowski, J. (2003). Biocidal polymers active by contact. V. Synthesis of polysiloxanes with biocidal activity. *Journal of Applied Polymer Science*, *75*(8), 1005–1012. doi:10.1002/(SICI)1097-4628(20000222)75:8<1005::AID-APP5>3.0.CO;2-W

Schartel, B. (2010). Phosphorus-based Flame Retardancy Mechanisms-Old Hat or a Starting Point for Future Development. *Materials (Basel)*, *3*(10), 4710–4745. doi:10.3390/ma3104710 PMID:28883349

Selvakumar, N., & Barshilia, H. C. (2012). Review of physical vapor deposited (PVD) spectrally selective coatings for mid- and high-temperature solar thermal applications. *Solar Energy Materials and Solar Cells*, *98*, 1–23. doi:10.1016/j.solmat.2011.10.028

Shen, G. X., Chen, Y. C., & Lin, C. J. (2005). Corrosion protection of 316 L stainless steel by a TiO_2 nanoparticle coating prepared by sol-gel method. *Thin Solid Films*, *489*(1-2), 130–136. doi:10.1016/j.tsf.2005.05.016

Steinfeldt, M., Gleich, A. V., Petschow, U., Haum, R., Chudoba, T., & Haubold, S. (2004) Nachhaltigkeitseffekte durch Herstellung und Anwendung nano-technologischer Produkte. Schriftenreihe des IÖW 17704.

Su, S. H., Conroy, S., Lin, T. L., Sheu, M. S., & Loh, I. H. (2004). Surface coatings. Encyclopedia of Biomaterials and Biomedical Engineering, 2, 1412-1431.

Sun, G., Xu, X., Bickett, J. R., & Williams, J. F. (2001). Durable and Regenerable Antibacterial Finishing of Fabrics with a New Hydantoin Derivative. *Industrial & Engineering Chemistry Research*, *40*(4), 1016–1021. doi:10.1021/ie000657t

Tavares, J., Swanson, E. J., & Coulombe, S. (2008). Plasma Synthesis of Coated Metal Nanoparticles with Surface Properties Tailored for Dispersion. *Plasma Processes and Polymers*, *5*(8), 759–769. doi:10.1002/ppap.200800074

Textor, T., Schollmeyer, H., Bahners, T., & Schollmeyer, E. (2004). Modification of low energy polymer surfaces by immobilization of fluorinated carboxylates with zirconium-based coupling agents. *Journal of Applied Polymer Science*, *94*(2), 789–795. doi:10.1002/app.20685

Thomas, F. A. P., Stoks, J. A. W., & Buegman, A. (2001). *US Patent 6291054*. Washington, DC: US Patent Office.

Thorvaldsson, A., Edvinsson, P., Glantz, A., Rodriguez, K., Walkenstrom, P., & Gatenholm, P. (2012). Superhydrophobic behaviour of plasma modified electro spun cellulose nanofiber-coated microfibers. *Cellulose (London, England)*, *19*(5), 1743–1748. doi:10.100710570-012-9751-z

Twardowski, A., Makowski, P., Malachowski, A., Hrynyk, R., Pietrowski, P., & Tyczkowski, J. (2012). Plasma treatment of thermoactive membrane textiles for superhydrophobicity. *Journal of Materials Science*, *18*(2), 163–166.

van Lente, H., & van Til, J. I. (2008). Articulation of sustainability in the emerging field of nanocoatings. *Journal of Cleaner Production*, *16*(8-9), 967–976. doi:10.1016/j.jclepro.2007.04.020

Vasiljevic, J., Gorjanc, M., Tomšic, B., Orel, B., Jerman, I., Mozetic, M., Vesel, A., & Simoncic, B. (2013). The surface modification of cellulose fibres to create super-hydrophobic, oleophobic and self-cleaning properties. *Cellulose (London, England)*, *20*(1), 277–289. doi:10.100710570-012-9812-3

Vihodceva, S., & Kukle, S. (2013). Improvement of UV Protection Properties of the Textile from Natural Fibres by the Sol-gel Method. *IOP Conf. Series: Materials Science and Engineering*, *49*, 1-4. 10.1088/1757-899X/49/1/012022

Wang, D., Zhang, Z. B., Li, Y. M., & Xu, C. H. (2014). Highly Transparent and Durable Superhydrophobic Hybrid Nanoporous Coatings Fabricated from Polysiloxane. *ACS Applied Materials & Interfaces*, *6*(13), 10014–10021. doi:10.1021/am405884x PMID:24955659

Windsor, B. (2005, March). *Hygienic Coatings & Surfaces*. PRA Coatings Technology Centre.

Wu, L., Zhang, J., Lia, B., & Wang, A. (2013). Mimic nature, beyond nature: Facile synthesis of durable superhydrophobic textiles using organosilanes. *Journal of Materials Chemistry. B, Materials for Biology and Medicine*, *1*(37), 4756–4763. doi:10.1039/c3tb20934b PMID:32261159

Xi, J. Q., Schubert, M. F., Kim, J. K., Schubert, E. F., Chen, M., Lin, S. Y., Liu, W., & Smart, J. A. (2007). Optical thin-film materials with low refractive index for broadband elimination of Fresnel reflection. *Nature Photonics*, *1*(3), 176–179. doi:10.1038/nphoton.2007.26

Xing, Y., Yang, X., & Dai, J. (2007). Antimicrobial finishing of cotton textile based on water glass by sol-gel method. *Journal of Sol-Gel Science and Technology*, *43*(2), 187–192. doi:10.100710971-007-1575-1

Xu, B., & Cai, Z. (2008). Fabrication of a superhydrophobic ZnO nanorod array film on cotton fabrics via a wet chemical route and hydrophobic modification. *Applied Surface Science*, *254*(18), 5899–5904. doi:10.1016/j.apsusc.2008.03.160

Xu, L., & He, J. (2012). Antifogging and Antireflection Coatings Fabricated by Integrating Solid and Mesoporous Silica Nanoparticles without Any Post-Treatments. *ACS Applied Materials & Interfaces*, *4*(6), 3293–3299. doi:10.1021/am300658e PMID:22663342

Xu, Q. F., Wang, J. N., & Sanderson, K. D. (2010). Organic–Inorganic Composite Nanocoatings with Super hydrophobicity, Good Transparency, and Thermal Stability. *ACS Nano*, *4*(4), 2201–2209. doi:10.1021/nn901581j PMID:20302323

Xu, T., & Xie, C. S. (2003). Tetrapod-like nano-particle ZnO/acrylic resin composite and its multi-function property. *Progress in Organic Coatings*, *46*(4), 297–301. doi:10.1016/S0300-9440(03)00016-X

Xue, C.-H., Chen, J., Yin, W., Jia, S.-T., & Ma, J.-Z. (2012). Superhydrophobic conductive textiles with antibacterial property by coating fibers with silver nanoparticles. *Applied Surface Science*, *258*(7), 2468–2472. doi:10.1016/j.apsusc.2011.10.074

Xue, C.-H., Jia, S.-T., Chen, H.-Z., & Wang, M. (2008). Superhydrophobic cotton fabrics prepared by sol-gel coating of TiO_2 and surface hydrophobization. *Science and Technology of Advanced Materials*, *9*(3), 1–5. doi:10.1088/1468-6996/9/3/035001 PMID:27877998

Xue, C.-H., Jia, S.-T., Zhang, J., & Tian, L.-Q. (2009). Superhydrophobic surfaces on cotton textiles by complex coating of silica nanoparticles and hydrophobization. *Thin Solid Films*, *517*(16), 4593–4598. doi:10.1016/j.tsf.2009.03.185

Yadav, R., Anubhav, V., Mathur, P., & Sheikh, J. (2019). Antibacterial, UV protective and antioxidant linen obtained by natural dyeing with henna. *Cellulose Chemistry and Technology*, *53*(3-4), 357–362. doi:10.35812/CelluloseChemTechnol.2019.53.36

Yang, H., Zhu, S., & Pan, N. (2004). Studying the mechanisms of titanium dioxide as ultraviolet-blocking additive for films and fabrics by an improved scheme. *Journal of Applied Polymer Science*, *92*(5), 3201–3210. doi:10.1002/app.20327

Yang, W., Thordarson, P., Gooding, J. J., Ringer, S. P., & Braet, F. (2007). Carbon nanotubes for biological and biomedical applications. *Nanotechnology*, *18*(41), 412001. doi:10.1088/0957-4484/18/41/412001

Yazhinia, K. B., & Prabu, H. G. (2015). Study on flame-retardant and UV-protection properties of cotton fabric functionalized with ppy–ZnO–CNT nanocomposite. *RSC Advances*, *5*(61), 49062–49069. doi:10.1039/C5RA07487H

Yeh, J. M., & Chin, C. P. (2003). Structure and properties of poly (*o*-methoxyaniline)-clay nanocomposite materials. *Journal of Applied Polymer Science*, *88*(4), 1072–1078. doi:10.1002/app.11829

Yousaf, S., Alhnan, M. A., Abdallah, A., Abdallah, B., Khan, I., & Ahmed, W. (2015). *Nanocoatings in medicine: antiquity and modern times*. Elsevier. doi:10.1016/B978-0-323-28990-0.00016-6

Yu, S., Guo, Z., & Liu, W. (2015). Biomimetic transparent and superhydrophobic coatings: From nature and beyond nature. *Chemical Communications*, *51*(10), 1775–1794. doi:10.1039/C4CC06868H PMID:25406877

Yu, Z., Lv, L., Ma, Y., Di, H., & He, Y. (2016). Covalent modifications of graphene oxide by metronidazole for reinforced anti-corrosion properties of epoxy coatings. *RSC Advances*, *6*(22), 18217–18226. doi:10.1039/C5RA23595B

Zhao, Y., Tang, Y., Wang, X., & Lin, T. (2010). Superhydrophobic cotton fabric fabricated by electrostatic assembly of silica nanoparticles and its remarkable buoyancy. *Applied Surface Science*, *256*(22), 6736–6742. doi:10.1016/j.apsusc.2010.04.082

Zhou, C.-E., Kan, C., Matinlinna, J. P., & Tsoi, J. K. (2017). Regenerable Antibacterial Cotton Fabric by Plasma Treatment with Dimethylhydantoin: Antibacterial Activity against S. aureus. *Coatings*, *7*(11), 1–14. doi:10.3390/coatings7010011

Zhu, J., Hsu, C. M., Yu, Z. F., Fan, S. H., & Cui, Y. (2010). Nanodome Solar Cells with Efficient Light Management and Self-Cleaning. *Nano Letters*, *10*(6), 1979–1984. doi:10.1021/nl9034237 PMID:19891462

Zimmermann, J., Reifler, F. A., Fortunato, G., Gerhardt, L.-C., & Seeger, S. (2008). A Simple, One-Step Approach to Durable and Robust Superhydrophobic Textiles. *Advanced Functional Materials*, *18*(22), 3662–3669. doi:10.1002/adfm.200800755

Chapter 2
Principles and Applications of Thermal Spray Coatings

John Henao
https://orcid.org/0000-0002-8954-6039
CONACYT-CIATEQ A.C., Mexico

Astrid L. Giraldo-Betancur
https://orcid.org/0000-0002-5056-7270
CONACYT-CINVESTAV, Mexico

Carlos A. Poblano-Salas
CIATEQ A.C., Mexico

Jorge Corona-Castuera
CIATEQ A.C., Mexico

Fabio Vargas
https://orcid.org/0000-0003-4484-3950
University of Antioquia, Colombia

Oscar Sotelo-Mazón
Universidad Autónoma del Estado de Morelos (UAEM), Mexico

ABSTRACT

The goal of the chapter is to address the fundamental theory of thermal spraying and its modern industrial applications, in particular, those involving flame spray, HVOF, plasma spray, and cold spray processes. During the last 30 years, thousands of manuscripts and various book chapters have been published in the field of thermal spray, displaying the evolution of thermally sprayed coatings in many industrial applications. Thermal spray coatings are currently interesting for different modern applications including prosthesis, thermal barriers, electrochemical catalysis, electrochemical energy conversion devices, biofouling, and self-repairing surfaces. The chapter will explain the fundamental principles of the aforementioned thermal spraying processes and discuss the effect of different controlling parameters on the final properties of the produced coatings. This chapter will also explore current and future industrial applications of thermal spray coatings.

LIST OF COMMON ABBREVIATIONS

APS: Atmospheric Plasma Spray
ASI: Adiabatic Shear Instabilities
CAPS: Controlled Atmosphere Plasma Spray

DOI: 10.4018/978-1-7998-4870-7.ch002

CGS: Cold Gas Spray
dB: Decibels (dBA is noise power measured in dB)
DC: Direct Current
D-Gun: Detonation Spray
HPCS: High Pressure Cold Gas Spray
HV: Hardness Vickers
HVAF: High Velocity Air Fuel
HVOLF: High Velocity Oxygen Liquid Fuel
HVOF: High Velocity Oxygen Fuel
HRC: Hardness Rockwell C
HVSFS: High Velocity Suspension Flame Spray
LPCS: Low Pressure Cold Gas Spray
LPPS: Low Pressure Plasma Spray
PS-PVD: Plasma Spray Physical Vapor Deposition
SPS: Shrouded Plasma Spray
TS: Thermal Spray
TWA: Twin Wire Arc
VPS: Vacuum Plasma Spray

INTRODUCTION TO THERMAL SPRAY

General Background

Surface modification of materials is a topic that has been intensively studied for the last hundred years to improve, transform, or provide specific functions to a surface originally designed to perform a definite task. Different methods have been employed to modify the surface of materials; for instance, the deleterious effects of corrosion on steels are controlled by applying a layer of a reactive and low melting point metal such as Zn (Marder et al., 2000). A similar protective effect is produced on the surface of iron-based materials when a polymer is deposited on their surface (Ates et al., 2016); however, the pursued application and the specific surface requirements dictate the procedure and more suitable materials for protecting or modifying the surface. Large numbers of materials and processes have been developed for providing surface modification to different substrate materials, from polymers, metals, and ceramics, to electronics, composites, biomaterials, and a myriad of possible combinations in the middle. Thermal spraying (TS) occupies a very specific niche among the technologies developed for providing specific surface features to metallic, ceramic, and composite materials. TS encompasses a group of coating technologies employed for applying metallic and nonmetallic materials on different types of substrates in order to provide them with specific features such as elevated corrosion resistance, biocompatibility, lubrication, high abrasion resistance, electrical and thermal insulation, etc. (Knight et al., 1998). Raw materials employed in such technologies can be in the form of powders, wire, or rods, depending on the selected TS process. All TS processes encompass thermal and kinetic contributions to the overall energy balance between the energy source and the sprayed material (Pawlowski, 2008). Depending on the TS process configuration, thermal and kinetic contributions to that energy balance can be modified; for instance, in high energy processes such as plasma, raw materials receive a larger amount of thermal energy than that arising from particle acceleration (Fauchais et al., 2014). In TS processes, raw materials are introduced into the energy source and converted to molten or semi-molten particles (with the exception of cold gas spray in which raw materials are introduced in powdered form and often remain in solid state), which ultimately undergo successive collisions on the substrate surface and produce a layer-by-layer coating. Upon impact, particles get bonded on the substrate surface, forming thin structures known as "splats"; then, coating build-up continues by the accumulation of subsequent splats, forming a typical structure that includes pores, molten and semi-molten particles, limits among particles, lamellaes, and cracks. The cooling rates associated with the formation of splats are typically in the order of 10^6 Ks^{-1} for metals (Pasandideh-Fard et al., 2002). The physical, chemical, and mechanical properties of coating produced by thermal spray are intimately related to the coating architecture, which in turn is influenced by the selected processing parameters (stand-off distance, feeding rate, raster speed, fuel/oxygen ratio, gas pressure, and flow, substrate temperature, etc.) Hence, repeatability of TS processes and final coating properties are a challenge since a large interplay occurs among processing variables and final coating properties (Davis, 2004).

Classification of Thermal Spray Processes

TS processes are typically classified into three main categories: i) combustion spraying, ii) electric discharge arc spraying, and iii) compressed gas expansion. This classification is stablished according to the source of energy used to provide thermal and kinetic energy to the sprayed material (Fauchais et al., 2014).

Combustion Spray Processes

These processes are based on the combustion produced by mixing a fuel (liquid or gaseous) and an oxidizing gas. Combustion spray processes can be classified as low and high velocity systems depending on operating pressures and gas velocities, both pressure and velocity are strongly influenced by the gun design. Low velocity combustion systems (<100 ms^{-1}) use either powder or wire feedstocks, which are fed into the oxyfuel flame by employing feeder machines with different setups such as vibrating devices for powders or push/pull systems for wires. When powders are used as feedstock (Figure 1a), they are injected into the oxyfuel flame, melted, and accelerated towards the substrate surface by the flame and air jets. The density and bond strength of low velocity combustion sprayed coatings are normally lower than those measured for its high velocity counterparts. These processes offer a wide range of feeding rates, ranging from 0.5 to 9 kgh^{-1}, with exception of low melting point metals which can be processed even at higher feeding rates. Such flexibility makes combustion spray processes attractive for manufacturing coatings on large areas requiring specific finishing and/or functionality. Figure 1 shows typical configurations for a combustion flame system.

Figure 1. a) conventional flame spray, adapted from (Dcleng, n.d.), and b) wire combustion spray, adapted from (Benxxoglobal, n.d.).

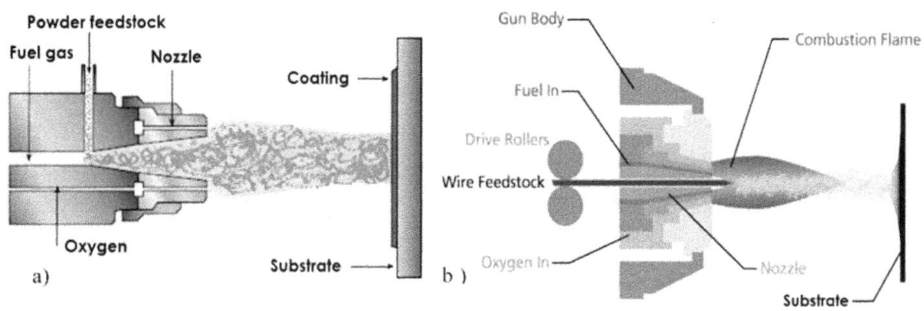

Figure 1b shows another setup of the flame spray process known as "wire combustion spray"; this process is very similar to the conventional flame spray system (Figure 1a), as both configurations inject the feedstock material axially into the flame to allow its complete melting. Feeding rates of materials sprayed in the wire combustion process are comparable to those values reported for conventional flame spray. However, higher substrate temperatures are reached in the wire combustion spray as the energy input required for melting wires or rods is higher than that employed to promote melting of powders. Combustion spray processes have low efficiency as only about 10% of the energy input is employed for melting the feedstock material, the rest of the energy is either dissipated in the flame or transmitted to the substrate. Hence, high substrate temperatures are achieved when these processes are employed. Substrate cooling is then sometimes recommended when combustion sprayed coatings are produced, among various reasons, mainly to avoid fast oxidation of the coating/substrate system (i.e. to reduce the oxidation kinetics process upon impact), although this procedure must be performed carefully as uncontrolled cooling rates may result in spalling or delamination of the coatings (Pawlowski, 2008).

There are currently various high-velocity combustion processes available in the industry; those processes include the high velocity oxyfuel (HVOF), high velocity air-fuel (HVAF), high velocity liquid fuel (HVOLF), high velocity suspension spray (HVSFS) and detonation spray (D-Gun). Currently, detonation spray is much less employed industrially than HVOF systems and HVOF-based processes such as HVAF and HVOLF. HVOF is a process in which gaseous fuels (propylene, propane, hydrogen) are combined with pure oxygen for producing a high-speed flame. When liquid fuel is used (Kerosene, Jet fuel), this process is known as HVOLF. In HVOF, a larger kinetic energy input is provided to the feedstock material in comparison to conventional flame spray systems; such conditions are achieved by employing either a barrel-type or a convergent/divergent de Laval nozzle, which accelerates the gas mixture up to $\sim 1000\ ms^{-1}$. Depending on the fuel and fuel/oxygen ratio employed in the process, flame temperature values can range from 2500 to 3100 °C. Combustion takes place in a high-pressure chamber from which the mixture exits and combines with the feedstock powder, the gas-powder mixture is then accelerated within the nozzle, traveling at high velocities (> Mach 1). HVOF guns are cooled by air or water; in some gun designs, air is also introduced to the chamber to complete the combustion reaction. Due to the high particle velocities and temperatures, dense and well-bonded coatings are produced by employing this technology. Typical feeding rates for this system ranges from 2 to 14 kgh^{-1} (Fauchais et al., 2014).

High velocity air-fuel (HVAF) process offers important versatility to produce coatings as it employs large volumes of air to complete the combustion reaction (instead of pure oxygen). HVAF also works with different fuels such as propane, propylene, natural gas, or hydrogen. The combustion of air and fuel often produces flame temperatures between 1900-1950 °C. This temperature range lies between the temperatures reported for HVOF and cold gas spray (CGS); therefore, HVAF is considered as a "warm" spray process. HVAF-sprayed coatings often show reduced oxidation when these are compared to HVOF-sprayed counterparts. HVAF process is mainly employed for the fabrication of metallic and cermet-based coatings, although some low melting point ceramics may also be produced by this method. Typical HVAF process has efficiency values ranging from 50 to 85%, with higher values reported for metals than those found for cermets (HVAF, n.d.). Maximum feeding rates are 30 kgh^{-1} for cermets and up to 23 kgh^{-1} for metals and alloys. Particle velocity values range from 1100 to 1200 ms^{-1} and depend on the particle size distribution and selected gun design (size, nozzle profile, etc.). Figure 2 shows a general scheme of both HVOF and HVAF processes.

Flame Spray Coatings and Modern Industrial Applications

Figure 2. a) HVOF process, adapted from (Advanced-Coating, n.d.), and b) HVAF gun, adapted from (Science direct, n.d.).

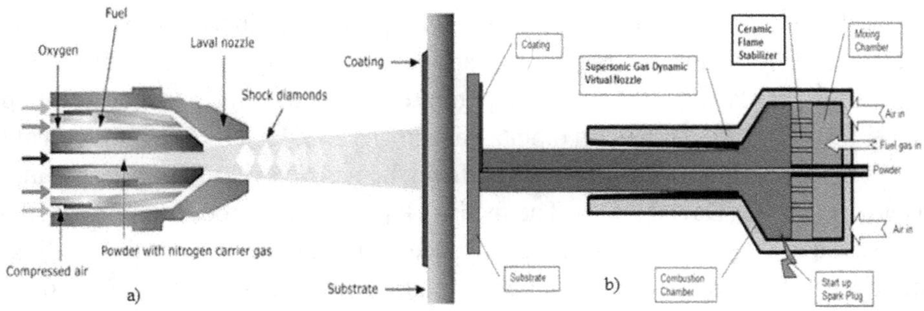

Electric Discharge Arc Spraying

Among electric discharge arc processes, that include all variations of plasma spraying processes, electric arc spray, also known as twin wire arc (TWA), is the simplest process. It employs two consumable wire electrodes that are connected to a high-current DC power unit; they are introduced to a gun where an electric arc is produced between both electrodes, the electrodes are then melted, atomized, and propelled by a carrier gas (normally dry air but any inert gas can be used) to the substrate (see Figure 3). TWA has high efficiency, as most of the energy input is consumed to promote melting of the electrodes; hence, a low substrate temperature can be maintained as a hot jet gas is not employed. A large range of feeding rates can be configured depending on the electrical current employed and the electrical conductivity and melting point of the wire. In TWA, typical feeding rates range between 4 to 55 kgh^{-1}. The process can be employed for producing metallic coatings employing aluminum, copper, iron, molybdenum, nickel, tin, titanium, and zinc as feedstock. The benefits of TWA process are high spraying rates and low-cost consumables (AMTAG, n.d.).

Figure 3. Schematic of TWA process, adapted from (Tocalo Company, n.d.).

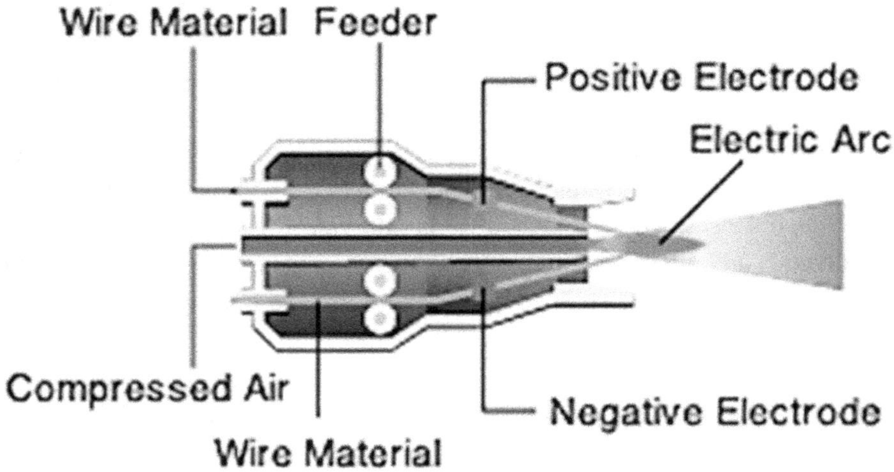

There are two main types of plasma spray processes available, i.e. atmospheric and vacuum assisted plasma spray. In both systems, the plasma is generated by passing an inert gas, or a mixture, between a high energy dc electric arc. Typically, high purity argon, nitrogen, hydrogen, or mixtures of them are employed for producing the plasma. Maximum temperatures achieved in the non-transferred plasma process can be up to ~ 19,000 °C depending on the plasmagen gas (He, H_2, N_2, Ar), although at the powder injection site, which is located few millimeters away from the nozzle exit, temperature values can range between 6,000 and 10,000 °C (Davis, 2004). The feedstock powder is introduced into the plasma plume by inert carrier gases suffering important heating and acceleration. Different powder feeding configurations can be employed in this process such as pneumatic, mechanical, or a combination of both. Due to the high temperatures reached in the process, the melting point of the feedstock material is not an issue; therefore, plasma spray is flexible in terms of the materials that can be sprayed, i.e., ceramics, cermets, and metals. The power outputs for commercial plasma guns range from 20 to 200 kW, and different gun configurations are available for different spraying requirements (internal and external surfaces, cascade arc, high and low power, triple-cathode, etc.) (Oerlikon Metco, n.d.). Spraying rates can vary depending on the feedstock material, gun type, powder injection system, selected gas mixture, etc.

Vacuum assisted plasma systems operate in a confined environment in which a close control of the atmosphere can be attained. In fact, the system works at a reduced pressure and, therefore, is also known as low pressure plasma spray (LPPS); operating pressure values are in the range from 10 to 50 kPa. At such low pressures, the plasma plume becomes larger in diameter and in length and, by employing specific nozzles, is accelerated towards the substrate. Because the amount of oxygen is drastically reduced, coatings with a low oxide content are produced by LPPS. High substrate temperatures are often recommended to produce LPPS coatings with high adhesion strength, as splashing of particles at impact can be significantly reduced (sometimes, this recommendation is also valid for other TS processes).

Compressed gas Expansion Spraying

Compressed gas expansion processes comprise TS techniques operating by employing compressed gases without the intervention of electrical energy and combustion reactions to induce melting of the feedstock material. These processes are also known as Cold gas spray (CGS) processes, in which a coating is formed by intensive deformation of particles upon impact at temperatures below the melting point of the sprayed material. The CGS process was developed in Russia in the mid-80´s at the Institute of Theoretical and Applied Mechanics. CGS employs a highly pressurized hot gas (Air, N_2 or He) to carry and accelerate powder particles through a convergent-divergent nozzle; particles can achieve velocities in the range from 300 to 1200 ms^{-1}(Papyrin et al., 2006). Due to the high kinetic energy transferred to the feedstock powder, particles can undergo severe plastic deformation upon impact, leading to high particle/substrate bonding and coating densification. In particular, metallic particles get bonded to the substrate when they reach a minimum velocity, also known as critical velocity, above which specific thermomechanical conditions are achieved at the particle/substrate interface at impact, allowing deformed particles to get high adhesion strength. Critical velocities have been estimated for several metallic materials and defined as a function of feedstock properties and particle sizes. The range of critical velocity reported for metals deposited by cold spray to date lies between 150 and 900 ms^{-1}(Viscusi et al., 2019). Figure 4 shows a schematic of the CGS process (Papyrin et al., 2006). Expansion of the carrier gas at the nozzle diverging zone accelerates particles and reduces their temperature; since gas temperatures (< 1000°C) are often below melting temperatures for most of metallic and ceramic materials, the feedstock particles are usually in solid state before impacting the substrate; in this way, this fact helps to minimize oxidation and phase transformations or microstructural undesired changes in the coatings obtained. CGS has attracted world-wide interest due to its high deposition efficiency, low cost, and feasibility to produce highly compact deposits with unlimited thickness.

Figure 4. Cold gas spray schematic, adapted from (Perez-Andrade, 2020).

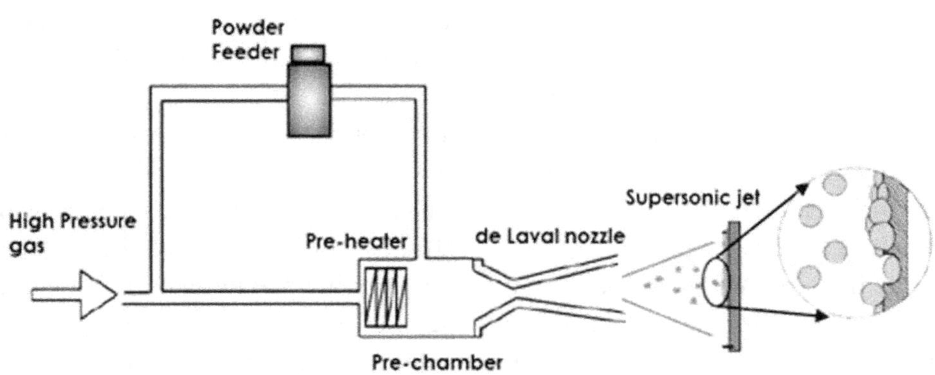

Stress State in Thermal Spray Coatings

Adherence is one of the most important aspects to control when thermal spray coatings are produced. Different factors interplay in defining the final adherence of a thermal spray coating, i.e., thermal expansion coefficient of substrate and coating materials, substrate temperature, particle velocity and temperature, spray distance, etc. Controlling residual stresses in TS coatings is paramount in order to fine tune adherence. Different residual stresses arise during coating fabrication, i.e., during coating build up and cooling. There are two different kinds of stresses arising during coating build up depending on the TS process employed; for instance, peening stresses are developed on the substrate when high-speed processes are employed, such as HVOF, HVAF or CGS. On the other hand, when using processes involving high temperatures such as atmospheric plasma or combustion flame spray, this kind of stresses are not generated. Peening stresses develop compressive residual stress state in coatings, which can be regarded as beneficial for some specific applications. Once molten particles impact the substrate, they quickly solidify with an associated contraction, which is restricted by adherence to the substrate (Matejicek, J., & Sampath, 2001); this process generates tensile stresses in the coating also known as quenching stresses. During coating-substrate cooling down, and due to the difference in thermal expansion coefficients between substrate and coating, thermal mismatch stresses are generated; such stresses can be either tensile or compressive depending on the thermo-physical properties of the materials employed. Hence, the final stress state of a thermal sprayed coating is dictated by the contribution of each stress state generated during coating build up and cooling.

Flame Spray Coatings and Modern Industrial Applications

Flame spray was first introduced in 1909 by Max Ulrich Schoop, being the first thermal spray process invented (Siegmann & Abert, 2013). In flame spray, the feedstock material (in powder or in wire shape) is fed into a flame produced by the combustion of a mixture of gases (acetylene or propane mixed with oxygen or air). As the feedstock particles flight through the flame, they are completely or partially melted and simultaneously accelerated towards the substrate. Upon impact, completely and partially molten particles are flattened producing splats, which adhere mechanically onto the substrate surface thanks to its controlled roughness. In this manner, flame sprayed coatings can have thicknesses that are typically in the range from 100 and 500 μm and are characterized for having a lamellar structure, in which pores are produced by failures in the stacking of lamellae or by the deposition of partially molten particles (Fauchais et al., 2014; Pawlowski, 2008). The flame spray process is known as the simplest and most economical thermal spray technique at present time.

The flame spray process was developed to manufacture metallic coatings; however, it is currently also used to spray ceramic particles, polymers, and cermets. Due to the thermal and kinetic characteristics of the flames employed in this process, flame sprayed coatings are usually highly porous and have low adhesion strength (Fauchais et al., 2014; Proner, 1999). It is worth mentioning that coating properties can significantly be improved by the optimization of processing parameters (i.e. it requires a proper selection of fuel and oxidizing gases, stand-off distances, feeding rates etc.). Among various oxy-fuel flames, the oxy-acetylene flame is the most energetic, delivering an enthalpy value up to 1300 $kJmol^{-1}$ and reaching a maximum flame temperature and flame velocity of ≈3170 °C and ≈12 ms^{-1}, respectively (Glassman et al., 2014; AGA company, n.d.). Currently, the oxy-acetylene flame is one of the most preferred thermal sources in flame spray because of the properties mentioned above. Nevertheless, in some cases, propane-oxygen (Brogan, 2000; Palathai et al., 2008)or propane-air mixtures are also used to prepare flame sprayed coatings, mainly, when low melting temperature feedstock materials are sprayed (*i.e.* polymers and low-melting point metals) (Leivo et al., 2004).

The combustion reaction of an oxy-acetylene flame is carried out in two stages. Firstly, an equimolar mixture of oxygen and acetylene reacts under the action of a spark (i.e. ignition source) with a temperature close to 300 °C (Eddings, 2005). This reaction is carried out in the primary zone of the flame, in which carbon monoxide and hydrogen are formed according to the following equation (1):

$$2C_2H_2 + 2O_2 \rightarrow 4CO + 2H_2 \qquad (1)$$

The primary zone of the flame is shaped like a blue cone, being the region with the highest temperature of the whole flame. The second stage of the combustion reaction takes place when carbon monoxide and hydrogen, produced during the formation of the primary zone, react with oxygen in the surrounding air, yielding what is known as the secondary zone of the flame. At this zone, the reaction products are carbon dioxide and water vapor as described in equation (2). The secondary zone has a plume shape and is the coldest and darkest region of the flame.

$$4CO + 2H_2 + 3O_2 \rightarrow 4CO_2 + 2H_2O \qquad (2)$$

Then, the overall reaction corresponding to the complete transformation of carbon into carbon dioxide, as well as, hydrogen into water, is shown in the following equation (3):

$$2C_2H_2 + 5O_2 \rightarrow 4CO_2 + 2H_2O + 1300\text{kJ/mol} \qquad (3)$$

In practical terms, an oxygen volume 2.5 times higher than that of acetylene is required to produce a neutral (stoichiometric) flame. When the combustion is the result of an acetylene and air mixture, the volume of air supplied to the process must be 11.9 times greater than that of acetylene (Eddings, 2005). The oxygen volume that must be supplied from the oxygen cylinders to the spraying torch in order to produce the primary zone in a neutral flame must be 1.7 times higher than the volume of acetylene. The volume of oxygen required to produce the chemical reaction corresponding to the secondary zone of a neutral flame is taken from the air of the surrounding atmosphere (Fauchais et al., 2014).

When a higher volume of oxygen is provided to the process, an oxidizing flame is produced; otherwise, a higher amount of acetylene produces a fuel rich flame (Fauchais et al., 2014; AGA Company, n.d.). In either case, the volume of gas that does not participate in the chemical reaction, does not contribute to heat generation. Instead, it is heated with the energy produced by the combustion reaction, causing the reduction of the flame temperature (Fauchais et al., 2014). On the other hand, as the oxygen content supplied to the torch increases, the length of the primary zone of the flame also increases (i.e., the length of the region with the highest temperature of the whole flame is enlarged due to the excess of oxygen) see Figure 5 (AGA Company, n.d.; Cadavid et al., 2016).

Figure 5. Thermal and physical characteristics of an oxy-acetylene flame obtained with different volumetric ratios of C_2H_2:O_2. a) Flame temperature, adapted from (AGA Company, n.d.); b) Length of the primary and secondary zone. (Thermal Spray booth at University of Antioquia, Courtesy of Edwin Cadavid, adapted from (Cadavid et al., 2016)).

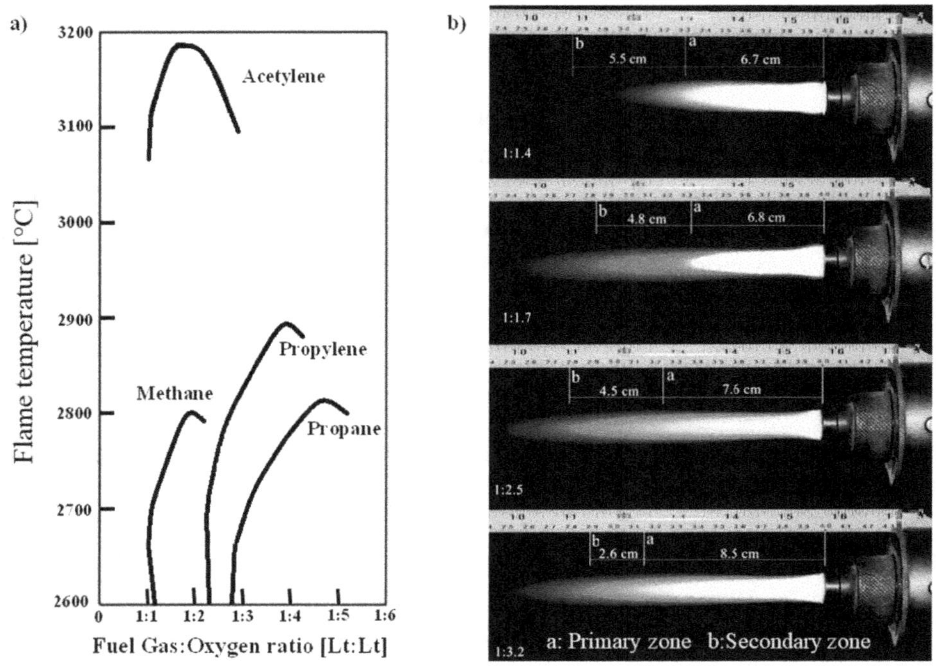

Typical flow rates of 30 lmin^{-1} of oxygen and 18 lmin^{-1} acetylene produce a neutral flame with a power level of about 40 kW (Fauchais et al., 2014); these flow rates are widely used to manufacture metallic coatings. However, the use of an oxidizing flame increases the residence time of particles in the primary zone of the flame, which could be helpful to favor melting of particles with low thermal conductivity and high melting points. As a result, ceramic powders are often sprayed by using oxidizing flames despite the lower temperature of these flames with respect to the neutral one (Ferrer et al., 2017; Pinzón et al., 2018). The combustion of a stoichiometric mixture of acetylene and oxygen with a 1:1.7 C_2H_2:O_2 volume ratio produces a flame whose primary zone has a maximum temperature ≈3170 °C and 6.8 cm in length. As the oxygen added increases up to a 1:2.5 C_2H_2:O_2 ratio, the maximum temperature decreases to approximately 3080 °C, but the length increases to 7.6 cm, and with an even greater oxygen enrichment generating a 1:3.2 ratio a maximum temperature of ≈2980 °C and a length of 8.5 cm are produced (AGA Company, n.d.; Cadavid et al., 2016).

Flame sprayed coatings often have greater porosity values than 10% and adhesion strength values generally below 30 MPa (Fauchais et al., 2014; Proner, 1999). In this manner, optimizing the flame spray process has been a key task in recent years to obtain the lowest porosity and the highest adhesion strength values in coatings produced by this method. This fact is fundamental to improve the performance of these coatings, mainly, when they are manufactured for some specific applications, including corrosion protection and mechanically aggressive structural applications. For instance, a previous study (Redjdal et al., 2013) reported the fabrication of FeCr coatings by using a neutral flame. In order to optimize the residence time of FeCr particles in the neutral flame, the feedstock powder was injected into the torch by setting a controlled flow. As a result, the FeCr coatings deposited by flame spray had a good structural quality, presenting porosity values between 1.7% and 3%; however, adhesion strength values only achieved a range between 5 and 11 MPa (Redjdal et al., 2013). Similarly, flame spraying of self-fluxing alloys, including NiCrBSi, promotes almost a complete elimination of the pores in the coatings obtained. Interestingly, a metallurgical bonding can be promoted on the coating-substrate interface by performing a melting treatment on the self-fluxing coatings (Fauchais et al., 2014; Rodriguez et al., 2003; Herman et al., 2000). On the other hand, alumina-titania coatings have also been prepared by flame spray and by using a feeding rate of 11 g/min of alumina ≈13 wt.% titania powder, a particle size distribution between 5-30 μm, an oxidizing flame with a 1:3.2 C_2H_2:O_2 ratio, and a stand-off distance of 9.5 cm. The coatings obtained under those spraying conditions had a good structural quality and low porosity values < 1%, see Figure 6 (Pinzón et al., 2018).

Figure 6. Flame sprayed coatings with low porosity values. a) Sprayed and fused NiCrBSiFe alloy, Courtesy of Bonem S.A.; b) Al_2O_3≈13 wt.% TiO_2, adapted from (Pinzón et al., 2018).

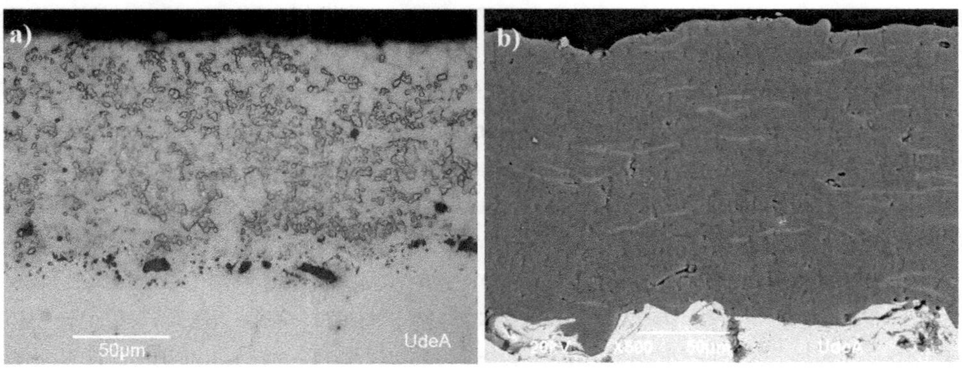

Overall, self-fluxing alloys are nickel- or cobalt-base alloys containing boron and silicon and, in some cases, chromium, carbon and iron. Boron and silicon elements serve as fluxing agents (hence the term self-fluxing) limiting oxidation of nickel, cobalt, chromium, and iron; these elements also control surface tension and fluidity of the alloy. Boron is also a potent hardener for both nickel and cobalt-based materials and combined with silicon reduces the melting temperature of the alloy by several hundred degrees due to the formation of eutectic alloy compositions. The chromium in these alloys is added to provide wear and corrosion resistance (Fauchais et al., 2014; Herman et al., 2000; TWI-global, n.d.).

After spraying, self-fluxing alloys are generally fused by heat treating, which is usually carried out by using an oxyacetylene or propane torch or a furnace, heating the coated surface over 900 °C. The result is a fully dense coating, metallurgically bonded to the substrate, with high hardness values ranging from 20 to 70 HRC (depending on boron and carbon content), and good impact resistance. These properties, attributed to exceptional coating integrity and metallurgical bonding, also offer good corrosion resistance and high hardness even in hot environments, allowing its service up to 650 °C. During the fusing process, a borosilicate glass phase is formed in the flame sprayed coating, which promotes the formation of a compact and smooth surface (Davis, 2014; TWI-global, n.d.).

Although flame spraying was developed more than one hundred years ago, it is still employed in modern industry for the fabrication of coatings for various applications. In recent years, the flame spray process has also been used for the functionalization of microparticles. Some examples of flame sprayed coatings applications are described in this section. For instance, NiCrBSiFe and NiCrBSi-WC self-fluxing alloys are widely used as flame sprayed coatings for agriculture and mining industry thanks to the hardness values achieved, usually higher than 60 HRC, and low friction coefficients, both giving excellent coating resistance against low-pressure abrasion and erosion (Castolin-Eutectic Eutalloy SF, n.d.; González et al., 2007). NiCrBSi also are used to coat paper machine rolls (Fauchais et al., 2014). Additionally, these self-fluxing coatings have exceptional impact resistance and a low friction coefficient. Flame sprayed NiBSi self-fluxing coatings are used to protect or repair cast iron and steel molds for plastic and glass industries, as well as, in gear teeth and pump bodies, valve seats, shafts, eccentrics, bearings, soldering tungsten carbide on drilling stabilizers and other parts for petroleum and gas industry requiring high metal-to-metal wear resistance (Franco et al., 2016; Castolin-Eutectic Eutalloy SF, n .d.).

Alternatively, flame sprayed zinc and aluminum coatings are widely used for corrosion protection of steel structures as bridges and offshore platforms. Although aluminum powder is expensive, it is used as coating due to its corrosion resistance to saltwater and acidic gaseous atmospheres. In addition, zinc coatings have resistance to alkaline corrosion (TWI-global, n.d.). These coatings are commonly used as an alternative to galvanizing.

Meanwhile, flame sprayed WC-Co cermet powders result in hard (> 60 HRC), tough, and dense coatings, which are mainly used for sliding wear, erosion, fretting, and abrasion applications at service temperature up to 480 °C. These types of coatings are often suitable for pump housings, capstans, rollers, exhaust fans, pipes, elbows dust collectors, rotating valves, and wire drawing (Saint Gobain Coating Solutions-TI-ELITE, n.d.; Oerlikon Metco, n.d.).

On the other hand, flame sprayed ceramic coatings are mainly employed for textile, steel, and glass industries. For instance, coatings manufactured from fine particles of Al_2O_3-TiO_2 (13 to 40 wt.% TiO_2) are used to protect textile friction parts as thread guides, guide bars, feed separators, as well as, capstans, pumps bodies, pump sleeves, shafts, pistons, cylinder rods, and pin guides. The high hardness of these coatings (between 700 and 1000 HV_{2N}) and their low porosity result in excellent sliding wear resistance (Saint Gobain Coating Solutions-TITEX, n.d.). Al_2O_3 – 13wt % TiO_2 coatings are harder than those with 40 wt.% TiO_2, but they are also more fragile (Vargas et al., 2010). Alternatively, flame sprayed Al_2O_3 -25 wt.% ZrO_2 coatings, despite their moderate hardness (≈600 HV_{3N}), have good high temperature fracture toughness, which allows them to be used for applications requiring abrasive wear resistance, thermal shock resistance, and thermal cycling resistance up to 700 °C. In fact, these coatings are employed in parts being in contact with molten glass and refractory funnels (AZ Supra Thermal Spray Flexicord; n.d.). Additionally, flame sprayed Al_2O_3 -25 wt.% MgO spinel coatings have a low wettability in contact with aluminum and zinc, as well as good stability up to 1400 °C. These coatings have good corrosion resistance when they are exposed to most acids and alkalis and a medium abrasion resistance, which allows them to be used as thermal barrier coatings for crucibles, chutes, molds, and funnels, as well as in feed rollers for galvanized wires manufacturing, steel converter chimneys, and other parts in contact with molten aluminum and zinc (Spinel Supra Thermal Spray Flexicord, n.d.).

Significant amounts of polymer feedstock powders have been sprayed by the flame spray process due to the excellent properties of these type of materials. Flame sprayed polymer coatings have been employed as protective coatings for ladder racking, bridges and municipal handrails, river tugboat hulls and barges, wastewater clarifiers, wastewater surge tanks, lighting poles, and coating of exhaust systems for submarines, among others (TWI-global, n.d.; Petrovicova & Schadler, 2002).

In recent years, the flame spray process has also been employed for the atomization of microparticles; the advantage of flame-atomization is related to the modification of the original physical properties of the processed powder. For instance, during the atomization process, as the particles are melted by the flame, they rotate on their own axis, producing enough superficial shear forces and leading to spheroidization. In some cases, these microparticles may consist of elements or compounds that volatilize at the atomization temperature. The gases generated during the process escape towards the surface of the particles, producing different superficial porosity patterns, see Figure 7.

Figure 7. Atomization of microparticles. a) Stages of the process; b) Spheroidization scheme. Courtesy of Freddy Vargas, Óscar Iván Rojas and Luis Guillermo Angarita.

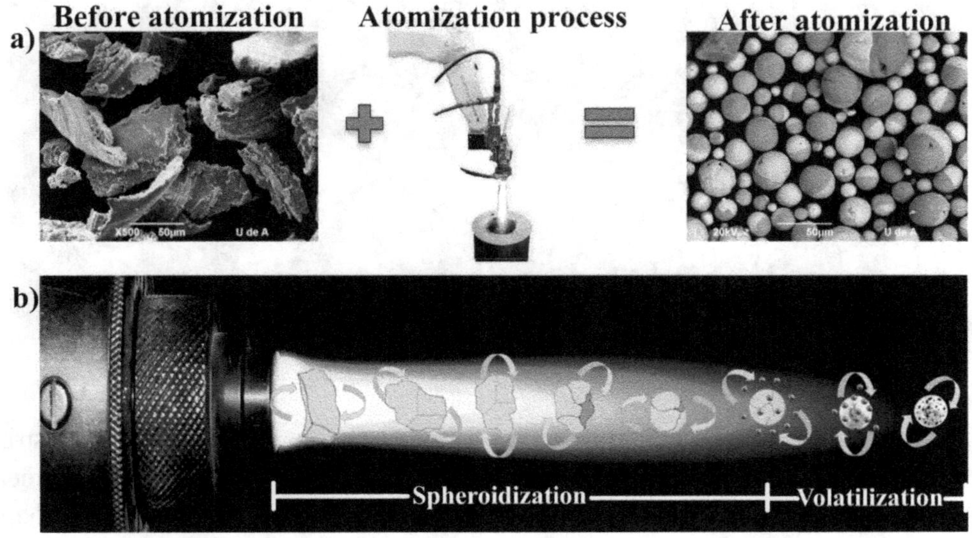

Irregular-shaped microparticles of alumina as well as slags produced by melting of metals, machining shavings of copper alloys, and bioactive glasses have been atomized by employing an oxyacetylene flame, achieving a high degree of spheroidization and improving their flowability, see Figure 8. For instance, the high surface area of spheroidized alumina microparticles can improve the thermal conductivity and hardness of various types of resins and rubbers used in the electronics industry, when the microparticles are added to these materials (Denka, n.d.). Additionally, the low density of highly porous microspheres obtained from slag atomization makes them useful as extenders and lightening materials in oil well drilling (Poirier & Quercia, 2015). Likewise, the porosity obtained in bioactive glass microspheres can allow them to be used as nano-carriers in biomedical applications (Hossain et al., 2018), while the microspheres obtained from copper shavings could be used as feedstock powder to produce coatings for wear applications.

Figure 8. Microspheres obtained by flame atomization. a) Alumina, courtesy of Freddy Vargas; b) Fusion slags, courtesy of Freddy Vargas; c) Bioactive glass, courtesy of Óscar Iván Rojas; d) Machining shaving of a CuAlZn alloy, courtesy of Efraín Gómez and Daniela Soto.

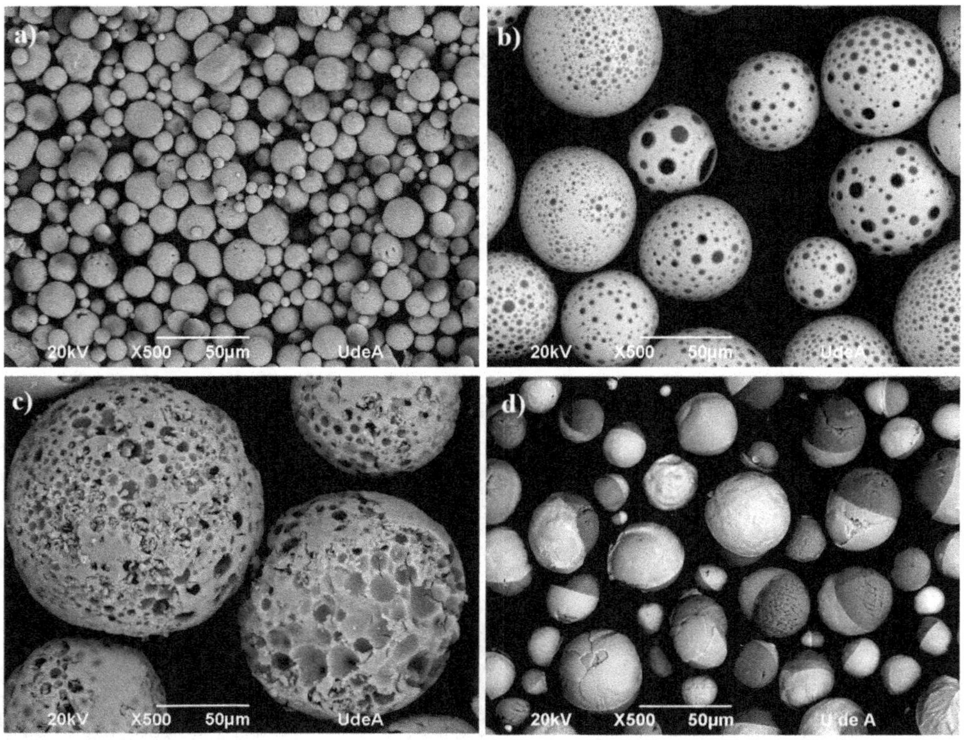

Overall, the versatility of the flame spray process has allowed its usage for different industrial applications. The flame spray process is often used to produce metallic, cermet, ceramic, and polymer coatings, including applications involving parts exposed to different conditions and aggressive environments. The flame spray process is also useful for functionalizing microparticles and re-incorporating solid waste into productive processes, offering an added value to the functionalized products.

High Velocity Oxygen Fuel (HVOF) Spray Coatings and Modern Industrial Applications

The high velocity oxygen fuel (HVOF) spray technique is classified as one of the combustion-based deposition processes. HVOF was first discovered by Union Carbide (Thorpe & Richter, 1992), nowadays known as Praxair Tech. HVOF was developed after great efforts to produce TS coatings with better mechanical properties than those obtained by conventional flame spray processes. The HVOF process was commercially introduced in the 1980s by JetKote (Deloro Stellite, Inc.) (Li et al., 1999) and, after more than 40 years in the market, is currently one of the most preferred TS processes for industrial applications. The basic operating principle of the HVOF process is based on the combustion of gases, where propane, hydrogen, and kerosene are the most used fuels; oxygen and compressed air are often employed as oxidizing agents. High pressure fuel and oxidizing gases are injected into a combustion chamber, producing a combustion flame that heat and accelerate the feedstock material (spherical or

Figure 9. Schematic representation of a DJH gun design and its operation, adapted from (HVOF Guns Oerlikon Metco, n.d.).

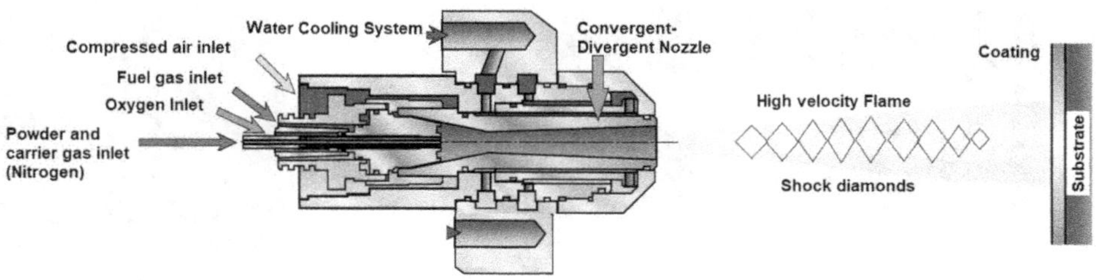

irregular powder) through a supersonic nozzle. The particles travel into the hot gases and can achieve supersonic velocities; heat transfer makes them to melt, partially or completely. Particles flatten plastically, cool down, and solidify upon impact to form the coating. HVOF coatings show often high density and a homogeneous microstructure due to the high kinetic energy of the particles at impact and the short residence time of the particle in the process (Fauchais et al., 2014).

HVOF systems are classified into 3 types, divided according to the combustion chamber pressure employed. The first, second, and third type of guns are designed to support pressures between 3 and 5 bar, 5 and 10 bar, and up to 25 bar, respectively. This classification is also known as the first, second and third generation of HVOF guns and has emerged due to the necessity of increasing the velocity of the particles and reducing their temperature. The first generation of HVOF guns was developed by using a barrel-type nozzle (maximum particle velocities between 350 and 450 ms^{-1}), whereas the second and third generations were developed by using a convergent-divergent nozzle (maximum particle velocities between 500 and 700 ms^{-1}). Most of HVOF guns use a water-cooling system (there are some devices only using air-cooling), which is intended to improve the thermal efficiency of the process (max. power delivered 100-300 KW) (Fauchais et al., 2014).

Figure 10. DJH2700 Gun operating with Propane and Oxygen, the figure displays the combustion flame obtained (Thermal Spray booth at CIATEQ A.C.; Courtesy of Juanita Hincapie Bedoya).

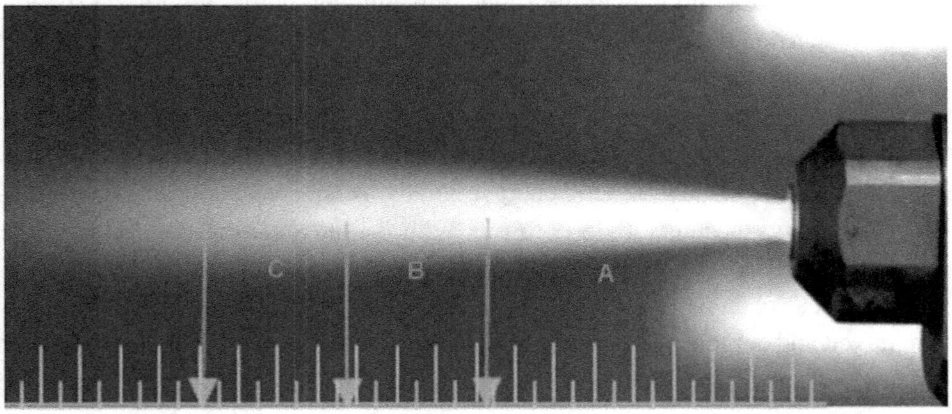

Figure 11. Relationship between coating microstructure and processing parameters in the HVOF process.

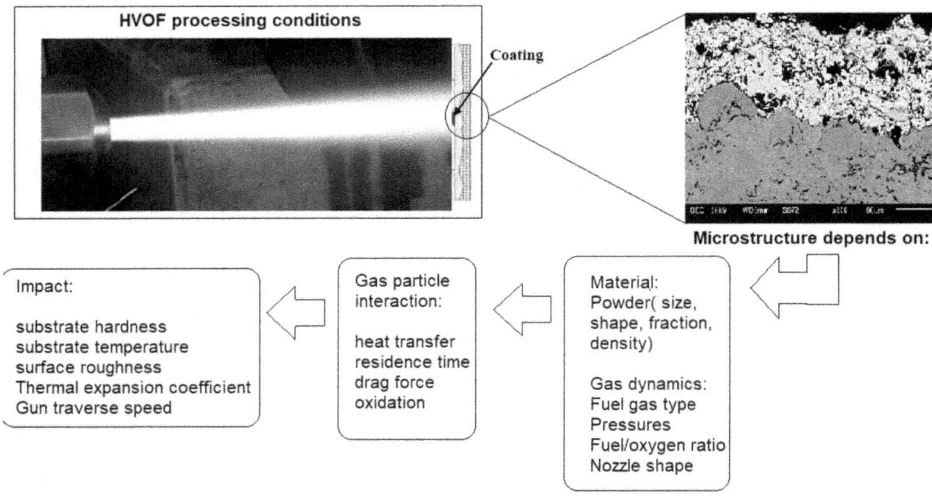

One of the most successful HVOF guns in industry is the Diamond Jet Hybrid (DJH) gun with axial powder injection. Figure 9 shows a schematic representation of a HVOF-DJH system. In this process, premixed fuel gas and oxygen are injected into the gas mixing zone, where they react to produce a combustion flame. The hot gases and the air injected from the annular inlet orifice are conducted by a convergent-divergent nozzle where they are compressed and expanded to reach supersonic velocities. The gun air cap is cooled by using air and water to prevent any damage. The feedstock powder is injected at the central inlet nozzle using nitrogen as the carrier gas. Consequently, rapid momentum and heat transfer between the gas and the powder results in the acceleration and heating of particles, which later deform and quickly solidify upon impact on the substrate surface (HVOF Guns Oerlikon Metco, n.d.).

Most HVOF systems are characterized by the formation of "shock" diamond patterns in the high velocity flame just at the exit of the nozzle (Fauchais et al., 2014). The diamonds are a characteristic of the supersonic nature of the HVOF-flame, as shown in Figures 9 and 10. The HVOF jet presents 3 main zones, see Figure 10; zone A is characterized by the formation of shock diamonds. One can observe that there is a certain number of diamonds in the flame, this number can vary with the fuel/oxygen ratio. Therefore, diamonds can be a good indication of how much kinetic energy is delivered by the jet. A similar phenomenon is observed in high performance jet engines exhausts. In fact, HVOF guns can be as noisy as a jet engine exhaust, reaching noise levels ranging from 125 to 133 dBA (permissible exposure limit for humans 90 dBA) (United States Department of Labor, n.d.). Zone B represents the disappearance of the shock diamonds and the loose of kinetic energy. Finally, zone C is the region where the flame loses kinetic and thermal energy because it starts to mix with the surrounding atmosphere. Then, particle temperature at impact will be a function of the zone A length and stand-off distance.

The physicochemical phenomena occurring in the HVOF process can be divided as follows: first, when gases are mixed and combustion begins, a conversion from chemical energy into heat occurs; then, when the combustion flame is compressed and expanded within the convergent-divergent nozzle, there is a conversion of thermal energy into kinetic energy. Once the feedstock particles are injected into the hot gas stream, momentum and heat transfer from the gas to the powder occurs. Finally, particles travel a short time into the flame, melt, and impact on the substrate surface. These processes occur almost simultaneously and are vital when optimizing the spraying process.

Table 1. Summary of main HVOF equipment available in the market and suppliers.

Liquid Fuel Systems	Gas Fuel Systems	Supplier	Ref.
Model JP-5000® SystemModel JP-8000™ System	Model 7700GFHV-2000	Praxair	(Praxair HVOF Guns, n.d.)
WokaStar-610-Sz®WokaJet-410-Sz ® Series	Diamond Jet™ Series	Oerlikon Metco	(HVOF Guns Oerlikon Metco, n.d.)
	Jet Kote® II and III Series	Deloro Stellite	(Delloro Stellite Guns, n.d.)
GTV K2 gun	GTV GLC GUN	GTV	(GTV Guns, n.d.)
CJS K4.2 (kerosene + H_2 fuel)ID CoolflowID Coolflow Mono	-	Thermico	(Thermico Guns, n.d.)
eGun system - ultra high pressure HVOF	-	Flame Spray Technologies	(Flame Spray Guns, n.d.)
MET-JET4L	-	Metallisation	(Metallisation Spray Guns, n.d.)
CastoJet® - CJK5		Castolin Eutectic	(Eutectic Guns, n.d.)
XPOJET® 5000 MJP-6000™ GUN	HIPOJET® 2700 HIJET® 9610	The Metallizing Equipment Company Pvt. Ltd. (MEC)	(Metallizing Company Guns, n.d.)
Convertable HVOF-HVAF Multipurpose HVAF	HVAF	Kermetico	(Kermiko HVAF Guns, n.d.)
-	M2™ Supersonic HVAF	Unicoat	(Unicoat Guns, n.d.)

Final properties of HVOF-sprayed coatings depend on several adjustable processing parameters, including particle size distribution, feeding rate, stand-off distance, fuel/oxygen ratio, substrate type, and cooling. Particle size is an important parameter in the HVOF process since particles can interact in a different way with the gas stream depending on their size. For instance, small particles can be accelerated and heated up, reaching high velocities and temperatures in a short time. However, when they leave the main jet core of the flame, their temperature and velocity decay rapidly. For particles of large size, plume velocity and heating rates are more stable than for small ones. The stand-off distance is defined as the distance between the nozzle exit and the substrate surface; this parameter is also important since it determines the particle conditions (speed and temperature) at which they impact the substrate. Substrate type and cooling will influence the solidification rate and the level of transformation of kinetic energy into viscoplastic deformation of the particles upon impact. Overall, the microstructure of the HVOF coatings results from the deformation, solidification, and sintering of the deposited particles. The way how particles arrive to the substrate surface and how they behave upon impact depends on the behavior of in-flight particles and on the substrate properties. It means that substrate hardness and temperature as well as particle temperature, velocity, and melting ratio determine the properties of the coating (Fauchais et al., 2014). Figure 11 summarizes the relationship between the coating microstructure and the main parameters involved in the HVOF process.

Figure 12. Turbine blade coated with a TBC system. The bond coat was prepared by using HVOF. (Thermal Spray booth at CIATEQ A.C.; courtesy of Angel Cámara).

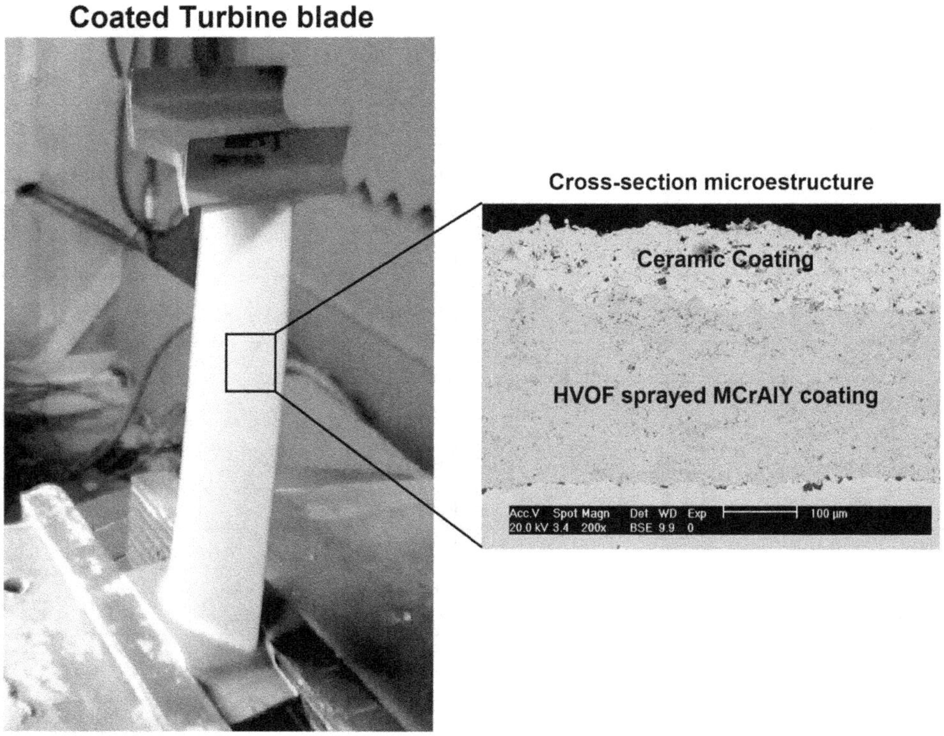

HVOLF systems employing liquid fuels are an alternative to fuel gas-based high velocity techniques; in fact, guns with general designs for gas fuel were also developed for liquid fuel. HVOLF often employs kerosene as liquid fuel instead of gas and oxygen instead of air. The use of kerosene and oxygen allows very high levels of power (up to 300 kW) and high hot gas velocities (2000-2200 ms^{-1}). With these guns, powders can also be injected radially or axially. The use of a liquid fuel simplifies the spraying process and improves operational safety (Kamnis & Gu, 2006).

Recently, Wire-HVOF (W-HVOF) has been developed by Metallizing Equipment Company Ltd (patent number 214843 2529/DEL/1997) as an alternative technique for producing thin coatings. Usually, HVOF guns are able to produce coatings above 50 microns; however, W-HVOF can also produce coatings with thickness values ranging from 10 to 40 µm. This system provides supersonic spray velocities and allows an improved heating and melting of feedstock material in wire shape. W-HVOF has a good potential when dense, high adhesion strength, and thin coatings (<50 µm) are required. This process is also economically and industrially attractive compared with physical vapor deposition (PVD)-based processes (Tailor et al., 2019).

The HVAF process is also based on HVOF systems, whereas air is employed instead of oxygen to complete the combustion process. HVAF coatings are similar to those produced by HVOF; however, its main advantage lies in the possibility of producing less oxidation of particles during the spraying process. HVAF is recommended when minimum oxidation and high bonding strength are required. One characteristic of HVAF coatings is their high fatigue and delamination resistance. This fact allows the fabrication of very thick coatings (from tens of microns to several millimeters) (Verstak & Baranovski, 2006; Matikainen et al., 2020).

Figure 13. Stainless-steel customized hip implant manufactured by 3D-printing (additive manufacturing department at CIATEQ A.C.). The multilayer coating was deposited by HVOF to provide bioactive properties to the hip prosthesis.

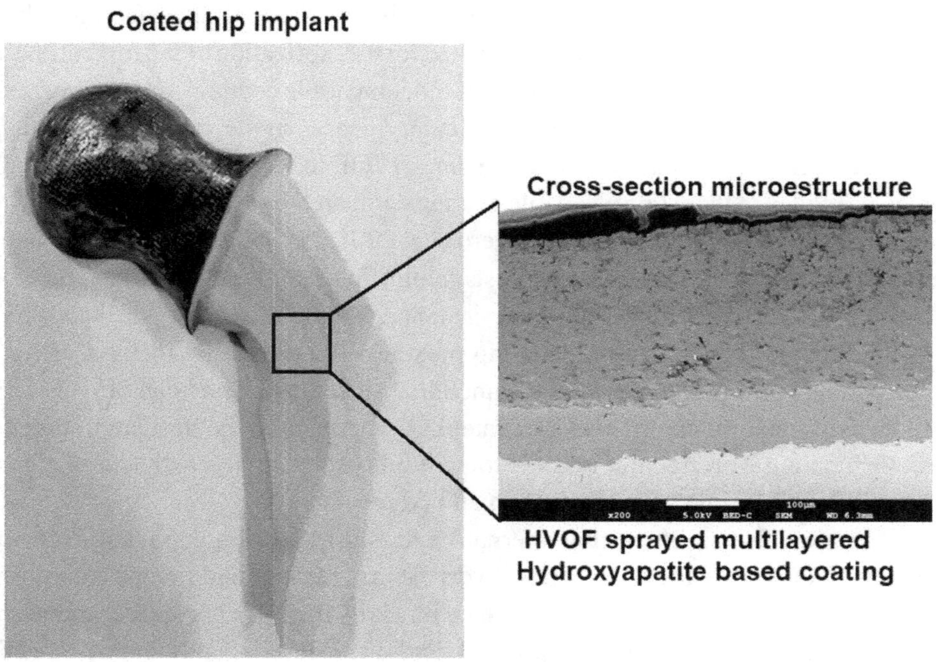

Nowadays, there is a wide range of HVOF guns in the market, the variety of commercial guns involves the use of gas or liquid fuel, water or air cooling, various nozzle designs, and axial or radial feedstock injection. Main suppliers of commercial high-velocity combustion spray systems are summarized in Table 1.

It is well known that the major advantage of HVOF is the high kinetic energy of the process, which allows the fabrication of coatings with high density. Other benefits include smooth surface finishes and low oxide levels. HVOF processes are suitable for metals and intermetallic compounds; some of the most recognized materials deposited by this technique are Tungsten Carbide-Cobalt (Diamalloy and Woka), Nickel-Aluminum, MCrAlY type-alloys (where the M is a metal of either Nickel, Cobalt or a combination of both), Cobalt-Molybdenum-Chromium (Triballoy), and Nickel-Chromium-Molybdenum (Hastelloy). The mentioned alloys have been traditionally applied as coatings for corrosion, wear, oxidation/sulfidation resistance, high temperature service, and restoration.

Industrial applications of HVOF coatings include the preparation of coatings for turbines in the aerospace industry. In this market, hot corrosion resistance, oxidation resistance, thermal insulation, and abradability are crucial for the operation of the turbine. With recent advances in turbine design, materials require to support higher temperatures than in the past. In addition, HVOF coatings are applied to support the application of heat-insulated resistant ceramics, typically called Thermal Barrier Coatings (TBCs).

The ceramic coating is usually applied by other techniques such as plasma spray on top of an oxidation resistant coating applied by HVOF. The benefits are improved mechanical properties and service life of the turbine component. The oxidation resistance HVOF coating is often an MCrAlY alloy. Figure 12 shows a TBC coating prepared by HVOF (bond coat) and plasma spray prepared on a turbine blade.

Currently, HVOF coatings are used in many hot section areas of a turbine, including combustors, compressors, liners, combustor cans or heat shields, and various stationary and rotating blades. Some sections require HVOF coatings for sealing. In this sense, abradable coatings are a good choice; they must be readily abradable and must resist impact erosion by abrasive particles moving at high velocities.

In recent years, HVOF coatings have been employed for the fabrication of biocompatible coatings. The HVOF process is advantageous for the fabrication of biocompatible coatings since temperatures involved in the process do not favor the precipitation of secondary phases in the sprayed material. In particular, hydroxyapatite coatings have been deposited by using HVOF in recent years (Henao et al., 2018). Although there is no evidence that hydroxyapatite coatings prepared by HVOF are currently commercially available, numerous studies (Li et al., 2004; Melero et al., 2018; Bolelli et al., 2014; Henao et al., 2018) have suggested their advantages after *in-vitro* tests, indicating that these coatings can be commercially available in the biomedical industry soon. For instance, Figure 13 displays a hydroxyapatite-based coating deposited by HVOF on a customized hip prosthesis fabricated by metal additive manufacturing. This type of applications is considered as functional since the coating can provide various surface properties to the base material in order to integrate the component to the host environment. In the case of protheses, the coating material provides biocompatibility, bioactivity, and acts as a barrier to avoid metal ion dissolution when implanted in the human body.

HVOF coatings are also an interesting alternative for slurry erosion applications in various industrial scenarios, such as processing of minerals, hydroelectric plants, mud pumps etc. Slurry damage to metallic parts represents an industrial problem since it results in costly preventive and corrective maintenance, performance reduction of the components, and increase of operational costs. HVOF coatings can be an excellent alternative to produce dense coatings employed when high mechanical resistance is required for slurry erosion damage. In particular, Tungsten Carbide-Cobalt-Chromium HVOF coatings have demonstrated to be effective in the production of dense coatings with high erosive wear resistance (Santacruz et al., 2019).

Similarly, air jet erosion is another recurrent problem occurring in the power generation industry; this phenomenon is usually observed in steel surfaces by the presence of fly ash in the air. Air jet erosion damage is a problem that results in the failure of steel components and increases operational costs. Recent works have demonstrated that Chromium Carbide- Nickel Chromium alloys display good erosive wear resistance in a wide interval of temperatures (200°C-600°C) (Nautiyal & Tyagi, 2020).

HVOF coatings are currently considered as an alternative for hard Chrome plating (electroplating). When seamless steel tubes are severely worn due to the flow of a liquid component, electroplating is a conventional method to produce protective films and to restore the functionality of the damaged tube. However, this process is limited to produce thin coatings, is time consuming and hazardous. In particular, manufacturers have been searching for new alternatives due to new environmental regulations and to ensure promptness in the restoration of worn parts. The HVOF process is an interesting option for replacing electroplating since it does not produce any hazardous material and can be employed to fabricate coatings in few hours (even minutes). Recently, Chromium-Molybdenum, Chromium Carbide-Nickel Chromium, and Nickel-Chromium HVOF coatings have demonstrated to be excellent options to hard chrome plating repair of worn surfaces; these coatings often possess high hardness (800-900 HV) and low sliding wear rates (Karuppasamy et al., 2019).

Overall, modern industrial applications of HVOF coatings are associated with the functionality of the coating material and the capability of the HVOF process to provide a coating on a required part. Coating functionalities include (and are not limited to) abrasive wear, sliding wear, erosion cavitation wear, hot corrosion resistance, bioactive behavior, magnetic behavior, and dielectric properties. The HVOF process is currently present in various industrial sectors, solving different industrial problems; for instance, in aeronautics, petrochemical, electronics, nuclear, agriculture, maritime, military, diesel industries, railroad, biomedical, and automotive.

Figure 14. Plasma Spray gun with axial particles injection, adapted from (Mostaghimi et al., 2017).

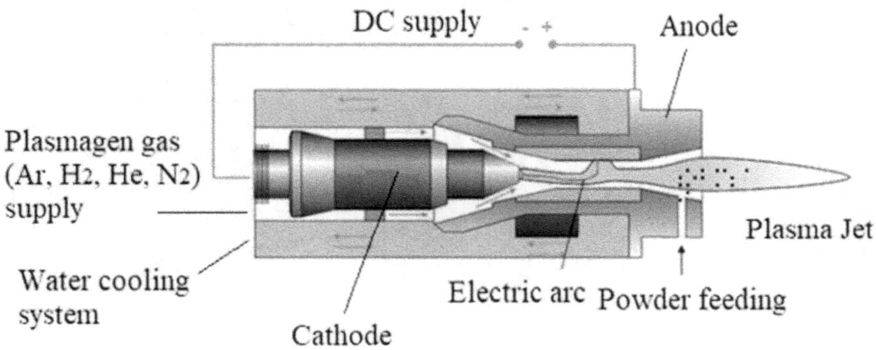

Plasma Spray Coatings and Modern Industrial Applications

Plasma spraying technology was developed after Second World War. In the fifties, arc plasma engineering was employed to develop systems for processing materials. The development of a small direct current (DC) non-transferred arc system gave rise to the plasma spray process (Berndt et al., 1992). Currently, plasma spray is one of the leading and most versatile technologies employed for manufacturing coatings, i.e., a large variety of materials can be processed from hundreds of micrometers up to few millimeters in thickness. The available sprayable materials range from superalloys to refractory intermetallic compounds and ceramics (Vardelle et al., 2015).

The thermal source employed to melt the feedstock material in plasma spray consists of a plasma jet that is created by the flow of gas through a high energy electric arc. Plasma gases such as nitrogen, hydrogen, helium, and argon (or a mixture of them) flow around a thoriated tungsten cathode and a hollow water-cooled copper anode. The plasma is initiated by a high voltage discharge, which causes localized ionization and a conductive path for a DC arc to form between cathode and anode. The resistance heating from the arc causes the gas to reach extreme temperatures, dissociate, and ionize to form a plasma, Figure 14. The plasma exits the anode nozzle as a free or neutral plasma jet (Pawlowski, 2008). As it happens for all thermal spray processes, the feedstock powder is rapidly heated and accelerated towards the substrate surface (particle velocity values are in the range between 100 and 350 ms^{-1} depending on particle size, plasma power, and other processing parameters). Stand-off distances employed in this process have values from 40 to 180 mm. Plasma spray is commonly used in mass production of different components since this process can be easily automated. In some specific cases, such as vacuum assisted plasma, plasma sprayed coatings can be much denser, cleaner, and with higher adherence than other thermally sprayed coatings (Apelian et al., 1983; Aruna et al.; 2017).

There are multiple key aspects in the plasma spray process to bear in mind; for instance, any variation in the plasma jet stability can influence the overall performance of the torch, affecting the particle heating and final coating quality (Duan & Heberlein, 2002).

Figure 15. Feedstock entry in a system radial and axial injection of particles, adapted from (Ganvir, 2016).

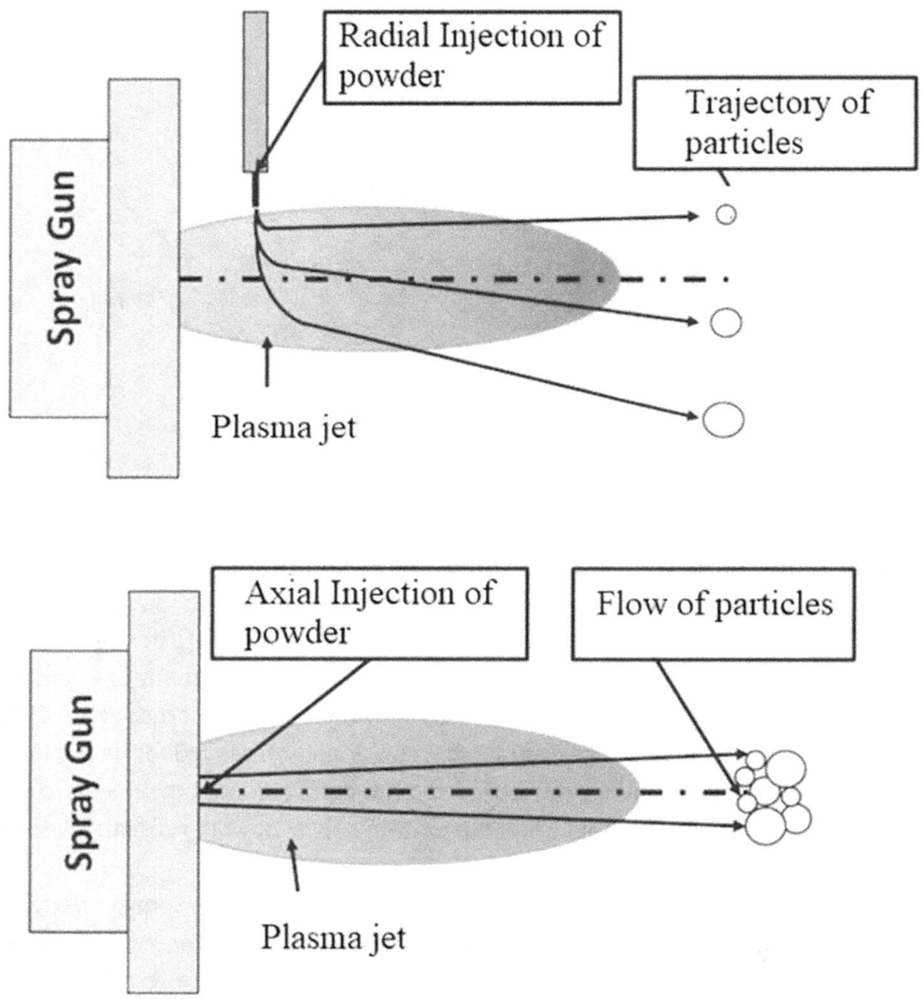

The feedstock powder injection is another key parameter in this process since the way particles are injected into the plasma jet can cause remarkable differences in their trajectory and thermal history. Feedstock powder can be injected either radially or axially in the plasma jet, as shown in Figure 15. Particle features such as size distribution, shape, and density are also important since they determine the particle path followed in the plasma jet. Due to the high thermal gradients achieved in the plasma spray process, size distribution and feedstock injection configuration are very important to obtain homogeneous coatings since particle melting and deposition strongly depends on their thermal and kinetic history in the plasma jet (Berndt et al., 1992; Fauchais et al. 2014).

The plasma spray process, as described in Figure 14, is most employed at normal atmospheric conditions and referred as Atmospheric Plasma Spraying (APS). In such conditions, plasma operation promotes an interaction between the fast-flying hot spray powder particles and the surrounding gas atmosphere, i.e., some air is entrapped in the hot plasma jet. When metals and intermetallic materials are sprayed by APS, the interaction of the plasma jet with the surrounding atmosphere can result in oxidation of the particles. Consequently, this type of coatings can have some oxide content depending on the oxidation kinetics attained under the processing conditions employed. For this reason, the APS process is mostly employed in the fabrication of ceramic coatings, although an important number of metallic and intermetallic coatings are also processed by this method. Environmental air entrainment into the plasma jet can also affect the coatings building up process. Entrained air often moves in opposite direction from the nozzle to the plasma gas injection channel and promotes a reduction in the thermal efficiency of the plasma jet, contributing to reduce thermal energy transferred from the plasma jet to the particles and to increase in-flight particles oxidation. Oxidation is generally not desirable in TS coatings as it dramatically reduces their performance (Fauchais et al., 2014; Pawlowski, 2008).

Ideal processing conditions to produce homogeneous plasma spray coatings include a uniform and controllable velocity of particles to produce splat formation upon impact, a uniform and controllable particle heating to attain a viscous state at impact without vaporization or undesired reactions, isolation of the particles from the environment, or at least, a minimum interaction with the surrounding environment, and a good stability of the plasma jet. Some additional processing conditions are feeding rate, stand-off distance, substrate temperature, particle size distribution, type and angle of powder injection, primary and secondary gas ratio, type of primary and secondary gas, and power level (Berndt et al., 1992; Fauchais et al., 2014).

Another plasma spray process is the Vacuum Plasma Spraying (VPS), which is based on the same operation principle than APS; however, VPS is carried out inside a controlled low-pressure chamber, whereas a small amount of inert gas is present. The process starts when the chamber is evacuated at pressures $<10^{-1}$ mbar; then the chamber is loaded with an inert gas. A pump system is constantly working to remove the injected plasma gases. In the VPS process, properties of the raw materials are maintained in the coating, as virtually there is not interaction between feedstock materials and oxygen. Metallic coatings, such as titanium-based systems, can be produced by VPS. VPS coatings often have higher bonding strength and higher density than their APS counterparts. VPS is also employed for the fabrication of functional coatings including various advanced technologies such as bioactive implants, advanced electronics, and self-cleaning surfaces. One disadvantage of VPS is the high fabrication costs involved; this process is then employed to produce coatings for high added value products (Pawlowski, 2008; Mauer et al., 2013).

The Shrouded Plasma Spray (SPS) process has been developed as an alternative to VPS; SPS is a process created to avoid the use of specialized pump systems as those employed in the VPS process. In SPS, the plasma jet is isolated from the surrounding atmosphere by using a shielded nozzle attached to the plasma torch anode. That is, the plasma jet is enveloped with an inert gas, which is not ionized and prevents the introduction of oxygen into the plasma jet; SPS is often employed to deposit oxidization susceptible materials and is a lower cost option than VPS.

The controlled atmosphere plasma spraying (CAPS) is another plasma process carried out in a chamber containing an inert gas, usually He, Ar or N_2. Spraying materials in a controlled atmosphere ensures the production of oxide-free coatings with high purity and density. CAPS is also employed in applications requiring a stringent contamination control; CAPS coatings are then produced for biomedical implants, solid oxide cells, airfoils, and combustion chambers for aeronautical applications (Singh et al., 2007).

In recent decades, another plasma spray process known as Plasma Spray-Physical Vapor Deposition (PS-PVD) was developed. The PS-PVD technique has emerged as an extension of the VPS process, seeking to combine the advantages of APS (high deposition rates and cost-efficiency) with the special features provided by PVD, such as the deposit of columnar structures and thin coatings (Mauer et al., 2013). In PS-PVD, the vaporization degree is a function of the powder feeding rate and plasma gas composition. For example, a very low vaporization degree is obtained by using a cold plasma gas ratio (100Ar/10H_2 slpm) combined with a high powder feeding rate (40-80 g/min). The coatings obtained under these spraying conditions can be used as insulation layers, gas-tight electrolytes, and gas separation membranes when they are deposited on porous substrates (Mauer et al., 2013).

Currently, plasma sprayed coatings are employed in various industrial applications; the most common applications are corrosion protection, temperature isolation, and abrasion resistance. Plasma spray processes have also been employed to produce glassy ceramic coatings. Moreover, these processes have been used to deposit high temperature superconductive materials. Other uses include reconditioning of worn and corroded components, as well as the production of aerospace and nuclear power components and biomedical implants (Singh et al., 2007).

Plasma sprayed hydroxyapatite-based and bioglass-based coatings have been proposed to provide bioactive characteristics to titanium-based biomedical implants. Usually, plasma sprayed coatings are applied to hip and dental implants. This type of coatings are commercially available in some biomedical devices and are accepted by international controlling organisms such as the Food and Drug Administration (FDA, USA) due to their outstanding *in-vitro* and *in-vivo* biocompatibility and bioactivity (Aruna, 2017).

On the other hand, plasma spray coatings have also been employed in the aerospace and energy sectors. Thermal barrier coatings (TBCs) are widely used for the protection of gas turbine components in aircraft engines; these coatings can work as thermal insulators, and are able to reduce the temperature of internally cooled metallic substrates in the order of 200 and 250 °C with respect to the operating temperature at the coating surface. Yttria-stabilized zirconia (YSZ) is the most common material employed as a thermal barrier coating; however, rare earth oxide-based powders have also been used, including $La_2Zr_2O_7$, $La_2Ce_2O_7$, and $Gd_2Zr_2O_7$. TBCs provide the capacity of increasing the operational temperature of the engines and help to increase their efficiency.

Overall, TBCs are formed at least by two layers; a metallic layer known as bond coat and a top ceramic layer. The bond coat is normally a MCrAlY alloy, in which M can be either Ni or Co, and performs two main functions: i) it inhibits substrate oxidation by the formation of an Al_2O_3 oxide scale at the bond coat/top coat interface, and ii) it works as an intermediate layer between top coat and substrate, which helps to decrease thermal expansion mismatch. Moreover, the main function of the ceramic top coat layer is to minimize the thermal conductivity of the array; since plasma sprayed ceramic coatings can contain some features such as pores and cracks, top ceramics layers can have some degree of stress-compliance when exposed to thermal cycling conditions. Top ceramic coatings often show good adhesion and have chemical compatibility with the oxide layer formed at the bond coat/top coat interface (Strangman, 1985; Vaßen et al., 2010). Since the 1980's, yttria-stabilized zirconia (YSZ) has been the preferred ceramic topcoat material due to its excellent thermal insulation performance. Yttria additions to zirconia (6-8 wt%) are crucial as they help to prevent a deleterious phase transformation of the later from tetragonal to monoclinic, which involves a large volume change (Vaßen et al., 2010). Moreover, the performance of a TBC coating also depends on the chemical composition of the feedstock powder, particle size distribution, and morphology; it also depends on the selected spraying conditions such as powder feed rate, plasma gas composition, electrical power, and plasma gun type. TBC performance is evaluated by measuring the coating thermal conductivity, thermal shock resistance, oxidation resistance, porosity and distribution of pores, residual stresses, and mechanical properties (adhesion and bending strength, hardness, fracture toughness, fatigue, etc.) (Kucuk et al., 2000).

Figure 16. Simple configuration of CGS process, adapted from (Villafuerte, 2015; Cavaliere, 2017).

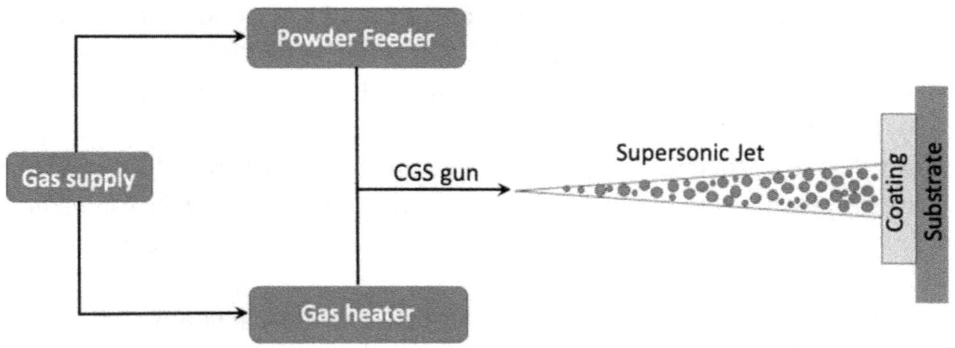

Currently, there are several ways to reduce the thermal conductivity of TBCs. The first one is by increasing the thickness of the ceramic coating; however, this option promotes the increment of residual stresses in the system, which may lead to premature failure. The use of multilayer ceramic coating architecture to reduce thermal conductivity is also attractive. This type of architectures often employ YSZ and zirconate-type ceramics in an alternated manner; alternatively, the introduction of pores and cracks (by a rigorous control of the feedstock powder morphology and optimization of processing parameters) is another alternative and effective way to reduce thermal conductivity in TBCs (Wang et al., 2011).

Plasma spray processes have also been employed in the energy sector for the development of metal-supported solid oxide fuel cells (SOFCs); in fact, plasma spraying has been the most suitable method for manufacturing this class of devices. For instance, some SOFCs prepared by plasma spray comprise electrodes manufactured from Ni-YSZ, Sr-doped $LaMnO_3$, and Sr-doped $LaCrO_3$ powders (Aruna, 2017). It is worth mentioning that these electrodes are often produced by VPS because of the flexibility of producing thin and dense ceramic coatings (Mauer et al., 2013).

Other industrial applications of plasma spraying involve the development of high-performance and high-functional coating materials. For instance, the development of plasma sprayed polyester/TiO_2 coatings on wood surfaces is one example of the innovative industrial applications of plasma sprayed coatings. Plasma sprayed coatings on wood products is an emerging field; this new application is intended to improve protection of wood-based products, especially when they are exposed to natural environmental conditions. Under these conditions, uncoated wood-based products can be affected by the environment on their natural durability (reducing their mechanical stability) and aesthetic appearance (Jnido et al., 2020; Oukach et al., 2020). Other functional applications include plasma sprayed photocatalytic coatings

(Zn/ZnO, porous TiO_2), electrical-heating coatings ($TiO_2/MgAl_2O_4$) and environmental multilayered barrier coatings ($Gd_2Zr_2O_7$, $Y_3Al_2(AlO_4)_3$) (Cheng et al., 2020; Li et al., 2020; Oluwafunmilade et al., 2020; Jordan et al., 2020; Wang et al., 2020).

Cold Spray Coatings and Modern Industrial Applications

Cold-gas spray (CGS) is a compressed gas expansion-based technology. CGS was developed in the mid-1980's by Dr. Anatolii Papyrin and colleagues at the Institute of Theoretical and Applied Mechanics in Russia. The CGS process is based on the application of high kinetic energy to a powdered feedstock and on the consolidation of coatings by plastic deformation of those particles at impact. In fact, CGS is regarded as a solid-state deposition process, wherein particles can achieve supersonic-velocities at impact (600-1200 ms^{-1}); usually, particles temperatures are lower (< 1000°C) than those achieved in conventional TS (Pawlowski, 2008; Cavaliere, 2017; Villafuerte, 2015).

A straightforward scheme of the CGS process is shown in Figure 16. The high-pressure gas (helium, nitrogen, or air) flows through the system following two different paths. The first one is used to drive the gas through a powder feeder to transport powder particles to the gun, whereas the second one conduces the gas to an electric heater where it is preheated. Both paths converge inside the CGS gun, commonly, equipped with a converging-diverging nozzle (De-Laval-type) (Pawlowski, 2008; Cavaliere, 2017; Villafuerte, 2015).

CGS is classified into two main categories according to the operating pressure range and the injection location of the powders into the nozzle. These categories are high-pressure (HPCGS) and low-pressure (LPCGS) CGS.

HPCGS is schematically represented in Figure 17a; it operates at maximum gas pressures between 50 and 55 bar. The working temperature of the gas is in the range of 20 to 1100 ºC, and particle velocities are often in the supersonic regime (Table 2). In HPCGS, particles are injected axially into the gas stream in a pre-chamber. Then, the high-pressure gas and particles are driven through the converging part of the nozzle, where the gas is compressed. Acceleration of the particles is obtained through the diverging part of the nozzle, where the gas is expanded. Gas heating is not only performed to increase the kinetic energy of the gas and to provide momentum to the particles, but also to increase the thermal energy of the system. The increase of gas temperature can promote a softening effect in some materials at impact. Alternatively, high gas pressures often influence particle velocity at impact, and, for some materials, this condition is required to induce formation of the coatings (Koivuluoto et al., 2012; Cavaliere, 2017).

LPCGS is schematically represented in Figure 17b. Unlike HPCGS, LPCGS operates at a maximum gas pressure of ~ 10 bar, working temperatures of the gas are in the range of 20 to 650 ºC, and particles are radially injected to the gas stream at the diverging part of the nozzle. In LPCGS, particle velocities are lower than those obtained in HPCS, being usually below 600 ms^{-1} (Table 2). LPCS is a low-cost system, and its operation is cost/effective because it employs compressed air as the propellant gas; then, it is portable and easy to manipulate (Assadi et al., 2003; Villafuerte, 2015; Kuivuluoto et al., 2012; Burkalov et al., 2007; Xu et al., 2006; Alhulaifi et al.; 2012; Bush et al., 2017; Cinca et al.; 2016; Chun et al., 2008; Dosta et al.; 2013).

Figure 17. Schematic representation of the two types of CGS systems. a) HPCGS system; b) LPCGS system. Adapted from (Cavaliere, 2017; Yin et al., 2018).

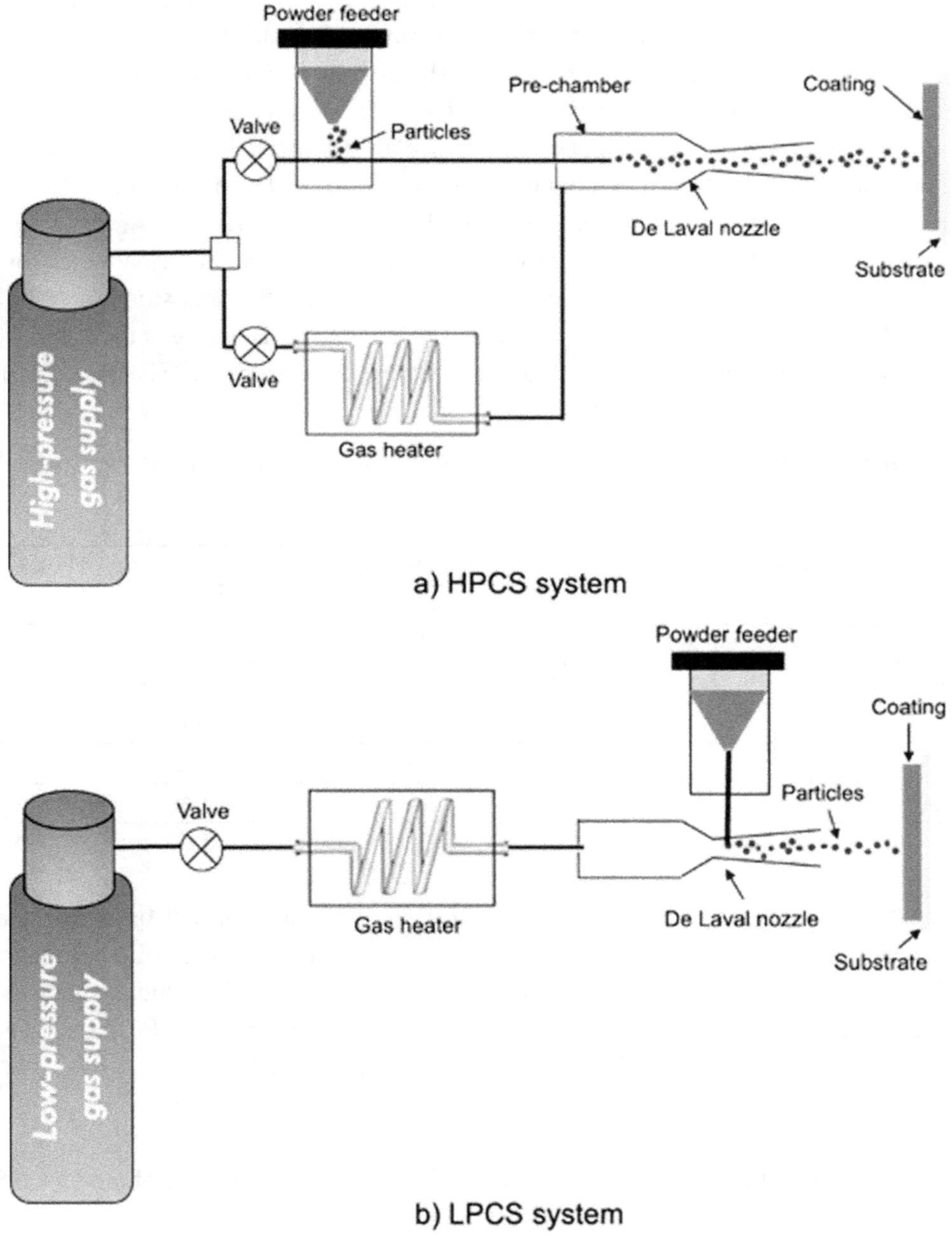

Both LPCGS and HPCGS techniques have similar mechanisms to explain the formation of metallic coatings; however, for ceramics, polymers, and cermets, the bonding mechanisms are still poorly understood. The most accepted bonding mechanism for metallic particles is associated with adiabatic shear instabilities (ASI) (Assadi et al., 2003; Cavaliere, 2017). This mechanism consists of severe localized shear deformation in the interfacial area between particle/particle and/or particle/substrate upon impact. At impact, breaking of the outer oxide layer of particles takes place, producing a strong particle/

substrate contact. This phenomenon, combined with the high compressive stresses developed during the collision of the particles at the substrate surface, favors bonding of particles and the formation of coatings (Assadi et al., 2003; Cavaliere, 2017). Although ASI is the most accepted bonding mechanism in CGS, some authors have proposed different approaches. Bae et al. (Bae et al., 2009) proposed that ASI is not required for a successful bonding mechanism. Meanwhile, Hussain et al. (Hussain et al., 2009) proposed another similar bonding mechanism associated with the mechanical interlocking of particles. Particle adhesion also depends on the contact surface area, plastic deformation, combination of processing parameters, and particle and substrate temperature (Assadi et al., 2003; Moridi et al., 2014; Cavaliere, 2017). Other bonding mechanisms have been proposed for metallic glass alloys (amorphous metallic alloys) deposited by CGS. Metallic glasses have different mechanical and thermal behavior than their crystalline counterparts. In fact, one of the accepted mechanisms for the deposition of metallic glasses by CGS is based on the impact of the particles under a viscous state. Bonding is then achieved when a non-Newtonian flow is promoted at the substrate/particle interface (Henao & Sharma, 2018).

Table 2. Comparison between HPCS and LPCS (Kuivulouto et al., 2012; Villafuerte, 2015; Cavaliere, 2017).

Parameter	HPCGS	LPCGS
Process gas	N_2, He, mixtures	Compressed air
Pressure (bar)	7-55	6-10
Pre-heating temperature (ºC)	RT-1100	RT-650
Powder feed rate (kg/h)	4.5-13.5	0.3-3
Spraying distance (mm)	10-50	5-15
Particle size (µm)	5-50	5-30

Depending on the required final coating properties, a post-treatment can be performed, such as thermal treatment, internal oxidation, reaction sintering, etc. It is important to point out that bonding of particles in LPCGS is weaker compared to HPCGS due to the differences in particle velocities at impact (Villafuerte, 2015; Cavaliere, 2017). The main features of both LPCGS and HPCGS sprayed metallic coatings are summarized as follows (Villafuerte, 2015; Cavaliere, 2017):

1. HPCGS coatings are denser than LPCS counterparts.
2. The flattening ratio of particles is smaller in HPCGS than in LPCGS; therefore, LPCGS coatings present a higher strain distribution than HPCGS ones.
3. LPCGS presents lower deposition efficiency compared to HPCGS. This fact can be promoted due to the differences in the shot peening effect related to high particle deformation and pressures at impact.
4. ASI is often achieved in HPCGS for most crystalline metals; this is the main mechanism associated with the consolidation and density of HPCGS coatings.

Some advantages of CGS deposition processes are listed as follows (Pawlowski, 2008; Villafuerte, 2015; Cavaliere, 2017):

- The main difference between CGS and conventional thermal spray techniques are the gas temperature and particle velocity attained. CGS induces high particle velocities instead of extremely high gas temperatures to produce the coatings; this fact has many advantages for processing oxygen-sensitive and temperature-sensitive materials.
- During CGS process, grain growth, chemical reaction, cracks formation, evaporation, residual stresses, thermal shrinkage, phase transformations, and oxidation are avoided.
- CGS sprayed coatings can be obtained with thickness values below 1 mm without delamination or damage due to the high compressive stress state in the coatings. It is also possible to produce coatings with thickness values above 1 mm, especially, in cylindrical surfaces.
- Properties of the feedstock materials can be maintained in the coatings (this point can be achieved depending on the optimization of the process).
- Higher deposition efficiency than that obtained in conventional thermal spray can be obtained (50-90%) depending on the optimization of the process.
- It is possible to manufacture hybrid structures by using highly dissimilar materials.

Some limitations of CGS deposition processes are listed as follows (Pawlowski, 2008; Villafuerte, 2015; Cavaliere, 2017):

- Nozzle clogging can take place, causing changes in deposition rates and having undesired effects on coating properties. This situation occurs for long spraying periods.
- Due to the nature of the technique, it requires ductility at impact. It is then used for ductile materials or combinations of ductile/hard materials. There are some reports though related to the feasibility of using this technique to deposit ceramic materials (Kumar et al., 2020).
- Microstructural changes of the sprayed material can be produced due to work-hardening and recrystallization processes occurring upon impact at the particle/particle interface.

Several processing parameters influence the quality of coatings produced by CGS. Some of them are listed as follows (Villafuerte, 2015; Cavaliere, 2017):

- *Nature of propellant gas:* Gas velocity is directly proportional to its molecular weight; thus, the nature of the gas in CGS has a significant effect on the dynamics of the particles. The use of helium results in higher velocities than employing nitrogen and air.
- *Nozzle:* The dimensions of the nozzle establish the Mach number of the gas at any axial location. This number is often below Mach 3. The geometry and dimensions of the nozzle can influence the residence time of the particles in the gas stream; therefore, the nozzle design has an influence on temperature and velocity of particles at impact. The type of material used to manufacture the CGS nozzle can also influence the gas dynamics and particles behavior. Some phenomena can take place within the nozzle including wearing and clogging due to the interaction of particles with the nozzle walls.
- *Traverse velocity:* traverse velocity influences heat transfer from the gas stream to the substrate; it also has an effect on the coating microstructure.
- *Stand-off distance:* This parameter has an effect on the condition of the particles attained at impact. The stand-off distance must be optimized because is one of the parameters responsible to establish the thermal and kinetic energy of the particles at impact.

- *Feedstock powder:* Each type of material: metal, ceramic, polymer, composite, etc. has a particular kinetic behavior in the CGS process. Differences in the mechanical behavior of materials strongly affects the particle/substrate and particle/particle interactions at impact. Materials with low plasticity are less suitable for CGS. Flowability of the powder is also important and is a parameter affected by the morphology of the particles and their particle size distribution.

It is important to highlight that, in CGS, optimization of processing parameters creates a specific environment to generate the conditions to deposit a specific material. Some important factors resulting from a combination of specific processing parameters are:

- *Critical velocity:* V_{crit} is defined as the minimum particle velocity necessary for a metallic material to adhere to the substrate (Papyrin, 2006; Assadi et al., 2003). The value of the V_{crit} depends on many factors, such as the substrate and feedstock powder composition, morphology of particles, particle size distribution, and substrate surface properties (Papyrin, 2006; Gärtner et al., 2006).
- *Gas temperature and pressure:* Both factors strongly affect the particle impact velocity (Gärtner et al., 2006; Papyrin, 2006; Maev & Leshchynsky, 2009; Meng et al., 2011; Yin et al.; 2018). High gas temperatures and high gas pressures generate high particle impact velocities, improving deposition efficiency and producing high quality coatings, *i.e.* coatings with high density and high adhesion strength. Thermal energy is beneficial for the deposition of metallic particles because it intensifies plastic deformation by promoting dislocations mobility (Gärtner et al., 2006; Maev & Leshchynsky, 2009; Meng et al., 2011; Yin et al., 2018). Sometimes high gas temperatures and pressures do not favor metallic particle deposition since impact conditions can be extremely aggressive and produce erosion. Then, it is important to deposit each specific material under its own combination of processing parameters to promote bonding instead of erosion.
- *Deposition efficiency:* It can be defined as the relationship between the weight of particles adhered to the substrate and the total weight of the initial feedstock particles without overspray. For metallic particles, the deposition efficiency increases as the particle velocity significantly exceeds the critical velocity (Gärtner et al., 2006; Cavaliere, 2017).

Some studies have shown that CGS is a favorable and cost-effective technology. CGS is used to produce protective coatings and to recover dimensions of different parts. Some industrial applications of CGS coatings are summarized in Table 3. In the last decades, new materials have also been deposited by CGS; these materials are also listed in Table 3. This diversification demonstrates the versatility of the CGS process. Metallic glasses are part of the diversity of new materials sprayed by CGS, and due to their unique physical and chemical properties the production of functional coatings from them is positively foreseen. Hydroxyapatite is another material of interest, mainly for biomedical applications, that can be deposited by means of CGS. Cold sprayed hydroxyapatite coatings can have better bioactive behavior than conventional ones since deterioration of crystalline phase can be avoided by CGS (Cavaliere, 2017; Vilardel et al., 2019; Sun et al., 2019). Overall, numerous studies have claimed that cold sprayed coatings can have good corrosion and wear resistance and have good mechanical integrity in the as-sprayed condition. These characteristics are attractive as these coatings can be employed in various applications in different industries (Cavaliere, 2017).

Table 3. Some applications of the CGS process in different industrial sectors (Villafuerte, 2015; Cavaliere, 2017).

Application	Coating materials	Industry sector
Corrosion resistance	Metals and alloys, metal/polymer, metal matrix composites (M/ceramic, M/metal, cermets), intermetallics, nanomaterials.	Aerospace Automotive Electronics Defense Oil and gas Petrochemical Power generation
Oxidation resistance	Metals and alloys, metal/polymer, metal matrix composites (M/metal, cermets), intermetallics, oxides,	
High specific strength	Metals and alloys, metal/polymer, metal matrix composites (M/ ceramic), intermetallics, ceramics, nanomaterials.	Various industrial sectors
Wear-resistant coatings	Metals and alloys, metal matrix composites (M/ceramic, M/metal, cermets), intermetallics, oxides, ceramics, WC cermets, nanomaterials.	Automotive Aerospace Machining
Biocompatibility, Biodevices, and Antibacterial	Metals and alloys, metal/polymer, oxides, ceramics, polymers nanomaterials.	Biomedical (devices and surgical implants)
Photocatalytic performance	Metals and alloys, oxides, ceramics, nanomaterials	Energy generation industry
Restoration	Metals and alloys, metal/polymer	Aerospace Defense Petrochemical

SUMMARY

Thermally sprayed coatings have been used in various industries since the invention of the flame spray process in the early 1900s. Afterwards, the development of different thermal spray technologies, such as APS, HVOF, HVAF, VPS, CGS, etc., has constantly increased the use of this type of coatings in different industrial sectors. Thermally sprayed coatings have been widely used in many applications due to the versatility of thermal spray technologies to operate under different engineering environments (aerospace, biomedical, energy, chemistry, agriculture, textile, automotive, marine, etc.). Overall, thickness values of thermally sprayed coatings lie in the range from 50 μm to few millimeters, depending on the spraying technique and on the processing parameters employed. In addition, density values of thermally sprayed coatings are in the range between 65% to 99%, depending on the spraying process. Advantages of thermal spray processes to produce coatings include their versatility to spray a variety of materials such as crystalline metals, metallic glasses, ceramics, cermets, intermetallic materials, and polymers. These materials can be sprayed under different shapes such as wire, rod, powder, and suspension. Another advantage is the capacity of thermal spray techniques to coat a wide range of substrate materials and to operate on different substrate shapes.

Nowadays, various industries have recognized multifaceted capacity and inherent economics of some thermal spray processes. Thermally sprayed coatings are currently positioned as crucial coating systems in various conventional and modern applications. Most contemporary applications of thermally sprayed coatings are associated with wear, corrosion, oxidation, and high temperature environments. Currently, some industrial fields of applications include biomedical, electronics, dimensional restoration, thermal barriers, and energy storage. It is envisioned that, in the near future, thermal spray techniques can offer new opportunities for emergent applications in healthcare (bioactive, antibacterial, and antivirus surfaces), electronics (sensors), energy (battery and fuel cells), environmental (self-cleaning and antifouling surfaces), aerospace (environmental barriers) and automobile (energy-absorbing surfaces) industries.

ACKNOWLEDGMENT

The authors would like to thank the support provided by the Mexican National Council of Science and Technology (CONACYT), specially the Cátedras program, projects number 848 and 881. The preparation of this book chapter was possible due to the collaboration among CIATEQ A.C., CINVESTAV Querétaro Unit, The Thermal Spray National Laboratory (CENAPROT), and University of Antioquia (Faculty of Engineering). The support from the authorities of University of Antioquia and CIATEQ A.C., in particular, Prof. John Jairo Arboleda, Prof. Esperanza López, Mrs. Luz Maritza Areiza, and Prof. Luis Gerardo Trapaga Martinez is greatly appreciated.

REFERENCES

Advanced, A. M. T. A. G. (n.d.). *Materials Technology, Twin Wire Arc Spraying*. Retrieved from: https://www.amt-ag.com/equipment/spray-processes/twin-wire-arc

Advanced-Coating. (n.d.). Retrieved from: http://www.advanced-coating.com/english/spraying-hvof.htm.

Alhulaifi, A. S., Buck, G. A., & Arbegast, W. J. (2012). Numerical and experimental investigation of cold spray gas dynamic effects for polymer coating. *Journal of Thermal Spray Technology*, *21*(5), 852–862. doi:10.100711666-012-9743-4

Apelian, D., Paliwal, M., Smith, R. W., & Schilling, W. F. (1983). Melting and solidification in plasma spray deposition—Phenomenological review. *International Metals Reviews*, *28*(1), 271–294. doi:10.1179/imtr.1983.28.1.271

Aruna, S. T. (2017). Thermal Spray Coatings Using Solution Combustion Synthesized Powders. In Concise Encyclopedia of Self-Propagating High-Temperature Synthesis (pp. 382-384). Elsevier. doi:10.1016/B978-0-12-804173-4.00155-1

Assadi, H., Gärtner, F., Stoltenhoff, T., & Kreye, H. (2003). Bonding mechanism in cold gas spraying. *Acta Materialia*, *51*(15), 4379–4394. doi:10.1016/S1359-6454(03)00274-X

Ates, M. (2016). A review on conducting polymer coatings for corrosion protection. *Journal of Adhesion Science and Technology*, *30*(14), 1510–1536. doi:10.1080/01694243.2016.1150662

Bae, G., Kumar, S., Yoon, S., Kang, K., Na, H., Kim, H. J., & Lee, C. (2009). Bonding features and associated mechanisms in kinetic sprayed titanium coatings. *Acta Materialia*, *57*(19), 5654–5666. doi:10.1016/j.actamat.2009.07.061

Benxxonglobal. (n.d.). Retrieved from: http://www.bexxonglobal.com/what-is-wire-and-powder-flame-spray-.html

Berndt, C. C., Brindley, W., Goland, A. N., Herman, H., Houck, D. L., Jones, K., & Smith, M. (1992). Current problems in plasma spray processing. *Journal of Thermal Spray Technology*, *1*(4), 341–356. doi:10.1007/BF02647162

Bolelli, G., Bellucci, D., Cannillo, V., Lusvarghi, L., Sola, A., Stiegler, N., & De Nardo, L. (2014). Suspension thermal spraying of hydroxyapatite: Microstructure and in vitro behaviour. *Materials Science and Engineering C*, *34*, 287–303. doi:10.1016/j.msec.2013.09.017 PMID:24268261

Brogan, J. A. (2000). Thermal-spraying of polymers and polymer blends. *MRS Bulletin*, *25*(7), 48–53. doi:10.1557/mrs2000.124

Burlacov, I., Jirkovský, J., Kavan, L., Ballhorn, R., & Heimann, R. B. (2007). Cold gas dynamic spraying (CGDS) of TiO2 (anatase) powders onto poly (sulfone) substrates: Microstructural characterisation and photocatalytic efficiency. *Journal of Photochemistry and Photobiology A Chemistry*, *187*(2-3), 285–292. doi:10.1016/j.jphotochem.2006.10.023

Bush, T. B., Khalkhali, Z., Champagne, V., Schmidt, D. P., & Rothstein, J. P. (2017). Optimization of cold spray deposition of high-density polyethylene powders. *Journal of Thermal Spray Technology*, *26*(7), 1548–1564. doi:10.100711666-017-0627-5

Cadavid, E., Parra-Velásquez, C., & Vargas, F. (2016). Estudio de llamas oxiacetilénicas usadas en la proyección térmica. *Revista Colombiana de Materiales*, (9), 15-26.

Cavaliere, P. (2017). *Cold-Spray Coatings: Recent Trends and Future perspectives*. Springer.

Chen, K., Liu, Y., Fu, Y., Huang, J., Feng, X., Wang, J., & Li, H. (2020). Loading of Zn/ZnO particles in the precursor feedstock affects the characteristics of liquid plasma sprayed nano-ZnO coatings for photocatalytic applications. *Nanotechnology*, *31*(18), 185301. doi:10.1088/1361-6528/ab6cda PMID:31945757

Chun, D. M., Kim, M. H., Lee, J. C., & Ahn, S. H. (2008). TiO2 coating on metal and polymer substrates by nano-particle deposition system (NPDS). *CIRP Annals*, *57*(1), 551–554. doi:10.1016/j.cirp.2008.03.111

Cinca, N., Vilardell, A. M., Dosta, S., Concustell, A., Garcia Cano, I., Guilemany, J. M., & Peiró, F. (2016). A new alternative for obtaining nanocrystalline bioactive coatings: Study of hydroxyapatite deposition mechanisms by cold gas spraying. *Journal of the American Ceramic Society*, *99*(4), 1420–1428. doi:10.1111/jace.14076

Company, A. G. A. (n.d.). *There is no better one*. Retrieved from: https://www.linde-gas.ee

Davis, J. R. (Ed.). (2004). *Handbook of thermal spray technology*. ASM International.

Dcleng. (n.d.). Retrieved from: https://www.dcleng.com.au/index.php?f=flame-spray-ceramic

Deloro Stellite Guns. (n.d.). Retrieved from: https://www.stellite.com/us/en/products/stellite-equipment/hvof-jet-kote/jet-kote-III.html

Denka spherical aluminal Denka Company Limited. (n.d.). Retrieved from: https://www.denka.co.jp/eng/product/detail_00040/

Dosta, S., Couto, M., & Guilemany, J. M. (2013). Cold spray deposition of a WC-25Co cermet onto Al7075-T6 and carbon steel substrates. *Acta Materialia*, *61*(2), 643–652. doi:10.1016/j.actamat.2012.10.011

Duan, Z., & Heberlein, J. (2002). Arc instabilities in a plasma spray torch. *Journal of Thermal Spray Technology*, *11*(1), 44–51. doi:10.1361/105996302770348961

Eddings, E. G. (2005). *Mechanical Engineers' Handbook*. John Wiley & Sons, Inc.

Eutectic Castolin Eutalloy, S. F. (n.d.). *Eutalloy - Eutalloy Spray and Fusing ®*. Retrieved from: https://www.castolin.com/sites/default/files/Eutalloy-powder-spray-fusing

Fauchais, P. L., Heberlein, J. V., & Boulos, M. I. (2014). *Thermal spray fundamentals: from powder to part*. Springer Science & Business Media. doi:10.1007/978-0-387-68991-3

Ferrer, M. Y., Vargas, F., Vera López, E., Moreno Téllez, C. M., & Peña Rodríguez, G. (2017). Evaluación de la morfología, resistencia al desgaste ya la corrosión de recubrimientos de ZrO2-Al2O3/Ni depositados sobre sustratos de acero al carbono mediante proyección térmica por llama. *Ingeniare. Revista Chilena de Ingeniería*, *25*(4), 721–732.

Flame Spray Guns. (n.d.). Retrieved from: https://www.fst.nl/systems/thermal-spray-guns-and-torches/egun-hvof-spray-gun/

Franco, D., Vargas, F., & López Gómez, M. E. (2016). Study of thermal shock resistance of flame sprayed coatings manufactured to protect molds used in glass containers industry. *Revista Chilena de Ingeniería*, *24*(2), 239–248. doi:10.4067/S0718-33052016000200007

Ganvir, A. (2016). *Microstructure and Thermal Conductivity of Liquid Feedstock Plasma Sprayed Thermal Barrier Coatings* (Doctoral dissertation). University West.

Gärtner, F., Stoltenhoff, T., Schmidt, T., & Kreye, H. (2006). The cold spray process and its potential for industrial applications. *Journal of Thermal Spray Technology*, *15*(2), 223–232. doi:10.1361/105996306X108110

Glassman, I., Yetter, R. A., & Glumac, N. G. (2014). *Combustion*. Academic press.

González, R., Garcia, M. A., Penuelas, I., Cadenas, M., del Rocío Fernández, M., Battez, A. H., & Felgueroso, D. (2007). Microstructural study of NiCrBSi coatings obtained by different processes. *Wear*, *263*(1-6), 619–624. doi:10.1016/j.wear.2007.01.094

GunsE. (n.d.). Retrieved from: https://www.castolin.com/product/hvof-castojet-cjk5

GunsG. T. V. Online. (n.d.). Available: https://www.gtv-mbh.com/_old/gtv-mbh-englisch/www.gtv-mbh.de/cms/front_content5d9e.html?idcat=84

GunsT. (n.d.). Retrieved from: http://thermico.de/EN/products/engineering/spray-gun-technology.html

GunsU. (n.d.). Retrieved from: http://www.uniquecoat.com/m2-hvaf-spray-gun.html

Guns Oerlikon MetcoH. V. O. F. (n.d.). Retrieved from: https://www.oerlikon.com/metco/en/products-services/coating-equipment/thermal-spray/spray-guns/spray-guns-hvof/hvof-diamond-jet/

HardKarb 12 Co Thermal Spray Flexicord | Saint-Gobain Coating Solutions. (n.d.). Retrieved from: https://www.coatingsolutions.saint-gobain.com/materials/flexicord/carbide-series/hardkarb-12-co

Henao, J., Cruz-bautista, M., Hincapie-Bedoya, J., Ortega-Bautista, B., Corona-Castuera, J., Giraldo-Betancur, A. L., & Poblano-Salas, C. A. (2018). HVOF hydroxyapatite/titania-graded coatings: Microstructural, mechanical, and in vitro characterization. *Journal of Thermal Spray Technology*, *27*(8), 1302–1321. doi:10.100711666-018-0811-2

Henao, J., & Sharma, M. M. (2018). Characterization, Deposition Mechanisms, and Modeling of Metallic Glass Powders for Cold Spray. In *Cold-Spray Coatings* (pp. 251–272). Springer. doi:10.1007/978-3-319-67183-3_8

Herman, H., Sampath, S., & McCune, R. (2000). Thermal spray: Current status and future trends. *MRS Bulletin*, *25*(7), 17–25. doi:10.1557/mrs2000.119

Hossain, K. M. Z., Patel, U., Kennedy, A. R., Macri-Pellizzeri, L., Sottile, V., Grant, D. M., & Ahmed, I. (2018). Porous calcium phosphate glass microspheres for orthobiologic applications. *Acta Biomaterialia*, *72*, 396–406. doi:10.1016/j.actbio.2018.03.040 PMID:29604438

Hussain, T., McCartney, D. G., Shipway, P. H., & Zhang, D. (2009). Bonding mechanisms in cold spraying: The contributions of metallurgical and mechanical components. *Journal of Thermal Spray Technology*, *18*(3), 364–379. doi:10.100711666-009-9298-1

HVAF. High Velocity Air-Fuel. (n.d.). Retrieved from: http://www.hvaf.com/index.html

Jnido, G., Ohms, G., & Viöl, W. (2020). One-Step Deposition of Polyester/TiO2 Coatings by Atmospheric Pressure Plasma Jet on Wood Surfaces for UV and Moisture Protection. *Coatings*, *10*(2), 184. doi:10.3390/coatings10020184

Jordan, E. H., Gell, M., Kumar, R., & Jiang, C. (2020). *U.S. Patent Application No. 16/584,820*. Washington, DC: US Patent Office.

Kamnis, S., & Gu, S. (2006). 3-D modelling of kerosene-fuelled HVOF thermal spray gun. *Chemical Engineering Science*, *61*(16), 5427–5439. doi:10.1016/j.ces.2006.04.005

Karuppasamy, S., Sivan, V., Natarajan, S., Babu, S. K., Duraiselvam, M., & Dhanuskodi, R. (2019). Establishment of Wear Resistant HVOF Coatings for 50CrMo4 Chromium Molybdenum Alloy Steel as an Alternative for Hard Chrome Plating. *Journal of The Institution of Engineers (India): Series C*, *100*(4), 635-645.

Kermiko HVAF Guns. (n.d.). Retrieved from: https://kermetico.com/

Knight, R., & Smith, R. W. (1998). *Thermal Spray Forming of Materials, Powder Metal Technologies and Applications* (Vol. 7). ASM Handbook. ASM International.

Koivuluoto, H., Coleman, A., Murray, K., Kearns, M., & Vuoristo, P. (2012). High pressure cold sprayed (HPCS) and low pressure cold sprayed (LPCS) coatings prepared from OFHC Cu feedstock: Overview from powder characteristics to coating properties. *Journal of Thermal Spray Technology*, *21*(5), 1065–1075. doi:10.100711666-012-9790-x

Kucuk, A., Berndt, C. C., Senturk, U., Lima, R. S., & Lima, C. R. C. (2000). Influence of plasma spray parameters on mechanical properties of yttria stabilized zirconia coatings. I: Four point bend test. *Materials Science and Engineering A*, *284*(1-2), 29–40. doi:10.1016/S0921-5093(00)00799-1

Kumar, S., Kumar, M., & Jindal, N. (2020). *Overview of cold spray coatings applications and comparisons: a critical review. World Journal of Engineering.*

Leivo, E., Wilenius, T., Kinos, T., Vuoristo, P., & Mäntylä, T. (2004). Properties of thermally sprayed fluoropolymer PVDF, ECTFE, PFA and FEP coatings. *Progress in Organic Coatings*, *49*(1), 69–73. doi:10.1016/j.porgcoat.2003.08.011

Li, C. J., Yang, H., & Li, H. (1999). Effect of gas conditions on HVOF flame and properties of WC-Co coatings. *Materials and Manufacturing Processes*, *14*(3), 383–395. doi:10.1080/10426919908914834

Li, H., Ng, B. S., Khor, K. A., Cheang, P., & Clyne, T. W. (2004). Raman spectroscopy determination of phases within thermal sprayed hydroxyapatite splats and subsequent in vitro dissolution examination. *Acta Materialia*, *52*(2), 445–453. doi:10.1016/j.actamat.2003.09.046

Li, X., Yang, K., Wang, D., Deng, C., & Zhou, Y. (2020). Property of TiO2-15MgAl2O4 Electrical-Heating Coating Prepared by Atmospheric Plasma Spraying and Hydrogen Heat Treatment. *Coatings*, *10*(2), 177. doi:10.3390/coatings10020177

Maev, R. G., & Leshchynsky, V. (2009). *Introduction to low pressure gas dynamic spray: physics and technology*. John Wiley & Sons.

Marder, A. R. (2000). The metallurgy of zinc-coated steel. *Progress in Materials Science*, *45*(3), 191–271. doi:10.1016/S0079-6425(98)00006-1

Matejicek, J., & Sampath, S. (2001). Intrinsic residual stresses in single splats produced by thermal spray processes. *Acta Materialia*, *49*(11), 1993–1999. doi:10.1016/S1359-6454(01)00099-4

Matikainen, V., Koivuluoto, H., & Vuoristo, P. (2020). A study of Cr3C2-based HVOF-and HVAF-sprayed coatings: Abrasion, dry particle erosion and cavitation erosion resistance. *Wear*, *446-447*, 203188. doi:10.1016/j.wear.2020.203188

Mauer, G., Hospach, A., & Vaßen, R. (2013). Process development and coating characteristics of plasma spray-PVD. *Surface and Coatings Technology*, *220*, 219–224. doi:10.1016/j.surfcoat.2012.08.067

Melero, H. C., Sakai, R. T., Vignatti, C. A., Benedetti, A. V., Fernández, J., Guilemany, J. M., & Suegama, P. H. (2018). Corrosion resistance evaluation of HVOF produced hydroxyapatite and TiO2-hydroxyapatite coatings in Hanks' solution. *Materials Research*, *21*(2). Advance online publication. doi:10.1590/1980-5373-mr-2017-0210

Meng, X., Zhang, J., Zhao, J., Liang, Y., & Zhang, Y. (2011). Influence of gas temperature on microstructure and properties of cold spray 304SS coating. *Journal of Materials Science and Technology*, *27*(9), 809–815. doi:10.1016/S1005-0302(11)60147-3

Metallisation Spray Guns. (n.d.b). Retrieved from: https://www.metallisation.com/product-category/hvof/

Metallizing Company Guns. (n.d.a). Retrieved from: https://www.mecpl.com/products/coating-equipments/hvof-spray-coatings.php

Metco, O. (n.d.). *Plasma Guns for Thermal Spray*. Retrieved from: https://www.oerlikon.com/metco/en/products-services/coating-equipment/thermal-spray/spray-guns/coating-equipment-plasma/

Moridi, A., Hassani-Gangaraj, S. M., Guagliano, M., & Dao, M. (2014). Cold spray coating: Review of material systems and future perspectives. *Surface Engineering*, *30*(6), 369–395. doi:10.1179/1743294414Y.0000000270

Mostaghimi, J., Pershin, L., & Yugeswaran, S. (2017). Heat Transfer in DC AND RF Plasma Torches. In F. Kulacki (Ed.), *Handbook of Thermal Science and Engineering*. Springer. doi:10.1007/978-3-319-32003-8_27-1

Nautiyal, H., & Tyagi, R. (2020). High temperature erosive wear behavior of high velocity oxy-fuel sprayed Cr3C225 (Ni20Cr) coating on (AISI-316) austenitic steel. *Journal of Tribology*, *142*(7), 1–16. doi:10.1115/1.4046290

Oluwafunmilade, A. K., Abidine, Z., Ando, Y., Noda, Y., & Iino, M. (2020). Deposition of Porous Photocatalytic TiO 2 Film for Dye-Sensitized Solar Cells by Low-Powered Atmospheric Plasma Spray Equipment. In *Renewable Energy and Sustainable Buildings* (pp. 945–953). Springer. doi:10.1007/978-3-030-18488-9_79

Oukach, S., Hamdi, H., El Ganaoui, M., & Pateyron, B. (2020). Protective Plasma Sprayed Coating forThermo-Sensitive Substrates. In *MATEC Web of Conferences* (*Vol. 307*, p. 01039). EDP Sciences. 10.1051/matecconf/202030701039

Palathai, T., Tharajak, J., & Sombatsompop, N. (2008). Hardness, adhesion index and microstructure of PEEK coating on Al or Fe substrate by LVOF flame spray. *Materials Science and Engineering A*, *485*(1-2), 66–73. doi:10.1016/j.msea.2007.07.049

Papyrin, A., Kosarev, V., Klinkov, S., Alkhimov, A., & Fomin, V. M. (2006). *Cold spray technology*. Elsevier.

Pasandideh-Fard, M., Pershin, V., Chandra, S., & Mostaghimi, J. (2002). Splat shapes in a thermal spray coating process: Simulations and experiments. *Journal of Thermal Spray Technology*, *11*(2), 206–217. doi:10.1361/105996302770348862

Pawlowski, L. (2008). *The science and engineering of thermal spray coatings* (2nd ed.). John Wiley & Sons. doi:10.1002/9780470754085

Perez Andrade, L. (2020). *Additive Manufacturing by Cold Spray of Nickel-based Superalloys Inconel 718* (Doctoral dissertation). Cinvestav, Queretaro, México.

Petrovicova, E., & Schadler, L. S. (2002). Thermal spraying of polymers. *International Materials Reviews*, *47*(4), 169–190. doi:10.1179/095066002225006566

Pinzón, A. V., Urrego, K. J., González-Hernández, A., Rincon Ortiz, M., & Vargas, F. (2018). Corrosion protection of carbon steel by alumina-titania ceramic coatings used for industrial applications. *Ceramics International*, *44*(17), 21765–21773. doi:10.1016/j.ceramint.2018.08.273

Poirier, T., & Quercia, G. (2015). Formation of hollow vitreous and semi-crystalline microspheres in slag flame spraying. *Ceramics International*, *41*(1), 369–377. doi:10.1016/j.ceramint.2014.08.080

Praxair HVOF Guns. (n.d.). Retrieved from: https://www.praxairsurfacetechnologies.com/en/components-materials-and-equipment/coating-equipment/thermal-spray-coating-systems/high-velocity-oxygen-fuel-hvof

Proner, A. (1999). Revêtements par projection thermique : Conclusion, perspectives d'avenir | Techniques de l'Ingénieur. *Techniques de l'ingénieur.* Retrieved from: https://www.techniques-ingenieur.fr/base-documentaire/materiaux-th11/traitements-de-surface-des-metaux-par-voie-seche-et-en-milieu-fondu-42360210/revetements-par-projection-thermique-m1645/conclusion-perspectives-d-avenir-m1645niv10008.html

Redjdal, O., Zaid, B., Tabti, M. S., Henda, K., & Lacaze, P. C. (2013). Characterization of thermal flame sprayed coatings prepared from FeCr mechanically milled powder. *Journal of Materials Processing Technology*, *213*(5), 779–790. doi:10.1016/j.jmatprotec.2012.11.018

Rodrıguez, J., Martın, A., Fernández, R., & Fernández, J. E. (2003). An experimental study of the wear performance of NiCrBSi thermal spray coatings. *Wear*, *255*(7-12), 950–955. doi:10.1016/S0043-1648(03)00162-5

Saint-Gobain Coating Solutions. (n.d.). *Coatings Al_2O_3, TI-ELITE Ceramic Supra series-ref : 98210 4447 000 Typical composition (%)-T TIP*. Retrieved from: https://www.coatingsolutions.saint-gobain.com/sites/imdf.coatingsolutions.com/files/

Saint-Gobain Coating Solutions. (n.d.). *Al_2O_3 and TiO_2, TI-TEX Ceramic Supra series-ref : 98210 4547 000.* Retrieved from: https://www.coatingsolutions.saint-gobain.com/sites/imdf.coatingsolutions.com/files/

Santacruz, G., Shigueaki Takimi, A., Vannucchi de Camargo, F., Pérez Bergmann, C., & Fragassa, C. (2019). Comparative study of jet slurry erosion of martensitic stainless steel with tungsten carbide HVOF coating. *Metals*, *9*(5), 600. doi:10.3390/met9050600

Science Direct. (n.d.). Retrieved from: https://www.sciencedirect.com/topics/engineering/spray-process

Siegmann, S., & Abert, C. (2013). 100 years of thermal spray: About the inventor Max Ulrich Schoop. *Surface and Coatings Technology*, *220*, 3–13. doi:10.1016/j.surfcoat.2012.10.034

Singh, H., Sidhu, B. S., Puri, D., & Prakash, S. (2007). Use of plasma spray technology for deposition of high temperature oxidation/corrosion resistant coatings–a review. *Materials and Corrosion*, *58*(2), 92–102. doi:10.1002/maco.200603985

Spinel Supra Thermal Spray Flexicord. (n.d.). *Ceramic Series | Saint-Gobain Coating Solutions*. Available: https://www.coatingsolutions.saint-gobain.com/materials/flexicord/ceramic-supra-series/spinel-supra

Spray, T. (n.d.). Thermal Spray Materials Guide, Metco, Sulzer. *Oerlikon metco*. Retrieved from: https://www.oerlikon.com/ecomaXL/files

Strangman, T. E. (1985). Thermal barrier coatings for turbine airfoils. *Thin Solid Films*, *127*(1-2), 93–106. doi:10.1016/0040-6090(85)90215-9

Sun, C., Zhou, X., & Xie, C. (2019). Effect of processing conditions on Al-based amorphous/nano-crystalline coating by cold-spraying. *Surface and Coatings Technology*, *362*, 97–104. doi:10.1016/j.surfcoat.2019.01.096

Supra Thermal Spray Flexicords, A. Z. Ceramic Series | Saint-Gobain Coating Solutions. (n.d.). Retrieved from: https://www.coatingsolutions.saint-gobain.com/materials/flexicord/ceramic-supra-series/az-supra

Tailor, S., Modi, A., & Modi, S. C. (2019). Thermally Sprayed Thin Copper Coatings by W-HVOF. *Journal of Thermal Spray Technology*, *28*(1-2), 273–282. doi:10.100711666-018-0770-7

Thorpe, M. L., & Richter, H. J. (1992). A pragmatic analysis and comparison of HVOF processes. *Journal of Thermal Spray Technology*, *1*(2), 161–170. doi:10.1007/BF02659017

Tocalo Company. (n.d.). Retrieved from: https://www.tocalo.co.jp/english/technical/arc.html

TWI-global. (n.d.). *What are the most common applications of flame spraying?* Retrieved from: https://www.twi-global.com/technical-knowledge/faqs/faq-what-are-the-most-common-applications-of-flame-spraying. (Accessed March 9th 2020).

United States Department of Labor. (n.d.). Retrieved from: https://www.osha.gov/SLTC/noisehearing-conservation/

Vardelle, A., Moreau, C., Themelis, N. J., & Chazelas, C. (2015). A perspective on plasma spray technology. *Plasma Chemistry and Plasma Processing*, *35*(3), 491–509. doi:10.100711090-014-9600-y

Vargas, F., Ageorges, H., Fournier, P., Fauchais, P., & López, M. E. (2010). Mechanical and tribological performance of Al2O3-TiO2 coatings elaborated by flame and plasma spraying. *Surface and Coatings Technology*, *205*(4), 1132–1136. doi:10.1016/j.surfcoat.2010.07.061

Vaßen, R., Jarligo, M. O., Steinke, T., Mack, D. E., & Stöver, D. (2010). Overview on advanced thermal barrier coatings. *Surface and Coatings Technology*, *205*(4), 938–942. doi:10.1016/j.surfcoat.2010.08.151

Verstak, A., & Baranovski, V. (2006). AC-HVAF sprayed tungsten carbide: properties and applications. *Thermal Spray 2006: Building on 100 Years of Success*, 15-18.

Vilardell, A. M., Cinca, N., Dosta, S., Cano, I. G., & Guilemany, J. M. (2019). Feasibility of using low pressure cold gas spray for the spraying of thick ceramic hydroxyapatite coatings. *International Journal of Applied Ceramic Technology*, *16*(1), 221–229. doi:10.1111/ijac.13088

Villafuerte, J. (Ed.). (2015). *Modern cold spray: materials, process, and applications*. Springer. doi:10.1007/978-3-319-16772-5

Viscusi, A., Astarita, A., Gatta, R. D., & Rubino, F. (2019). A perspective review on the bonding mechanisms in cold gas dynamic spray. *Surface Engineering*, *35*(9), 743–771. doi:10.1080/02670844.2018.1551768

Wang, D., Dong, S., Zeng, J., Liang, P., Liao, H., Wang, Y., & Cao, X. (2020). Influence of doping Mg2+ or Ti4+ captions on the microstructures, thermal radiation and thermal cycling behavior of plasma-sprayed Gd2Zr2O7 coatings. *Ceramics International*, *46*(9), 13054–13065. doi:10.1016/j.ceramint.2020.02.076

Wang, L., Wang, Y., Sun, X. G., He, J. Q., Pan, Z. Y., Zhou, Y., & Wu, P. L. (2011). Influence of pores on the thermal insulation behavior of thermal barrier coatings prepared by atmospheric plasma spray. *Materials & Design*, *32*(1), 36–47. doi:10.1016/j.matdes.2010.06.040

Xu, Y., & Hutchings, I. M. (2006). Cold spray deposition of thermoplastic powder. *Surface and Coatings Technology*, *201*(6), 3044–3050. doi:10.1016/j.surfcoat.2006.06.016

Yin, S., Cavaliere, P., Aldwell, B., Jenkins, R., Liao, H., Li, W., & Lupoi, R. (2018). Cold spray additive manufacturing and repair: Fundamentals and applications. *Additive Manufacturing*, *21*, 628–650. doi:10.1016/j.addma.2018.04.017

Chapter 3
Cyclic Oxidation of Combined LTA/YSZ and Alumina Thermal Spray Coatings

Pritee Deshpande-Purohit
https://orcid.org/0000-0002-5741-8497
Army Institute of Technology, India

Shashikant Vagge
College of Engineering, Pune, India

Bhavana Shrigadi
Cummins India Ltd., Pune, India

ABSTRACT

Thermal barrier coatings protect the substrate from thermal diffusion, oxidation, phase transformations, elastic deformation, plastic deformation, creep deformation, thermal expansion, thermal radiation. It allows parts and components of gas turbines to withstand high temperature upto1650 °C. Cylic oxidation behavior of alumina incorporated, lanthanum titanium aluminum oxide (LaTi2Al9O19), and yttria stabilized zirconia (YSZ), that is LTA/YSZ top ceramic layer coating, was investigated. Two coating combinations, L 100 having top LTA layer thickness of 100 μm and L 150 having top layer of LTA having thickness 150 μm, were tested for thermal cycles at the temperature of 1100°C. The performances of these coatings were compared with conventional YSZ coatings. Microstructure studies, EDX, and XRD analysis demonstrated the formation of mainly LTA, LaAlO3, Al2TiO5, Al2O3, and TiO2 at 1100°C in both coatings. But in L 150 coating, the rate of oxidation was found slower than L 100 coating. Annealed L 150A and L 100A specimens show cyclic oxidation life of 272th and 250th cycles, respectively.

DOI: 10.4018/978-1-7998-4870-7.ch003

INTRODUCTION

The thermal efficiency and power generation of gas turbines and aero engines can be enhanced using different thermal barrier coatings (TBCs). TBCs comprises of ceramic top coat of yttria stabilized zirconia (YSZ) having low thermal conductivity with good insulation capability. The top coat is applied over an oxidation resistant MCrAlY bond coat (Gok, & Goller, 2016). At higher temperatures for longer duration of operation or in case of cyclic thermal loading conditions, the spallation failure of TBC occurs mainly due to sintering and phase transformation of YSZ (Padture et al., 2002). To overcome the challenges associated with higher inlet gas temperature in advanced gas turbine and aero engines, rigorous researches were carried out to find alternative to ceramic materials better than conventional YSZ (Cao, Vassen, Tietz, & Stoever, 2006). Few of them are aluminates, pyrochlores, doped zirconia, perovskites. Excellent high temperature performance and higher thermal stability has been revealed by lanthanum zirconate as a top coat material (Vaben, Cao, Tietz, Basu, & Stover, 2000).

Studies revealed that TGO growth can retards with reduction in the bond coat internal oxidation. It is observed that the NiCrAlY bond coat with incorporated alumina proved better hot corrosion resistant compared to the NiCrAlY bond coat with incorporated YSZ (Wei, Guo, Gong, & Xu, 2008). Oxidation barriers layer of Alumina suppresses the excess bond coat oxidation. The Al_2O_3/YSZ coatings revealed higher oxidation and spallation resistance and increased rate of phase transition and densification (Ma, et al., 2008). The composite coatings with the combination of YSZ/alumina top coat, showed much better resistance to oxidation and thermal cycling (Friedrich, Gadow, & Schirmer, 2001).

La_2Zr_2O7 (LZ) pyrochlore as a top coat material has revealed excellent high-temperature capability and high thermal stability. H. Dong et. al, developed $La_2Ce_2O_7$ (LC) coating using APS by using La2Ce2.5O8 powder. The LC/YSZ coating has thermal cycling life 40% more than YSZ coating at 1320°C. Thermal conductivity of DCL coatings having top layer of 50% $La_2Zr_2O_7$ is reduced and 50% $Gd_2Zr_2O_7$ increased. Pyrochlore LZ is having lower thermal conductivity and good sintering resistance compared to YSZ. But it is having short life due to thermal expansion mismatch and higher thermal stresses generating from it. The LZ coating has low thermal expansion coefficient (TEC) which leads to higher thermal stresses and very short life. Recently co-doping of rare earth material Sc_2O_3 in YSZ was studied. Materials for TBCs are more complex and they were introduced through many studies. $LaTi_2Al_9O_{19}$ (LTA) (Xie, 2011), Lanthanide Tantalate ($RETa_3O_9$), Nd (Neodymium), Sm (Samarium), Eu (Europium), Gd (Gadolinium), Dysprosium-Tantalum Oxide (DTO) (Wu, 2018), Dy (Dysprosium), Er (Erbium)) (Chen L, 2018), Lanthanide Niobate (Ln_3NbO_7) (LNO), Dy (Dysprosium), Er (Erbium), Y (Yttrium), Yb (Ytterbium)) (Yang, 2019), Magnesium-Silicon Oxide (MSO) (Chen S, 2019), Calcium-Magnesium Alumino-Silicate (CMAS) (Gildersleeve, 2019), Zirconium Lanthanate ($Zr_3Ln_4O_{12}$) where, Ln is La (Lanthanum), Gd (Gadolinium), Y (Yttrium), Er (Erbium), and Yb (Ytterbium)) (Zhao M, 2019), Magnetoplumbite ($LnMgAl_{11}O_{19}$), Pr (Praseodymium), and Gadolinium-Zirconium oxide (GZO) (Vaßen, 2020).

LTA/YSZ ceramic layer showed excellent oxidation performance (Xie, Guo, Gong, & Xu, 2011). However, no data on cyclic oxidation study of combined YSZ/alumina TBCs is available in open literature. In present work, the thermal cyclic behavior of the plasma sprayed coatings for oxidation was evaluated at 1100°C. Oxidation kinetics was studied. Phases involved were determined using field emission scanning electron microscopy (FESEM) and x-ray diffraction (XRD).

BACKGROUND

Thermal barrier coatings (TBCs) is a complex material systems, sustainable at very high temperatures above 1000 °C and provide the thermal insulation and with very less thermal expansion. It is sprayed on the metallic substrate to protect the underlying components at the hot sections of gas turbine and aero engines (Backman, 1992; Padture, 2002). To increase the performance and efficiency of gas turbines and aero engines thermal barrier coatings are used. Thermal barrier coatings (TBC) consist of an insulating ceramic top coat of yttria stabilized zirconia (YSZ) with lower thermal conductivity applied over a metallic MCrAlY bond coat (M= Ni and Co) as an oxidation resistant layer. At higher temperatures, due to sintering and phase transformation of YSZ the spallation failure of TBC occurs. Rigorous researches carried out to find new ceramic materials better than YSZ to meet the challenges associated with higher inlet gas temperature in advanced gas-turbine and aero engines. Few of them are aluminates, pyrochlores, doped zirconia, perovskites (Vasen, 2000; Cao, 2004; Tarasi, 2011).

Studies also have been done to search a method to reduce the internal oxidation of the bond coat which is the main reason for TGO growth. Alumina layer acts as an oxygen barrier and retards the further bond coat oxidation. Al_2O_3/YSZ coatings have shown good spallation and oxidation resistance and increased densification and phase transition rate. In oxidation test, the composite coatings of alumina as a top coat and the mixed YSZ alumina layer, showed better resistance. Studies proved that the alumina incorporated NiCrAlY bond coat have better hot corrosion resistance than the YSZ incorporated NiCrAlY bond coat.

$LaTi_2Al_9O_{19}$ (LTA) has been proposed as TBC material, which is having excellent phase stability. It is combination of rare earth element and ceramic that is Lanthanum oxide La_2O_3, titanium oxide TiO_2 and aluminium oxide Al_2O_3. But due to its relative low fracture toughness it is having short life. This difficulty can be solved by using a double L/YSZ ceramic layer.

MATERIALS AND METHODS

The Inconel 718 was used as substrate material for developing the coatings. Specimens of 10 mm x 10 mm x 4 mm were cut using wire cut EDM. The edge effect of the specimen was eliminated by grinding the edges. Using alumina powder the specimens were grit blasted before spraying. LTA powder was prepared using La_2O_3-99%, TiO_2-99%, and Al_2O_3-99% having size 325 µm. These powders were dried for removal of moisture at 200°C for 10 h, mixed together and then ball milled for 5 h. Mixed powder was dried at 200°C for 5 h and calcined at 1500°C for 24 h. It was compacted at 400 MPa and sintered at 1600°C for 72 h. Finally sintered mass crushed and sieved with 90 µm sieve. The fluidity was maintained around 38 to 50 gm/min.

The coatings were prepared at Metallizing Equipment, Jodhpur. A 40 kW Metco thermal plasma spray unit having plasma spray F4 gun was used for developing coating. All powders selected were thermal grade powders having fluidity in the range of 38 to 50 gm/min. The particle size of powders was kept 90 µm. The thickness of the coatings was maintained in the range of 320-370 µm. Before spraying the LTA powder was preheated to 150°C. Two combinations of coatings were finalized with varying topcoat thickness, 1. L 100 with 100 µm thickness and 2. L 150 with 150 µm thickness. It is shown in Fig. 1. Annealing of L 150 and L 100 coatings was done at 1050°C for 5 h in resistance heated muffle furnace for recrystallization of coating (Xie, 2011; Sreedhar, 2009).

Figure 1. Weight gain curves for coatings subjected to thermal cycles at 1100°C

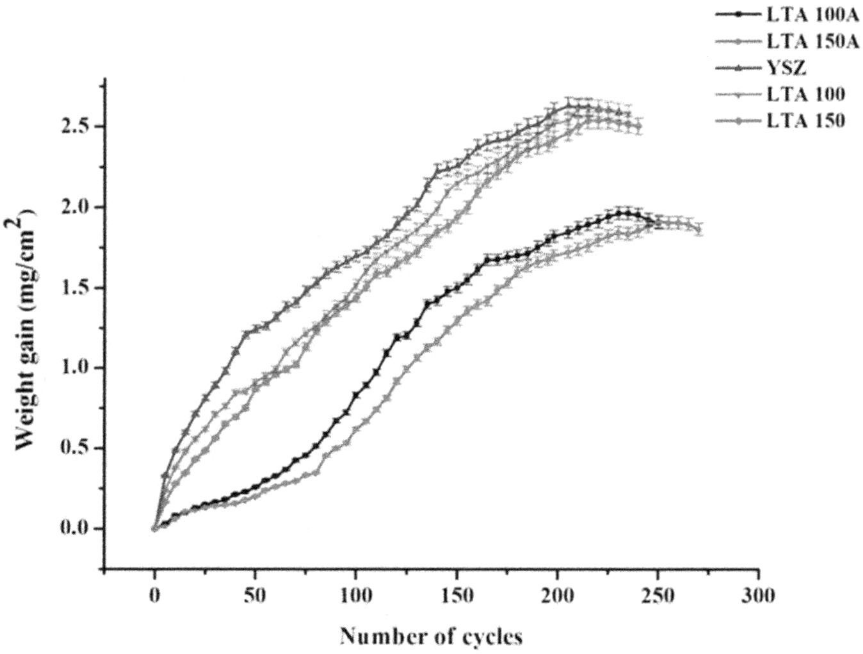

Cyclic oxidation test of coated specimens was conducted at 1100°C in a Therelek make muffle furnace. One cycle comprises of heating the specimens at 1100°C for 45 min and cooling in air for 15 min that is total 1h. The specimens were removed from the furnace after an interval of 45 min and cooled in a room for 15 min. Weight of the coated samples was taken using Afcoset electronic balance with least count of 0.001g. The cyclic oxidation test was discontinued after 20% failure of coated specimens. Oxidation kinetics of the specimens was studied using the mass gain data. The specimens were evaluated for various phases obtained using XRD, FESEM. Energy Dispersive X-Ray analysis (EDX) was done to do the elemental mapping and determine the elemental percentage. The porosity of the coating was checked by measuring the pores and the coating mass via image analysis technique on optical microscope.

RESULTS

Fig.1 depicts the relation between mass gain in mg/cm^2 and total number of cycles at 1100°C. The annealed specimens have shown least weight gain compared to as-received specimens. As-received L 150 and L 100 specimen sustained upto 242nd and 236th cycles. The mass gain due to oxidation of 2.49 mg/cm^2 and 2.57 mg/cm^2 is observed respectively. The traditional (YSZ) specimen resulted in the mass gain of 2.589 mg/cm^2 due to oxidation and was sustained upto 230th cycles. Annealed specimens L 150A and L 100A, remained intact till 272nd and 250th cycles. Thus L 150A and L 100A proved better with the least mass gain due to oxidation of 1.87 mg/cm^2 and 1.89 mg/cm^2 respectively. Most of the coatings do not exhibit parabolic behavior. It show accelerated oxidation behavior due to minor oxidation of ceramic constituents and impurities at the top layer.

Figure 2. XRD patterns of L150 coating subjected to thermal cycles at 1100°C

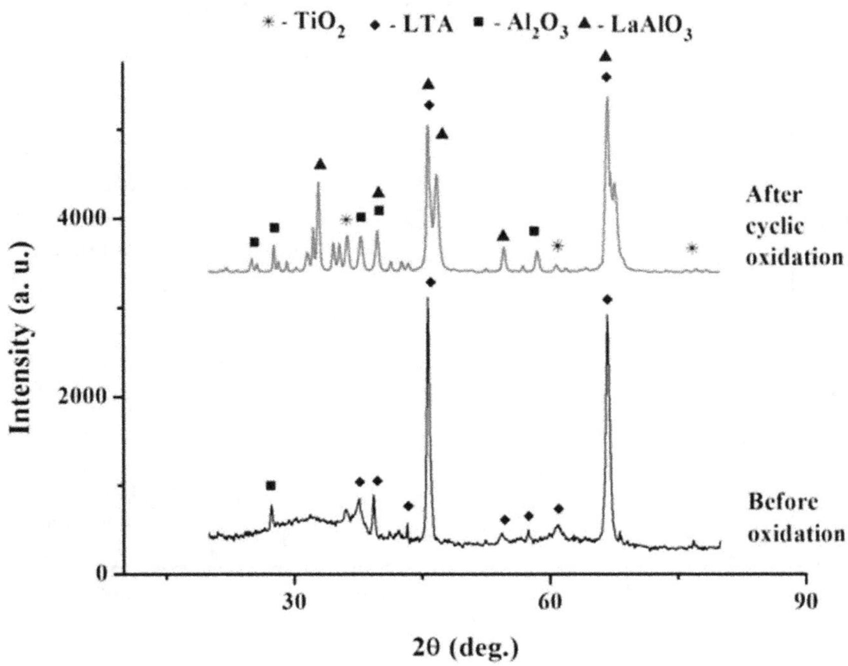

X-Ray Diffraction Analysis

The XRD peaks of L 150 and L 100 specimens prior to cyclic oxidation and post thermal cycles at a temperature of 1100°C are depicted in Fig. 2 and 3 respectively. The retained phases without many changes were observed in all specimens pre and post oxidation. It proves the negligible phase change in both L 100 and L 150 specimens. The XRD analysis of L 100 and L 150 samples after thermal cycling oxidized to $LaAlO_3$, Al_2O_3 and LTA. The $LaAlO_3$ phases were observed due to incorporated alumina. But multiple phases of Al_2TiO_5 were observed in L 100 specimens after cyclic oxidation. Presence of $LaAlO_3$, Al_2O_3 and almost LTA phases were detected in L 150 specimens proves the limited oxidation of L 150 coating. No drastic phase change was observed in the XRD pattern except few peaks broadening. An intact thermal barrier was created by L 150 coating.

Figure 3. XRD patterns of L100 coating subjected to thermal cycles at 1100°C

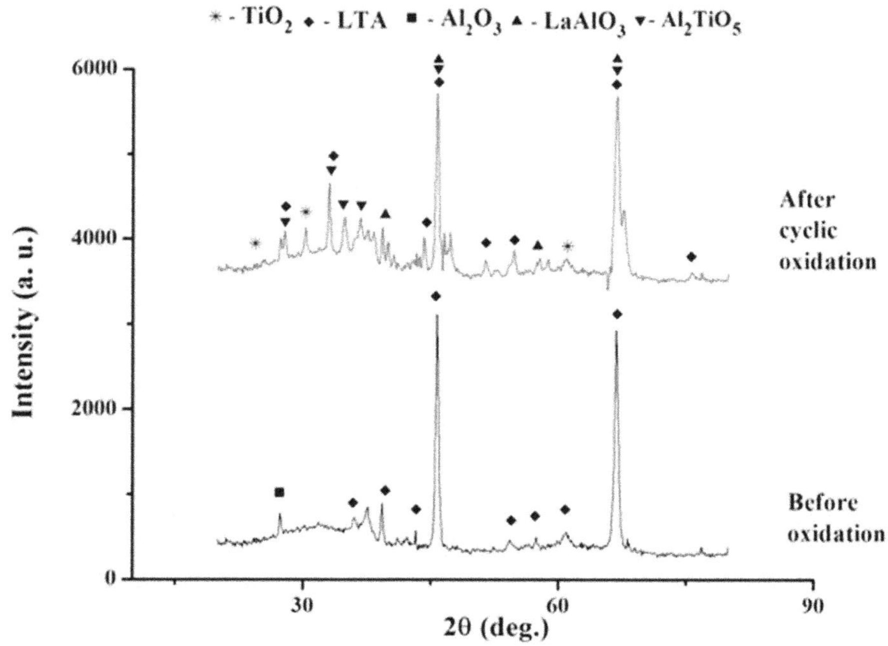

After subjected to thermal cycling and oxidation at 1100°C the phases of LTA are transformed from tetragonal t'-phase to monoclinic m-phase. Also the other LTA phases are dissolved to Al_2O_3 and other similar oxides in case of both the coatings. With the increase in the number of repetitive heating and cooling cycles at 1100°C the crystalline LTA phase decreases. Phases formed in XRD analysis are in good agreement with the elements observed in EDX. These phase formation is discussed later.

Cross Section Morphology

As shown in Fig. 4a and 4e the cross-section of as-received L 100 and as-received L 150 coatings before oxidation shows a clear and precise interface at the LTA/YSZ top layer and the incorporated alumina and bond coat respectively. Cross section morphology both annealed L 100 and L 150 before thermal cycling is shown in Fig 4c and 4g. Dense microstructure is observed in both annealed coating.

Cross section morphology of as-received L 100 coating after thermal cycling at 1100°C is shown in Fig. 4b. It shows, after cyclic oxidation, the cracks generated leads to wider opening in the coating, results in rapid bond coat oxidation near to the alumina and bond coat interface. The oxidation kinetics of coating is limited by the diffusion of O_2 and Al_2O_3 (Sreedhar, MasroorAlam, & Raja, 2009). Thus oxidized specimens show depletion of alumina layer at some places and hence the consistency in thickness of alumina is not observed. The transverse cracks were observed above the TGO layer. In the top layer parallel to the surface smaller cracks were generated. With the increase in number of thermal cycles the transverse cracks increased and resulted in the localized fracture and formation of chip of the top coat. Fig. 4b shows that major cracks were initiated and propagated from the coating surface and ended up at top coat alumina interface. The bi-axial stress due to compression at the surface is the predominant reason of coating failure. In L 100 coating, the top most coating surface has more number of cracks and also the crack width observed were large. It shows extensive cracking at the coatings interface. TGO formation leads to extensive stress and enhanced cracking.

Figure 4. Cross section images of samples subjected to thermal cycling: (a) as-received LTA 100 coating before oxidation, (b) as-received L100 coating after oxidation, (c) as-received L150 coating before oxidation, (d) as-received L150 coating after oxidation, (e) annealed L100 coating before oxidation, (f) annealed L100 coating after oxidation, (g) annealed LTA 150 coating before oxidation, (h) annealed L150 coating after oxidation

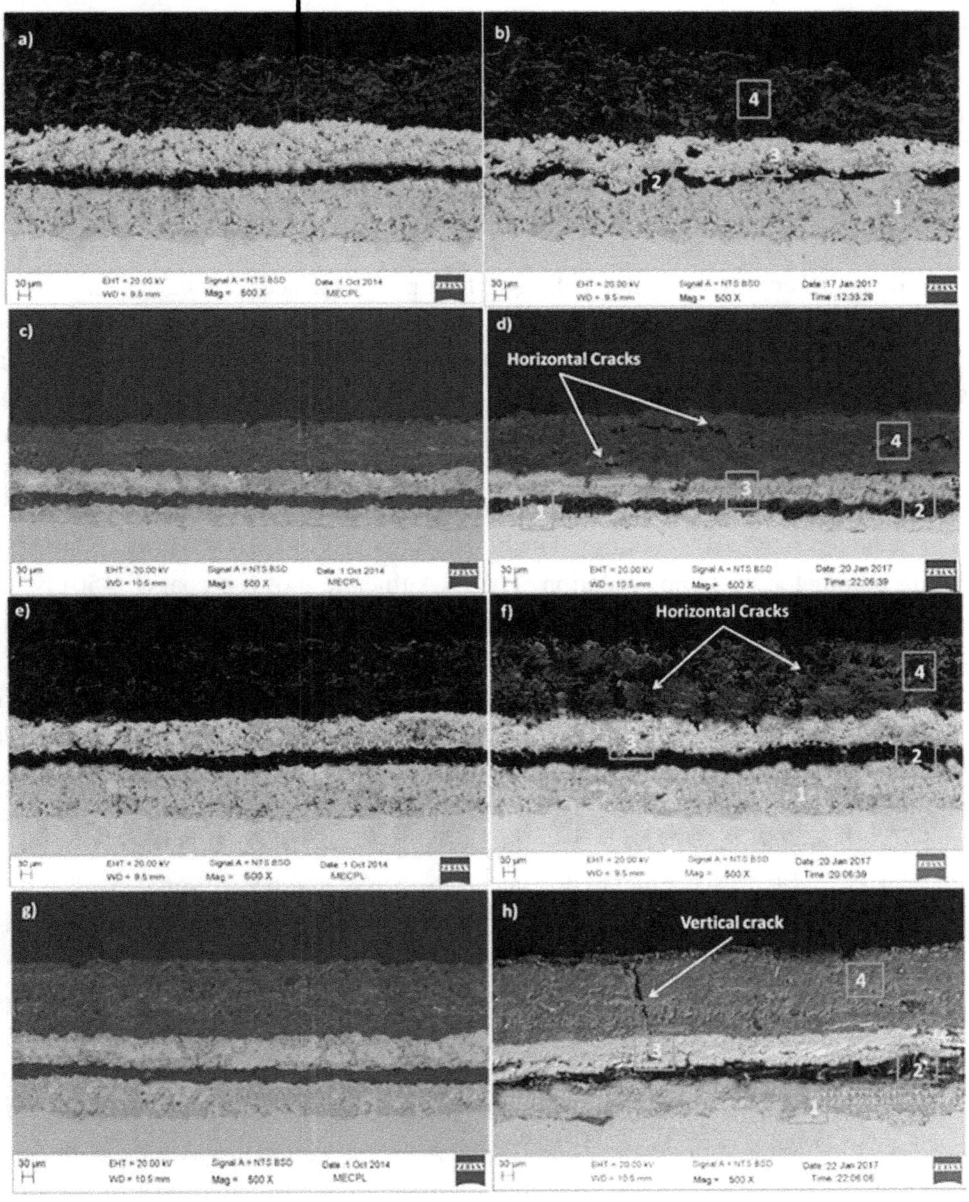

The cross-section of the annealed L 100 coated sample after thermal cycling is shown in Fig. 4d. After cyclic oxidation of annealed L 100 coatings no sharper interface were observed as a result of the mixed layers of the top coat and adjacent bottom layers as depicted in Fig. 4d. It shows a considerable split between the alumina-YSZ interfaces, caused by the stresses initiated as an effect of compression (compressive stresses) during cooling process. It shows immense cracking and loss of coating integrity due to cyclic oxidation. The cracks propagation and enhancement is also due to the thermal interfacial stress.

The micro-cracks with vertical as well as transverse orientation are observed in the L 100 coatings were propagated in the downward side of the protective TGO layer. Also, the large cracks acts as an oxygen diffusion path towards the surface of bond coat, and results in the accelerated oxidation of bond coat. Therefore, partially the transverse cracks also results in the failure due to spallation of coating. The early failure of L 100 coating is caused by the formation of longitudinal as well as transverse cracks as a by-product of the thermal shock generated during the thermal cycling. Cracks initiated due to thermal cycling appear parallel to bond coat and top coat interface.

The coatings when exposed to high temperature during prolonged cyclic oxidation results in the closure of pores at the top coat and thus increases its thermal conductivity and makes the coating rigid and very much prone to failure due to spallation. During heating, the LTA/YSZ top surface of the as-received and annealed L 100 coating was subjected to large bi-axial compressive stresses. These stresses cause the non-linear and continuous deformation of the top coat surface. During cooling, the radial stress converted to tensile stress as a result induced strain. The alteration of stresses from compressive to tensile with large magnitude was adequate to surpass the fracture durability of the top surface layer and results in the vertically oriented cracks. Also, major cracks most likely initiate within the bond coat-top coat interface a cleavage mode. The cracks and the existing imperfections in the ceramic top layer result in longer cracks and spallation failure of both bond coat as well as top coat.

The thermal cycling performance and the overall oxidation resistance of as-received L 150 coating were found to be better than L 100 coatings. The higher thickness of L 150 coating is one of the coating lives enhancing parameter in cyclic oxidation. The LTA thickness of top coat (L 150) is 150 μm and the overall coating thickness is 370 μm. This 150 μm LTA layer over the YSZ layer is responsible for greater thermal gradients.

Fig. 4f shows the cross section morphology of as-received L 150 coating after cyclic oxidation. In case of as-received L 150 specimen the LTA top coat which is 150 μm thick acts as a crack arresting barrier. Though no cracks were observed on the top LTA layer but, cracks were detected at the interface of YSZ/alumina and bond coat due to TGO formation. The transverse crack was more visible as depicted in Fig. 4f. The tensile stress at the top coat-alumina and alumina-bond coat interface was responsible in the initiation of cracks which are parallel to that interface, results in propagation of crack and an absolute delamination of coating structure. The oxidation of bond coat resulted in the higher compressive-stress which is in plane, producing tensile-stress which is out of plane and inducing cracks almost parallel to the entire coating interface. Under repeated thermal cycling, these cracks open, propagate and resulted in a complete spallation failure of coating. Thus, the L 150 as-received coatings were failed at the YSZ and bond coat interface.

The cyclic oxidation resistance of annealed L 150 coating was better than L 100 coating. The thicker top layer develops the higher strain energy and results in changing and passivation of the cracks direction. Cross section of annealed L 150 coating after thermal cycling test at 1100°C is shown in Fig. 4h. It shows longitudinal branched cracks and intra splat micro cracks that were propagated through the coatings surface. Also, multiple branched cracks originating from the large crack and longitudinal and transverse cracks were observed. As shown in Fig. 4h the transverse cracks perpendicular to the interface are the results of tensile stresses due to cyclic heating and cooling. An extensive vertically oriented crack initiates from the top most layer of the coating towards the TGO layer. From Fig. 4h, it is clear that the development of a longer longitudinal crack and its sub-cracks at the YSZ/alumina and bond coat interface resulted in the spallation failure of the annealed L 150 coating at the YSZ and bond coat interface.

From Fig. 4f and 4h it is observed that, the cracks developed in this region were discontinuous and were propagated randomly in entire region separately. The discontinuous crack present in the LTA layer liberates the thermal stress and hindered the spallation at the coatings interface. It results in the failure of the coating at another weak location that is at the YSZ-alumina-bond coat interface.

The EDX analysis of cross sections of L 100 and 150 coating after thermal cycling is summarized in Table 1. It evidences the occurrence of oxides of Al, La and Ti. Also the formation of Ni and Cr-spinel oxides of were observed. It may be because of the oxygen barrier nature and premature failure of L 100 coatings. At point 4, in both L 150 and L 100 coatings traces of O, La, Ti and Al were found. Higher percentage of O, Al and Ti reveals the formation of LTA, $LaAlO_3$, Al_2O_3 and Al_2TiO_5 in L 100 coating. Formation of LTA, $LaAlO_3$, Al_2O_3 in L 150 coatings was observed. But at the top layer traces of Cr and Ni were not observed in L 150 coating (point 4). It evidences that no Ni and Cr-spinel oxides were formed at the top surface of as-received as well as annealed L 150 coating. But they were formed at the surface of L 100 coatings. The results obtained were in well agreement with the results obtained from the XRD and EDX analysis done at the coating surface.

Table 1. Presence of different elements determined through EDX analysis (at. %) after thermal cycling

Specimen	Location	Elements Present on coating surface							
		O	Al	Ti	La	Zr	Cr	Y	Ni
L 100 Fig. 4b	1	43.05	11.11	4.55	0.02	0.00	10.34	0.00	35.32
	2	51.56	35.98	0.01	0.01	0.97	1.55	0.00	0.14
	3	30.16	19.18	0.01	0.00	12.27	0.08	0.42	0.66
	4	78.22	47.01	3.99	0.93	0.12	0.00	0.04	0.39
L 150 Fig. 4d	1	25.42	13.12	0.01	0.01	0.01	13.35	0.00	45.89
	2	40.98	31.47	0.24	0.00	0.19	1.08	0.00	0.81
	3	31.97	03.04	1.53	0.20	16.72	0.00	0.34	0.05
	4	61.98	27.02	2.77	0.87	0.35	0.00	0.00	0.02
L 100A Fig. 4f	1	44.86	8.65	1.79	0.05	0.00	9.88	0.00	41.93
	2	57.09	34.02	0.03	0.00	0.94	1.21	0.06	0.10
	3	24.22	21.20	0.01	0.14	4.72	1.91	0.43	0.03
	4	72.21	35.21	5.34	0.82	0.16	0.00	0.76	0.09
L 150A Fig. 4h	1	31.21	08.54	0.01	0.01	0.00	12.17	0.00	35.42
	2	50.24	34.93	0.02	0.02	0.32	1.06	0.32	0.52
	3	30.19	20.43	2.18	0.21	4.74	0.00	0.61	0.02
	4	65.88	26.01	3.24	0.89	0.09	0.00	0.00	0.00

Figure 5.Elemental mapping microstructure of as-received L100

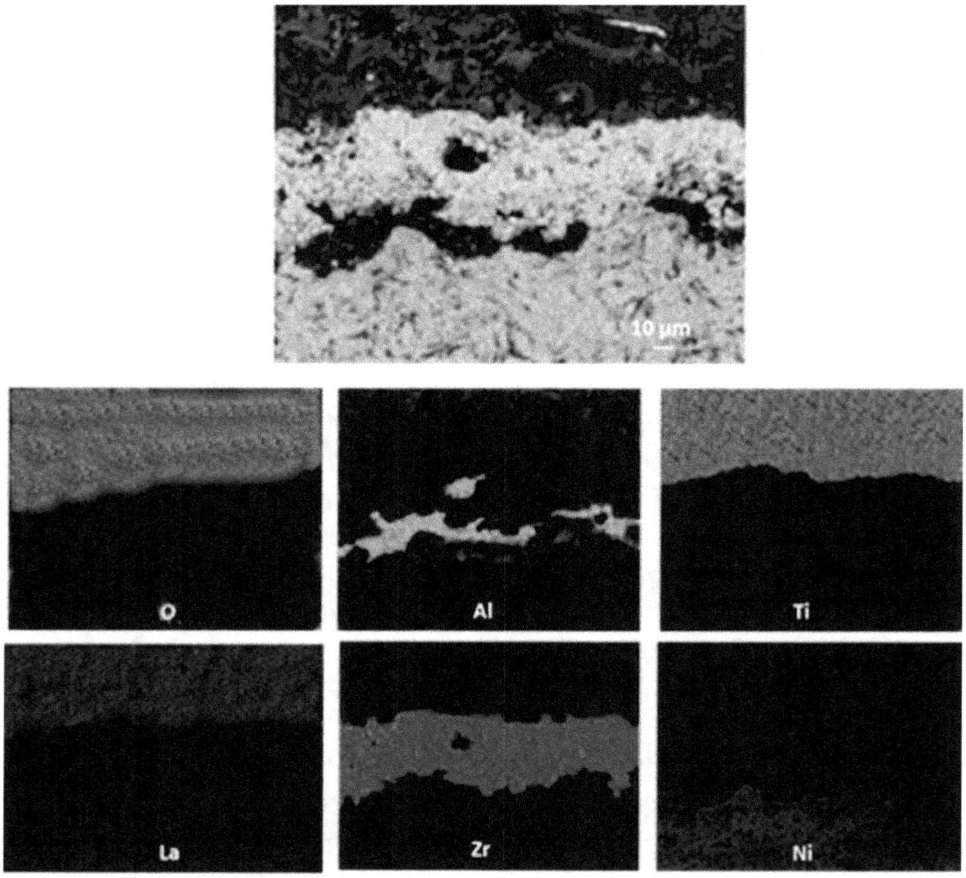

Presence of high percentage of O at the bond coat (point 1) in case of both as-received and annealed L 100 coatings evidences the accelerated oxidation of bond coat. In case of as-received and annealed L 150 coating the O percentage is much less compared to L 100 coating. The amount of oxygen available for formation of oxides was more in case of L 100 coating due to inadequate top coat thickness. It has been validated through XRD, as the phases of oxides having Al and Ti were comparatively more in case of L 100 specimen. the In both L 100 and L 150 coating at point 2 traces of O and Al were found which indicates the formation of protective oxide layer of Al_2O_3.

The elemental mapping of as-received and annealed L 100 is shown in Fig. 5 and Fig. 6 and elemental mapping of as-received and annealed L 150 is shown in Fig. 7 and Fig. 8. The photographs of coatings prior and post thermal cycling are shown in Fig. 9. In Fig. 9 specimen a and b are as-received L 100 and L 150 coatings before cyclic oxidation respectively. Fig. 9 specimen c and d are annealed L 100 and L 150 coatings before cyclic oxidation. Specimens from Fig. 9e and f are as-received L 100 and L 150 coatings after the heat cycling test. Fig. 9 g and h shows the annealed L 100 and L 150 coatings after thermal cycling. Visible cracks were observed in case of L 100 coating.

Figure 6.Elemental mapping microstructure of annealed L100

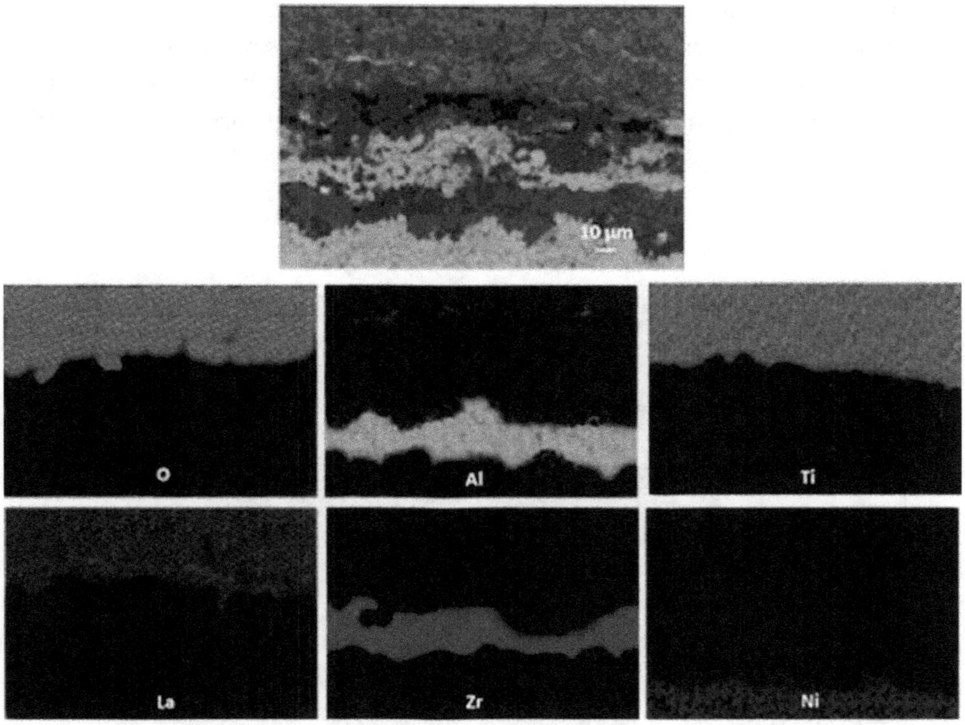

DISCUSSION

The present study is thermal cycling tests for traditional thermal barrier coatings having YSZ as top coat and compared with thermal cycling behavior of two different combinations of coating after incorporating a readily available alumina layer between bond coat and LTA/YSZ top coat layer at 1100°C. The coatings were developed by keeping parameters like coating layer thicknesses and coating constituents constant except top layer thickness of LTA. L 150 coatings showed higher resistance oxidation under thermal cycling and traditional YSZ showed less resistance to oxidation. The oxidation rate of TBC can be affected by change in the temperature, exposure timing, overall composition and thickness of top coat, surface roughness of bond coat and spraying method. The higher resistance to oxidation of L 150 coatings is due to the sufficient thickness of coating to act as a thermal and oxygen diffusion barrier. Also, due to negligible degradation of LTA contributed towards protection ability of coatings. Bond coat oxidation is prevented by alumina layer and resulted in better oxidation resistance in L 100 as well as L 150 coating as it acts as a diffusion barrier of oxygen. Although YSZ top coat thickness is more this thickness doesn't contribute to enhance the TBC life as the cohesion of YSZ top layer is stronger than adhesion with MCrAlY bond coat. Also, the interconnected porosities and microcracks of YSZ coating results in the formation of a strain tolerant coating, but permits the diffusion of oxygen and results in enhanced bond coat oxidation (Ramachandran, Balasubramanian, Ananthapadmanabhan & Viswabaskaran, 2012).

Thus after 230th oxidation cycle at 1100°C, the YSZ coating were shown spallation failure. One of the reasons for spallation of YSZ TBC is the stress generation due to thickening of the TGO. The reason for stress development in YSZ coating is thermal expansion coefficient mismatch (TEC). But in case of LTA coating the TEC is around 11.2 x 10^{-6} K^{-1}, that is very close to the values of TEC of YSZ at temperature above 1000°C. Thus initially the stress level is much lower to initiate cracks in the coating. However, with increasing time, crack formations due to other reasons resulted in further oxidation and hence thus formations of the mixed oxides in L 100 coating (Li et al., 2012). Also, the stress generated due to oxide formation is independent of the cycling temperature but varies with the cycle time (Ren, He & Wang, 2011). Figure 1 depicts higher mass gain in YSZ coating.

Figure 7. Elemental mapping microstructure of as-received L150

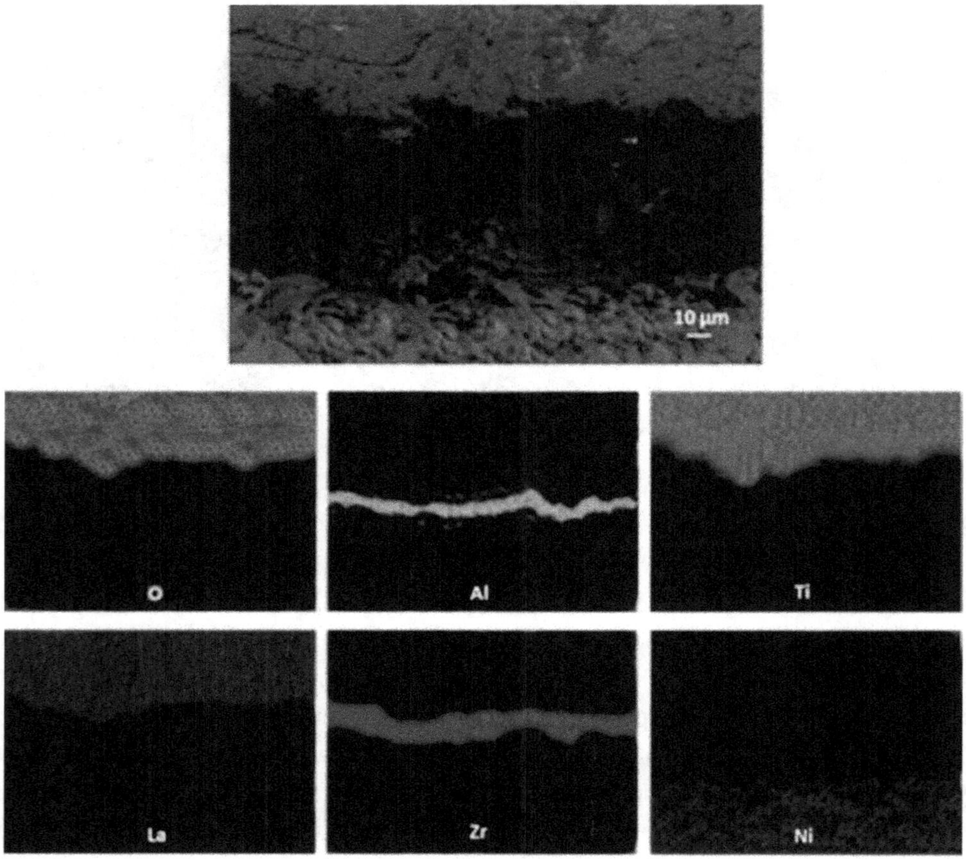

The cyclic oxidation of present coatings gave rise to various oxides and spinel. Also, LTA phase is the only one ternary compound found and mentioned in the combined Al_2O_3-La_2O_3-TiO_2 phase system. In the ternary phase diagram, 7 binary compounds are formed and $LaAlO_3$ is one of them. According to the ternary phase diagram of Al_2O_3-La_2O_3-TiO_2 and its chemical compositions and the chemical ratios it clear that the phases can be decomposed to different oxides rich in La and Ti. Therefore after thermal cycling the different oxides present on the coating surface can be the solid solution of $LaAlO_3$ and $La_{2/3}TiO_3$. LTA phase is a stable phase at high temperatures. The formation of $LaAlO_3$ is comparatively easy compared to different aluminates. It is very slow for temperature less than 1500°C and can be more than 132 days. Therefore, no other products of oxidation were generated after thermal cycling of LTA.

Figure 8. Elemental mapping microstructure of annealed L150

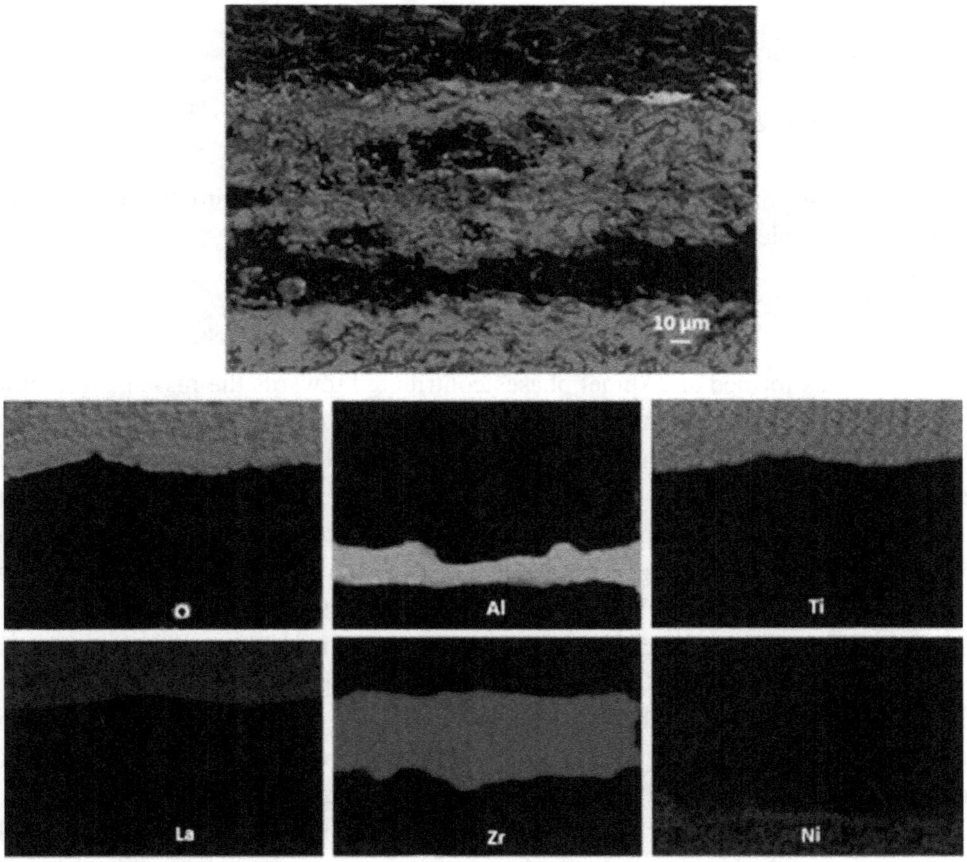

Figure 9. Macro photograph of specimen: (a) L100 and (e) L150 as-received coating before thermal cycling, (c) L100 and (g) L150 annealed coating before thermal cycling, (b) L100 and (f) L150 as-received coating after thermal cycling, (d) L100 and (h) L150 annealed coating after thermal cycling

Therefore LTA can be decomposed to only $LaAlO_3$, at temperature less than 1500°C (Skapin, Kolar & Suvorov, 1993).

Therefore, possible reactions are proposed here.

$$LaTi_2Al_9O_{19} \rightarrow LaAlO_3 + 2TiO_2 + 4Al_2O_3 \quad (1)$$

The available anatase phase that is α-Al_2O_3 and rutile phase that is TiO_2 simultaneously forms Al_2TiO_5. Equation (2) is the possible reaction of this phase transformation,

$$Al_2O_3 + TiO_2 \rightarrow Al_2TiO_5 \quad (2)$$

The different oxides formed and spinel phases contributed towards the mass gain in case of L 100 specimen than L 150 specimen. Formation of different oxides due to cyclic oxidation is confirmed through SEM, EDX and XRD analysis. From the mechanism of thermal cycling kinetics it is understood that, the scale growth of protective oxides in L 150 is steady, continuous and slower.

The alumina layer acts as an oxygen diffusion barrier and prevents further oxidation of immediate bond coat and protects the underlying substrate. The formation of oxides can be due to oxygen inward diffusion (Ropp, & Libowitz, 1978). In L 150 and L 100 coatings the diminished oxidation is observed due to lack of availability of oxygen compared to traditional porous YSZ coating. It resulted in the very slow rate of TGO growth in case of L 150A and L 100A coating (Nagano et al., 1999). Both, L 150A and L 100A specimens were intact till 272th and 250th cycles respectively in thermal cycling at 1100°C with negligible mass gain much compared to YSZ specimens. Thus, cyclic oxidation rates of TBC with various different ceramic layers (L 100 and L 150) are compared and studied. But in both L 100 and L 150, annealed coatings showed better cyclic oxidation life compared to as-received coatings. The results show that larger top coat thickness and double layer coatings can prove better oxidation resistant.

CONCLUSION

The cyclic oxidation of YSZ, DCL and alumina incorporated coatings of LTA/YSZ were tested at 1100°C. Following conclusions were made from the results obtained. Negligible mass gain of 1.86 mg/cm^2 is observed in alumina incorporated L 150A coating. Spinel oxides of Al, Cr and Ni were detected and contributed towards enhancement of cyclic oxidation lives of coatings. Comparatively optimum thickness of the coating layers contributed towards better cyclic oxidation life of L 150A coatings. Due to excellent chemical stability of LTA its decomposition is negligible after oxidation test. Better oxidation resistance is observed in case of L 150A coating. Cyclic oxidation of LTA/YSZ/Al_2O_3 top ceramic layer coating caused oxides formations like Al_2O_3, $LaAlO_3$, LTA, TiO_2, Al_2TiO_5 and also spinels of Ni and Cr in both L 100 and L 150 coatings.

REFERENCES

Backman, D. G., & Williams, J. C. (1992). Advanced Materials for Aircraft Engine Applications. *Science*, *255*(5048), 1082–1087. doi:10.1126cience.255.5048.1082 PMID:17817782

Cao, X. Q., Vassen, R., & Stoever, D. (2004). Ceramic Materials for Thermal Barrier Coatings. *Journal of the European Ceramic Society*, *24*(1), 1–10. doi:10.1016/S0955-2219(03)00129-8

Cao, X. Q., Vassen, R., Tietz, F., & Stoever, D. (2006). New double-ceramic-layer thermal barrier coatings based on zirconia-rare earth composite oxides. *Journal of the European Ceramic Society*, *26*(3), 247–251. doi:10.1016/j.jeurceramsoc.2004.11.007

Chen, L., Jiang, Y., Chong, X., & Feng, J. (2018). Synthesis and Thermophysical Properties of $RETa_3O_9$ (RE = Ce, Nd, Sm, Eu, Gd, Dy, Er) as Promising Thermal Barrier Coatings. *Journal of the American Ceramic Society*, *101*(3), 1266–1278. doi:10.1111/jace.15268

Chen, L., Jiang, Y., Chong, X., & Feng, J. (2018). Synthesis and Thermophysical Properties of $RETa_3O_9$ (RE = Ce, Nd, Sm, Eu, Gd, Dy, Er) as Promising Thermal Barrier Coatings. *Journal of the American Ceramic Society*, *101*(3), 1266–1278. doi:10.1111/jace.15268

Chen, S., Zhou, X., Song, W., Sun, J., Zhang, H., Jiang, J., Deng, L., Dong, S., & Cao, X. (2019). Mg_2SiO_4 as a Novel Thermal Barrier Coating Material for Gas Turbine Applications. *Journal of the European Ceramic Society*, *39*(7), 2397–2408. doi:10.1016/j.jeurceramsoc.2019.02.016

Friedrich, C., Gadow, R., & Schirmer, T. (2001). Lanthanum hexaluminate a new material for atmospheric plasma spraying of advanced thermal barrier coatings. *Journal of Thermal Spray Technology*, *10*(4), 592–598. doi:10.1361/105996301770349105

Gildersleeve, E., Viswanathan, V., & Sampath, S. (2019). Molten Silicate Interactions with Plasma Sprayed Thermal Barrier Coatings: Role of Materials and Microstructure. *Journal of the European Ceramic Society*, *39*(6), 2122–2131. doi:10.1016/j.jeurceramsoc.2019.01.023

Gok, M. G., & Goller, G. (2016). Production and characterization of GZ/CYSZ alternative thermal barrier coatings with multilayered and functionally graded designs. *Journal of the European Ceramic Society*, *36*(7), 1755–1764. doi:10.1016/j.jeurceramsoc.2016.01.036

Keyvani, A., Saremi, M., & Sohi, M. H. (2011). Oxidation resistance of YSZ-alumina composites compared to normal YSZ TBC coatings at 1100°C. *Journal of Alloys and Compounds*, *509*(33), 8370–8377. doi:10.1016/j.jallcom.2011.05.029

Li, Y., Xie, Y., huang, L., Liu, X., & Zheng, X. (2012). Effect of physical vapor deposited Al2O3 film on TGO growth in YSZ/CoNiCrAlY coatings. *Ceramics International*, *38*(6), 5113–5121. doi:10.1016/j.ceramint.2012.03.014

Ma, W., Mack, D., Malzbender, J., Vaben, R., & Stover, D. (2008). Yb2O3 and Gd2O3 doped strontium zirconate for thermal barrier coatings. *Journal of the European Ceramic Society*, *28*(16), 3071–3081. doi:10.1016/j.jeurceramsoc.2008.05.013

Nagano, M., Nagashima, S., Maeda, H., & Kato, A. (1999). Sintering behaviour of Al_2TiO_5 base ceramics and their thermal properties. *Ceramics International*, *25*(8), 681–687. doi:10.1016/S0272-8842(98)00083-2

Padture, N. P., Gell, M., & Jordan, E. H. (2002). Thermal barrier coatings for gas turbine engine applications. *Science*, *296*(5566), 280–284. doi:10.1126cience.1068609 PMID:11951028

Purohit, P., & Vagge, S. T. (2016). Evaluation of alumina incorporated combined ceramic layer thermal barrier coating. *Surface and Coatings Technology*, *307*, 871–878. doi:10.1016/j.surfcoat.2016.10.022

Ramachandran, C. S., Balasubramanian, V., Ananthapadmanabhan, P. V., & Viswabaskaran, V. (2012). Influence of the intermixed interfacial layers on the thermal cycling behavior of atmospheric plasma sprayed lanthanum zirconate based coatings. *Ceramics International*, *38*(5), 4081–4096. doi:10.1016/j.ceramint.2012.01.066

Ren, C., He, Y. D., & Wang, D. R. (2011). Cyclic oxidation behavior and thermal barrier effect of YSZ-(Al2O3/YAG) double-layer TBCs prepared by the composite sol–gel method. *Surface and Coatings Technology*, *206*(6), 1461–1468. doi:10.1016/j.surfcoat.2011.09.025

Ropp, R. C., & Libowitz, G. G. (1978). The nature of the alumina-rich phase in the System La_2O_3-$A1_2O_3$. *Journal of the American Ceramic Society*, *61*(11-12), 473–475. doi:10.1111/j.1151-2916.1978.tb16119.x

Skapin, S., Kolar, D., & Suvorov, D. (1993). X-ray diffraction and microstructural investigation of the Al_2O_3-La_2O_3-TiO_2 system. *Journal of the American Ceramic Society*, *76*(9), 2359–2362. doi:10.1111/j.1151-2916.1993.tb07777.x

Sreedhar, G., Alam, M. D. M., & Raja, V. S. (2009). Hot corrosion behavior of plasma sprayed YSZ/Al2O3 dispersed NiCrAlY coatings on Inconel-718 superalloy. *Surface and Coatings Technology*, *204*(3), 291–299. doi:10.1016/j.surfcoat.2009.07.026

Tarasi, F., Medraj, M., Dolatabadi, A., Oberste-Berghaus, J., & Moreau, C. (2011). High-Temperature Performance of Alumina-Zirconia Composite Coatings containing Amorphous Phases. *Advanced Functional Materials*, *21*(21), 4143–4151. doi:10.1002/adfm.201100471

Vaben, R., Cao, X., Tietz, F., Basu, D., & Stover, D. (2000). Zirconates as new materials for thermal barrier coatings. *Journal of the American Ceramic Society*, *83*, 2023–2028.

Vaßen, R., Bakan, E., Mack, D., Schwartz-Lückge, S., Sebold, D., Sohn, Y. J., Zhou, D., & Guillon, O. (2020). Performance of YSZ and $Gd_2Zr_2O_7$/YSZ Double Layer Thermal Barrier Coatings in Burner Rig Tests. *Journal of the European Ceramic Society*, *40*(2), 480–490. doi:10.1016/j.jeurceramsoc.2019.10.021

Vassen, R., Cao, X., Tietz, F., Basu, D., & Stöver, D. (2000). Zirconates as New Materials for Thermal Barrier Coatings. *Journal of the American Ceramic Society*, *83*(8), 2023–2028. doi:10.1111/j.1151-2916.2000.tb01506.x

Wei, Q. L., Guo, H. B., Gong, S. K., & Xu, H. B. (2008). Novel microstructure of EBPVD double ceramic layered thermal barrier coatings. *Thin Solid Films*, *516*(16), 5736–5739. doi:10.1016/j.tsf.2007.07.032

Wu, P., Hu, M. Y., Chong, X. Y., & Feng, J. (2018). The Glass-like Thermal Conductivity in ZrO_2-Dy_3TaO_7 Ceramic for promising Thermal Barrier Coating Application. *Applied Physics Letters*, *112*(13), 131903. doi:10.1063/1.5022610

Xie, X., Guo, H., & Gong, S. (2010). Mechanical properties of $LaTi_2Al_9O_{19}$ and thermal cycling behaviours of plasma-sprayed $LaTi_2Al_9O_{19}$/YSZ thermal barrier coatings. *Journal of Thermal Spray Technology*, *19*(6), 1179–1185. doi:10.100711666-010-9529-5

Xie, X., Guo, H., Gong, S., & Xu, H. (2011). Lanthanum-titanium-aluminum oxide: A novel thermal barrier coating material for applications at 1300°C. *Journal of the European Ceramic Society*, *31*(9), 1677–1683. doi:10.1016/j.jeurceramsoc.2011.03.036

Yang, J., Han, Y., Shahid, M., Pan, W., Zhao, M., Wu, W., & Wan, C. (2018). A Promising Material for Thermal Barrier Coating: Pyrochlore-related Compound Sm_2FeTaO_7. *Scripta Materialia*, *149*, 49–52. doi:10.1016/j.scriptamat.2018.02.005

Zhao, Y., Min, X., Fang, M., Wu, X., Liu, Y., & Huang, Z. (2019). Dependence of Crystal Structure on Mechanical and Thermophysical Properties of Magnetoplumbite-type $LnMgAl_{11}O_{19}$ Ceramics with Substitution of Ln^{3+} Ions. *International Journal of Applied Ceramic Technology*, *16*(4), 1596–1605. doi:10.1111/ijac.13216

Chapter 4
Advances in Low Thermal Conductivity Materials for Thermal Barrier Coatings

Rafael Vargas-Bernal
https://orcid.org/0000-0003-4865-4575
Instituto Tecnológico Superior de Irapuato, Mexico

Bárbara Bermúdez-Reyes
Universidad Autónoma de Nuevo León, Mexico

ABSTRACT

One of the areas of research that continue to attract researchers worldwide is the development of thermal barrier coatings (TBCs) especially associated with the design of new ceramic topcoats with low thermal conductivity and a high coefficient of thermal expansion. The purpose of this chapter is to present the advances that have been achieved regarding ceramic topcoats in the last decades, making a historical journey that culminates with the contributions of this decade. The introduction of new crystalline structures and chemical compositions have opened the door to the real possibilities of replacing yttria-stabilized zirconia (YSZ) to ensure the optimal thermomechanical-chemical properties required by TBCs. Future research directions associated with this topic are also provided.

INTRODUCTION

Thermal barrier coatings (TBCs) can be defined as advanced deposited material systems, operating at high temperatures to offer improved thermal stability and lower thermal conductivity These are placed on metal surfaces to protect the components of the hot section of gas turbine engines, and thereby, achieving higher fuel efficiency and lower emission objectives (Backman, 1992; Padture, 2002; Herzog, 2006). These coatings additionally improve erosion and impact resistance, which are crucial to increase engine durability and performance. Unfortunately, these coatings are susceptible to accelerated degradation due to deposition of silicates (known as CMAS) by environmental debris such as dust, sand, and ash that

DOI: 10.4018/978-1-7998-4870-7.ch004

adheres to them (Vaßen, 2012; Poerschke, 2017). The advanced materials used for this purpose are based on combinations of doped transition metal oxides with rare earth oxides, which reduce oxidation and thermal fatigue in the metal part. The most commonly used oxides are yttria stabilized zirconia (ZrO_2-Y_2O_3), mullite ($3Al_2O_3$-$2SiO_2$), alumina (Al_2O_3), ceria (CeO_2), lanthanum zirconate ($La_2Zr_2O_7$), lanthanum oxide (La_2O_3), oxide niobium (Nb_2O_5), and praseodymium oxide (Pr_2O_3) (Vasen, 2000; Cao, 2004; Tarasi, 2011). New and innovative materials for TBCs are being introduced such as $LaTi_2Al_9O_{19}$ (LTA) (Xie, 2011), lanthanide tantalate ($RETa_3O_9$) (where RE = Ce (Cerium), Nd (Neodymium), Sm (Samarium), Eu (Europium), Gd (Gadolinium), Dy (Dysprosium), Er (Erbium)) (Chen L, 2018), dysprosium-tantalum oxide (DTO) (Wu, 2018), magnesium-silicon oxide (MSO) (Chen S, 2019), lanthanide niobate (Ln-$_3NbO_7$) (LNO) (where Ln or L = Dy (Dysprosium), Er (Erbium), Y (Yttrium), Yb (Ytterbium)) (Yang, 2019), zirconium lanthanate ($Zr_3Ln_4O_{12}$) (where Ln = La (Lanthanum), Gd (Gadolinium), Y (Yttrium), Er (Erbium), and Yb (Ytterbium)) (Zhao M, 2019), magnetoplumbite ($LnMgAl_{11}O_{19}$) (where Ln = La (Lanthanum), Pr (Praseodymium), Nd (Neodymium), Sm (Samarium), Eu (Europium), Gd (Gadolinium)) (Zhao Y, 2019), and gadolinium-zirconium oxide (GZO) (Vaßen, 2020).

Materials for thermal barriers must meet five main requirements: 1) low thermal conductivity, 2) high thermal expansion coefficient (CTE), 3) high melting point, 4) excellent damage tolerance, and 5) moderate mechanical properties (Liu, B. 2019). Metal surfaces, where thermal barrier coatings are deposited, are based on high temperature superalloys. These are exposed to dangerous environments and hot combustion of engines and turbines used for propulsion and power generation (Zhao M, 2019).

In addition, thermal barrier coatings guarantee the phase stability of the materials to be coated, offer corrosion resistance and high fracture toughness. These coatings continue to be used on the hot parts of gas turbines with two purposes: 1) to increase turbine efficiency and 2) to extend the life of these metal parts (Zhou, 2014). A high coefficient of thermal expansion increases the service life of a thermal barrier coating (TBC) (Zhou, 2020). The global market for the manufacture of thermal barrier coatings, considering both materials and equipment, in 2015 was estimated in the order of 7.58 billion US dollars. Its compound annual growth rate is 7.79% and it is expected to reach 11.89 billion US dollars by 2021 (Vardelle, 2016).

Until now, vapor deposition techniques are the main techniques for the manufacture of thermal barrier coatings (Vaßen, 2010). The main techniques are based on electron beam physical vapor deposition (EB-PVD) and atmospheric plasma spray (APS). The first technique produces columnar microstructures and is used for thermomechanically loaded blades of gas turbines, and the second offers better robust operation and economic viability compared to the first.

In this chapter, a brief history describing the evolution that thermal barrier coatings have experienced is presented. Current directions in thermal barrier coatings research and recent worldwide progress are also analyzed. This study will provide reference to develop new lower thermal conductivity materials used as thermal barrier coatings. Also, potential challenges and opportunities are briefly highlighted in order to minimize the thermal conductivity of the materials used in the coatings in the coming decades.

This chapter has been divided as follows: In the section titled *Background*, a description of the basic concepts, materials and common problems found in thermal barrier coatings is presented. Furthermore, a brief history of the evolution of coatings is provided in this section. Subsequently, in the section entitled *Why use low thermal conductivity?*, a brief discussion on the importance of further decreasing the conductivity of the material used as a thermal barrier coating is presented with a view to introducing new materials. Subsequently, in the section titled *Solutions and Recommendations*, some alternative coating materials that have a lower thermal conductivity than those presented in the *Background* section are described. Moreover, in the section entitled *Future research directions*, a brief analysis of the strategies that should be promoted in the coming years to improve coatings are discussed. Finally, in the section entitled *Conclusions* some testaments are given to compile the importance of this research work.

BACKGROUND

Structural applications such as aircraft turbines, electricity generators, and marine thrusters must be protected through coatings from corrosion, wear, erosion, and must provide lubrication and thermal insulation (Padture, 2002). Thermal barrier coatings are ideal for protecting parts and components from temperatures up to 1650 °C. The use of these coatings allows parts and components to be manufactured. They must operate in high thermal atmospheres outside of less expensive substrates, without compromising performance. Different physical phenomena occur at the same time in a thermal barrier coating: thermal and matter diffusion, oxidation, phase transitions, elastic deformation, plastic deformation, creep deformation, thermal expansion, thermal conduction, thermal radiation, fracture, fatigue, and sintered (Padture, 2002). Thermal barriers must not cause distortion, warping, cracking, or any change to the base metals or substrate. The working conditions to which the coatings are subjected, and which are difficult to reproduce completely in a laboratory are: high temperatures, steep temperature gradients, extremely severe environments, rapid temperature transients, high hot gas pressures, as well as, mechanical loading and corrosion (Liu, B. 2019). These materials must meet the restrictive specifications of aircraft companies such as Pratt and Whitney, General Electric and Rolls Royce. Common metal substrates where thermal barrier coatings are applied are: white metals such as aluminum and titanium, red metals such as bronze, brass, and copper, gray steels such as low carbon steel, medium carbon steel, high carbon steel, heat treated steels, stainless steels (200, 300, 400 series), and tool steels (D2, M2, H13), castings such as cast iron, cast steels, and ductile iron, carbides, super alloys such as nickel-based superalloys, and cobalt-based steels. A wide variety of materials are used as substrates, which must be covered by thermal barrier coatings, among which are nickel-chrome-aluminum-yttrium (NiCrAlY) alloys, nickel-based superalloys, and nickel-chromium-silicon-boron (NiCrSiB) alloys (Herzog, 2006). Delivery time is typically 3-5 working days. These coatings are typically 100 microns to 2 millimeters thick (Clarke, 2005).

Thermal barrier coatings operate as thermal insulation materials to insulate parts and components from long-term (10,000-30,000 hr) and large thermal loads and can maintain a considerable temperature difference between the load-bearing alloy and the surface of the coating (Padture, 2002; Herzog, 2006). In addition, this enables coatings to withstand higher operating temperatures, extend part life by reducing oxidation and thermal fatigue, and limit thermal exposure of structural components (Backman, 1992). These coatings work in conjunction with the air ventilation system included in the turbines to allow working at fluid temperatures above the melting point of metal airfoils. In the design of aeroengines operating at high temperatures and with high efficiency, lighter moving and/or rotating components using very thin and durable coatings are required. Like thermal shields, thermal barriers must also have thermal emissivity properties. In addition, the use of the coating reduces the requirements of the cooling system, decreasing polluting emissions, and improves the overall performance of the system (Dong, 2013; Dhomne, 2019). That is why researchers worldwide have as a research topic the development of new and advanced thermal barrier coatings.

During heating and cooling of the parts, the coatings will experience thermal expansion stresses. These can be adequately controlled through the introduction of porosity and the appropriate matching of thermal expansion coefficients of the metal substrate and the thermal barrier (Herzog, 2006). As an additional requirement, the coating must maintain phase stability in order to avoid significant volume changes that occur with phase transitions, which consequently cause fracture or failure of the coating. Thermal barriers must also withstand tribological thermomechanical environments.

The main properties to be met by thermal barrier coatings (TBC) are (Cao, 2004): 1) high melting point, 2) no phase transition in the range between ambient temperature and operating temperature, 3) low thermal conductivity, 4) chemically inert, 5) matching the thermal expansion coefficient of the metal substrate, 6) good adhesion to the substrate, and 7) low sintering ratio to produce a porous structure. Among the desired microstructural characteristics for the coatings are: 1) distributed porosity, 2) vertical cracks through thickness, and 3) lack of cracks and/or boundaries due to large-scale detachment (Jadhav, 2006). Ceramic materials are the only engineering materials that meet the above requirements for the implementation of thermal barrier coatings.

Traditionally, thermal barriers consist of four layers namely: a metal substrate (MS), a metal bond coat (MBC), a thermally grown oxide (TGO), and a ceramic topcoat (CT), as shown in Figure 1. Traditionally, the ceramic topcoat is based on yttria-stabilized zirconia (YSZ) that has low thermal conductivity and is chemically stable (Padture, 2002; Jadhav, 2006). Furthermore, this layer creates the largest thermal gradient of the barrier and maintains the other layers at a lower temperature than on its surface. YSZ with a 7 to 8% ZrO_2 range is the most widely used thermal barrier coating because it provides the best performance for high temperature gas turbines. Above 1200 °C, the thermal stability of the YSZ is compromised since this material undergoes phase transitions going from monoclinic (from room temperature until 1173 °C) to tetragonal (from 1173 °C to 2370 °C), and finally to cubic (from 2370 °C to 2690 °C) before of its melting, as depicted in Figure 2 (Cao, 2004; Vassen, 2000). The main disadvantage of the YSZ for its long-term application is its operating temperature less than 1473 K or 1200 °C. At temperatures higher than 1200 °C, phase transitions from *t'*-tetragonal to tetragonal and cubic ($t + c$) and then to monoclinic (m) occur, leading to the formation of cracks in the coating. As indicated in the figure, two different cooling rates can lead to different phases depending on the temperature at which cooling starts. Depending on the percentage of yttria included in zirconia, the presence of either a monoclinic, tetragonal or cubic phase or their combinations is favored (Witz, 2007). These phase transitions lead to cracking within the topcoat and over time to coating failure. Recent advances in materials engineering have proposed novel ceramics to replace YSZ based on rare earth zirconates, which perform better at temperatures above 1200 °C, and have lower fracture toughness than YSZ (Clarke, 2005). Zirconates ($Y_4Zr_3O_{12}$) and hafnates ($Gd_2Hf_2O_7$, $Yb_4Hf_3O_{12}$) due to their high content of oxygen ion vacancies, these facilitate the transport of oxygen and promote the formation of TGO (Vassen, 2000; Poerschke, 2017). Due to excess TGO in the coating, the detachment of the ceramic topcoat occurs, and the thermal barrier fails catastrophically. Therefore, these coatings require additional ceramic layers that resist oxidation based on alumina or mullite (Padture, 2002; Clarke, 2012). Lanthanum zirconate can be doped with cerium (by replacing zirconium) to increase its coefficient of thermal expansion and decrease its thermal conductivity, making use of the change in structure pyrochlore to defective fluorite (Haoliang, 2020). Yttrium doped lanthanum zirconate can be used as a coating, but care must be taken during the synthesis process to avoid the formation of the fluorite phase which increases thermal conductivity (Jiang, 2019). Rare earth zirconates can be used as thermal barriers due to their relatively low thermal stability and thermal conductivity, but these suffer from low fracture toughness and susceptibility to react with TGO (Jiang, 2019). As previously described, the thermophysical properties of zirconates can be tuned thanks to their relatively open pyrochlore and fluorite-based crystalline structure, where rare-earth and Zr sites can be replaced by other chemical elements with similar properties.

Figure 1. The basic layered structure of a thermal barrier coating (TBC) used as thermal insulation

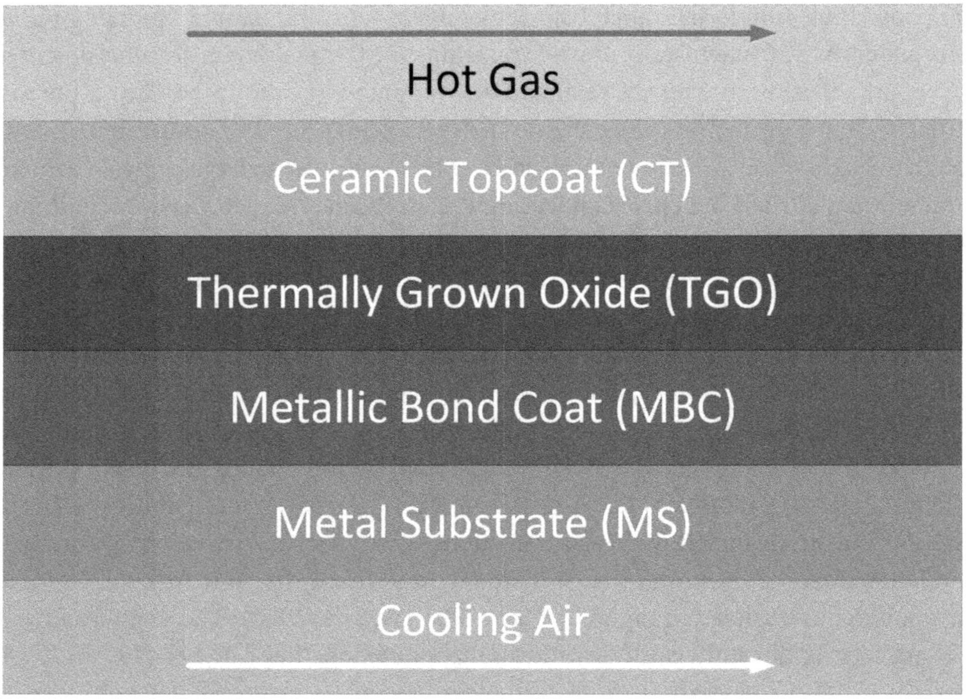

Strontium zirconate ($SrZrO_3$) as a thermal barrier coating material shows better cycling performance at surface temperatures greater than 1250 °C than barium zirconate ($BaZrO_3$) (Vaßen, 2010). However, this material undergoes the following phase transitions from orthorhombic to pseudo-tetragonal to 740 °C, then to tetragonal at 840 °C, and finally to cubic at 1130 °C, as established by Equation 1. To suppress such transformation, gadolinium (Gd) or ytterbium (Yb) doped is necessary to improve the thermophysical properties of the coatings at high temperatures (Ma, 2008).

$$\underset{Pnma}{\text{orthorhombic}} \xrightarrow{740°C} \underset{Imma}{\text{pseudo-tetragonal}} \xrightarrow{840°C} \underset{I4/mcm}{\textit{tetragonal}} \xrightarrow{1130°C} \underset{Pm\bar{3}m}{\textit{cubic}} \qquad (1)$$

The metal bondcoat is an oxidation resistant layer that is deposited directly onto the metal substrate. It is 75 to 150 microns thick and is made of nickel-chrome-aluminum-yttrium (NiCrAlY) or nickel-cobalt-chrome-aluminum-yttrium (NiCoCrAlY) alloys, or nickel (Ni) or platinum (Pt) aluminides (Padture, 2002; Herzog, 2006). The main function of this layer is to protect the metal substrate against oxidation and/or corrosion, particularly oxygen (O_2) and environmental corrosive elements that pass the ceramic topcoat (Clarke, 2012). Before ceramic topcoats were developed, metal bondcoats provided the function of high-temperature and environmental operation coatings (Pollock, 2012). Aluminide-type coatings such

Figure 2. Phase diagram of the YSZ used as a thermal barrier coating

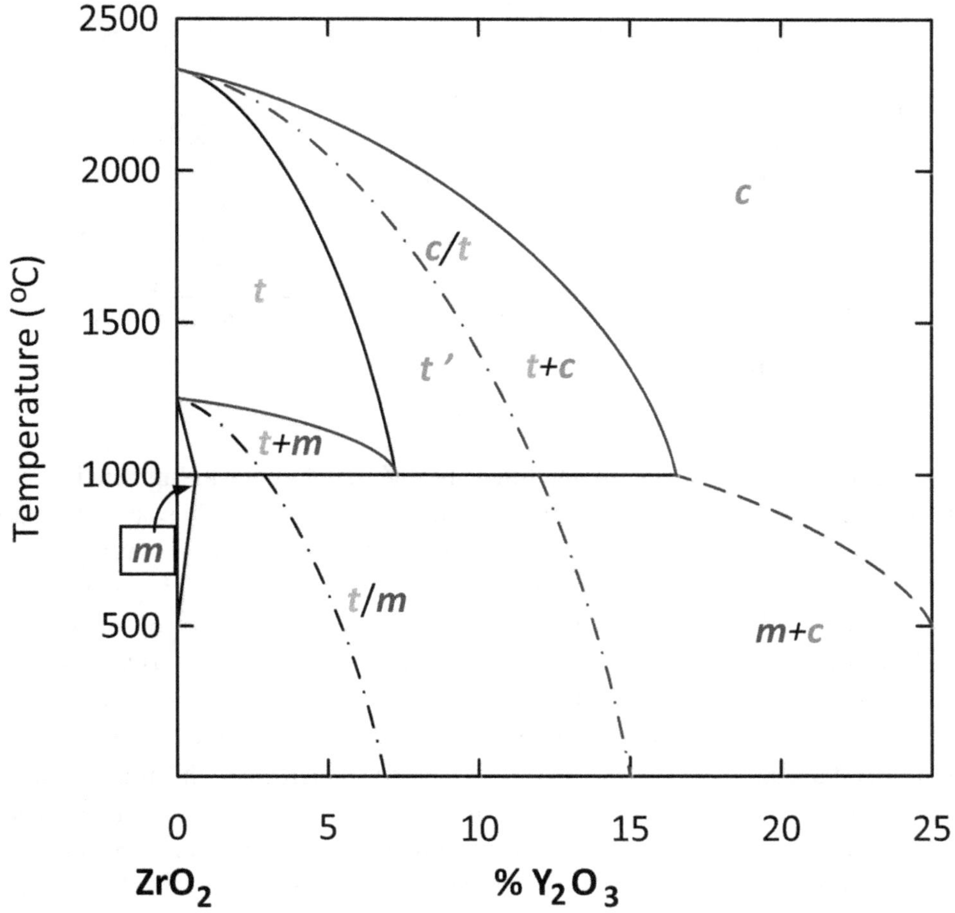

as AlNi and NiCoCrAl were materials to protect the metal substrate against oxidation, while chromides such as CoNiCrAl are used to protect the substrate material against hot corrosion (Vaßen, 2020). In addition to the functions described above, metal bondcoats should minimize interdiffusion of metals between layers and limit phase transitions at the interface with the metal substrate. This coating must be chemically complex to optimize the multifunctional thermomechanical-chemical properties. After deposition of the coating, it is vitally important to apply thermal treatments to avoid phase transitions considering the phase diagram of the associated metal system as well as the formation of intermetallic materials in the coating, as shown in Figure 3. The diagram shows the possible phases found in an Al-Ni-Cr alloy used as a substrate for the deposition of the thermal barrier coating. As can be seen, the composition and the operating temperature affect the present phases, so if the coating performs its function, it will avoid phase transitions. A more exhaustive study of these coatings can be found in (Pollock, 2012).

Figure 3. Ternary phase diagram of the Al-Ni-Cr system used as a substrate

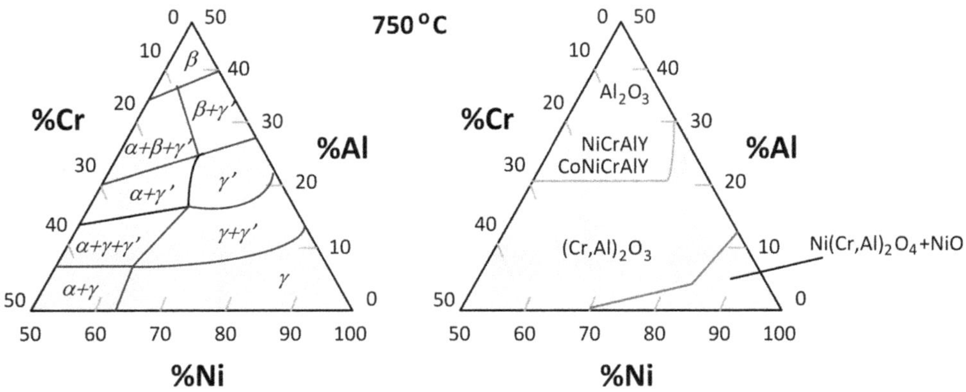

Both the presence of oxygen and temperatures above 700 °C in gas turbine engines cause the top bondcoat to oxidize resulting in the thermally grown oxide layer (TGO). This layer inevitably forms in high temperature applications, so coating designers make it grow slowly and evenly. In addition, this layer should have a crystalline structure with low oxygen diffusivity. Thus, it is possible controlling the diffusion of metals from the bondcoat rather than oxygen diffusion from the topcoat (Clarke, 2012).

The metal substrate is regularly a nickel or cobalt based superalloy, which is air cooled from within or through internal hollow channels that allow a temperature gradient to be established across the component walls (Padture, 2002). The superalloy-based component is cast either in the form of a single or polycrystalline crystal, and regularly contains 5 to 12 chemical elements. This provides properties such as high temperature resistance, ductility, oxidation resistance, resistance to hot corrosion and castability.

Three main mechanisms degrade coatings in aircraft engines leading to failure: 1) accelerated oxidation, 2) hot corrosion, and 3) molten deposit degradation (Backman, 1992; Vaßen, 2012). Oxidation degrades the life of the metal substrate, due to the thermal fatigue, to which it is subjected. Because the different layers of a thermal barrier coating have different thermal expansion coefficients, a deformation is introduced that causes cracking. Therefore, if its effect continues, then the catastrophic failure of the coating occurs (Poerschke, 2017). The most common failure mechanism involves the cracking of the thermally grown oxide layer and the metal bondcoat. This is due to that the growth of the TGO introduces a stress associated with the volumetric expansion occurred by the difference in thermal expansion coefficients (Cao, 2004). Furthermore, the growth of the TGO produces a depletion of the aluminum in the metal bondcoat, which leads to the formation of undesirable phases that contribute to mismatch stresses of thermal expansion (Clarke, 2012).

A summary of the main properties of common thermal barrier materials (TBCs) that have been widely used so far is presented in Table 1 (Cao, 2004; Dong, 2013; Dhomne, 2019). A graph illustrating the values of thermal conductivities of ceramic materials commonly used as thermal barrier coatings is shown in Figure 4. Among the main properties from the thermal point-of-view are the thermal expansion coefficient, thermal conductivity and thermal shock resistance. The data was collected from different references, and therefore, these may not be complete. The advantages and disadvantages of coating materials are compared to YSZ. In Table 2, the advantages and disadvantages of traditional coating materials are summarized (Dhomne, 2019).

Table 1. Main properties of common thermal barrier materials (TBCs)

Material	Chemical formula	Structure	Properties	Thermal conductivity
Yttria-stabilized Zirconia (YSZ)	ZrO_2-Y_2O_3	Fluorite (Cubic-Monoclinic)	YSZ has low thermal conductivity, high thermal expansion coefficient, low thermal shock resistance, can be deposited as a thick film, but is unstable at 1200 °C, and it is corroded due to its transparency to oxygen.	2.2-2.9 W/m·K
Mullite	$3Al_2O_3$·$2SiO_2$ or $2Al_2O_3$·SiO_2	Fluorite (Cubic-Monoclinic)	Mullite has low density, high thermal stability, low thermal conductivity, corrosion and oxidation resistance, suffers from shrinkage above 800 °C, and therefore, fractures and delaminates, and exhibits wide temperature variations on its surface.	3.5-11 W/m·K
Alumina	Al_2O_3	Monoclinic	Its alpha phase is a unique stable one, it has high hardness, it is chemically inert, it has high thermal conductivity, low thermal expansion coefficient, it improves resistance to oxidation and corrosion, it increases the hardness and resistance to bonding, and its ease of phase transition leads to the formation of microcracks.	14-30 W/m·K
Ceria+Yttria-stabilised Zirconia	CeO_2+ ZrO_2-Y_2O_3	Fluorite (Cubic-Monoclinic)	Ceria has a high thermal expansion coefficient and a low thermal conductivity for YSZ, when added to YSZ improves resistance to thermal shock, provides better thermal insulation and thermal expansion coefficient, but decreases hardness, and it is less porous.	3.23-3.78 W/m·K
Lanthanum zirconate	$La_2Zr_2O_7$	Pyrochlore (Cubic-Monoclinic)	Stable phase until it reaches its melting point. It has a huge number of vacancies that allow it to tune its thermal properties, low thermal conductivity for YSZ, low thermal expansion coefficient, and low hardness.	0.50-0.66 W/m·K
Rare earth oxides	La_2O_3, Nb_2O_5, Pr_2O_3, CeO_2, etc.	Fluorite or Pyrochlore (Cubic, Monoclinic)	Used in combinations of 2 or more of them, they have low thermal conductivity and high thermal expansion coefficient with respect to the YSZ, they have a polymorphic nature and instability of their phases, which results in low thermal shock resistance.	2.1-6 W/m·K, 1 W/m·K, 1 W/m·K, 2.2 W/m·K, etc.

The chemical elements whose oxides can be applied in thermal barrier coatings belong to groups IIIB (rare earth elements), IVB (Ti, Zr, and Hf), IIIA (Al) and IV A (Si) (Cao, 2004). Magnesium (Mg) and calcium (Ca) are used as stabilizers for zirconia. Rare earth oxides are considered promising materials for coatings due to their low thermal conductivity, high thermal expansion coefficient, and being chemically inert. Lanthanum zirconate can be doped to increase its coefficient of thermal expansion and be applied at high temperatures. The MCrAlY/YSZ system achieves its optimal performance as a TBC, however, researchers around the world continually continue to propose different material combinations as well as deposition processes. Due to its excellent thermal performance, zirconia stabilized with 8% yttria by weight or (8YSZ) has been the ceramic topcoat material used in recent decades (Cao, 2019). The multi-layer concept was introduced in order to improve the performance of coatings against thermal shock, since no single material can satisfy this requirement (Cao, 2004). The multi-layer scheme includes an erosion resistant layer as the outer layer, a ceramic coat, a corrosion-oxidation resistant metal layer, a thermal stress control layer and a diffusion resistant layer, as is shown in Figure 5. Despite decades-long advances in the development of thermal barrier coatings (TBCs), the results show that when all the necessary properties are considered, the YSZ is difficult to replace.

Figure 4. Thermal conductivities of ceramic materials commonly used as thermal barrier coatings

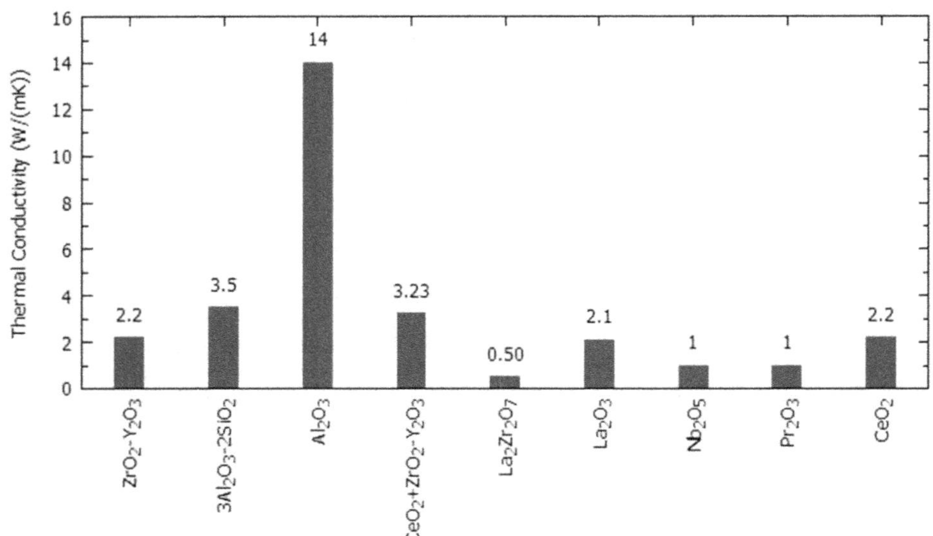

Table 2. Advantages and disadvantages of common thermal barrier coatings (TBCs)

Material	Advantages	Disadvantages
Alumina (Al_2O_3)	• High corrosion resistance • High hardness • Not oxygen-transparent	• Phase transformation (1000 °C) • High thermal conductivity • Very low expansion coefficient
Mullite ($3Al_2O_3{\cdot}2SiO_2$)	• High corrosion resistance • Low thermal conductivity • Good thermal-shock resistance below 1000 °C • Not oxygen transparent	• Crystallization (800-1000 °C) • Very low thermal expansion coefficient
Yttria-stabilized Zirconia (YSZ) (ZrO_2-Y_2O_3)	• High thermal expansion coefficient • Low thermal conductivity • High thermal shock resistance	• Sintering above 1200 °C • Phase transition (1170 °C) • Corrosion • Oxygen-transparent

Figure 5. The layered structure of a thermal barrier subjected to tribological conditions

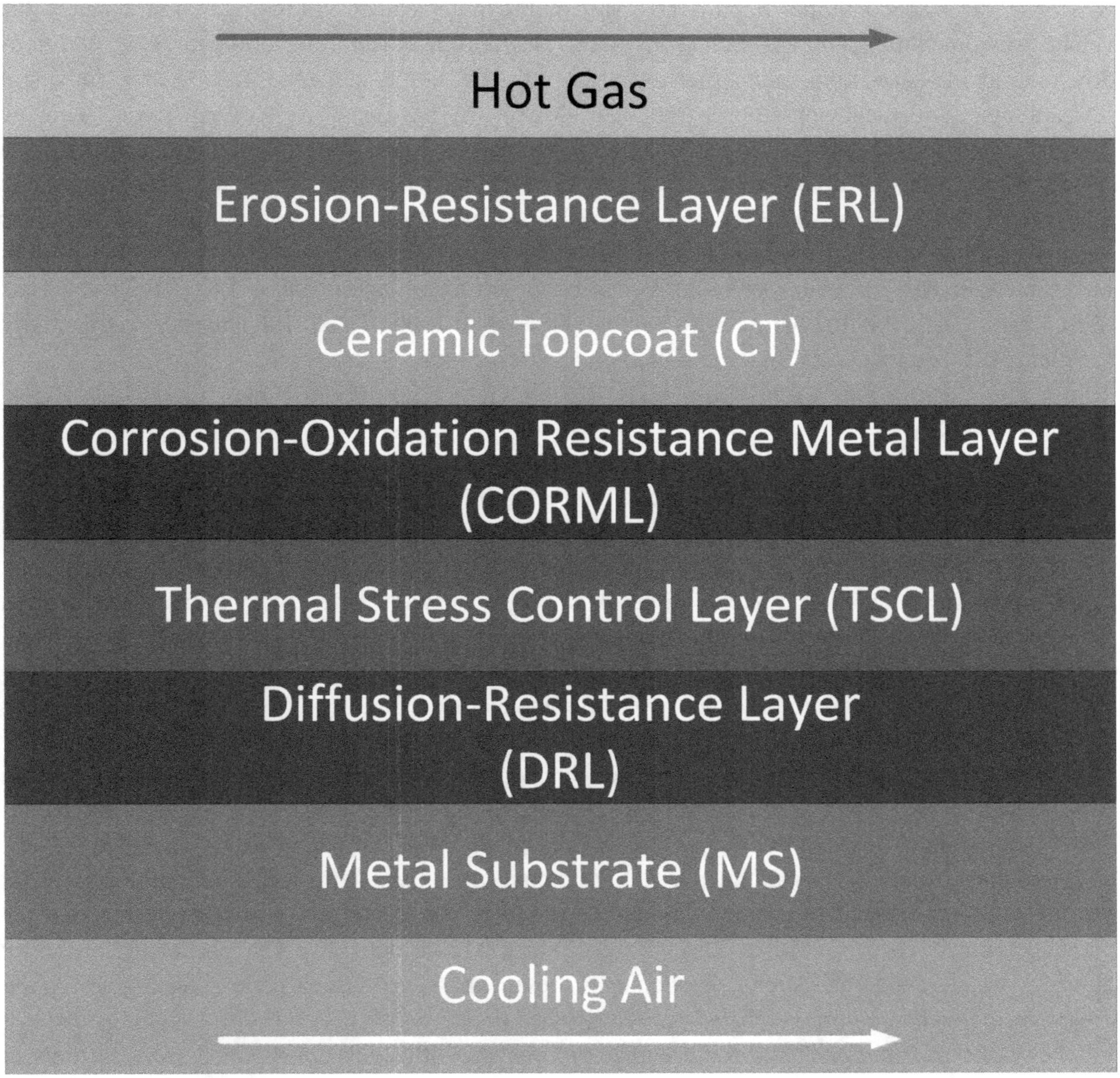

Among the material options that have been proposed as thermal barriers so far are fluorite oxides, pyrochlore oxides, and other oxides with low thermal conductivity (Clarke, 2005). Fluorite structured materials such as hafnia (HfO_2), ceria (CeO_2) and thoria (ThO_2) have thermal properties like zirconia (ZrO_2) especially when these are co-doped and have low conductivity due to the inclusion of porosity. $A_2^{3+}B_2^{4+}O_7$ pyrochlore materials have lower thermal conductivity than YSZ (Guo, 2012). These materials can be viewed as eight fluorite unit cells containing on average a single oxygen vacancy. For example, the structure of $Y_2Zr_2O_7$ is unstable, but by replacing yttrium with lanthanum or gadolinium, a stable structure is obtained up to at least 1500 °C, or by replacing the zirconium ion with ions such as titanium or molybdenum, its structure is also stabilized (Gildersleeve, 2019). These materials with stable structure have low thermal conductivity in the temperature range of 700 to 1200 °C, however, plumbates may have lower conductivities but are more susceptible to detachment. In the case of zirconates, this can be overcome through doping with neodymium (Nd), europium (Eu), or gadolinium (Gd). Additionally, other oxide-based compounds such as garnets ($Y_3Al_xFe_{5-x}O_{12}$), monazite ($LaPO_4$) and lanthanum hexaaluminate with magnetoplumbite structure ($LaMgAl_{11}O_{19}$) have been proposed as materials with low thermal conductivity for TBCs (Zhao, 2019). Unfortunately, although they have a thermal conductivity less than 3 W/m·K, they do not offer a lower conductivity than pyrochlore zirconates. Another possible alternative of materials for TBCs are ABO_3 perovskites, which are stable at high temperatures, but their thermal conductivity is not less than that of pyrochlore zirconates. Of special interest are perovskites in layers interspersed with cations with Ruddlesden-Popper structure such as $BaLa_2Ti_3O_{10}$ (BLT) which has been proposed as a thermal barrier with a stable phase up to 1500 °C and with a thermal conductivity less than 20% than that of 8YSZ (Guo, 2012).

A Brief History

Probably the first ceramic coating designed to protect a turbine blade was reported in 1947 by Harrison, Moore, and Richmond in the National Advisory Committee (NACA) for Aeronautics, published in the United States (Miller, 2009). The ceramic mixes attempted to protect SAE 1020 steel at a temperature of 667 °C were alumina (Al_2O_3), chromium oxide (Cr_2O_3), cobalt oxide (Co_2O_3), manganese dioxide (MnO_2), nickel oxide (NiO), magnesium oxide (MgO), calcium silicate ($2CaO{\cdot}SiO_2$), calcium chromate ($CaCrO_4$), and calcium fluoride (CaF_2). The first three materials offered promising options. For the year 1953, also at NACA, ceramic-based air-cooled turbine blade coatings for non-strategic steels and cast satellites were experimentally developed to protect them from oxidation and corrosion (Miller, 2009). For the year 1961 in NASA the thermal barrier concept was started using a topcoat based on zirconia (ZrO_2) and a bondcoat based on a chrome-nickel alloy. For the year 1973 at NASA ceramic and metal-based coatings were proposed to improve the thermal conductivity of hydrogen-oxygen cooled rockets. These coatings were based on alumina (Al_2O_3), zirconia (ZrO_2), and hafnia (HfO_2). For the year 1987, yttrium, ytterbium and hafnium doped alumina and zirconia-based coatings were proposed for gas turbine parts operating at 1040 °C (Miller, 2009). For the year 1994, it is noted that TBCs have been in use for almost 3 decades, and yttria stabilized zirconia is presented as a coating for parts of jet engines. In the 1990s, bismuth and barium plumbates were introduced as alternative materials for designing coatings. For the year 2000, strontium, barium and lanthanum zirconate with high melting points, excellent thermal stability and thermal shock are proposed as thermal barriers. By 2010, materials such as aluminates and pyrochlore oxides and perovskites are already in use. In addition, the concept of double ceramic layer is used. In the 2000s the use of calcium magnesium aluminosilicates (CMAS) was applied in testing at high temperature with YSZ coating (Gildersleeve, 2019). These silicates do not withstand temperatures greater than 1200 °C, infiltrate the TGO and fracture the TBC, and therefore, it experiences detachment and fails when the ceramic cools (Poerschke, 2017). Gadolinium zirconate (GZO) has resistance to CMAS infiltration through control of its porosity (Vaßen, 2020). In this decade, ceramic materials with Ruddlesden-Popper structure such as $BaLa_2Ti_3O_{10}$ for thermal barrier coatings are applied more extensively (Guo, 2012). Neodymium zirconate is an alternative coating material to YSZ thanks to the good quality and homogeneity of the layer, a porosity of less than 5% which reduces its thermal diffusivity, and with it, its thermal conductivity (Moskal, 2012). A thermal barrier coating strategy is the use of ceramic double coatings such as that based on neodymium (III) oxide (Nd_2O_3)-doped yttria-stabilized zirconia (YSZ) and boron nitride whiskers (BNws)-doped yttria-stabilized zirconia (YSZ) (Haoliang, 2020).

WHY USE LOW THERMAL CONDUCTIVITY?

For turbines to operate more efficiently, it is necessary that them to operate at higher temperatures, which is only achieved using the best thermal barrier coatings (TBCs) (Clarke, 2005). A 50% reduction in thermal conductivity has been estimated to decrease the temperature of the alloy by about 55 °C. The thermal conductivity of many oxides as a function of temperature and the effect of the level of porosity. These problems have been the subject of research interest in the area of materials engineering since the 1950s. For most oxides, thermal conductivity decreases as the inverse of temperature, which is completely necessary for TBCs (Liu, Q. 2019). However, for some oxides such as YSZ, UO_{2-x}, and $Th_{0.7}U_{0.3}O_{2+x}$, they show temperature-independent thermal conductivity at high temperatures. The reason for this independence stems from the fact that these materials contain very high concentrations of point defects which disperse the phonons. The inclusion of defects on purpose in the materials used in the TBC is the main object of study and experimentation by professionals and researchers of this type of coatings.

At low temperatures, the thermal conductivity of the oxides is independent of the presence of defects such as dislocations, individual vacancies, and long-range deformations due to inclusions and dislocations (Clarke, 2005). This is because defects at these temperatures only affect phonon transport on length scales greater than interatomic spacing. Phonons are collective thermal excitations of atoms or molecules in solids when the material is heated (Liu, B. 2019). Depending on the temperature to which a material is subjected, the frequency of oscillation of the phonons will be. At low temperatures, thermal conductivity is high because all phonons are traceable, making their contribution high. However, at high temperatures the phonons are not traceable and their contribution to thermal conductivity is low. The thermal insulation ability of thermal barrier coatings comes from their low thermal conductivity (Zhao, 2017). Ceramic materials thanks to their inherent porosity can be designed with low thermal conductivity. It can be achieved through the defects of their crystalline lattice and the dispersion of phonons produced by the vacancies, as well as by substitutions of the oxygen contained in the oxides of the thermal barrier. However, high mobility oxygen vacancies in the coatings can lead to rapid diffusion of oxygen at high temperatures. Therefore, these accelerate the failure of the coating due to effects such as grain coarsening, sintering, and simultaneous oxidation of the underlying metal bondcoat.

Additionally, the properties of the coatings are highly dependent on the deposition process and the chemical compositions (Liu, Q. 2019). With the reduction of thermal conductivity, it is not only desired to reduce the temperature to which the metal substrate will be exposed, but also to increase the service life of the coating, and thus, that of the turbine. Practical estimates indicate that the life of coated elements can be extended three to four times, while fuel consumption is reduced (Moskal, 2012).

In the design of alternative materials to YSZ for thermal barrier coatings, it is necessary to consider the decrease in thermal conductivity, phase stability and durability at high temperatures (Zhao, 2017). The required mechanical properties and thermal cycling time are achieved through control of porosity and microstructure. Pyrochlore or fluorite structures are regularly chosen, because they decrease thermal conductivity to less than 2 W/m·K and reach operating temperatures of 1600 °C. Thermal stresses are improved by increasing the thickness of the coating; however, this complicates the thermal matching capabilities with the metal substrate. However, the most avant-garde strategy for reducing thermal conductivity and that has provided better results, it is the inclusion of defects in the crystal lattice, such as vacancies and substitutions. A combination of oxygen vacancies and cation substitutions can produce a much more effective route to reduce thermal conductivity, and thereby, increase efficiency of the other required properties.

SOLUTIONS AND RECOMMENDATIONS

In the search for new materials to be used as thermal barriers, the engineer should not only consider the use of refractory materials. It is necessary look for crystalline materials that satisfy the technical requirements from the thermal point of view using atomistic simulations and crystalline chemistry (Clarke, 2005). In the following paragraphs, different material proposals that have been recently proposed either theoretically or experimentally are analyzed.

The 17 rare-earth elements that can be used in thermal barrier coatings are cerium (Ce), dysprosium (Dy), erbium (Er), europium (Eu), gadolinium (Gd), holmium (Ho), lanthanum (La), lutetium (Lu), neodymium (Nd), praseodymium (Pr), promethium (Pm), samarium (Sm), scandium (Sc), terbium (Tb), thulium (Tm), ytterbium (Yb), and yttrium (Y).

A recent ceramic material option for TBC is $Ba(Sr_{1/3}Ta_{2/3})O_3$ (BST) called strontium barium titanate which has showed a stable hexagonal perovskite structure up to 1600 °C and anisotropic growth of grains above 1400 °C (Cao, 2019). This has an elastic modulus low, high fracture toughness, high damage tolerance, moderate coefficient of thermal expansion, and higher thermal insulation capacity.

Another promising candidate for TBCs are $RETa_3O_9$ compounds (RE = Ce, Nd, Sm, Eu, Gd, Dy, Er) with a defective perovskite structure exhibiting low thermal conductivity (Chen, 2018). This conductivity is less than for YSZ and $La_2Zr_2O_7$ and a coefficient of thermal expansion close to them. Due to their high melting point and phase stability at high temperature it makes them candidate materials for TBCs. Their thermophysical properties for TBCs are derived from the asymmetric TaO_6 octahedra and the number of vacant cations in the rare earth oxide structure.

A novel TBC material based on magnesium silicate (Mg_2SiO4) showed excellent phase stability between room temperature and 1300 °C, showed low thermal conductivity at high temperature that is lower by 20% compared to 8YSZ, moderate coefficient of thermal expansion comparable to that of 8YSZ, resistance to higher thermal shock than for the 8YSZ, excellent thermal cycling durability, high fracture toughness, and slightly less Young's modulus than for the 8YSZ. The release of this material is due to the recrystallization of the coating, and the mismatch by thermal expansion with the TGO. It is expected that the thermal life cycle can be improved by controlling the process parameters and using a double ceramic layer coating (Chen, 2019).

$La_2Hf_2O_7$ ceramics are thermal barrier coating materials with phase stability up to 1600 °C, with a coefficient of thermal expansion less than that of 8YSZ and thermal conductivity less than that of 8YSZ. These can be used as thermal insulators and for double ceramic layer coatings (Liang, 2019).

Ceramic matrix composites based on yttrium titanate ($Y_2Ti_2O_7$) and nickel or silicon carbide have been proposed as thermal barrier coatings (Nguyen, 2016). This pyrochlore-structured titanate has high temperature chemical stability, a lower coefficient of thermal expansion than YSZ, and a lower thermal conductivity than YSZ.

Entropy engineering is used to design rare earth zirconate based coatings with excellent mechanical properties, large coefficients of thermal expansion, and excessively low thermal conductivity, without sacrificing other properties (Ren, 2020). Multi-component zirconates ($(Sm_{0.2}Eu_{0.2}Tb_{0.2}Dy_{0.2}Lu_{0.2})_2Zr_2O_7$ or $5RE_2Zr_2O_7$) were synthesized through co-precipitation, followed by sintering by solid state reaction and sintering using spark plasma. These coatings achieved ultra-low thermal conductivity and a large coefficient of thermal expansion, which represented an advance with scientific and technological significance.

An ytterbium aluminate-based coating ($Yb_3Al_5O_{12}$) exhibited phase stability, low thermal conductivity, good thermal stability, a high coefficient of thermal expansion which makes it a potential candidate for designing a thermal barrier (Wang, 2014). This material has higher mechanical and thermal properties as well as good resistance to steam corrosion for high temperature structural applications.

$RETaO_4$ ceramics have low thermal conductivity for high temperatures, good thermal tolerance, significant porosity, good fracture toughness, and lower Vickers hardness (Wang, 2017). Furthermore, these tantalates have low hardness at high temperature, and the operating temperature is expected to exceed 1600 °C, which makes them good candidates for thermal barrier coatings in high-temperature applications.

Ceramics based on $(ZrO_2)_x$-$(Dy_3TaO_7)_{1-x}$ were synthesized with an orthorhombic system and showed lower thermal conductivities than YSZ and glass-like caused by oxygen vacancies in the structure (Wu, 2018). These ceramics also show phase stability at high temperature and a coefficient of thermal expansion comparable to YSZ.

In the search for materials for thermal barrier coatings for high temperature applications, lanthanum titanium aluminum oxide or LTA is one of the technological possibilities. This material showed phase stability up to 1600 °C, low thermal conductivity, high coefficient of thermal expansion, desirable thermal cyclability, YSZ-like microhardness but less fracture toughness than YSZ (Xie, 2011). Its crystalline structure is monoclinic centered on the faces, that is, a huge unit cell when compared to the magnetoplumbite phase or for interleaved with layers of pseudobrookite.

Ceramic with pyrochlore structure Sm_2FeTaO_7 has also been proposed as a coating material because it has better thermal insulation capabilities such as low thermal conductivity, phase stability at high temperature, improved fracture toughness and low Young's modulus. (Yang, 2018). With a monoclinic structure, its phase stability is guaranteed due to its complex and distorted crystal lattice, high concentration of defects and high differences in the atomic mass of its cations.

$TaZr_{2.75}O_8$ ceramic can be used as a thermal barrier for high temperature applications since it has a lower thermal conductivity than YSZ, has a moderate coefficient of thermal expansion, shows high microhardness and fracture toughness (Yuan, 2019). Its structure presents a high phase stability because the presence of Ta^{5+} ions increase the stability of the tetragonal phase even up to 1500 °C and its oxide increases the corrosion resistance of zirconia.

$LnMgAl_{11}O_{19}$ magnetoplumbite materials (LnMA, Ln = La, Ce, Pr, Nd, Sm, Eu, Gd) are considered promising coating materials because they offer high melting points, low thermal conductivity, high coefficient of thermal expansion, and stability. phase and thermal up to at least 1600 °C (Zhao, 2019). These materials have hexagonal symmetry and consist of spinel blocks separated by mirror planes with rare earth cations coordinated with 12 oxygen atoms, which lose their coordination by the other magnesium and aluminum-based oxide structures resulting in disordered structures. This disorder can be exploited to obtain the thermal properties required for thermal barriers.

Among the coating alternatives, $Y_4Al_2O_9$ (YAM) ceramics with monoclinic phase have phase stability, low thermal conductivity, improved coefficient of expansion, and lower density (Zhou, 2014). This material is stable up to at least 1373 °C, its thermal conductivity can be reduced by doping it, or introducing more oxygen vacancies.

$SmTaO_4$ ceramics have phase stabilities and mechanical properties for high temperature applications, with a monoclinic phase structure when doping them with ceria (CeO_2), the thermal conductivity is decreased, and the coefficient of thermal expansion becomes slightly higher than that of YSZ (Zhou, 2020). The introduction of ceria as a chemical dopant in samarium tantalate introduces defects, which reduce the thermal conductivity of the material.

A special coating case for high temperature applications where silicon carbide is used as a substrate is known as an environmental barrier coating or EBC, it will be described below. Silicon-based ceramics show excellent oxidation and slip resistance for high temperature applications such as gas turbine engines, but unfortunately experience degradation of their stability to water vapor in high speed combustion environments (Vaßen, 2012). The addition of rare earth elements to the silicates gives them increased corrosion resistance, high temperature durability, low coefficient of thermal expansion, excellent phase and chemical stability, and low thermal conductivity. In order to achieve optimal balanced properties, researchers have chosen to chemically dope the silicates to improve their thermal and mechanical properties through rare earth elements. A rare earth based multifunctional silicate has been prepared with hot pressing and shows an interesting balance of physical properties to be used as a thermal barrier coating such as high Young's modulus, low thermal conductivity and high coefficient of thermal expansion (Ren, 2019).

A summary of the novel thermal barrier coatings that have been introduced in this decade are presented in Table 3. A graph illustrating the values of the thermal conductivities of the ceramic materials recently introduced and used as thermal barrier coatings is depicted in Figure 6. Furthermore, a graph illustrating the values of the coefficients of thermal expansion of these materials is shown in Figure 7.

Thin-Film Thermal Conductivity Characterization

To measure thermal conductivity there are direct methods and indirect methods (Borca-Tasciuc, 2004). In the case of direct measurements, they are regularly based on electrical characterization techniques. In direct measurements, it is necessary to know the thermal flux, the area and thickness of the sample, and the temperature drop between two points of the sample, as illustrated in Figure 8 (Çengel, 2015). Because thin films are anisotropic and/or polycrystalline, it is necessary to determine the thermal conductivity in different directions (the direction of the transverse plane, which is perpendicular to the plane of the thin film, or the direction in plane which is parallel to the plane thin film). For indirect measurements, these are commonly based on optical characterization techniques. These methods use radiation energy as the heat source to determine the dynamic response of the sample because it is difficult to determine exactly the amount of heat absorbed in the sample. Variation of temperature in the sample under varying heating conditions over time generates different signatures that can be detected and used to determine thermal conductivity. Examples of these signatures are thermal emission, refractive index, and thermal expansion, as illustrated in Figure 8 (Keppler, 2008).

Table 3. Thermal properties of materials recently proposed for TBCs

Material	Coefficient of thermal expansion (CTE)	Thermal Conductivity	Melting point	Reference
$Ba(Mg_{1/3}Ta_{2/3})O_3$ (BMT)	11×10^{-6}/K @ 1150 °C	2.5 W/m·K @ 1200 °C	3100 °C	(Cao, 2019)
$RETa_3O_9$ (RE = Ce, Nd, Sm, Eu, Gd, Dy, Er)	4-10.2×10^{-6}/K @ 1000 °C	1.33-2.37 W/m·K (100-800 °C)	--	(Chen, 2018)
Mg_2SiO_4	8.6-11.3×10^{-6}/K @ (200-1350 °C)	3.6 W/m·K (25 °C) to 1.68 W/m·K (800 °C)	1910 °C	(Chen, 2019)
$BaLa_2Ti_3O_{10}$ (BLT)	9.5-11.3×10^{-6}/K (*a-b* plane) 10.4-12.1×10^{-6}/K (*c*-axis)	1.41-1.71 W/m·K (*a-b* plane) 1.31-1.60 W/m·K (*c*-axis)	--	(Guo, 2012)
$La_2Hf_2O_7$ (LH)	8.76×10^{-6}/K	1.34 W/m·K @ 1000 °C	> 2000 °C	(Liang, 2019)
$Y_2Ti_2O_7$ (YT)	8.6×10^{-6}/K	2.25-2.6 W/m·K @ 500 °C	1827 °C	(Nguyen, 2016)
$(Y_{1/4}Ho_{1/4}Er_{1/4}Yb_{1/4})_2SiO_5$ (YHoErYbS)	5.5-6.5×10^{-6}/K @ (400-1200 °C)	2.05 W/m·K @ 0 C 1.6 W/m·K @ 600 C 2.2 W/m·K @ 1000 C	2200 °C	(Ren, 2019)
$(Sm_{0.2}Eu_{0.2}Tb_{0.2}Dy_{0.2}Lu_{0.2})_2Zr_2O_7$ SmEuTbDyLuZr	10.5×10^{-6}/K @ 1000 °C	0.86 W/m·K @ 1000 C	--	(Ren, 2020)
$Yb_3Al_5O_{12}$	7.83-23.74×10^{-6}/K @ (25-100 °C)	4.67-2.05 W/m·K @ (27-1127 °C)	2000 °C	(Wang, 2014)
$RETaO_4$ (RE = Nd, Eu, Gd, Dy, Er, Yb, Lu)	11.2×10^{-6}/K @ 1200 °C	1.38-1.4 W/m·K 800 °C	1400 °C	(Wang, 2017)
$(ZrO_2)_x$-$(Dy_3TaO_7)_{1-x}$ (x = 0, 0.02, 0.04, 0.06, 0.08, 0.1)	9.7-11×10^{-6}/K	1.3-1.8 W/m·K	--	(Wu, 2018)
$LaTi_2Al_9O_{19}$ (LTA)	8.0-11.2×10^{-6}/K @ 200-1400 °C	1.0-1.3 W/m·K @ 300-1500 °C	--	(Xie, 2011)
Sm_2FeTaO_7 (SFT)	6.4-10×10^{-6}/K @ 300-1600 °C	1.65-1.25 W/m·K @ 0-1000 °C	--	(Yang, 2018)
$TaZr_{2.75}O_8$	9.14×10^{-6}/K @ 25-1300 °C	1.65 W/m·K @ 800 °C	> 2000 °C	(Yuan, 2019)
$GdMgAl_{11}O_{19}$	9.6×10^{-6}/K @ 1600 °C	1.91-1.78 W/m·K @ 25-800 °C	--	(Zhao, 2019)
$Y_4Al_2O_9$ (YAM)	8.91×10^{-6}/K @ 27-1273 °C	1.81 W/m·K @ 1000 °C	> 2000 °C	(Zhou, 2014)
CeO_2-doped $SmTaO_4$	10.3×10^{-6}/K @ 1200 °C	1.09-2.75 W/ m·K @ 100-800 °C	> 2000 °C	(Zhou, 2020)

Figure 6. Thermal conductivities of the ceramic materials recently introduced as thermal barrier coatings

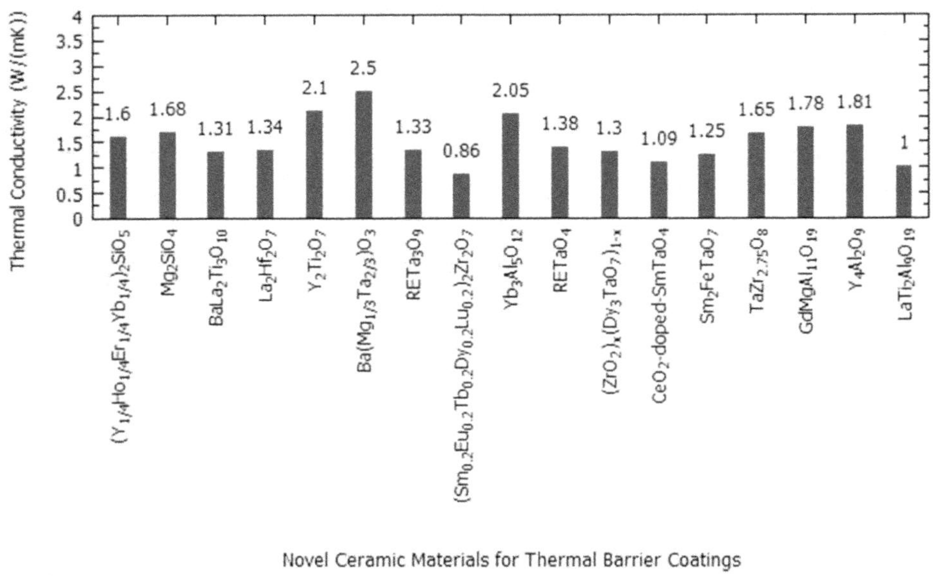

Figure 7. Coefficients of thermal expansion of novel ceramic materials used as thermal barrier coatings

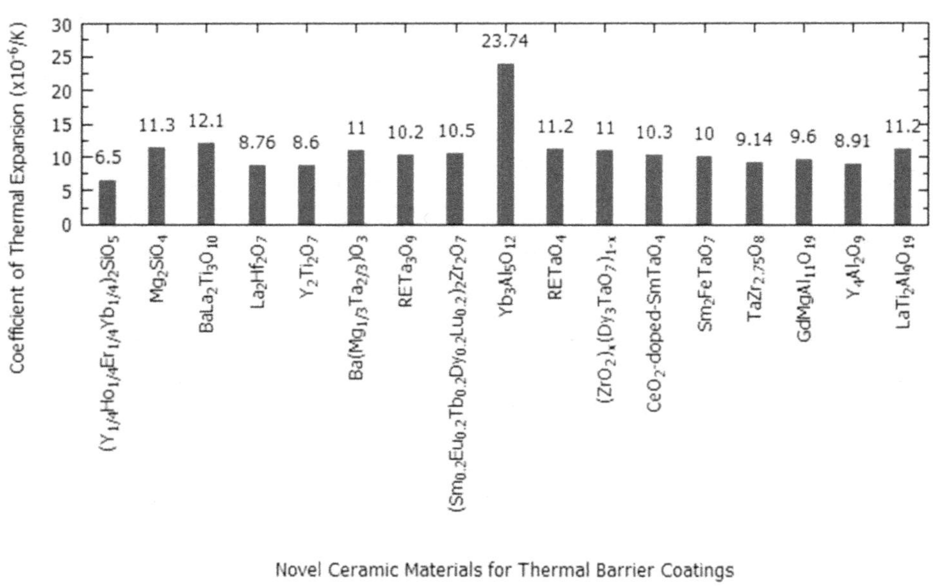

FUTURE RESEARCH DIRECTIONS

The search for materials with low thermal conductivity for thermal barrier coatings is still in its infancy. The range of possible chemical compositions as well as crystalline structures is unlimited, so the search will continue in this century as it will consume too much time and cost. Atomistic simulations of both the crystal structure and its thermal properties will help predict the results that can be achieved in experiments, so a lot of work and time can be saved by researchers worldwide. According to the results achieved so far, the most suitable materials for thermal barriers must have structures with high atomic weights, highly distorted and distorted, and with loose bonds. At an experimental level, doping and/or alloying will continue to offer the possibility of reducing the thermal conductivity of coatings. The main strategy for the design of future thermal barrier coatings (TBCs) must combine the theoretical calculations developed in computer simulation with advances at the experimental level. The challenge of the future design of thermal barriers is to identify a multifunctional strategy that optimizes all the required properties, either increasing or decreasing their values, but jointly not independently for only one of the requirements. This implies a crystal structure design of the coatings that considers the contributions of the strong and weak chemical bonds, completely linked on their hard and soft substructures, which give rise to the required mechanical, thermal and chemical properties. Furthermore, modeling of the aging and durability of the thermal barrier coating should be a research activity that should be undertaken because degradation of properties is directly related to theoretically predicted knowledge and experimentation gained from years of experience that practitioners can share. From a microstructural point of view, it is necessary that the new thermal coating materials have as a base a multi-functional operation, a high segmentation crack density, and that the nanostructured qualities be exploited to optimize the overall performance of the coating.

Figure 8. Thermal conductivity measurement of thin films using electrical and optical characterization techniques

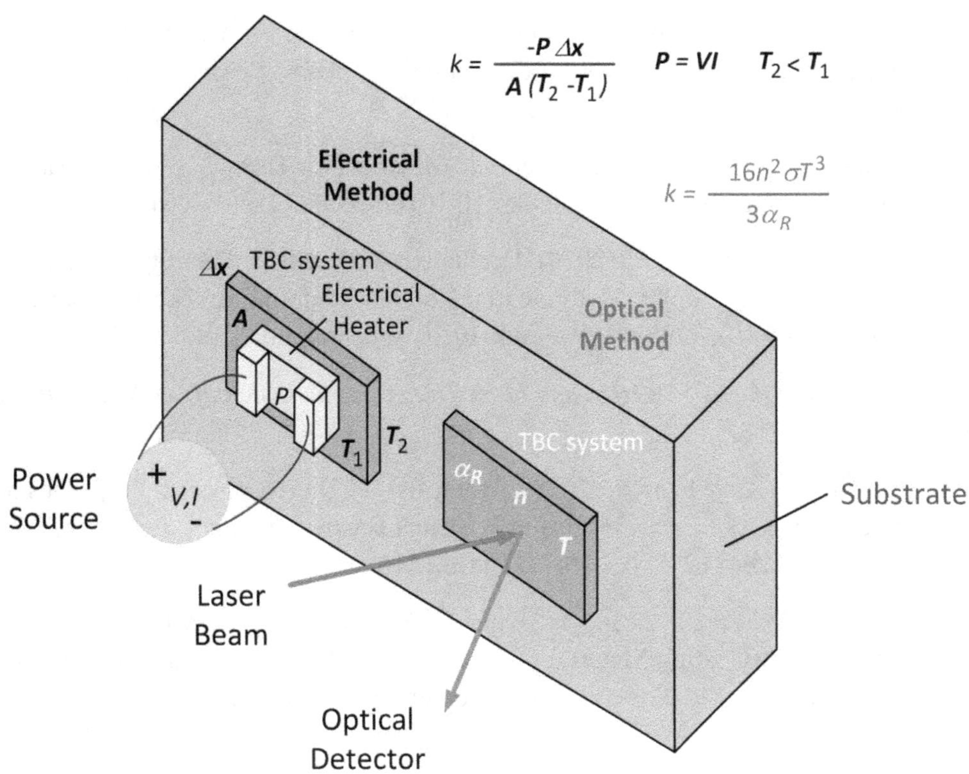

CONCLUSION

The advances in scientific research and technological development of thermal barrier coatings (TBCs) based on ceramic materials for their topcoats in recent decades has been presented in this chapter. The results achieved so far show a marked trend towards the use of crystalline structures that take advantage of defect engineering to decrease thermal conductivity, increase the coefficient of thermal expansion, increase the melting point, guarantee thermal and phase stability, and improve mechanical properties. Although the results show significant advances in both chemical, thermal and mechanical properties, researchers must continue to develop novel materials that increasingly satisfy the requirements to make the application of TBCs more reliable and with a longer lifetime.

ACKNOWLEDGMENT

The author thanks the access to scientific articles supplied by CONRICYT. It is economically sponsoring by CONACYT in Mexico. Besides, he wants to thank his wife and son for their support in time to conduct this research.

REFERENCES

Backman, D. G., & Williams, J. C. (1992). Advanced Materials for Aircraft Engine Applications. *Science, 255*(5048), 1082–1087. doi:10.1126cience.255.5048.1082 PMID:17817782

Borca-Tasciuc, T., & Chen, G. (2004). Experimental Techniques for Thin-Film Thermal Conductivity Characterization. In T. M. Tritt (Ed.), *Thermal Conductivity: Theory, Properties and Applications* (pp. 205–237). Kluwer Academic. doi:10.1007/0-387-26017-X_9

Cao, X. Q., Vassen, R., & Stoever, D. (2004). Ceramic Materials for Thermal Barrier Coatings. *Journal of the European Ceramic Society, 24*(1), 1–10. doi:10.1016/S0955-2219(03)00129-8

Cao, Y., Wang, Q., Ning, X., & Liu, Y. (2019). Evaluation of the Phase Stability, and Mechanical and Thermal Properties of $Ba(Sr_{1/3}Ta_{2/3})O_3$ as a Potential Ceramic Material for Thermal Barrier Coatings. *Ceramics International, 45*(10), 12989–12993. doi:10.1016/j.ceramint.2019.03.227

Çengel, Y. A., & Ghajar, A. J. (2015). *Heat and Mass Transfer: Fundamentals & Applications* (5th ed.). McGraw-Hill.

Chen, L., Jiang, Y., Chong, X., & Feng, J. (2018). Synthesis and Thermophysical Properties of $RETa_3O_9$ (RE = Ce, Nd, Sm, Eu, Gd, Dy, Er) as Promising Thermal Barrier Coatings. *Journal of the American Ceramic Society, 101*(3), 1266–1278. doi:10.1111/jace.15268

Chen, S., Zhou, X., Song, W., Sun, J., Zhang, H., Jiang, J., Deng, L., Dong, S., & Cao, X. (2019). Mg_2SiO_4 as a Novel Thermal Barrier Coating Material for Gas Turbine Applications. *Journal of the European Ceramic Society, 39*(7), 2397–2408. doi:10.1016/j.jeurceramsoc.2019.02.016

Clarke, D. R., Oechsner, M., & Padture, N. P. (2012). Thermal-Barrier Coatings for More Efficient Gas-Turbine Engines. *MRS Bulletin*, *37*(10), 891–898. doi:10.1557/mrs.2012.232

Clarke, D. R., & Phillpot, S. R. (2005). Thermal Barrier Coating Materials. *Materials Today*, *8*(6), 22–29. doi:10.1016/S1369-7021(05)70934-2

Dhomne, S., & Mahalle, A. M. (2019). Thermal Barrier Coating Materials for SI Engine. *Journal of Materials Research and Technology*, *8*(1), 1532–1537. doi:10.1016/j.jmrt.2018.08.002

Dong, H., Wang, D., Pei, Y., Li, H., Li, P., & Ma, W. (2013). Optimization and Thermal Cycling Behavior of $La_2Ce_2O_7$ Thermal Barrier Coatings. *Ceramics International*, *39*(2), 1863–1870. doi:10.1016/j.ceramint.2012.08.034

Gildersleeve, E., Viswanathan, V., & Sampath, S. (2019). Molten Silicate Interactions with Plasma Sprayed Thermal Barrier Coatings: Role of Materials and Microstructure. *Journal of the European Ceramic Society*, *39*(6), 2122–2131. doi:10.1016/j.jeurceramsoc.2019.01.023

Guo, L., Guo, H., Ma, G., Abbas, M., & Gong, S. (2012). Ruddlesden-Popper Structured $BaLa_2Ti_3O_{10}$, a Highly Anisotropic Material for Thermal Barrier Coatings. *Ceramics International*, *38*(5), 4345–4352. doi:10.1016/j.ceramint.2012.02.017

Haoliang, T., Guo, J., Erbao, L., Fuyuan, W., Changliang, W., Mengqiu, G., Junguo, G., & Yongjing, C. (2020). Preparation and Performance of Thermal Barrier Coatings made of BNw-containing modified Nd_2O_3-doped Yttria-stabilized Zirconia. *Ceramics International*, *46*(1), 500–507. doi:10.1016/j.ceramint.2019.08.289

Herzog, R., Warnken, N., Steinbach, I., Hallstedt, B., Walter, C., Müller, J., Hajas, D., Münstermann, E., Schneider, J. M., Nickel, R., Parkot, D., Bobzin, K., Lugscheider, E., Bednarz, P., Trunova, O., & Singheiser, L. (2006). Integrated Approach for the Development of Advanced, Coated Gas Turbine Blades. *Advanced Engineering Materials*, *8*(6), 535–562. doi:10.1002/adem.200500277

Jadhav, A. D., Padture, N. P., Jordan, E. H., Gell, M., Miranzo, P., & Fuller, E. R. Jr. (2006). Low-Thermal-Conductivity Plasma-Sprayed Thermal Barrier Coatings with Engineered Microstructures. *Acta Materialia*, *54*(12), 3343–3349. doi:10.1016/j.actamat.2006.03.024

Jiang, D., Wang, Y., Wang, S., Liu, R., & Han, J. (2019). Thermal Conductivity of Air Plasma Sprayed Yttrium Heavily-doped Lanthanum Zirconate Thermal Barrier Coatings. *Ceramics International*, *45*(3), 3199–3206. doi:10.1016/j.ceramint.2018.10.222

Keppler, H., Dubrovinsky, L. S., Narygina, O., & Kantor, I. (2008). Optical Absorption and Radiative Thermal Conductivity of Silicate Perovskite to 125 Gigapascals. *Science*, *322*(5907), 1529–1532. doi:10.1126cience.1164609 PMID:19056982

Liang, P., Dong, S., Zeng, J., Yuan, J., Jiang, J., Deng, L., Zhou, X. & Cao, X. (2019). $La_2Hf_2O_7$ Ceramics as Potential Top-Coat Materials for Thermal/Environmental Barrier Coatings. *Ceramics International*, *45*(17), 22432-22436. doi:10.1016/j.ceramint.2019.07.235

Liu, B., Liu, Y., Zhu, C., Xiang, H., Chen, H., Sun, L., Gao, Y., & Zhou, Y. (2019). Advances on Strategies for Searching for Next Generation Thermal Barrier Coating Materials. *Journal of Materials Science and Technology*, *35*(5), 833–851. doi:10.1016/j.jmst.2018.11.016

Liu, Q., Huang, S., & He, A. (2019). Composite Ceramics Thermal Barrier Coatings of Yttria Stabilized Zirconia for Aero-Engines. *Journal of Materials Science and Technology*, *35*(12), 2814–2823. doi:10.1016/j.jmst.2019.08.003

Ma, W., Jarligo, M. O., Mack, D. E., Malzbender, J., Vaßen, R., & Stöver, D. (2008). New Generation Perovskite Thermal Barrier Coating Materials. *Journal of Thermal Spray Technology*, *17*(5-6), 831–837. doi:10.100711666-008-9239-4

Miller A.R. (2009). *History of Thermal Barrier Coatings for Gas Turbine Engines: Emphasizing NASA's Role From 1942 to 1990*. NASA/TM-2009-215459.

Moskal, G., Swadźba, L., Hetmańczyk, M., Witala, B., Mendala, B., Mendala, J., & Sosnowy, P. (2012). Characterization of the Microstructure and Thermal Properties of $Nd_2Zr_2O_7$ and $Nd_2Zr_2O_7$/YSZ Thermal Barrier Coatings. *Journal of the European Ceramic Society*, *32*(9), 2035–2042. doi:10.1016/j.jeurceramsoc.2011.12.004

Nguyen, S. T., Nakayama, T., Suematsu, H., Suzuki, T., Takeda, M., & Niihara, K. (2016). Low Thermal Conductivity $Y_2Ti_2O_7$ as a Candidate Material for Thermal/Environmental Barrier Coatings. *Ceramics International*, *42*(9), 11314–11323. doi:10.1016/j.ceramint.2016.04.052

Padture, N. P., Gell, M., & Jordan, E. H. (2002). Thermal Barrier Coatings for Gas-Turbine Engine Applications. *Science*, *296*(5566), 280–285. doi:10.1126cience.1068609 PMID:11951028

Poerschke, D. L., Jackson, R. W., & Levi, C. G. (2017). Silicate Deposit Degradation of Engineered Coatings in Gas Turbines: Progress toward Models and Materials Solutions. *Annual Review of Materials Research*, *47*(1), 297–330. doi:10.1146/annurev-matsci-010917-105000

Pollock, T. M., Lipkin, D. M., & Hemker, K. J. (2012). Multifunctional Coating Interlayers for Thermal-Barrier Systems. *MRS Bulletin*, *37*(10), 923–931. doi:10.1557/mrs.2012.238

Ren, K., Wang, Q., Shao, G., Zhao, X., & Wang, Y. (2020). Multicomponent High-Entropy Zirconates with Comprehensive Properties for Advanced Thermal Barrier Coating. *Scripta Materialia*, *178*, 382–386. doi:10.1016/j.scriptamat.2019.12.006

Ren, X., Tian, Z., Zhang, J., & Wang, J. (2019). Equiatomic Quaternary $(Y_{1/4}Ho_{1/4}Er_{1/4}Yb_{1/4})_2SiO_5$ Silicate: A Perspective Multifunctional Thermal and Environmental Barrier Coating Material. *Scripta Materialia*, *168*, 47–50. doi:10.1016/j.scriptamat.2019.04.018

Tarasi, F., Medraj, M., Dolatabadi, A., Oberste-Berghaus, J., & Moreau, C. (2011). High-Temperature Performance of Alumina-Zirconia Composite Coatings containing Amorphous Phases. *Advanced Functional Materials*, *21*(21), 4143–4151. doi:10.1002/adfm.201100471

Vardelle, A., Moreau, C., Akedo, J., Ashrafizadeh, H., Berndt, C. C., Berghaus, J. O., Boulos, M., Brogan, J., Bourtsalas, A. C., Dolatabadi, A., Dorfman, M., Eden, T. J., Fauchais, P., Fisher, G., Gaertnet, F., Gindrat, M., Henne, R., Hyland, M., Irissou, E., ... Vuoristo, P. (2016). The 2016 Thermal Spray Roadmap. *Journal of Thermal Spray Technology*, *25*(8), 1376–1440. doi:10.100711666-016-0473-x

Vaßen, R., Bakan, E., Mack, D., Schwartz-Lückge, S., Sebold, D., Sohn, Y. J., Zhou, D., & Guillon, O. (2020). Performance of YSZ and $Gd_2Zr_2O_7$/YSZ Double Layer Thermal Barrier Coatings in Burner Rig Tests. *Journal of the European Ceramic Society*, *40*(2), 480–490. doi:10.1016/j.jeurceramsoc.2019.10.021

Vassen, R., Cao, X., Tietz, F., Basu, D., & Stöver, D. (2000). Zirconates as New Materials for Thermal Barrier Coatings. *Journal of the American Ceramic Society*, *83*(8), 2023–2028. doi:10.1111/j.1151-2916.2000.tb01506.x

Vaßen, R., Jarligo, M. O., Steinke, T., Mack, D. E., & Stöver, D. (2010). Overview on Advanced Thermal Barrier Coatings. *Surface and Coatings Technology*, *205*(4), 938–942. doi:10.1016/j.surfcoat.2010.08.151

Vaßen, R., Kagawa, Y., Subramanian, R., Zombo, P., & Zhu, D. (2012). Testing and Evaluation of Thermal-Barrier Coatings. *MRS Bulletin*, *37*(10), 911–916. doi:10.1557/mrs.2012.235

Wang, J., Chong, X., Zhou, R., & Feng, J. (2017). Microstructure and Thermal Properties of $RETaO_4$ (RE = Nd, Eu, Gd, Dy, Er, Yb, Lu) as Promising Thermal Barrier Coating Materials. *Scripta Materialia*, *126*, 24–28. doi:10.1016/j.scriptamat.2016.08.019

Wang, X., Xiang, H., Sun, X., Liu, J., Hou, F., & Zhou, Y. (2014). Thermal Properties of a Prospective Thermal Barrier Material: $Yb_3Al_5O_{12}$. *Journal of Materials Research*, *29*(22), 2673–2681. doi:10.1557/jmr.2014.319

Witz, G., Shklover, V., Steurer, W., Bachegowda, S., & Bossmann, H.-P. (2007). Phase Evolution in Yttria-Stabilized Zirconia Thermal Barrier Coatings Studied by Rietveld Refinement of X-Ray Powder Diffraction Patterns. *Journal of the American Ceramic Society*, *90*(9), 2935–2940. doi:10.1111/j.1551-2916.2007.01785.x

Wu, P., Hu, M. Y., Chong, X. Y., & Feng, J. (2018). The Glass-like Thermal Conductivity in ZrO_2-Dy_3TaO_7 Ceramic for promising Thermal Barrier Coating Application. *Applied Physics Letters*, *112*(13), 131903. doi:10.1063/1.5022610

Xie, X., Guo, H., Gong, S., & Xu, H. (2011). Lanthanum-Titanium-Aluminum Oxide: A Novel Thermal Barrier Coating Material for Applications at 1300 °C. *Journal of the European Ceramic Society*, *31*(9), 1677–1683. doi:10.1016/j.jeurceramsoc.2011.03.036

Yang, J., Han, Y., Shahid, M., Pan, W., Zhao, M., Wu, W., & Wan, C. (2018). A Promising Material for Thermal Barrier Coating: Pyrochlore-related Compound Sm_2FeTaO_7. *Scripta Materialia*, *149*, 49–52. doi:10.1016/j.scriptamat.2018.02.005

Yuan, J., Song, W., Zhang, H., Zhou, X., Dong, S., Jiang, J., Deng, L., & Cao, X. (2019). $TaZr_{2.75}O_8$ Ceramics as a Potential Thermal Barrier Coating Material for High-Temperature Applications. *Materials Letters*, *247*, 82–85. doi:10.1016/j.matlet.2019.03.102

Zhao, M., Pan, W., Wan, C., Qu, Z., Li, Z., & Yang, J. (2017). Defect Engineering in Development of Low Thermal Conductivity Materials: A Review. *Journal of the European Ceramic Society*, *37*(1), 1–13. doi:10.1016/j.jeurceramsoc.2016.07.036

Zhao, Y., Min, X., Fang, M., Wu, X., Liu, Y., & Huang, Z. (2019). Dependence of Crystal Structure on Mechanical and Thermophysical Properties of Magnetoplumbite-type $LnMgAl_{11}O_{19}$ Ceramics with Substitution of Ln^{3+} Ions. *International Journal of Applied Ceramic Technology*, *16*(4), 1596–1605. doi:10.1111/ijac.13216

Zhou, X., Xu, Z., Fan, X., Zhao, S., Cao, X., & He, L. (2014). $Y_4Al_2O_9$ Ceramics as a Novel Thermal Barrier Coating Material for High-Temperature Applications. *Materials Letters*, *134*, 146–148. doi:10.1016/j.matlet.2014.07.027

Zhou, Y., Gan, G., Ge, Z., Song, P., & Feng, J. (2020). Microstructure and Thermophysical Properties of CeO_2-doped $SmTaO_4$ Ceramics for Thermal Barrier Coatings. *Journal of Materials Research*, *35*(3), 242–251. doi:10.1557/jmr.2020.15

ADDITIONAL READING

Basu, D., Funke, C., & Steinbrech, R. W. (1999). Effect of Heat Treatment on Elastic Properties of Separated Thermal Barrier Coatings. *Journal of Materials Research*, *14*(12), 4643–4650. doi:10.1557/JMR.1999.0628

Kimura, T., Moriya, S.-I., & Takahashi, M. (2019). Effects of Heat Treatments to Inner Liner Material, Thermal Barrier Coating, and Outer Shell Material on Lifetime of a Combustion Chamber. *Acta Astronautica*, *158*, 244–252. doi:10.1016/j.actaastro.2018.05.019

Liang, J. J., Wei, H., Hou, G. C., Zheng, Q., Sun, X. F., Guan, H. R., & Hu, Z. Q. (2008). Thermal Stability of Phases in a NiCoCrAlY Coating Alloy. *Journal of Materials Research*, *23*(8), 2264–2274. doi:10.1557/JMR.2008.0282

Mahade, S., Ruelle, C., Curry, N., Holmberg, J., Björklund, S., Markocsan, N., & Nylén, P. (2019). Understanding the Effect of Material Composition and Microstructural Design on the Erosion Behavior of Plasma Sprayed Thermal Barrier Coatings. *Applied Surface Science*, *488*, 170–184. doi:10.1016/j.apsusc.2019.05.245

Qian, G., Nakamura, T., Berndt, C. C., & Leigh, S. H. (1997). Tensile Toughness Test and High Temperature Fracture Analysis of Thermal Barrier Coatings. *Acta Materialia*, *45*(4), 1767–1784. doi:10.1016/S1359-6454(96)00257-1

Xu, R., Fan, X., & Wang, T. J. (2016). Mechanisms Governing the Interfacial Delamination of Thermal Barrier Coating System with Double Ceramic Layers. *Applied Surface Science*, *370*, 394–402. doi:10.1016/j.apsusc.2016.02.180

KEY TERMS AND DEFINITIONS

Gas Turbine: Combustion engine that converts natural gas or other liquid fuels into mechanical energy.

Metal Bondcoat: Oxidation resistant metal layer that is deposited directly on the metal substrate or topcoat.

Metal Substrate: Base material to which a coating is applied or the material on which a process is conducted.

Rare Earth Metal: One of the seventeen chemical elements on the periodic table specifically including the fifteen lanthanides as well as scandium and yttrium.

Thermal Barrier Coating (TBC): Advanced materials system applied to metal surfaces operating at high temperatures for the management of heat exhaust in gas turbines or parts of aero engines.

Thermal Conductivity: A measure of the ability of a material to conduct heat.

Thermal Grown Oxide (TGO): Layer formed at the interface of the bond coat and/or topcoat to protect the metal substrate from high temperatures.

Thermal Insulation: Reduction of heat transfer between objects of different temperatures in thermal contact or the radiative influence range.

Chapter 5
Application of Thermal Spraying Techniques Used for the Surface Protection of Boiler Tubes in Power Plants:
Thermal Spraying to Combat Hot Corrosion

Gaurav Prashar
https://orcid.org/0000-0002-3549-5838
Lovely Professional University, Phagwara, India

Hitesh Vasudev
https://orcid.org/0000-0002-1668-8765
Lovely Professional University, Phagwara, India

ABSTRACT

Distinct methods of depositing the coatings are available according to the intended area of application with a sole objective to protect the surface of structural component like boiler, boiler tubes, and heat exchangers from any mechanical or chemical damage. The main benefit is to minimize the manufacturing cost of a new component and also its fabrication. Thermal spraying is the commonly used technique to tailor the surface properties with a cost-effective approach. The different techniques of thermal spraying such as plasma, HVOF, cold spraying, etc. were investigated and introduced. Although each of these processes has advantages, there are also certain disadvantages associated with them, which limit their application. In the chapter, different methods of depositing coating by thermal spray are discussed and compared. By developing advanced techniques and new coating materials, the life and efficiency of power plants can be enhanced in the future.

DOI: 10.4018/978-1-7998-4870-7.ch005

1. INTRODUCTION

Materials are lifeline or backbone of the modern industry and nothing exist without them (Prashar & Vasudev, 2020). Hot corrosion or high temperature corrosion of boiler tubes is the serious problem faced by the power plant industry (Beltran & Shores, 1972). According to a survey annual loss due to all types of corrosion is US$ 6500 million (Kumar et al., 2018). Corrosion mainly occurs due to aggressive gases such as HCl, H_2O, etc in combination with the metal and alkali chlorides that are generated during combustion process. Low grade fuel used in boilers contains sulphur and during combustion produces SO_2 which get oxidized to SO_3. This SO_3 by further reacting with sodium chloride (NaCl) and water vapor forms Na_2SO_4 having melting point 884°C and get condensed onto the boiler tubes together with fly ash (Natesan, 1976). Vanadium in small quantities is also present in coal and during combustion forms V_2O_5 having melting point 670°C. This V_2O_5 further reacts with Na_2SO_4 and forms sodium vanadate. This sodium vanadate is extremely corrosive in nature to the materials that operates at high temperatures (Hwang & Rapp, 1989; Khanna & Jha, 1998). According to (Shih et al., 1989) when surface of a metal or alloy got covered with liquefied salt (Na_2SO_4) at high temperatures, then oxidation occur at higher rates. This accelerated form of oxidation is termed as hot corrosion. Hot corrosion is also shown to be caused by sulfates, vanadates and carbonates (Pettit, 2011).

The reactions taking place at high temperatures for formation of sulphates, vanadates, and carbonates are as following:

SO_3 reacts with sodium chloride (NaCl) and water vapour to form Na_2SO_4; equation (1)

$$2(Na, K)Cl + SO_2 + ½ O_2 + H_2O = (Na, K)_2SO_4 + 2HCl \quad [1]$$

When temperature is approximately around 600°C, compounds of sodium sulfate and vanadium pentoxide are formed, which have more corrosive properties. This vanadium pentoxide further reacts with sodium sulfate and forms sodium vanadate which is extremely corrosive ; equation (2)

$$Na_2SO_4 + V_2O_5 = 2NaVO_3 + SO_3 \quad [2]$$

Corrosion results from two main impurities: carbon dioxide and oxygen. Carbonates and bicarbonates in fresh water are the major source of carbon dioxide contamination. Natural water contains sodium, calcium and magnesium bicarbonate. When these minerals enter into the boiler, heat causes them to break down as follow: equation (3)

$$Ca(HCO_3)_2 + Heat = CaCO_3 + H_2O + CO_2 \quad [3]$$

Calcium bicarbonate calcium carbonate sludge

One alternate is to protect the boiler tubes with a suitable coating material and is considered to be a preferred approach (Eliaz et al., 2002). High temperature performance of the coatings not only relies on coating composition but also depend upon the deposition process used and its parameters. Thermal spraying is the commonly used method to tailor the surface properties with a cost effective approach. In addition to this thermal spray is capable of depositing any type of coating powder (ceramics, polymer, composite and metallic) onto any kind of substrate material (Davis, 2004; Gil & Staia, 1999, 2002; Pawlowski, 2008; Singh & Prakash, 2009). Thermal spraying is considered to be one of the most adaptable hard facing techniques feasible for applying coating materials on the surface of structural components to combat high-temperature oxidation and wear. In general, the materials that don't decompose, vaporize or dissociate on heating can be thermally sprayed. Thermal spraying can deposit metallic as well as for non-metallic materials. Different thermal spraying techniques are available but choosing the best one depends on certain parameters such as functional requirements, availability, and cost of equipment, coating material adaptability to the selected coating technique, etc. Furthermore, thermal spraying techniques cause minor or negligible harm to environment in comparison with other techniques like chemical vapour deposition (CVD), physical vapour deposition (PVD) and electroplating (Choy, 2003; Koropchak & Roychowdhury, 1990; Tsai et al., 2002; Young, 1965).

To date, different techniques of thermal sprsaying such as plasma, HVOF, cold spraying etc were investigated and introduced. Although each of these processes have advantages but there are also certain disadvantages like presence of pores and weak inter-splat bonding associated with them which limit their application. (Richer et al., 2002) examined the oxidation behavior of plasma-sprayed CoNiCrAlY coatings at 1000ºC. It was observed that the most visible defects in developed coatings were porosity (5.3%) and oxide content, which leads to the poor performance of the coating. The main reason for the presence of pores was the shrinkage of molten droplets, as they strike on the surface and rapidly got solidified and as a result insufficient plastic deformation of the particles resulted in equiaxed pores. However, these types of porosities were not observed in moderate temperature techniques such as high-velocity oxy-fuel and cold spraying. The minimum shrinkage upon cooling and high velocities sufficiently deforms the particles and produce dense coatings. (Sadeghimeresht et al., 2016b) in their experimentation studied the corrosion demeanour and microstructural properties of Ni, NiAl and NiCr coatings deposited by HVOF, APS and HVAF process. Authors concluded that HVAF coating microstructure was dense with less porosity levels and minimum oxide content leading to high corrosion resistance among all. Nonetheless, thermal spraying field is continuously evolving and advancements have led to improvements in the coating quality. This chapter briefly elaborates the advancements made in the field of thermal spray coatings including role of composite coatings, microstructural characteristics and quality of the surface developed with regard to high temperature corrosion performance.

2. HOT CORROSION AND OXIDATION PROCESS IN BOILER TUBES

Degradation process of boiler tubes used in power plants is categorized into three types depending upon the range of temperature shown in fig. 1:

a) Hot corrosion type-1
b) Hot corrosion type-2
c) Oxidation

2.1 Hot Corrosion Type-1

This type occurs above the melting point of Na_2SO_4 in temperature range of 850°C-950°C (Edris et al., 1997). At elevated temperatures condensation of the fused alkali salts like sodium acetate, sodium carbonate and sodium sulfide initiates and they deposit onto the exposed surface of boiler tubes. Then chemical reaction occurs and depletes the Cr content of the base materials by attacking the layer of protective oxide. Due to this depletion in Cr content, rapid oxidation of base metal occurs which leads to the formation of porous and non-protective scale.

The formation of oxide layer/ scale can be observed in the oxidized coatings in microscopic observation, whereas, macroscopic appearance is characterized by color changes and severe peeling of coating. Microscopically, is exhibited by sulphidation followed by depletion region beneath the porous scale (Rapp, 1986).

2.2 Hot Corrosion Type-2

This type occurs below the melting point of Na_2SO_4 in temperature range of 650°C-850°C having minor content of SO_3 (Stringer, 1987). It is characterized by pitting attack due to formation of eutectic mixture (low melting point) of sodium sulphate and cobalt sulphate. Microscopically, in type-2 hot corrosion sulphidation attack followed by depletion region beneath the porous scale is not observed. In this type localized failure of scale takes place due to erosion, chemical reaction and thermal cycling (Kamal et al., 2017; Meier, 1989).

2.3 Oxidation

In this type degradation of material occurs when temperature rises above 950°C. Oxidation rate depends on transportation of the cations/anions through the oxide layer. Dissociation pressure of the oxide will determine most stable oxide from the set of elements. It is lower lowest for aluminum and chromium in comparison with iron, cobalt and nickel. If chromium content is >10 wt. % in Ni-Cr alloy, protective oxide layer of Cr will form. But chromium content will be 25 wt% in cobalt based super alloys due to difference in diffusion behavior of chromium in cobalt alloys.

Furthermore, it is recommended to add aluminum in chromium when temperature is raised above 850°C to increase the oxidation resistance because above 850°C chromium sublimates and forms volatile 'Cr_2O_3'. Another advantage of using $Al_2O_3$is that the formation rate of alumina will be slower than chromium oxide at same temperature. Reason being the higher activity of chromium than Al_2O_3 alumina (Cr will be enriched with oxygen rather than Al) and formation of Al will reduce in lower alumina activity in Cr-Al mixture. These types of coating will be referred for isothermal oxidation conditions.

During thermal cycling conditions severe spalling of the oxide scale can take place due to thermal stresses. For such conditions yttrium and rare earth elements can be added that segregate into grain boundaries along with the alumina scale and cause reduction in the transportation rates of alumina and oxygen and promoting oxidation resistance. Overlay coatings that contain 18-22 wt. % Cr, upto 12 wt. %Al, NiCrAlY and NiCoCrAlY will be candidate choice for oxidation resistance (Shourgeshty et al., 2016).

3. HIGH TEMPERATURE COATINGS

The main aim of coating is to protect the surface or specific part from aggressive environment in various fields, ranging from automotive industry to power plants and also to tiny implants inside human body. Early methods to combat high temperature corrosion are the development of steels for boiler components (Fukuda, 2017). But with increase in further electricity demand power plants have suffered by two problems ie; low thermal efficiency and increased level of emissions. This problem can be solved only if the temperature of power plant is increased above 700°C but conventional steel used in boilers will not be able to operate at higher temperatures because of low yield strength above 600°C. Therefore keeping in mind higher efficiencies and lower emission levels solution is high temperature corrosion resistant coatings.

Figure 1. Temperature range for hot corrosion-I, hot corrosion-II, and oxidation (Shourgeshty et al., 2016).

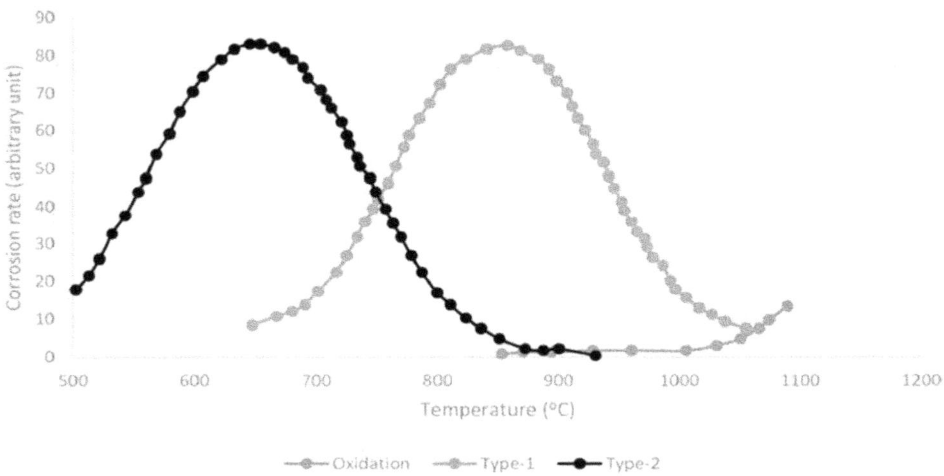

Table 1. Comparison of the properties for coatings used in waste to energy and biomass boiler tubes.

Coatings professes	Weld-overlay	Thermal spray	Laser cladding
Application records	Many	Many	Only testing
Coating materials	SS, Alloy-622 & 625 etc.	Cermets, ceramics & Ni-based alloys	Alloy-625 etc.
Dilution of base material	Large, 5-20%	Almost nothing	Medium, <5% controllable
Defect formation	Small	Small pores & oxides	Small
Thickness of the deposited coating	2-3mm	0.1-0.8mm	0.5-4mm
Corrosion life time	Long	Medium	Long
Coating sites	On-site & In-shop	On-site & In-shop	In-shop only
Feedstock materials	Powder, cored wire & solid wire	Powder, cored wire & solid wire	Powder, cored wire & solid wire
Cost	Economical	More than weld over lay but less than laser cladding	High cost

Presently the most suitable, economical and technical method to resist high temperature corrosion is thermal spray coatings for boiler components. Table.1 shows a comparison of the properties for coatings used in waste to energy and biomass boiler tubes (Kawahra, 2016).

3.1. Thermal Spray Coating

Thermal spray technique utilizes the heat energy generated from electric, chemical combustion or plasma jet to melt and deposit any kind of material (metallic or non-metallic) onto the substrate surface forming a protective layer (Davis, 2004). Fully melted or un-melted particles entrained in a high velocity jet flame strikes onto the substrate surface with impact and deform in a shape called "splats". The bonding of the coating with substrate takes place by the method of mechanical interlocking when particles strikes with impact and solidifies. Thermal spray can produce reliable corrosion and wear resistant coatings for power plants and other applications. Thickness of the coating ranges from 20µm to a several millimeters can be achieved and is higher than that achieved in CVD, PVD and electroplating (Pawlowski, 2008). Earliest records for thermal spray were found in patents of M.U. Schoop from a period of 1882 to 1889 (Davis, 2004; Pawlowski, 2008). Historic developments in the field of thermal spray coatings up-to 2019 are shown in fig.2. Thermal spray processes can be classified into different categories based on characteristics and specifications. However, most popular types are classified on the basis of velocity and temperature like high velocity oxy fuel (HVOF), flame, warm/cold spray, plasma, high velocity air fuel (HVAF), wire arc spraying and detonation gun spray (Sadeghi et al., 2019) shown in fig. 3.

It should also important to mention that before designing advanced coating for boiler components complete understanding is must among relationship between process, microstructure properties and coating performance. Therefore, for selecting a right technique for a candidate application, thorough understanding of the complexities and capabilities of thermal spray is required. In next section commonly used thermal spray techniques for boiler applications were discussed along with their features.

3.1.1. Plasma Spray Coating

The schematic view of plasma spray coating is shown in fig. 4. This process can be classified into two types based on its operating conditions ie; vacuum or atmospheric. This technique has a capability of spraying any type of feed stock material such as ceramics, refractory alloys or cermets owning to its high melting temperature (15000°C) that is generated by DC/induction plasma (Fauchais, 2004). Feed stock material is fed into this hot stream of plasma and due to high temperature in stream material gets melted. To achieve supersonic velocities convergent-divergent nozzle is preferred so that particles entrained in plasma stream attain high velocities and strikes the substrate surface with impact. Flexible nature of plasma spray allows depositing any kind of feedstock like slurry, liquid, powder and suspensions (Karthikeyan et al, 1997). High corrosion resistance of plasma spray coatings was reported by many authors for chromium oxide and nickel chromium alloys (Bulloch & Callagy, 1999; Knuuttila et al., 1999). (Singh et al., 2007) already reported that plasma spray technique is a cost effective solution to combat hot corrosion/oxidation (Fauchais et al., 1997; Singh et al., 2007). But its major drawback is its high level of porosities and large number of un-melted particles that makes them unsuitable for corrosion applications (Hussain et al., 2013; Odhiambo et al., 2019; Singh et al., 2005).

Vacuum plasma spraying (VPS) takes place in confined chamber and uses low pressure ($<10^{-1}$ mbar). This process is suitable for those materials that are reactive when sprayed in open air. Velocity attained in this is Mach 3 resulting in less porosity and better adhesion (Sidky & Hocking, 1999). It is most suitable for surface modification of plastics, polymers, metals, rubbers and fibres (Betancourt-Dougherty & Smith, 1998; Nemani et al., 2018; Padture et al., 2002). Disadvantage of vacuum spraying process is its high cost and skill involving in attaining vacuum conditions.

Figure 2. Historic advancements in field of thermal spray coatings.

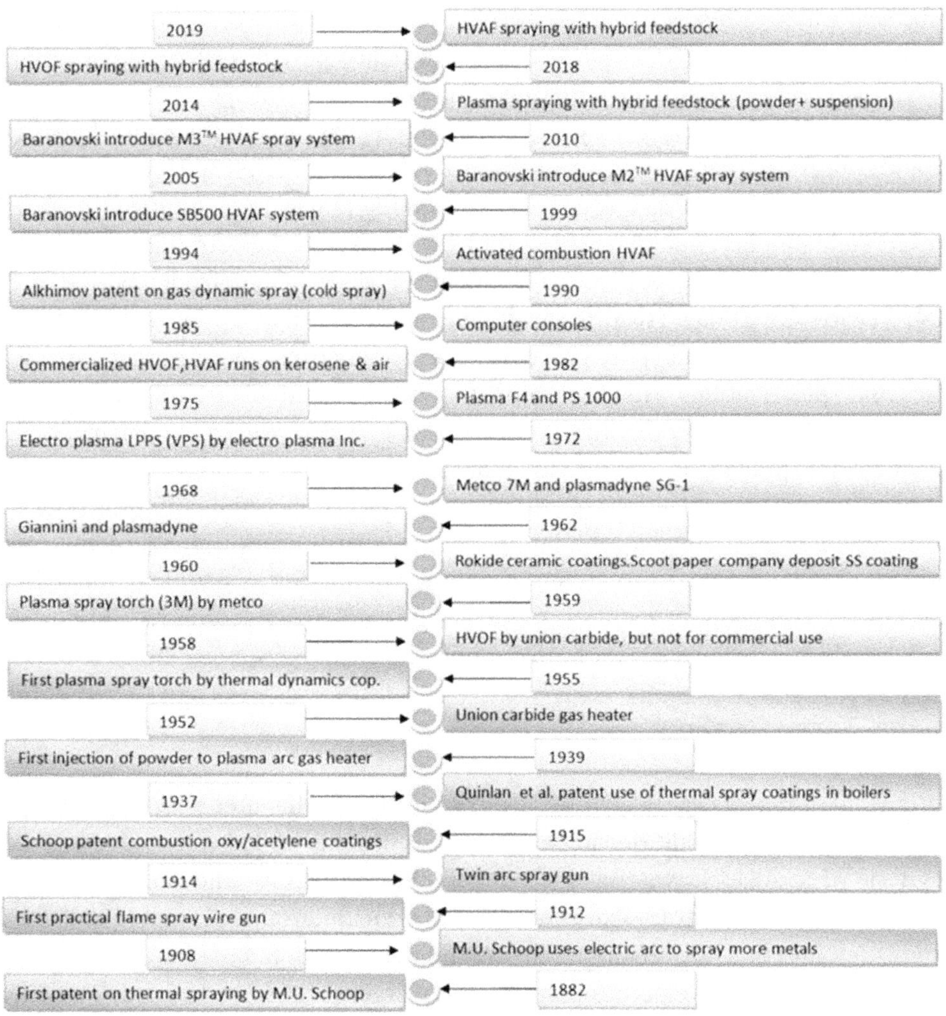

3.1.2. High Velocity Oxy Fuel (HVOF)

HVOF is rapidly developing technique to deposit surface coatings and is also challenging VPS that is costly due to its setup (US$2 million) and vacuum procedures. HVOF third generation setup even allows the spraying of materials that are prone to oxidation in atmosphere (Kaur et al., 2008). HVOF technique makes use of thermal plus kinetic energy to deposit coatings onto the substrate surface. The housing unit makes use of oxygen-fuel mixture that consists of hydrogen, propylene or propane which in turn depends upon the requirements of the user to produce best quality coatings. Mixture of fuel and gas is mixed in combustion chamber of the gun and is ignited once it comes out from the nozzle having a velocity of 1000m/s (Thorpe & Richter, 1992). The temperature of the combustion chamber ranges from 2760°C to 3315°C and it depends upon fuel used. Coating material is injected into this hot stream and this stream pushes the particles in semi-solid state towards the substrate (Lugscheider et al., 1998). The developed coating has the following characteristics;

Figure 3. Thermal spray processes categorized in terms of velocity and temperature of produced jet stream (Sadeghi et al., 2019).

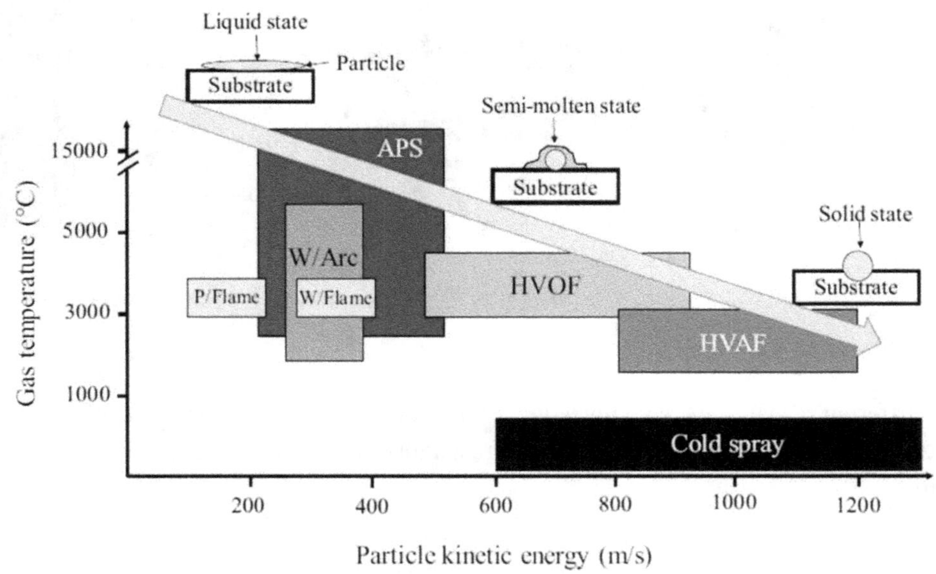

Figure 4. The plasma spraying process.

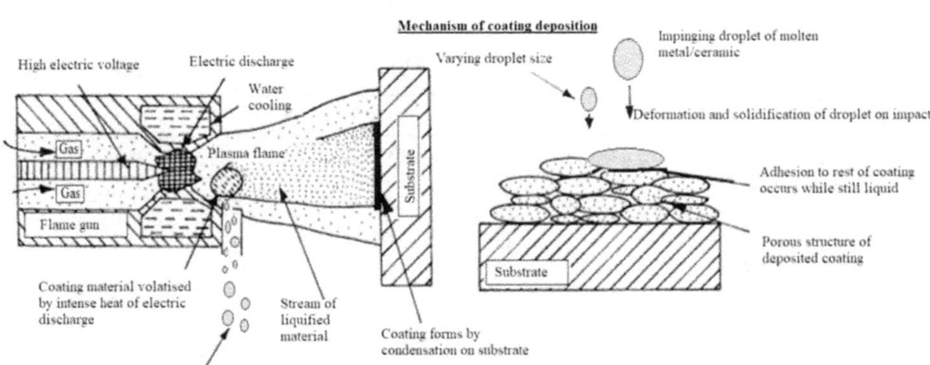

a) High density coatings due to high kinetic energy generated in this process.
b) High bond strength (> 8.27 X 10^7 N/m^2).
c) Low levels of porosities (due to no interconnecting network of porosities).

Figure 5. Cross section of typical HVOF system.

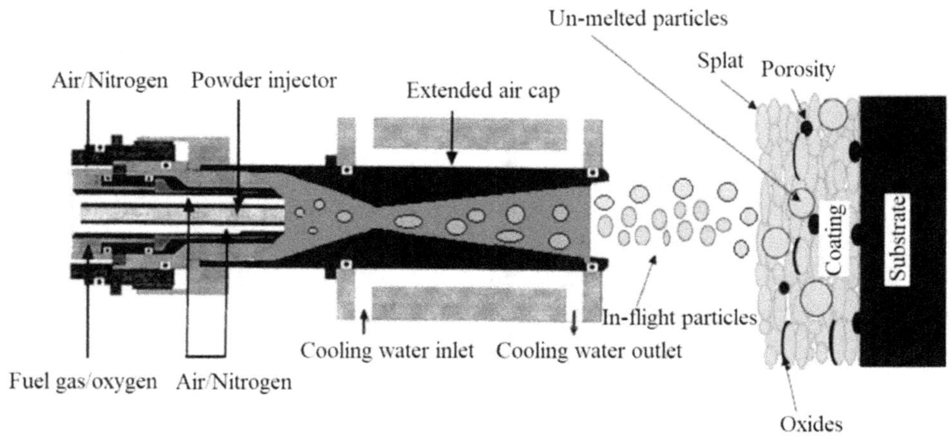

To date, different researchers have investigated the corrosion resistance of HVOF coatings and based on studies coatings exhibits improved corrosion resistance performance. Summary of some studies are listed in Table2.

Table 2. Weight gain data and percentage reduction in hot corrosion rate of Ni-based coatings.

Coating composition	Hot Corrosion rate mg/cm^2		Percentage reduction in hot corrosion after coating (%)
	Un-coated	Coated	
Ni-20Cr (Kaushal et al., 2011b)	3	0.7	76.67
Cr_3C_2 -NiCr (Kaur et al., 2009)	6.174	2.925	52.62
75Cr_3C_2-25NiCr (Bhatia et al., 2015)	33.33	6.89	79.33
Ni20-Cr (Kaushal et al., 2011a)	55	23	58.18
Cr_3C_2 -NiCr (Kaur et al., 2012)	10.57	6.89	34.82

3.1.3. Cold Spray Coating

Cold spray process depends more on particle impact rather than temperature (< 1000°C) to deposit coatings (Cavaliere, 2018). The working mechanism depends on size of the particle (1 to 100μm), temperature (< 1000°C), critical velocity and properties of the coating particles (Moridi et al., 2014). Feedstock in powder form is injected into the laval nozzle by using a carrier gas (helium or nitrogen). The bonding of the coating with substrate takes place by particle-substrate impact, when particles strikes with impact and energy deforms the particles. Cold spraying is considered to be spraying of suitable for soft metals such as aluminum and copper but in literature studies on hard materials such as W and Ti also reported (Li et al., 2007; Tsui et al., 1998). Cold spray process offers several benefits like no or little oxidation occurs during spraying, phase changes does not takes place, grain size remains intact and high deposition efficiency. But on the other hand cost involving in this process makes him unsuitable for its use in boiler applications (Sadeghi et al., 2019). Furthermore, micro-pores within the cold spray coatings are unavoidable due to the bonding nature of the particles (Barmouz et al., 2011; Vardelle et al., 2016). Interfacial strength between the cold spray particles are generally less than the strength of bulk material due to limited atomic diffusion between the splats. Ductility of as sprayed coatings is also limited that restricts its use for wide range of applications. Schematic diagram of cold spray process is shown in fig. 6.

Figure 6. Schematic diagram of cold spray process (Sadeghi et al., 2019).

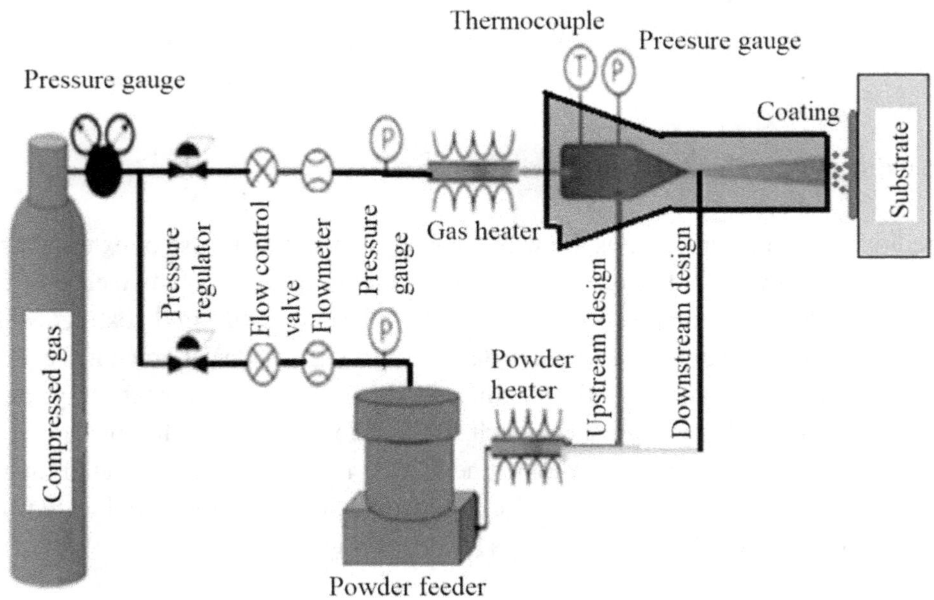

3.1.4. High Velocity Air Fuel Coating Process (HVAF)

HVAF has gained attention in last decade or so in many applications including power plant industry. This is relatively latest technology in the group of thermal spray processes. HVAF makes use of a spark plug to ignite the mixture of compressed air and fuel gas as shown in fig. 7. As soon as the temperature of catalytic ceramic wall exceeds the auto-ignition temperature of mixture, it assists in further combustion and taking over the role of spark plug (Sadeghimeresht et al., 2016a; Verstak & Baranovski, 2003). Microstructure of the coatings developed by HVAF process is lamellar, with less porosity levels and has little or no oxide content as compared to HVOF and APS thermal spraying techniques [Eklund et al., 2018, 2019; Sadeghimeresht et al., 2016c, 2018a, 2018b, 2019). High velocity (1200m/s) leads to better plastic deformation of the in-flight particles and produce coatings with minimum voids or pores at splat boundaries [65-66]. Less temperature (<1950°C) along with less residence time makes them suitable candidate for spraying of materials such as Al or Cr that are sensitive to oxidation and high temperatures (Davis, 2004). (Sadeghimeresht et al., 2016b) in their experimentation studied the corrosion demeanour and microstructure -properties-performance of Ni, NiAl and NiCr coatings deposited by HVOF, APS process and compared it with HVAF process. Authors concluded that HVAF coating microstructure was dense with less porosity levels and minimum oxide content leading to high corrosion resistance among all.

Figure 7. Schematic diagram of HVAF process (Sadeghi et al., 2019).

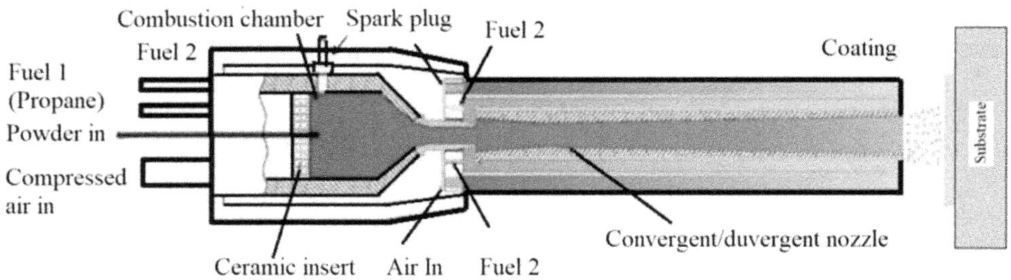

3.1.5 Arc Wire Spray Coating Method

Another method of thermal spraying is arc wire spray. It does not make use of external heat source for melting of feed stock powder but utilizes an electric arc that occurs between two electrode wires which melts the consumable wire. Composition of the wires is close to the desired coating. Molten particles are directed towards the substrate by compressed air or inert gas stream as shown in fig. 8. This process has the advantage of achieving high bond strength, less porosity levels and higher spray rates in comparison with flame spraying (Champagne & Helfritch, 2013). Moreover, it also harms the environment and health of human beings by producing fumes, ozone and arc light. Both wire arc and flame spraying are cost effective approaches but cannot be used for high temperature corrosion applications due to higher porosity levels and in situ oxides in the coatings (Sadeghi et al., 2019).

Figure 8. Schematic of wire arc spraying setup.

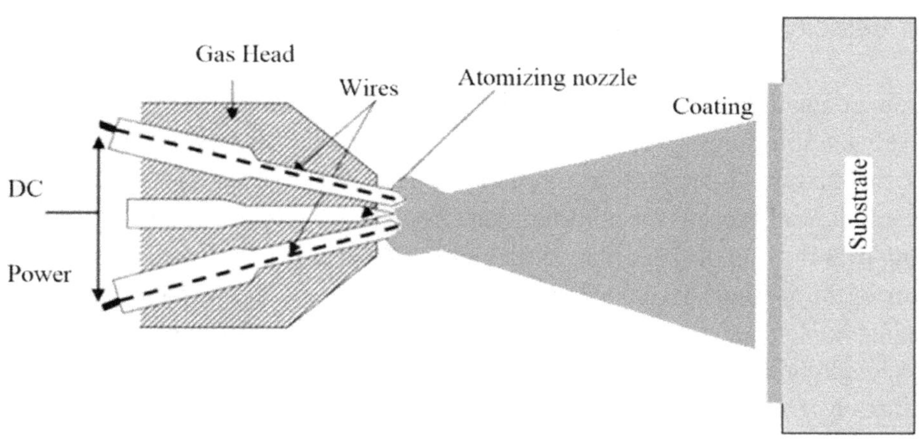

4. COMPARISON BETWEEN DIFFERENT THERMAL SPRAY TECHNIQUES

Table 2 shows the comparison between different thermal spray techniques. As discussed in previous section it is clear that microstructure of the deposited coatings varies according to the technique used. Micro-structural properties such as porosities, bond strength and in situ oxide content can be controlled by varying thermal and kinetic energy of the spraying technique. As in case of cold spray or HVAF techniques oxide free and dense coatings were obtained by high velocity rather than high thermal energy which show that for obtaining oxide free and dense coatings kinetic energy is more important.

Table 3. Comparison between different thermal spraying techniques (Sadeghi et al., 2019).

Parameters	Plasma	HVOF	Cold spray	HVAF	Wire arc
Jet temperature(°C)	12000-15000	2800	<1000	1950	4000
Jet velocity(m/s)	200-600	900	1200	1200	50-100
Particle temperature (°C)	>3800	3000	>600	<1600	>3800
Particle velocity(m/s)	600	1000	1000	1200	200
Porosity level	High	Moderate	low	Low	High
Oxide level	Moderate to coarse	Moderate to low	none	Low	Moderate to high
Bond strength (Mpa)	40-70	70-100	80-100	100-200	10-40
Materials	ceramics, refractory alloys, cermets	Nickel based alloys	Soft metals	Al or Cr	High alloy Fe based coatings

5. ROLE OF MICROSTRUCTURE AND COATING COMPOSITION IN HIGH TEMPERATURE CORROSION

Practically in boilers erosion-corrosion can act simultaneously and can cause premature failure of boiler tubes. Alloying elements like nickel can be combined with protective scale forming elements such as Al or Cr to improve the erosion-corrosion resistance [68-69].Therefore, erosion-corrosion can be easily and effectively controlled with proper surface treatment, in recent years by use of thermal spray coatings (Fantozzi et al., 2017; Oksa et al., 2013, 2014; Zhou et al., 2010). Furthermore, in these harsh environments, coating composition and microstructure will decide its performance. Hence need arises for the coating materials that have properties like hardness, strong bond strength, low porosity, good thickness as well as appropriate phase composition to provide corrosion resistance against the salt contaminants such as Na_2SO_4 and V_2O_5.

Recently (Galedari et al., 2019) summarizes the corrosion behavior of materials and concluded that super alloy (Inconel-625) must be considered for providing resistance to corrosion due to its composition. Although super alloys has good corrosion resistant properties, but it is must to enhance its wear resistance in order to prolong the service life of parts that are partially damaged during operation. Hence, metal matrix composite coatings consisting of super alloys and thermally stable reinforcing particles are the candidate choice to provide protective coatings against wear and corrosion for aggressive environments (Dooley & Wiertel, 2009). Hard phases used as reinforcement additives include WC, Al_2O_3, Cr_3C_2, Cr_2O_3, TiC, SiC, TiN, CeO_2, TiO_2 etc. (Buytoz et al., 2013; Cai et al., 2013; Grewal et al., 2013; Harsha et al., 2008; He et al., 2014; Kim et al., 2003; Koivuluoto & Vuoristo, 2009; Martin et al., 2001; Maatta et al., 2011; Ramesh et al., 2010; Tu et al., 1997; Tobar et al., 2006; Wang et al., 1996; Yin et al., 2011).

(Vasudev et al., 2019) integrated a bi-layer of 718-alloy in NiCrAlY coating by HVOF process. High temperature corrosion and erosion behavior of both coated and bare samples was studied. The authors reported that NiCrAlY coating with a bi-layer of alloy-718 has a better corrosion and erosion resistance than substrate. Enhanced erosion-corrosion resistance is attributed with high hardness and formation of stable oxides (TiO_2, Al_2O_3 and Cr_2O_3) on the coating surface. (Sevillano et al., 2013) deposited Ni-$Al_2O_3$by cold spray technique to examine the oxidation behavior at 520°C. Authors reported that Ni-Al_2O_3founds to be suitable to candidate to combat oxidation/corrosion in power plant applications due to absence of interconnected porosities and reduced porosity levels (<3%). (He et al. 2014) in their investigation nickel base alloy was reinforced with $Al_2O_3$and CeO_2 and deposited it on Al alloy by using plasma spray technique. It was found that mechanical and micro structural features of as deposited coatings like hardness, fracture toughness, bonding strength and porosity levels improved after addition of $Al_2O_3$and CeO_2.

Apart from coating composition, microstructure of the as deposited coatings such as porosity levels, splat boundaries also effect erosion-corrosion performance and these features varies according to the spraying process and spraying parameters (Rodriguez et al., 2000; Souza & Neville, 2003; Saaedi et al., 2010). High porosity levels leads to formation of discontinuous oxide scale and effect oxidation resistance of the coating (Uusitalo et al., 2002). Literature reveals that the presence of these features in coatings effect the uniform supply of Al or Cr and ultimately results in discontinuous oxide scale (Hussain et al., 2013; Oksa et al., 2014). Hence next section will focus mainly on splat boundaries and porosity.

5.1. Splat Boundaries

'Splat' as discussed earlier is the name given to single impacted particle. Large number of splats overlapped, solidifies and form coating as shown in fig. 9. Splats are the building block of coating formation. Molten or semi-molten particles are spherical in shape and on impacting the substrate surface their structure changes to disc like shape known as splat. Irrespective of the spraying technique splat and splat boundaries exists. Area between the two splats indicates the plastic deformation achieved on impact and also the degree of particle melting. It is through these splat boundaries corrosive species will travel and affect the corrosion performance of the coating if particles were not deformed completely on impact (Fauchais et al., 2014). Severe corrosion can occur along the rounded un-melted particles and splat boundaries (Durul & Nurbas, 2010). Therefore, these regions or areas need attention in corrosion protection due to several reasons. First, improper contact between the splats paved way for corrosion to trigger (Zhou et al., 2009; Xu et al., 2011). Secondly, due to in situ oxidation of feed stock particles during spraying, oxides formation will occur along the splats and this oxidation damage the protective oxide layer by depleting Al and Cr (Tao et al., 2010; Zhang et al., 2012).

Figure 9. Mechanism of the coating by splat formation.

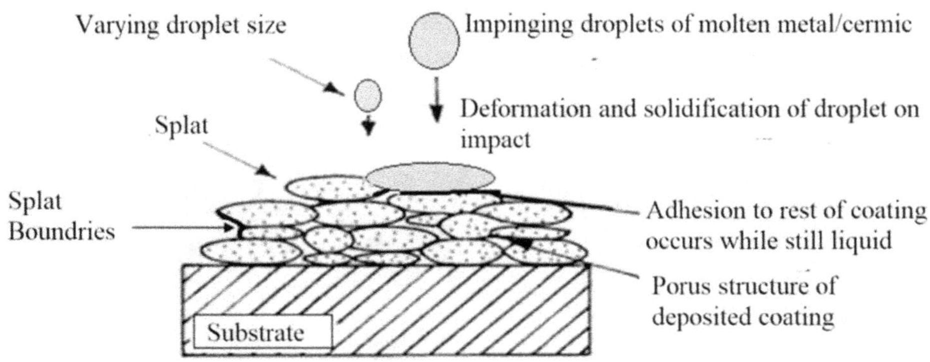

5.2. Pores

Un-melted particles during spraying process are considered to be as the common source for coating porosity. Un-melted particles require higher impact velocity for better plastic deformation and also for better closer of the voids. If the velocity will be less than the solid particles will remain trapped inside the coating and due to this proper contact between the splats will not take place and voids occur. It is clear from vast technical literature that presence of pores significantly reduces the corrosion resistance (Guilemany et al., 2002; Camp et al., 2009; Monticelli et al., 2004; Bolelli et al., 2008; Wang et al., 2014; Milanti et al., 2015; Zhang et al., 2016). Porosity can be categorized differently by authors. (Shanmugam et al., 2014) categorized pores according to the formation mechanism and size variation as type A (0 to 10µm) and type B (10 to 25µm). Type A is the result of interaction among gaseous medium and particles whereas type B is developed by splashing on impact with material deposited. Other researchers classified pores as open pore, closed pores and micro cracks. They reported that these types of pore and micro cracks occur due to enclosed gases and shrinkage of splats as a result of fast solidification [Ctibor et al., 2006; Konyashin & Chukalovskaya, 1997). Furthermore, others in addition to this categorization pores are classified as globular pore, flat pore and vertical pore. Globular pores are convex in shape whereas inter-lamellar flat pores are like thin voids perpendicular to the direction of spraying (Radka et al., 2004).

Furthermore, porosities give rise to oxide content resulting in mass gain of coatings. Mass gain not only depends upon the porosity but also on to the pore structure morphology. Large porosity levels (open porosity) forms voids and results in forming of interconnected networks of pores which communicate with coating surface. This structure allows the oxygen to enter deep into coating and reach cavities to form oxides enhancing the mass gain of coatings (Sadeghimeresht, 2018).

Figure 10. Effect of different types of pores on the corrosion behavior of coatings and long term exposure to corrosion leads the closed pores to interconnected pores.

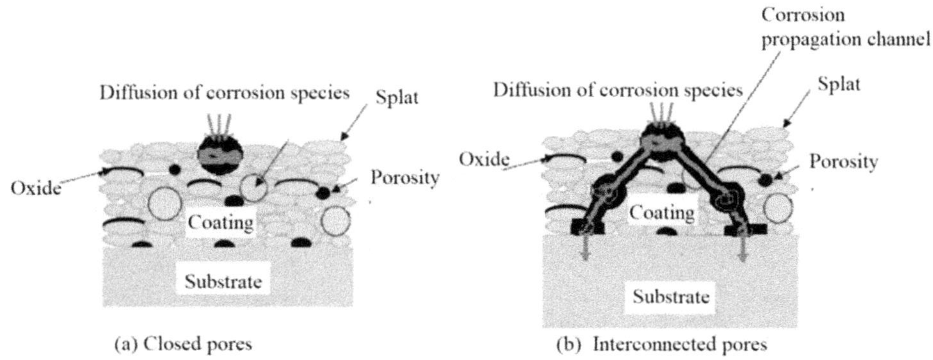

6. CONCLUSION

With rapid surging in industrialization and global warming threats, the power plants have to operate under high temperatures above 700°C. Above this temperature range, load bearing components made up of conventional alloys (like super heater tubes and water walls) have to suffer from high temperature corrosion and they degrade very easily. Thermal spray processes then comes into picture and are increasingly being used in these sectors for the protection of structural components against corrosion with an ideal combination of metal matrix coatings consisting of Ni-based super alloys plus various reinforcements. The composite coating provides effective results in minimizing the rate of corrosion only if microstructure obtained from thermal spraying techniques. Therefore, it should be important to mention that, before designing advanced coating for boiler components complete understanding is must amongst the structure-property relationship. Hence, for selecting a right technique for a candidate application thorough understanding of the complexities and capabilities of thermal spray process is required. While depositing nickel based coatings, HVOF is recommended and has shown promising approach to increase the efficiency and lifetime of components in boilers.

REFERENCES

Barmouz, M., & Givi, M. K. B. (2011). Fabrication of in situ Cu/SiC composites using multi-pass friction stir processing: Evaluation of microstructural, porosity, mechanical and electrical behavior. *Composites. Part A, Applied Science and Manufacturing*, *42*(10), 1445–1453. doi:10.1016/j.compositesa.2011.06.010

Beltran, A. M., & Shores, D. A. (1972). Hot Corrosion. In The superalloys (pp. 4289-4296). Academic Press.

Betancourt-Dougherty, L., & Smith, R. (1998). Effects of load and sliding speed on the wear behaviour of plasma sprayed TiC NiCrBSi coatings. *Wear*, *217*(1), 147–154. doi:10.1016/S0043-1648(97)00212-3

Bhatia, R., Sidhu, H. S., & Sidhu, B. S. (2015). High Temperature Behavior of Cr3C2-NiCr Coatings in the Actual Coal-Fired Boiler Environment. *Metallurgical and Materials Transactions E*, *2*(1), 1011–1022.

Bolelli, G., Lusvarghi, L., & Giovanardi, R. (2008). A Comparison between the Corrosion Resistances of Some HVOF-Sprayed Metal Alloy Coatings. *Surface and Coatings Technology, 202*(19), 4793–4809. doi:10.1016/j.surfcoat.2008.04.056

Bulloch, J., & Callagy, A. (1999). An in situ wear-corrosion study on a series of protective coatings in large induced draft fans. *Wear, 233-235*, 284–292. doi:10.1016/S0043-1648(99)00229-X

Buytoz, S., Ulutan, M., Islak, S., Kurt, B., & Celik, O. N. (2013). Microstructural and wear characteristics of high velocity oxygen fuel (HVOF) sprayed NiCrBSi–SiC composite coating on SAE 1030 steel. *Arabian Journal for Science and Engineering, 38*(6), 1481–1491. doi:10.100713369-013-0536-y

Cai, B., Tan, Y., He, L., Tan, H., & Gao, L. (2013). Tribological properties of TiC particles reinforced Ni-based alloy composite coatings. *Transactions of Nonferrous Metals Society of China, 23*(6), 1681–1688. doi:10.1016/S1003-6326(13)62648-5

Camp, M., Carboneras, M., Lopez, M. D., Torres, B., Rodrigo, P., Otero, E., & Rams, J. (2009). Corrosion Resistance of Thermally Sprayed Al and Al/SiC Coatings on Mg. *Surface and Coatings Technology, 203*(20-21), 3224–3230. doi:10.1016/j.surfcoat.2009.03.057

Cavaliere, P. (2018). *Cold-Spray Coatings: Recent Trends and Future perspectives*. Springer. doi:10.1007/978-3-319-67183-3

Champagne, V. K., & Helfritch, D. J. (2013). A demonstration of the antimicrobial effectiveness of various copper surfaces. *Journal of Biological Engineering, 7*(8), 1–7. doi:10.1186/1754-1611-7-8 PMID:23537176

Choy, K. L. (2003). Chemical vapor deposition of coatings. *Progress in Materials Science, 48*(2), 57–170. doi:10.1016/S0079-6425(01)00009-3

Ctibor, P., Lechnerova, R., & Benes, V. (2006). Quantitative analysis of pores of two types in a plasma-sprayed coating. *Materials Characterization, 56*(4-5), 297–304. doi:10.1016/j.matchar.2005.11.016

Davis, J. R. (2004). *Handbook of Thermal Spray Technology*. ASM International.

Dooley, R., & Wiertel, E. (2009). A survey of erosion and corrosion resistant materials being used on boiler tubes in waste to energy boilers. *Proceedings of the 17the Annual North American Waste-to-Energy Conference, NAWTEC17*. 10.1115/NAWTEC17-2334

Durul, E. N. A., & Nurbas, M. (2010). Corrosion Behavior of Different Thermal Spray Coatings and Hard Chromium Electroplating on A286 Super Alloy. *Advanced Materials Research, 154-155*, 226–229. doi:10.4028/www.scientific.net/AMR.154-155.226

Edris, H., Mc-Cartney, D. G., & Sturgeon, A. J. (1997). Micro-structural Characterization of HVOF Sprayed Coatings of Inconel 625. *Journal of Materials, 32*, 863–868.

Eklund, J., Phother, J., Sadeghi, E., Joshi, S., & Liske, J. (2019). High-Temperature Corrosion of HVAF-Sprayed Ni-Based Coatings for Boiler Applications. *Oxidation of Metals, 91*(5), 729–747. doi:10.100711085-019-09906-0

Eklund, J., Sadeghimeresht, E., Simon, J. P., Liske, J., Markocsan, N., & Joshi, S. (2018). Effect of Water Vapor on the Oxidation Behavior of HVAF-Sprayed NiCr and NiCrAlY Coatings. *Mater. Corros.*, *69*(10), 1431–1440. doi:10.1002/maco.201710005

Eliaz, N., Shemesh, G., & Latanision, R. M. (2002). Hot Corrosion in Gas Turbine Components. *Engineering Failure Analysis*, *9*(1), 31–43. doi:10.1016/S1350-6307(00)00035-2

Fantozzi, D., Matikainen, V., Uusitalo, M., Koivuluoto, H., & Vuoristo, P. (2017). Chlorine induced high temperature corrosion of Inconel 625 sprayed coatings deposited with different thermal spray techniques. *Surface and Coatings Technology*, *318*, 233–243. doi:10.1016/j.surfcoat.2016.12.086

Fauchais, P. (2004). Understanding Plasma Spraying. *Journal of Physics. D, Applied Physics*, *37*(9), 86–108. doi:10.1088/0022-3727/37/9/R02

Fauchais, P., Coudert, J. F., & Vardelle, M. (1997). Transient Phenomena in Plasma Torches and in Plasma Sprayed Coating Generation. *Journal de Physique. IV*, *7*(C4), 187–198. doi:10.1051/jp4:1997414

Fauchais, P. L., Heberlein, J. V. R., & Boulos, M. I. (2014). Industrial Applications of Thermal Spraying Technology. In *Thermal Spray Fundamentals* (pp. 1401–1566). Springer. doi:10.1007/978-0-387-68991-3_18

Fukuda, M. (2017). 22 - Advanced USC technology development in Japan. *Materials for Ultra-Supercritical and Advanced Ultra-Supercritical Power Plants*, 733-754.

Galedari, S. A., Mahdavi, A., Azarmi, F., Huang, Y., & McDonald, A. (2019). A comprehensive review of corrosion resistance of thermally-sprayed and thermally-diffused protective coatings on steel structures. *J. Therm. Spray Tech.*, *28*(4), 645–677. doi:10.100711666-019-00855-3

Gil, L., & Staia, M. H. (1999). Microstructure and properties of HVOF thermal sprayed NiWCrBSi coatings. *Surface and Coatings Technology*, *120–121*, 423–429. doi:10.1016/S0257-8972(99)00494-6

Gil, L., & Staia, M. H. (2002). Influence of HVOF parameters on the corrosion resistance of NiWCrBSi coatings. *Thin Solid Films*, *420–421*, 446–454. doi:10.1016/S0040-6090(02)00815-5

Grewal, H. S., Singh, H., & Agrawal, A. (2013). Microstructural and mechanical characterization of thermal sprayed nickel–alumina composite coatings. *Surface and Coatings Technology*, *216*, 78–92. doi:10.1016/j.surfcoat.2012.11.029

Guilemany, J. M., Fernández, J., Delgado, J., Benedetti, A. V., & Climent, F. (2002). Effects of Thickness Coating on the Electrochemical Behaviour of Thermal Spray Cr_3C_2-NiCr Coatings. *Surface and Coatings Technology*, *153*(2-3), 107–113. doi:10.1016/S0257-8972(01)01679-6

Harsha, S., Dwivedi, D., & Agarwal, A. (2008). Influence of CrC addition in NiCrSiB flame sprayed coatings on microstructure, microhardness and wear behaviour. *International Journal of Advanced Manufacturing Technology*, *38*(1-2), 93–101. doi:10.100700170-007-1072-2

He, L., Tan, Y., Wang, X., Xu, T., & Hong, X. (2014). Microstructure and wear properties of Al_2O_3–CeO_2/Ni-base alloy composite coatings on aluminum alloys by plasma spray. *Applied Surface Science*, *314*, 760–767. doi:10.1016/j.apsusc.2014.07.047

Hussain, T., Dudziak, T., Simms, N. J., & Nicholls, J. R. (2013). Fireside Corrosion Behavior of HVOF and Plasma-Sprayed Coatings in Advanced Coal/Biomass Co-fired Power Plants. *J. Therm. Spray Technol., 22*(5), 797–807. doi:10.100711666-013-9887-x

Hwang, Y. S., & Rapp, R. A. (1989). Thermo-Chemistry and Solubility of Oxides in Sodium Sulfate-Vanadate Solutions. *Corrosion, 45*(11), 933–937. doi:10.5006/1.3585003

Kamal, S., Sharma, K. V., Pedapati, S. R., & Mamat, O. (2017). Thermal spray coatings for hot corrosion resistance, engineering applications of nanotechnology, Topics in Mining. *Metallurgy and Materials Engineering*, 235–268.

Karthikeyan, J., Berndt, C. C., Tikkanen, J., Reddy, S., & Herman, H. (1997). Plasma spray synthesis of nanomaterial powders and deposits. *Materials Science and Engineering A, 238*(2), 275–286. doi:10.1016/S0921-5093(96)10568-2

Kaur, M., Singh, H., & Prakash, S. (2008). A survey of the literature on the use of high velocity oxy-fuel spray technology for high temperature corrosion and erosion-corrosion resistant coatings. *Anti-Corrosion Methods and Materials, 55*(2), 86–96. doi:10.1108/00035590810859467

Kaur, M., Singh, H., & Prakash, S. (2009). High-temperature corrosion studies of HVOF-sprayed Cr_3C_2-NiCr coating on SAE-347H boiler steel. *Thermal Spray Technology, 18*(4), 619–632. doi:10.100711666-009-9371-9

Kaur, M., Singh, H., & Prakash, S. (2012). High-temperature behavior of a high-velocity oxy-fuel sprayed Cr_3C_2-NiCr coating. *Metallurgical and Materials Transactions. A, Physical Metallurgy and Materials Science, 43*(8), 2979–2993. doi:10.100711661-012-1118-4

Kaushal, G., Singh, H., & Prakash, S. (2011a). Comparative high temperature analysis of HVOF sprayed and detonation gun sprayed Ni–20Cr coating in laboratory and actual boiler environment. *Oxidation of Metals, 76*(3-4), 169–191. doi:10.100711085-011-9245-7

Kaushal, G., Singh, H., & Prakash, S. (2011b). High temperature corrosion behaviour of HVOF-sprayed Ni-20Cr coating on boiler steel in molten salt environment at 900°C. *International Journal of Surface Science and Engineering, 5*(5/6), 415–433. doi:10.1504/IJSURFSE.2011.044388

Kawahra, Y. (2016). An Overview on Corrosion-Resistant Coating Technologies in Biomass/Waste-to-Energy Plants in Recent Decades. *Coatings, 6*(3), 1–24.

Khanna, A. S., & Jha, K. S. (1998). Degradation of Materials under Hot Corrosion Conditions. *Transactions of the Indian Institute of Metals, 51*(5), 279–290.

Kim, H. J., Hwang, S. Y., Lee, C. H., & Juvanon, P. (2003). Assessment of wear performance of flame sprayed and fused Ni-based coatings. *Surface and Coatings Technology, 172*(2-3), 262–269. doi:10.1016/S0257-8972(03)00348-7

Knuuttila, J., Ahmaniemi, S., & Mantyla, T. (1999). Wet abrasion and slurry erosion resistance of thermally sprayed oxide coatings. *Wear, 232*(2), 207–212. doi:10.1016/S0043-1648(99)00147-7

Koivuluoto, H., & Vuoristo, P. (2009). Effect of ceramic particles on properties of cold sprayed Ni–20Cr + Al_2O_3 coatings. *J. Therm. Spray Technol., 18*(4), 555–562. doi:10.100711666-009-9345-y

Konyashin, I. Y., & Chukalovskaya, T. V. (1997). A technique for measurement of porosity in protective coatings. *Surface and Coatings Technology*, *88*(1-3), 5–11. doi:10.1016/S0257-8972(95)02758-0

Koropchak, J. A., & Roychowdhury, S. B. (1990). Evidence for aerosol ionic redistribution within aerosols produced by chrome electroplating. *Environmental Science & Technology*, *24*(12), 1861–1863. doi:10.1021/es00082a012

Kumar, S., Handa, A., & Kumar, M. (2018). Combating hot corrosion of boiler tubes- A study. *Engineering Failure Analysis*, *94*, 379–395. doi:10.1016/j.engfailanal.2018.08.004

Li, W. Y., Liao, H., Li, C. J., Bang, H. S., & Coddet, C. (2007). Numerical simulation of deformation behavior of Al particles impacting on Al substrate and effect of surface oxide films on interfacial bonding in cold spraying. *Applied Surface Science*, *253*(11), 5084–5091. doi:10.1016/j.apsusc.2006.11.020

Liu, S. L., & Zheng, X. P. (2009). Microstructure and Properties of AC-HVAF Sprayed Ni60/WC Composite Coating. *Journal of Alloys and Compounds*, *480*(2), 254–258. doi:10.1016/j.jallcom.2009.02.028

Lugscheider, E., Herbst, C., & Znao, D. (1998). Parameter Studies on High- Velocity Oxy-Fuel Spraying of MCrAlY Coatings. *Surface and Coatings Technology*, *108-109*, 16–23. doi:10.1016/S0257-8972(98)00630-6

Lyphout, C., Bjorklund, S., Karlsson, M., Runte, M., Reisel, G., & Boccaccio, P. (2014). Screening Design of Supersonic Air Fuel Processing for Hard Metal Coatings. *J. Therm. Spray Technol.*, *23*(8), 1323–1332. doi:10.100711666-014-0139-5

Maatta, A., Kanerva, U., & Vuoristo, P. (2011). Structure and tribological characteristics of HVOF coatings sprayed from powder blends of Cr_3C_2–25NiCr and NiCrBSi alloy. *J. Therm.Spray Technol.*, *20*(1-2), 366–371. doi:10.100711666-010-9579-8

Martin, A., Rodriguez, J., Fernandez, J. E., & Vijande, R. (2001). Sliding wear behaviour of plasma sprayed WC–NiCrBSi coatings at different temperatures. *Wear*, *251*(1-12), 1017–1022.

Meier, G. H. (1989). A review of advances in high-temperature corrosion. *Materials Science and Engineering A*, *120-121*(1), 1–11. doi:10.1016/0921-5093(89)90712-0

Milanti, A., Matikainen, V., Koivuluoto, H., Bolelli, G., Lusvarghi, L., & Vuoristo, P. (2015). Effect of Spraying Parameters on the Microstructural and Corrosion Properties of HVAF-Sprayed Fe-Cr-Ni-B-C Coatings. *Surface and Coatings Technology*, *277*, 81–90. doi:10.1016/j.surfcoat.2015.07.018

Monticelli, C., Frignani, A., & Zucchi, F. (2004). Investigation on the Corrosion Process of Carbon Steel Coated by HVOF WC/Co Cermets in Neutral Solution. *Corrosion Science*, *46*(5), 1225–1237. doi:10.1016/j.corsci.2003.09.013

Moridi, A., Hassani-Gangaraj, S. M., Guagliano, M., & Dao, M. (2014). Cold spray coating: Review of material systems and future perspectives. *Surface Engineering*, *30*(6), 369–395. doi:10.1179/17432 94414Y.0000000270

Natesan, K. (1976). Corrosion-Erosion behavior of materials in a coal-gasification environment. *Corrosion*, *32*(9), 364–370. doi:10.5006/0010-9312-32.9.364

Nemani, S. K., Annavarapu, R. K., Mohammadian, B., Raiyan, A., Heil, J., Haque, M. A., … Sojoudi, H. (2018). Surface Modification of Polymers: Methods and Applications. Adv. Mater. Interfaces, 1-26.

Odhiambo, J. G., Lei, W. G., Zhao, Y. T., & Lei, C. L. (2019). Porosity and its significance in plasma sprayed coatings. *Coatings*, *9*(460), 1–24. doi:10.3390/coatings9070460

Ohkubo, N., Miyakusu, K., Uematsu, Y., & Kimura, H. (1994). Effect of alloying elements on the mechanical properties of the stable austenitic stainless steel. *ISIJ International*, *34*(9), 764–772. doi:10.2355/isijinternational.34.764

Oksa, M., Auerkari, P., Salonen, J., & Varis, T. (2014). Nickel-based HVOF coatings promoting high temperature corrosion resistance of biomass-fired power plant boilers. *Fuel Processing Technology*, *125*, 236–245. doi:10.1016/j.fuproc.2014.04.006

Oksa, M., Tuurna, S., & Varis, T. (2013). Increased lifetime for biomass and waste to energy power plant boilers with HVOF coatings: High temperature corrosion testing under chlorine-containing molten salt. *J. Therm Spray Technol.*, *22*(5), 783–796. doi:10.100711666-013-9928-5

Padture, N. P., Gell, N., & Jordan, E. H. (2002). Thermal barrier coatings for gas-turbine engine applications. *Science*, *296*(5566), 280–284. doi:10.1126cience.1068609 PMID:11951028

Pawlowski, L. (2008). *The Science and Engineering of Thermal Spray Coatings*. John Wiley & Sons ltd. doi:10.1002/9780470754085

Pettit, F. (2011). Hot corrosion of metals and alloys. *Oxidation of Metals*, *76*(1-2), 1–21. doi:10.100711085-011-9254-6

Prashar, G., & Vasudev, H. (2020, February). *Hot corrosion behavior of super alloys*. Paper presented at the International conference on material processing and charcterziation. GLA University, Mathura.

Radka, J., Ctibor, P., & Benes, V. (2004). Analysis of porous structure in plasma-sprayed coating. *Image Analysis & Stereology*, *23*, 45–52.

Ramesh, M. R., Prakash, S., Nath, S. K., Sapra, P. K., & Venkataraman, B. (2010). Solid particle erosion of HVOF sprayed WC–Co/NiCrFeSiB coatings. *Wear*, *269*(3-4), 197–205. doi:10.1016/j.wear.2010.03.019

Rana, R. S., Purohit, R., & Das, S. (2012). Reviews on the influences of alloying elements on the microstructure and mechanical properties of aluminum alloys and aluminum alloy composites. *Int. J. Scientific Res. Publ.*, *2*(6), 1–7.

Rapp, R. A. (1986). Chemistry and electrochemistry of the hot corrosion of metals. *Corrosion*, *42*(10), 568–577. doi:10.5006/1.3583026

Richer, P., Yandouzi, M., Beauvais, L., & Jodoin, B. (2010). Oxidation behaviour of CoNiCrAlY bond coats produced by plasma, HVOF and cold gas dynamic spraying. *Surface and Coatings Technology*, *204*(24), 3962–3974. doi:10.1016/j.surfcoat.2010.03.043

Rodriguez, M., Staia, M., Gil, L., Arenas, F., & Scagni, A. (2000). Effect of Heat Treatment on Properties of Nickel Hard Surface Alloy Deposited by HVOF. *Surface Engineering*, *16*(5), 415–420. doi:10.1179/026708400101517404

Saaedi, J., Coyle, T. W., Arabi, H., Mirdamadi, S., & Mostaghimi, J. (2010). Effects of HVOF Process Parameters on the Properties of Ni-Cr Coatings. *J. Therm. Spray Technol.*, *19*(3), 521–530. doi:10.100711666-009-9464-5

Sadeghi, E., Markocsan, N., & Joshi, S. (2019). Advances in Corrosion-Resistant Thermal Spray Coatings for Renewable Energy Power Plants. Part I: Effect of Composition and Microstructure. *J Therm Spray Tech.*, *28*(8), 1749–1788. doi:10.100711666-019-00938-1

Sadeghimeresht, E. (2018). *Ni-Based Coatings for High Temperature Corrosion Protection.* (Doctoral dissertation). University West, Department of Engineering Science, Division of Subtractive and Additive Manufacturing.

Sadeghimeresht, E., Hooshyar, H., Markocsan, N., Joshi, S., & Nylen, P. (2016a). Oxidation Behavior of HVAF-Sprayed NiCoCrAlY Coating in H_2-H_2O Environment. *Oxidation of Metals*, *86*(3-4), 299–314. doi:10.100711085-016-9637-9

Sadeghimeresht, E., Markocsan, N., & Nylen, P. (2016b). A Comparative Study on Ni-Based Coatings Prepared by HVAF, HVOF, and APS Methods for Corrosion Protection Applications. *J. Therm.Spray Technol.*, *25*(8), 1604–1616. doi:10.100711666-016-0474-9

Sadeghimeresht, E., Markocsan, N., Nylen, P., & Bjorklund, S. (2016c). Corrosion Performance of Bi-Layer Ni/Cr_2C_3-NiCr HVAF Thermal Spray Coating. *Applied Surface Science*, *369*, 470–481. doi:10.1016/j.apsusc.2016.02.002

Sadeghimeresht, E., Reddy, L., Hussain, T., Huhtakangas, M., Markocsan, N., & Joshi, S. (2018a). Influence of KCl and HCl on HighTemperature Corrosion of HVAF-Sprayed NiCrAlY and NiCrMo Coatings. *Materials & Design*, *148*, 17–29. doi:10.1016/j.matdes.2018.03.048

Sadeghimeresht, E., Reddy, L., Hussain, T., Markocsan, N., & Joshi, S. (2018b). Chlorine-Induced High Temperature Corrosion of HVAF Sprayed Ni-Based Alumina and Chromia Forming Coatings. *Corrosion Science*, *132*, 170–184. doi:10.1016/j.corsci.2017.12.033

Sadeghimeresht, E., Zhang, P., Chen, S., Li, X. H., Markocsan, N., & Joshi, S. (2019). Effects of Surface Finish on the Initial Oxidation of HVAF-Sprayed NiCoCrAlY Coatings. *Surface and Coatings Technology*, *364*, 43–56. doi:10.1016/j.surfcoat.2019.02.068

Sevillano, F., Poza, P., Munez, C. J., Vezzu, S., Rech, S., & Trentin, A. (2013). Cold-Sprayed Ni-Al_2O_3 Coatings for Applications in Power Generation Industry. *Journal of Thermal Spray Technology, 22,* 772-782.

Shanmugam, K., Thirumalaikumarasamy, D., & Balasubramanian, V. (2014). Establishing empirical relationships to predict porosity level and corrosion rate of atmospheric plasma-sprayed alumina coatings on AZ31B magnesium alloy. *J. Magnes. Alloy*, *2*(2), 140–153. doi:10.1016/j.jma.2014.05.002

Shih, S., Zhang, Y., & Li, X. (1989). Sub-melting point hot corrosion of alloys and coatings. *Journal of Materials Science and Engineering: A, 120-121*(1), 277–282.

Shourgeshty, M., Aliofkhazraei, M., & Alipour, M. M. (2016). *Introduction to High-Temperature Coatings, High Temperature Corrosion.* IntechOpen. Available from: https://www.intechopen.com/books/high-temperature-corrosion/introduction-to-high-temperature-coatings

Sidky, P. S., & Hocking, M. G. (1999). Review of inorganic coatings and coating processes for reducing wear and corrosion. *British Corrosion Journal, 34*(2), 171–183. doi:10.1179/000705999101500815

Singh, H., & Prakash, S. (2009). *Plasma Spray Coatings for Superalloys: Characterization and High-Temperature Oxidation Behavior.* VDM Verlag.

Singh, H., Puri, D., & Prakash, S. (2005). High Temperature Oxidation Behaviour of Plasma Sprayed NiCrAlY Coatings on Ni-Based Superalloys in Air. *Transactions of the Indian Institute of Metals, 59,* 215–227.

Singh, H., Sidhu, B. S., Puri, D., & Prakash, S. (2007). Use of plasma spray technology for deposition of high temperature oxidation/corrosion resistant coatings – a review. *Materials and Corrosion, 58*(2), 92–102. doi:10.1002/maco.200603985

Souza, A. D., & Neville, A. (2003). Linking Electrochemical Corrosion Behaviour and Corrosion Mechanisms of Thermal Spray Cermet Coatings (WC-CrNi and WC/CrC-CoCr). *Materials Science and Engineering A, 352*(1-2), 202–211. doi:10.1016/S0921-5093(02)00888-2

Stringer, J. (1987). High temperature corrosion of super alloys. *Materials Science and Technology, 3*(7), 482–493. doi:10.1080/02670836.1987.11782259

Tao, Y., Xiong, T., Sun, C., Kong, L., Cui, X., Li, T., & Song, G. L. (2010). Microstructure and Corrosion Performance of a Cold Sprayed Aluminium Coating on AZ91D Magnesium Alloy. *Corrosion Science, 52*(10), 3191–3197. doi:10.1016/j.corsci.2010.05.023

Thorpe, M., & Richter, H. (1992). A pragmatic analysis and comparison of HVOF processes. J. *Therm. Spray Technol., 1*(2), 161–170. doi:10.1007/BF02659017

Tobar, M. J., Alvarez, C., Amado, J. M., Rodriguez, G., & Yanez, A. (2006). Morphology and characterization of laser clad composite NiCrBSi–WC coatings on stainless steel. *Surface and Coatings Technology, 200*(22-23), 6313–6317. doi:10.1016/j.surfcoat.2005.11.093

Tsai, W. T., Chen, H. P., & Hsien, W. Y. (2002). A review of uses, environmental hazards and recovery/recycle technologies of perfluorocarbons (PFCs) emissions from the semiconductor manufacturing processes. *Journal of Loss Prevention in the Process Industries, 15*(2), 65–75. doi:10.1016/S0950-4230(01)00067-5

Tsui, Y., Doyle, C., & Clyne, T. W. (1998). Plasma sprayed hydroxyapatite coatings on titanium substrates Part 1: Mechanical properties and residual stress levels. *Biomaterials, 199*(22), 2015–2029. doi:10.1016/S0142-9612(98)00103-3 PMID:9870753

Tu, T. P., Liu, M. S., & Mao, Z. Y. (1997). Erosion resistance of Ni–WC self-fluxing alloy coating at high temperature. *Wear, 209*(1-2), 43–48. doi:10.1016/S0043-1648(96)07457-1

Uusitalo, M. A., Vuoristo, P. M. J., & Mantyla, T. A. (2002). High Temperature Corrosion of Coatings and Boiler Steels in Reducing Chlorine-Containing Atmosphere. *Surface and Coatings Technology*, *161*(2-3), 275–285. doi:10.1016/S0257-8972(02)00472-3

Vardelle, A., Moreau, C., & Vuoristo, P. (2016). The 2016 Thermal Spray Roadmap. *J. Therm.Spray Technol.*, *25*(8), 1376–1440. doi:10.100711666-016-0473-x

Vasudev, H., Thakur, L., Bansal, A., Singh, H., & Zafar, S. (2019). High temperature oxidation and erosion behaviour of HVOF sprayed Inconel-718 coating. *Surface and Coatings Technology*, *362*, 366–380. doi:10.1016/j.surfcoat.2019.02.012

Verstak, A., & Baranovski, V. (2003). *Activated Combustion HVAF Coatings for Protection Against Wear and High Temperature Corrosion.* Paper presented at the International Thermal Spray Conference and Exposition, Orlando, FL.

Wang, H., Xia, W., & Jin, Y. (1996). A study on abrasive resistance of Ni-based coatings with a WC hard phase. *Wear*, *195*(1-2), 47–52. doi:10.1016/0043-1648(95)06762-0

Wang, Y., Normand, B., Mary, N., Yu, M., & Liao, H. (2014). Microstructure and Corrosion Behavior of Cold Sprayed SiC_p/Al 5056 Composite Coatings. *Surface and Coatings Technology*, *251*, 264–275. doi:10.1016/j.surfcoat.2014.04.036

Xu, C., Du, L., Yang, B., & Zhang, W. (2011). Study on Salt Spray Corrosion of Ni-Graphite Abradable Coating with 80Ni20Al and 96NiCr-4Al as Bonding Layers. *Surface and Coatings Technology*, *205*(17-18), 4154–4161. doi:10.1016/j.surfcoat.2011.03.007

Yin, B., Zhou, H. D., Chen, J. M., & Yan, F. Y. (2011). Microstructure and high temperature sliding wear behavior of laser remelted NiCrBSi/35(WC–12Ni) composite coating. *Surface Engineering*, *27*(6), 458–463. doi:10.1179/026708410X12687356948517

Young, M. A. (1965). Health hazards of electroplating. *Journal of Occupational and Environmental Medicine*, *7*(7), 348–352. doi:10.1097/00043764-196507000-00008 PMID:14321464

Zhang, C., Chan, K. C., Wu, Y., & Liu, L. (2012). Pitting Initiation in Fe- Based Amorphous Coatings. *Acta Materialia*, *60*(10), 4152–4159. doi:10.1016/j.actamat.2012.04.005

Zhang, S. D., Wu, J., Qi, W. B., & Wang, J. Q. (2016). Effect of Porosity Defects on the Long-Term Corrosion Behaviour of Fe-Based Amorphous Alloy Coated Mild Steel. *Corrosion Science*, *110*, 57–70. doi:10.1016/j.corsci.2016.04.021

Zhou, J., Walleser, J. K., Meacham, B. E., & Branagan, D. J. (2010). Novel in Situ transformable coating for elevated-temperature applications. *J. Therm. Spray Technol.*, *19*(5), 950–957. doi:10.100711666-010-9498-8

Zhou, Z., Wang, L., Wang, F., & Liu, Y. (2009). Formation and Corrosion Behavior of Fe-Based Amorphous Metallic Coatings Prepared by Detonation Gun Spraying. *Transactions of Nonferrous Metals Society of China*, *19*(3), 634–638. doi:10.1016/S1003-6326(10)60123-9

Chapter 6
A Review of the Cold Gas Dynamic Spraying Process

Sudesna Roy
https://orcid.org/0000-0003-3129-0891
KIIT University, India

Subhrasmita Tripathy
https://orcid.org/0000-0002-6633-6880
KIIT University, India

ABSTRACT

In this modern era, use of coatings on engineering materials has become highly inevitable. One such emerging coating method is the cold gas dynamic spraying. It is a solid-state process where deposition on to the surface of the material is done at high pressure and velocity. Adhesion of the powder to the substrate is possible because of the high amount of plastic deformation. This chapter introduces the CGDS system and discusses the types of set-ups and its modifications that are generally used. Further, the chapter delves into the process parameters in the spraying process and the correlation of these parameters with the coating properties. It also provides a comprehensive review of the current theories of bonding mechanism in cold spray. It aims to provide an overview of the material systems that have been investigated so far for cold spraying with an outline of the experimental and numerical simulation that have been researched.

1. INTRODUCTION

A deposition process is used to deposit a material (referred to as coating) on to a bulk material (referred to as substrate). Depending on the desired coating properties different methods may be used for the application of the coating such as physical vapor deposition (PVD), chemical vapor deposition (CVD), sputtering, laser cladding, electroplating, thermal spraying and cold gas dynamic spraying (Steyer2001), (Kalita2014), (Anisur2018) and (Zhou2019). The process may deposit the coating material atom by atom, as in the vapor deposition process or as molecule and ions as in the laser cladding and electroplating or

DOI: 10.4018/978-1-7998-4870-7.ch006

as aggregate powders as in thermal and cold gas dynamic spraying. It is the aim of this chapter to focus on the cold gas dynamic spraying method. Cold gas dynamic spraying (CGDS) or simply known as cold spraying, is an emerging technology where powders or particles are carried with a gas and impacted on the substrate at high velocities. CGDS was invented in 1980's at the Institute of Theoretical and Applied Mechanics of Siberian Branch Russian Academy of Science (Papyrin2007) and (Dykhuzein1998). CGDS has many advantages as to other deposition methods that include high deposition efficiency, low residual stress, less heat input, elimination/minimization of porosity, control of phase transformation and compositional stability (Papyrin2007) and (Koivluoto2010). Since the deposition temperatures are at room temperatures and much below the melting point, the particles cannot be melted in the gas flow. Being a solid state process, oxide inclusion and void formation can be avoided in cold spraying (Sobolov, 2004) and (Stoltenhoff,2001). Further it has also been reported that wide range of coating materials, e.g. pure metals like Al, Cu, Ti, Fe, Ag, Zn and Ni; metal alloys like Cu-Al, Cu-Zn, Ni-Cu, Ni-Al, CoNiCrAIY and stainless steel and composites like Cu+Al2O3, Ni+TiC, Al- Al_2O_3 have also been used in the cold spray system (Champagne,2007), (Maev, 2006) and (Dykhuzein,1998).

Figure 1. Schematic diagram of cold gas dynamic spraying

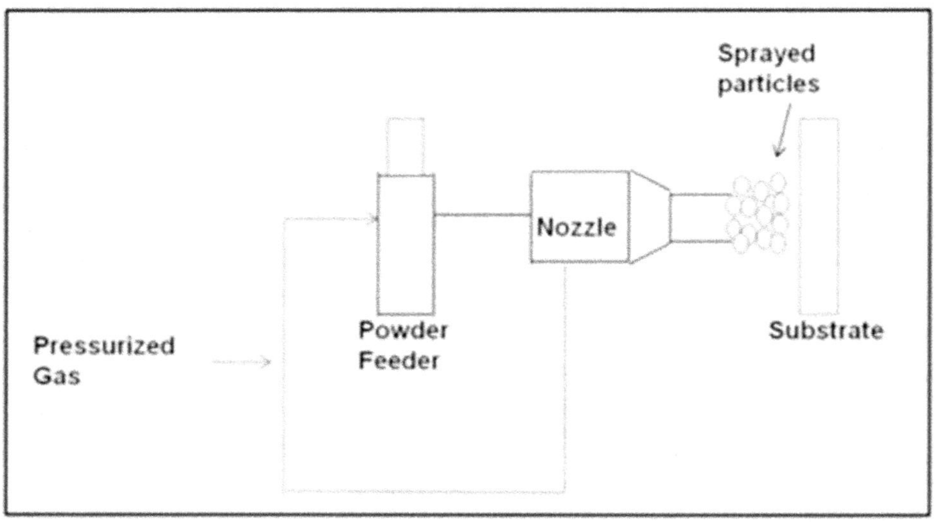

1.1 Principle of Cold Spraying

Cold spray being a solid state process is used to form a deposit by the means of ballistic impingement (Koivluoto2001). In this process powder particle strikes the substrate with a high velocity and leads to formation of plastic deformation or erosion of the substrate. A gas (nitrogen, helium or air) is accelerated to supersonic velocity by a converging diverging de Laval type nozzle. On impinging to the substrate the particles either adhere on to the substrate or bounce back (rebound). This phenomena of the particles is due to the drag effect of the gas as the particles flow in it (Maev2006). There is no sign of adhesion of the particles until a critical particle velocity is exceeded. The temperature of the powder particle (also same as the gas temperature) plays an important role during the impact. The particle size of the metals range from 5 to 100 μm. These particles are accelerated by high velocity stream of gas generated through expansion of pressurized, preheated gas through nozzle. The solid particles impact on the substrate above the critical velocity for the powder and substrate will deform and form a dense layer (Papyrin2007), (Dykhuzein1998) and (Stoltenhoff2001). The bonding mechanism depends on many factors and has been discussed in detail in section 3.1.

1.2 Types of Cold Spray

Cold spray is basically catagorised as high pressure cold spray (HPCS) and low pressure cold spray (LPCS). The LPCS uses gas pressure between 5-10 bars and gas temperatures are usually around 550° C. A schematic of the LPCS is shown in Figure 2. The system has a design improvement where the pressurized gas is passed through a heater before it interacts with the powder stream. In this type the gas contained in the reservoir flows through a pipeline to the heating system where it is heated and brought to the low pressure chamber. After heating the gas in spray gun it is forced through the convergent-divergent nozzle. The velocity of the particles reach between 300-600 m/s at exit of the nozzle. As the LPCS system is light weight, compact and portable it is used in industrial platforms without any difficulties. The life span of the nozzle is more because during spraying the particles do not flow through the throat of the nozzle. Moreover, LPCS system is more compatible and cheaper as compared to the other CGDS systems (Singh, 2013).

On the other hand, in the HPCS is bulky and heavy weight which can generally be used for hard materials that require pressures greater than 10 bar. A schematic of the general arrangement is shown in Figure 3. Here, generally helium or nitrogen is used as the carrier gas. The working temperature here can reach up to 1000°C and the gas pressure varies between 25-40 bar. In this process powder feed stock particles mix with the propellant gas in the pre-chamber zone and then fed into the gas stream axially through the upstream of the converging section of the nozzle at high pressure to prevent the backflow of the carrier gas to the powder feeder. The velocity of the particles reaches between 600-1200 m/s at the exit of nozzle (Powlowski, 2008). In the high pressure system a high pressure powder feeder is used for this high pressure in the main gas stream is required. But the challenge here is that the high pressure powder feeders are usually very big and expensive. In the HPCS system when hard particles are sprayed, often the throat of nozzle gets worn out due to particle erosion. Due to this, the function of the nozzle is affected, resulting in poor deposition quality. Another challenge in the HPCS system is clogging of nozzle occurs due to the increase in the temperature and particle velocity. On the other hand the LPCS type has a much simpler equipment (Figure 2). However, the inlet pressure is restricted (< 1MPa) to allow the atmospheric pressure to supply powder into the nozzle. Due to this powder feeding technique, the particle velocity is restricted in the LPCS type.

A number of investigators have experimentally investigated the CGDS process and consequently different modifications of the system have been explored. Vacuum cold spray is one such method where low atmospheric pressure is used to keep the specimen in a vacuum tank. A vacuum pump is combined with the vacuum tank in low pressure environment. As a result recovery of gas and collection of over sprayed powder is done via the vacuum tank (Muehlbrger, 2007). A similar design was used by Akedo et.al. (2008) named as aerosol deposition method. In this process a propellant gas flow of helium or air is used to spray nano particles in a vacuum chamber. This process is used to deposit very small particles because it reduces the presence of shock wave at the substrate. The propellant gas used is also below the atmospheric pressure. Yet another variation to the CGDS set up is called the pulsed gas dynamic spraying. In this process, particles are heated below the melting temperature, which is albeit higher than the temperature experienced in the CGDS process. In this process critical velocity is reduced with an subsequent increase in the gas temperature. This leads to higher levels of plastic deformation. Discontinuous nature of this process exploits non-stationary pressure waves to generate pressure & temperature simultaneously (Jodion, 2007).

Figure 2. Schematic diagram of LPCS

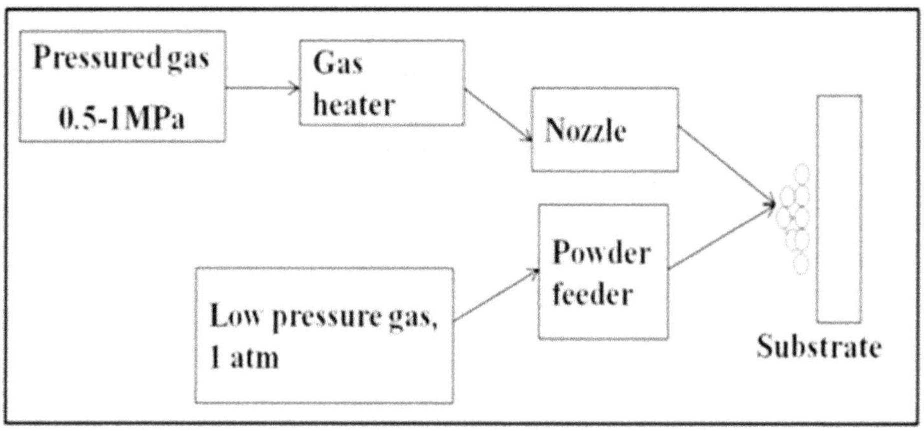

Figure 3. Schematic diagram of HPCS

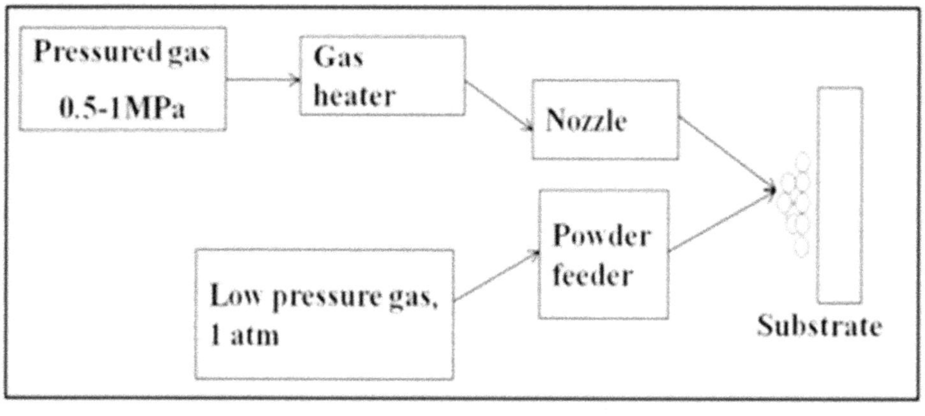

2. EXPERIMENTAL PARAMETERS FOR COLD SPRAY

The different process parameters involved in this method are particle velocity, critical particle velocity, characteristics of propellant gas, spraying angle, substrate to nozzle (stand-off distance) and most importantly the nozzle design.

2.1 Gas Parameters

In this deposition process air, nitrogen or helium are generally used as the major propellant gases. Though nitrogen gas is used widely for deposition, for higher velocity particles helium is used as it is lighter than nitrogen. This is because hard materials cannot be deposited without achieving higher velocity. Li et al. reported that particles accelerated to higher velocity in helium as compared to nitrogen (Singh2012). However, helium is costly than nitrogen making it economically unfavorable for many application unless it is recycled (Yekane). Therefore the most important gas characteristics that affects the deposition are the gas density, pressure and temperature. The lower the density of the gas, the drag effect on the particles is low. Therefore where hydrogen has a density of 0.09 g/l, helium has a density of 0.179 g/L, nitrogen has a density of 1.25 g/L, argon has a density of 1.78 g/L and air has a density of 1.205 g/L at STP. This implies that particles will move faster (higher particle velocity) in helium as compared to air (Singh 2012).

2.2 Spraying Nozzle

The spraying nozzle plays an important role during deposition by cold spraying specifically for HPCS. It has a high influence on particle velocity during spraying. A typical design of the de Laval nozzle is shown in Figure 4. The de Laval nozzle is a converging diverging nozzle that has the capability to achieve supersonic velocities at the outlet. The principle of the de Laval nozzle utilizes the thermal and pressure energy to kinetic energy. The geometry of the nozzle is such that as the gas is constricted to pass through the throat region, the velocity of the gas increases rapid. The length of the converging and diverging sections are denoted as L_{con} and L_{div}, respectively. The diameter of the constriction (or throat) is denoted as d_t, whereas the diameter of the inlet and outlet are denoted as d_i and d_e, respectively. The gas velocity depends on the ratio of the inlet and outlet diameter of the converging-diverging section. The throat diameter of the nozzle, the exit diameter or the inlet diameter influences the particle velocity. As the nozzle length has material and fabrication constraints, it limits its application in different sectors. Therefore new materials and design are optimized to improve the powder flow through the nozzle and optimize the design required to minimize the gas flow through the nozzle (W-Y Li2005). Li et al. (2005) had introduced a special internal diameter spray gun with 6.25 expansion ratio and 30 mm standoff distance for divergent nozzle of 40 mm length. Helium and nitrogen were used as propellant gas resulting in dense deposition of coating after cold spray (C-J Li2005). Similarly, Karthikeyan et al. (2007) designed a nozzle using tungsten carbide to coat GRCop-84 special alloy. In order to deposit aluminium powder, thermo plastic nozzle was designed by Champagne et al. which corrected the clogging effect of the steel nozzle (Champagne et al,2008) and (DeForce,2007).

Figure 4. Schematic diagram of the de Laval nozzle used in CGDS

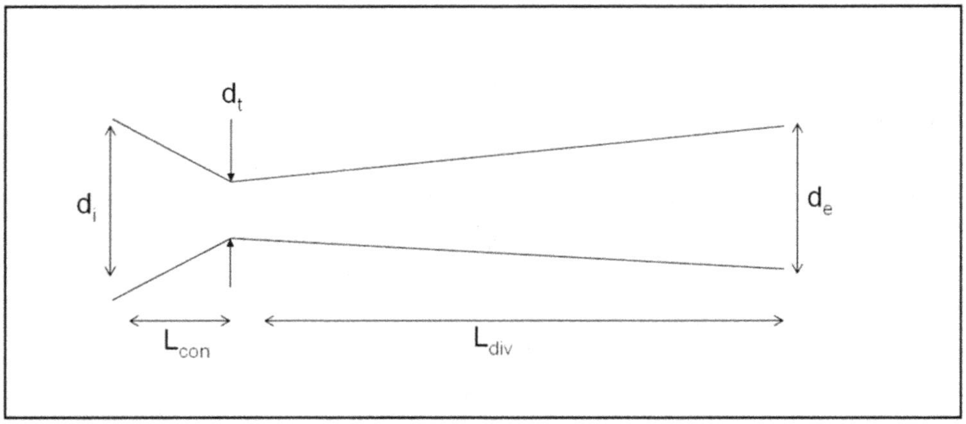

2.3 Standoff Distance

Stand off distance is the distance between the nozzle and the substrate. Typically a particle ejected from the nozzle accelerates some distance away from the exit and attains a maximum velocity after travelling some 50 to 100 mm towards the substrate. This is then the ideal standoff distance where the particle velocity is at its maximum. Ideally the particles at the center are considered, whereas the particles at the periphery diverge and rebound off. When the nozzle geometry is fixed (inlet-outlet diameter ratio and length of the converging-diverging chamber), the standoff distance substantially affect the particle velocity as well as the deposition efficiency. Li et al. (2008) have shown that an optimum standoff distance exists, where the deposition efficiency is the maximum (W-Y Li2008). The optimum standoff distance usually increases with increasing particle size. Moreover, their experiments have also conclusively shown that the optimum standoff distance (and particle velocity) is shorter for lighter materials (low density) than that for denser materials (higher density). At shorter stand-off distances, Pattison et al have shown that the occurrence of bow shock waves that impinge on the substrate to reduce the particle velocity as well as alter the coating characteristics (Pattison2008). When the optimum standoff distance is maintained, the deposition efficiency is at its maximum.

2.4 Powder Parameters

According to researchers powder characteristics such as powder particle size, particle morphology and purity level affect the deposition quality (Kairet2007). The particle size is typically in the range 1 to 100 μm. In order to achieve adhesive coating the used powder must be dry, free flowing and properly blended (Abdulaziz2012). It is also reported that powder mixtures can only be sprayed only when one of the powder mixture is metallic powder. Moreover, powder particles need to be deformable and ductile for intense adhesion through plastic deformation (Steenkiste2002), (Lee2005) and (Lee2004). Ductility of the metallic powder is essential for flattening of the particles by plastic deformation. Powder material must ensure plastic flow without fracturing in order to produce dense coating. (Alkhimov1995).

2.5 Role of Surface Preparations

The surface where deposition will be done must be cleaned and polished using different grits of emery papers. This will remove presence of any dirt, grease, grime, scale and oxides that may interfere with the bonding process. According to researchers there are there are three areas to differentiate i.e. effect of thermo-property, surface roughness and surface texture. Bonding of deposition with the substrate is fine only when the surface is free of oxidation and contaminants during cold spraying. It is also reported that grinding or polishing, grit blasting and sand blasting are used for metals whereas for glass and polymeric substrate cleaning and degreasing is followed before deposition (Temitope2019).

Bonding between the deformed particle the substrate was favorably affected on rough surface were reported by few investigators (Marrocco2006), (C-J Li2006) (Sutcliffe2003) and (Legoux2010). According to them various combinations of particle/substrate depend on different surface preparation for adhesive deposition. Yin et al. (2015) reported that there was a decrease in bonding strength for grit-blasted surface. But according to Wayne et al. impact of titanium particles on sapphire substrate improved the coating thickness which had roughness of 3 nm. Reduction of bonding strength of 0.05 μm was investigated by Hussain et al. (2011) for polished surface which had 3.9 μm surface roughness.

3. MECHANISMS INVOLVED

3.1 Bonding Mechanism

The coating characteristics like deposition efficiency and adhesion strength is dependent on the bonding mechanism, which occurs due to the impingement of the powder on the substrate. As the bonding between powders and substrate improves, the parameters like deposition efficiency and adhesion strength of the coating will improve. The mechanism that is involved in bonding is rather complex and there has been an attempt to address from various approaches. According to Grujicic (2003) adhesion of the coating depends on the size of the interfacial area between the particle and the substrate (Grujicic2003). One of the mechanism for bonding is mechanical interlocking, usually present in the interfacial bonding (Klinkov2005). In cold spraying due to the high kinetic energy of the particles on the substrate, there is a shot peening like effect on the coating formation in cold spraying, improves its strength and adhesion. Sticking of the particles becomes stronger following the particles repeated impact on the substrate, resulting in better adhesion between the particles and between the particle and substrate.

In cold spraying, combination of different bonding mechanism depend on bonding between particles and substrate and between particles. According to Hussain et al. (2009), a combination of metallurgical bonding and interlocking occurs in cold spraying. "Other studies reported a combination of mechanical blocking, metal-metal bonding, van der Waal's adhesion and diffusion in cold spraying (Hussain2009, Guetta2009). In cold sprayed Cu plates and Al substrate diffusion was found as the bonding mechanisms resulting in formation of a thin nano-crystalline layer of an intermetallic, at the interface (Guetta2009). Bonding is influenced by substrate softening during impact on the soft substrate materials e.g. when Cu impacts on Al substrate making strong adhesion between coating and substrate (Kairet2007). Grujicic et al. (2003) have reported that atomic diffusion at the particle-substrate interface is not the dominant bonding mechanism. On the other hand, Kairet (2007) reports the occurrence of slight diffusion at the interface (in a Cu-Al system). However, the most widely accepted mechanism for bonding is by far that due to the adiabatic shear instability (Assadi2003, Moridi2014).

3.2 Shear Instability as Bonding Mechanism

According to experimental studies and computational modeling for better understanding of CS investigator (Dykhuzein1998)] have reported that adhesion occurs only when the powder particles exceed a critical velocity (Stoltenhoff2002). When the particle impacts on the substrate a strong pressure field is formed on the substrate. As a result the material gets accelerated due to shear load. Due to this localized shear strain occurs, which under critical condition leads to adiabatic shear instability (Grijicic2003), (Assadi2003) and (Grijicic2004). Hence, the bonding mechanism is attributed to adiabatic shear instability, that occurs at and beyond critical velocity.

Adiabatic shear instability was first reported for the deformation mechanism in explosive welding where shock waves are created that causes shear loading that leads to deformation behaviour. It was adapted by Assadi et al (2003) to explain the crater like formation that clearly is related to the formation of bond between the substrate and particle. Their article describes that a critical velocity exists beyond which adhesion is markedly improved. Further they have explained through numerical analysis the deformation behavior of the particles and shown material jetting and splat flattening that occurs only at and beyond the critical velocity. This has been experimentally validated by impacting copper particles on copper substrate at high velocities, that show material jetting occurring at the splat-substrate interface. This was widely well received in the scientific community. Similarly Grujicic et al (2003) have independently published similar bonding mechanism supported by finite element numerical analysis and validated experimentally. Xie et al (2016) have shown that metallurgical bonding, by removal of the oxide layer by a shot-peening like effect, is responsible for the bonding mechanism. Hassani-Gangaraj (2018) have claimed that although material jetting is a necessary condition for the bonding mechanism, however, the absence of adiabatic shear bands does not purport the adiabatic shear instability as the bonding mechanism. They show numerically that the bonding mechanism may originate due to the metallurgical bonding at the particle-substrate interface initiated by the shock waves that are hydrodynamic (pressure driven) in nature. The proposed mechanism has met with some critique and Assadi et al (2019) commented that while adiabatic shear instability is not the necessary condition for jetting, and hence bonding, it is both experimentally and numerically supported. In fact they argue that high interfacial strain is necessary for bonding and jetting may or may not occur depending on the spraying conditions. Hassani-Gangaraj (2019), however, have reiterated that the jetting behaviour is experimentally validated and have explained the critical velocity as a function of the bulk speed of sound. They have also defined a function for the same, considering the effects of particle size, strength and temperature on it. There have been a string of recent publications in this area, that strengthen/demonstrate the jetting behavior and some refute material jetting. Nevertheless, material jetting and critical velocity still remain an integral part of cold spraying mechanism and all important for the bonding mechanism.

3.3 Particle Velocity

Deposition in cold spray is influenced by both gas velocity as well as particle velocity. The particle velocity, due to the effect of the drag force does not have the same value as the gas velocity. Walker (2018) show numerically that the values of the particle velocity and gas velocity differ by almost 100 m/s for the same system. The gas is forced to enter through the converging-diverging nozzle and acquires a supersonic velocity as it exits the nozzle. The powder-feeder feeds the powder to the gas stream at the exit of the nozzle only after it has acquired the required velocity. As the particles flow in the gas stream, the drag force that exists reduces the particle velocity by at least a 100 m/s. The particles as they hit the substrate after travelling some distance (optimized standoff distance) may either bounce back (erosion of the surface) or stick (adhesion to the substrate). This condition for adhesion has been determined to be dependent on the particle velocity. The particle velocity at which the particles start sticking to the surface without any erosion, is called the critical (particle) velocity (Papyrin2007). Without critical velocity of the powder particles formation of coatings is not possible in cold spraying. Critical velocity depends on the characteristics of the sprayed material like density, size and temperature (Dykhuizein1999) (Assadi2003) and (Schmidt2006). Particle adhesion is only possible when the particle velocity is at or beyond the critical velocity of the given material. Adhesion is usually preceded by material jetting, which is verified both experimentally and numerically when the particle velocity exceeds the critical velocity (Ghelichi2009). The simulated critical velocities for a few materials like Al, Ni, Fe & Cu are 680-700 m/s, 620-640 m/s, 560-580 m/s, respectively (Ghelichi2009). Karthikeyan et al. (2006) in their numerical simulation study with four particle system have shown that, there are four different stages in the deposition process. In the first stage, when the particle velocity is less than the critical velocity, a delay time occurs between the beginning of the particle flow and the particle adhesion. This leads to rebounding or subsequently erosion of the substrate. In the second stage when the particle velocity just exceeds critical velocity, a thin layer of material gets adhered. In the third and final stage, the subsequent layers of coating are deposited on the first layer (Schimidt2006) and (Singh2012).

3.4. Effect of Temperature

The effect of temperature in 'cold' spraying is rather an oxymoron. Temperature may either be applied pre-deposition to improve the adhesion mechanism or post deposition to remove residual stresses created by the impact. In both cases the temperatures involved are a small fraction of the melting point of the material. There are broadly two types of pre-deposition heating, i.e. gas heating and substrate heating, done to improve the adhesion.

Gas heating is probably the most common type of cold spraying set-up. Here, the gas is passed through a heater before entering the nozzle. The temperature of the gas, reduces its density which in turn increases its velocity. Therefore even at lower gas pressures it can attain supersonic velocities (Papyrin2007, Dyzhukein1998). This has been very easily accommodated in most CGDS set-ups. Walker (2018) show that as the temperature increases the critical velocity decreases, such that adhesion can take place at lower particle velocities. Assadi (2003) show that the critical velocity reduces by almost 40% of the gas temperature for copper particles on copper substrates.

Substrate heating on the other hand has only been partially explored since it can only be applied to substrates that are not oxygen sensitive. Moreover heating of the substrate may lead to variations in its mechanical properties. Pre-heating of substrate shows to improve both the deposition efficiency as well as the adhesion strength of the coating (Yin2015, Xie2015). Whereas, in some materials like zinc and tin, the deposition efficiency is not affected by the substrate temperature (Legoux2007). Moreover, Legoux (2007) have shown that although the critical velocity is not affected by the substrate pre-heating temperature, the adhesion and deposition efficiency improve with the increase in the temperature. This is probably related to the bonding mechanism where thermal softening contributes to the ASI at the interface. The thermal softening (referred to as impact temperature) has also been confirmed to occur even without substrate pre-heating (Legoux2007).

Post deposition heat treatment has been attempted by many to mainly remove any residual stresses that may arise due to cold spraying. Huang et al. (2015) have worked on heat treated Al, Cu, Ti and stainless steel 316 deposition. It was found that for Al coating, after heat treatment at 200°C, deposition became stronger but ductility was very poor. Similarly for cold sprayed Ti coatings, the tensile strength and elongation decreased as compared to bulk Ti. But for cold sprayed Cu coating, the tensile strength was found to be higher by heat treatment. After heat-treating at 300°C, the deposition became stronger, however, with poor ductility. When the temperature was increased to 400°C & 500°C, good mechanical properties were obtained. But when the temperature was raised to 700°C, it was found that tensile strength & elongation of the coating decreased as compared to heat treated at 400°C & 500°C.The tensile strength of cold sprayed stainless steel 316 coatings did not show much improvement as compared to heat treated below 600°C (Huang2015).

4. APPLICATIONS OF COLD GAS DYNAMIC SPRAYING.

This process is heavily used in automobiles, aeroplanes, medical, electronics & petrochemical industries. Cold spray is used as restoration procedure to promote the renewal of antique objects, such as cars, aeroplanes & metal sculptures (Villafuerte2010) and (Shkodin 2010). Space shuttle boosters are coated with aluminum and aluminum alloy by cold spraying for refurbishment. In anti-skidding cold spraying is used as corrosion reducing agent (Karthikeyan 2004). Solid surfaces such as bed rails, faucets, light switches, door knobs & other hardware can be coated using CGDS (Sova2013). High thermal insulators and high corrosion resistance components with less cost of manufacturing can be produced using CGDS (Yoon2007). Al-Sn alloy are used in ship building & automobile industries. Shear surface is provided by Sn and the friction properties in aluminium matrix during sliding helps in high temperature environment. Ning et al successfully coated Al-10Sn & Al-5Sn through CGDS process with well-bonded structures & low porosity (Ning2009). In power generation station, CGDS coating is used. In case of wear rings, impeller seal section, turbines blade, water pump housing the coating offer caution & protection from high temperature corrosion. At low temperature cold sprayings can be used to reconditioned wear rings by bronze coating (Assadi2016). GRCop-84 substrate was successfully coated with NiCrAly and CuCrAl particles using CGDS (Balani2005). A special tungsten carbide nozzle was designed for this deposition.

Although metals are the prevalent materials that are deposited (coating material) using cold spray, nevertheless, ceramic materials have widely been deposited. Due to the lack of deformability of ceramic particles only thin layers can be produced using cold spray. In biomedical implants titanium coatings have potential applications. Tungsten oxide and titanium oxide are semiconductor materials with a band gap at around 3 eV, with good chemical stability in aqueous solutions. Thin ceramic coatings are used in gas turbines, piston tops, cylinder heads, valves and diesel engines (Moridi2004). Hydroxyapatite (HAP) is commonly used to coat implant surfaces to improve osseointegration with the bone tissues. As bio-ceramic coating, induces high temperature cold spraying coating is used at low temperature. It was reported that ASB industries had deposited Ti-HAP composite coating using CGT kinetics 4000 cold gas dynamic system.

Air conditioning equipment in vehicle and Al-tube heat exchangers are fabricated by cold gas dynamic spraying. Copper components in transformer are protected by removing the electrochemical oxidation of the elements by using this coating method. Aluminum tips & electric mains in transformer can be protected by coating copper powder. Transmission gear box in helicopter is protected by barrier coating that helps to restore dimensional tolerances of heat sensitive materials such as ZE41A-T5 (Champagne2008).

5. CONCLUSION & FUTURE PROSPECTIVE

The future for cold spraying is opening up as more and more materials are being tested and used for spraying. Many companies have started developing and selling off-the-shelf cold spray systems with many built in and custom features. With the increase in investigations into the spraying mechanisms and the bonding mechanism, there is more and more control over the process-parameter-property tetrahedron. With the spike in recent investigations and subsequent publications, cold spraying seems to be a encouraging solid state processing method.

Many researchers have worked on various phenomena during deposition and development of coating during cold spraying. Basically the high drag co-efficient porous powder enables it to attain high deposition & good adhesion. This chapter has systematically introduced the different aspects of cold spraying which includes the process, process parameters, the equipment, the know-how, the metallurgical bonding aspect, the various advantages and disadvantages, the material system and the present status of the process.

REFERENCES

Akedo, Nakano, Park, Baba, & Ashida. (2008). The aerosol deposition method. *Synthesial. England., 1*, 121-130.

Alhulaifi, Buck, & Arbegast. (2012). Numerical and Experimental Investigation of Cold Spray Gas Dynamic Effects for Polymer Coating. *Journal of Thermal Spray Technology, 21*(5), 852.

Alkhimov, A., Papyrin, A., Kosarev, V., Nesterovich, N., & Shushpanov, M. (1995). *Method and Device for Coating*. European Patent 0488533B1.

Anisur, M., Chakrobarty, P., Easton, D., & Raman, R. (2018). Controlling hydrogen environment and cooling during CVD graphene growth on nickel for improved corrosion resistance. *Carbon, 127*, 131–140. doi:10.1016/j.carbon.2017.10.079

Assadi, H., Gärtner, F., Klassen, T., & Kreye, H. (2019). Comment on "Adiabatic shear instability is not necessary for adhesion in cold spray". *Scripta Materialia, 162*, 512–514. doi:10.1016/j.scriptamat.2018.10.036

Assadi, H., Gartner, F., Stoltenhoff, T., & Kreye, H. (2003). Bonding mechanism in cold gas spraying. *Acta Materialia, 51*(15), 4379–4394. doi:10.1016/S1359-6454(03)00274-X

Assadi, H., Kreye, H., Gärtner, F., & Klassen, T. (2016). Cold spraying – a materials perspective. *Acta Materialia, 116*, 382–407. doi:10.1016/j.actamat.2016.06.034

Balani, K., Agarwal, A., Seal, S., & Karthikeyan, J. (2005). Transmission electron microscopy of cold sprayed 1100 aluminum coating. *Scripta Materialia*, *53*(7), 845–850. doi:10.1016/j.scriptamat.2005.06.008

Champagne, V. (2007). *The cold spray Materials Deposition Process: Fundamentals and Application*. Woodhead Publishing Ltd.

Champagne, V. K. (2008). The repair of magnesium rotorcraft components by cold spraying. *Journal of Failure Analysis and Prevention*, *8*(2), 164–175. doi:10.100711668-008-9116-y

Champagne, V. K., Leyman, P. F., & Helfritch, D. (2008). Magnesium repair by cold spraying. *Plating and Surface Finishing*, *95*, 34.

Choudhuri, A., Mohanty, P. S., & Karthikeyan, J. (2009). *Bioceramic composite coatings by cold spraying technology*. Addit. Manuf. Process.

DeForce, Eden, Potter, Champagne, Leyman, & Helfritch. (2007). Application of aluminium coatings for the corrosion protection of magnesium by cold spraying. *TRI Serv. Corros.*, 1-16.

Dykhuizen & Smith. (1998). Gas dynamic principle of cold spray. *Thermal Spray Technology*, 205-212.

Dykhuizen, R. C., Smith, M. F., Glimore, D. L., Neiser, R. A., Jiang, X., & Samapth, S. (1999). Impact of high velocity cold spray particles. *Journal of Thermal Spray Technology*, *8*(4), 559–564. doi:10.1361/105996399770350250

Ghelichi, R., & Guaglians, M. (2009). Coating by the cold spraying process. Frat Ed Integrita Strut Ed, 3, 30-44.

Grujicic, M., Saylor, J., Beasley, D., DeRosset, W., & Helfritch, D. (2003). Computational analysis of the interfacial bonding between feed-powder particles and the substrate in the cold-gas dynamic spray process. *Applied Surface Science*, *219*(3-4), 211–227. doi:10.1016/S0169-4332(03)00643-3

Grujicic, M., Zaho, C. L., DeRosset, W. S., & Helfritch, D. (2004). Adiabatic shear instability based mechanism for particles substrate bonding in the cold-gas dynamic-spray process. *Materials & Design*, *25*, 68–81.

Guetta, S., Berger, M., Borit, F., Guipont, V., Jeandin, M., Bousite, M., Ichikawa, Y., Sakaguchi, K., & Ogawa, K. (2009). Influence of Particle Velocity on Adhesion of Cold-Sprayed Splats. *Journal of Thermal Spray Technology*, *18*(3), 331–342. doi:10.100711666-009-9327-0

Hassani-Gangaraja, Veysset, Champagne, Nelson, & Schuh. (2019). Adiabatic shear instability is not necessary for adhesion in cold spray. *Scripta Materialia, 162*, 515-51.

Huang, R., Sone, M., Ma, W., & Fukanuma, H. (2015). The effect of heat treatment on the mechanical properties of cold sprayed coatings. *Surface & Coating Technology*, *261*, 278–288.

Hussain, McCartney, Shipway, & Zhang. (2009). Bonding Mechanism in cold Spraying: The Contribution of Metallurgical and Mechanical components. *Journal of Thermal Spray Technology, 18*(3), 364-379.

Hussain, T., McCartney, P., & Shipway, P. H. (2011). Shipway, "Impact phenomenon in cold spraying of titanium on various ferrous alloys. *Surface and Coatings Technology*, *205*(21-22), 5021–5027. doi:10.1016/j.surfcoat.2011.05.003

Jodion, B., Richer, P., Berube, G., Ajdelszatajn, L., & Yanouzi, M. (2007). Pulsed gas dynamic spraying process analysis, development and selected coating examples. *Surface Coating Technology, 201*, 7544–7551.

Kairet, T. (2007). *A contribution to the study of cold gas dynamic spraying of copper: Influence of the powder characteristics on the mechanical properties of the coating* (Doctoral thesis). University Libre deBruxelles, Belgium.

Kalita, G., Muhammed, E., Ayhan, E., Sharma, S., & Shinde, M. (2014). Low temperature deposited graphene by surface wave plasma CVD as effective oxidation resistive barrier. *Corrosion Science, 78*, 183–187. doi:10.1016/j.corsci.2013.09.013

Karthikeyan, J. (2007). *Development of oxidation resistant coatings on GRCop-84 substrates by cold spraying processs*. NASA/CR.

Karthikeyan, J. (2004). Cold spraying Technology: International Status and USA Efforts. Academic Press.

Klinkov, S., Kosarev, V., & Rein, M. (2005). Cold spray deposition: Significance of particle impact phenomena. *Aerospace Sci. Technology, 9*, 582–591.

Koivluoto. (2010). *Microstructural characteristics and corrosion properties of cold sprayed coating*. Tampere University of Technology.

Lee, H., Jung, S., Lee, S., You, Y., & Ko, K. H. (2005). Correlation between Al2O3 particles and interface of Al- Al2O3 coatings by cold spray. *Applied Surface Science, 252*(5), 1891–1898. doi:10.1016/j.apsusc.2005.03.148

Lee, H., Yu, Y., Lee, Y., Hong, Y., & Ko, K. (2004). Cold Spray of SiC and Al2O3 with soft metal incorporation: A Technical Contribution. *Journal of Thermal Spray Technology, 13*(2), 184–189. doi:10.1361/10599630419355

Legoux, J.-G., Irissou, E., Desauluniers, S., Bobyn, J., Harvey, B., Wong, W., Gagnon, E., & Yue, S. (2010). *Characterization and performance evaluation of a helium recovery system designed for cold spraying. NRC Publ Arch*. NP Arc.

Li, C.-J., Li, W.-Y., & Liao, H. (2006). Examination of the critical velocity for deposition of particles in cold spraying. *Journal of Therm. Spray Technol., 15*(2), 212–222. doi:10.1361/105996306X108093

Li, C.-J., Li, W.-Y., Wang, Y.-Y., Yang, G.-J., & Fukanuma, H. (2005). A theoretical model for prediction of deposition efficiency in cold spraying. *Thin Solid Films, 489*(1-2), 79–85. doi:10.1016/j.tsf.2005.05.002

Li, W.-Y., & Li, C.-J. (2005). Optimal design of a novel cold spraying gun nozzle at a limited space. *Journal of Therm Spray Technol., 14*(3), 391–396. doi:10.1361/105996305X59404

Li, W.-Y., Zhang, C., Guo, X. P., Zhang, G., Liao, H. L., Li, C.-J., & Coddet, C. (2008). Effect of standoff distance on coating deposition characteristics. *Materials & Design, 29*(2), 297–304. doi:10.1016/j.matdes.2007.02.005

Maev, R., & Leshchynsky, V. (2006). Air gas dynamic spraying of powder mixtures: Theory and Applications. *Journal of Thermal Spray Technology, 15*(2), 198–205. doi:10.1361/105996306X108048

Marrocco, T., McCartney, D. G., & Sturgeon, A. J. (2006). Production of titanium deposits by cold-gas dynamic spray: Numerical modeling and experimental characterization. *Journal of Thermal Spray Technology*, *15*(2), 263–272. doi:10.1361/105996306X108219

Muehlbrger, E. (2007). *Method and apparatus for low pressure cold spraying*. US Patent 6759085.

Ning, X.-J., Kim, J.-H., Kim, H.-J., & Lee, C. (2009). Characteristics and heat treatment of cold-sprayed Al–Sn binary alloy coatings. *Applied Surface Science*, *255*(7), 3933–3939. doi:10.1016/j.apsusc.2008.10.074

Papyrin, A., Kosarev, V., Klinkov, S., Alkimov, A., & Fomin, V. (2007). Cold spray technology. Elsevier.

Pattison, Celotto, Khan, & O'Neill. (2008). Standoff distance and bow shock phenomena in the Cold Spray process. *Surface and Coatings Technology, 202*(8), 1443-1454.

Pawlowski, L. (2008). *The science and engineering of thermal spray coating*. John Wiley &Sons. doi:10.1002/9780470754085

Schmidt, T., Gartner, F., Assadi, H., & Kreye, H. (2006). Devlopement of a generalized parameter window for cold spray deposition. *Acta Materialia*, *54*(3), 729–742. doi:10.1016/j.actamat.2005.10.005

Shkodkin, Oklyuev, & Buzdygar. (2010). *Peculiarities of Gas Dynamic Spray Applications in Russia.* Academic Press.

Singh, Sidhu, Kalsi, & Karthikeyan. (2013). Development of Cold Spray from innovation to emerging future coating technology. *Journal of Brazilian Soc Mech*, 231-245.

Singh, H., Sidhu, T. S., & Kalsi, S. (2012). *Cold spray technology: Future of coating deposition processes*. Gurunaka Dev University.

Sobolev, V., Guilemany, J., & Nutting, J. (2004). *High velocity oxy-fuel spraying theory, structure-property Relationships and Application.* Maney Publishing.

Sova, Grigoriev, Okunkova, & Smurov. (2013). Potential of cold gas dynamic spray as additive manufacturing technology. *International Journal of Advance Manufacturing Technology, 69*, 2269-2278.

Steyer, P., Millet, J.-P., Anderbouhr, S., & Jacquot, P. (2001). Effect of Ni Interlayer on Corrosion Behaviour of Crn Pvd Coating Deposited onto Plastic Mould Steels. *Surface Engineering*, *17*(4), 1743–2944. doi:10.1179/026708401101517962

Stoltenhoff, T., Kreye, H., & Ritcher, H. (2001). An Analysis of the Cold Spray Process and its Coatings. *Journal of Thermal Spray Technology*, *11*(4), 542–550.

Stoltenhoff, T., & Ritcher, H. J. (2002). An analysis of the cold spray process and its coatings. *Journal of Thermal Spray Technology*, *11*(4), 542–550. doi:10.1361/105996302770348682

Sutcliffe, C. J., Papworth, A., Gallagher, C., Fox, P., Morgan, R., & O'Neill, W. (2003). Cold gas dynamic manufacturing a new approach to near-net shape metal component fabrication. *Mater.Res.Soc. Proc.*, *758*, 73–84.

Temitope & Jen. (2019). A comparative review on cold gas dynamic spraying processes and technologies. *Manufacturing Review*.

Van Steenkiste, T., Smith, J., & Teets, R. (2002). Aluminium Coatings via Kinetic Spray with Relatively Large particle powders. *Surface Coating Technology*, *154*, 237–252.

Villafuerte, J. (2010). Current and future applications of cold spraying technology. *Metal Finishing*, *108*(1), 37–39. doi:10.1016/S0026-0576(10)80005-4

Xie, Y., Yin, S., Chen, C., Planche, M.-P., Liao, H., & Lupoi, R. (2016). New insights into the coating/substrate interfacial bonding mechanism in cold spray. *Scripta Materialia*, *125*(December), 1–4. doi:10.1016/j.scriptamat.2016.07.024

Yekane, Deshmukh, & Suryavanshi. (n.d.). Corrosion Protection of Light Metal Using Cold Spray Technique. *Journal of Computer Science*, 72-78.

Yin, S., Xie, Y., Suo, X., Liao, H., & Wang, X. (2015). Interfacial bonding features of Ni coating on Al substrate with different surface pretreatments in cold spraying. *Materials Letters*, *138*, 143–147. doi:10.1016/j.matlet.2014.10.016

Yoon, S., Kim, H., & Lee, C. (2007). Fabrication of automotive heat exchange using kinetic spraying process. Surface Coating Technology, 201, 9524-9932.

Zhou, Liu, Li, Zhu, & Xiong. (2019). *Microstructure and properties of Fe-Cr-Ni alloy coating on T10steel by laser cladding*. Material Research Express.

Chapter 7
Fabrication of Functionally Graded Metal and Ceramic Powders Synthesized by Electroless Deposition

Onur Güler
Engineering Faculty, Department of Metallurgical and Materials Engineering, Karadeniz Technical University, Trabzon, Turkey

Temel Varol
https://orcid.org/0000-0002-1159-5383
Engineering Faculty, Department of Metallurgical and Materials Engineering, Karadeniz Technical University, Trabzon, Turkey

ABSTRACT

One of the most important factors in powder metallurgy is the powder properties that directly affect the final product properties. By using the functionally graded materials (FGMs) in powder metallurgy, the desired properties can be obtained by means of layers having microstructure having more than one feature in a single material structure. Similarly, by the production of functionally graded powders (FGPs), different properties can be obtained in a single powder structure and the materials that have different properties in the same structure are developed by integrating these powders with powder metallurgy. In this context, the FGMs synthesized from the FGPs produced by electroless deposition (ED) of metal or ceramic-based powder materials facilitates the production of advanced material. Therefore, the purpose of this chapter is on the fabrication of metal and ceramic-based FGPs by ED and to discuss of their advantages on the powder metallurgy parts.

DOI: 10.4018/978-1-7998-4870-7.ch007

INTRODUCTION

The emergence of delamination problems in the matrix and reinforcement material of composite materials developed as a solution to the problems such as corrosion and wear of conventional mono-component or alloy materials has encouraged researchers to produce a new type of material. These novel materials were called as functional graded materials (FGMs) whose features change gradually corresponding with the dimensions in the structure (Ning Zhang et al., 2019). These materials are widely used in electrical, aerospace, energy, automotive, bioengineering and many other applications. Since conventional engineering materials do not provide the desired properties, they are highly advantageous in terms of obtaining different material properties in a single structure. The application areas of FGMs are shown schematically in Figure 1.

Figure 1. Application areas of FGMs

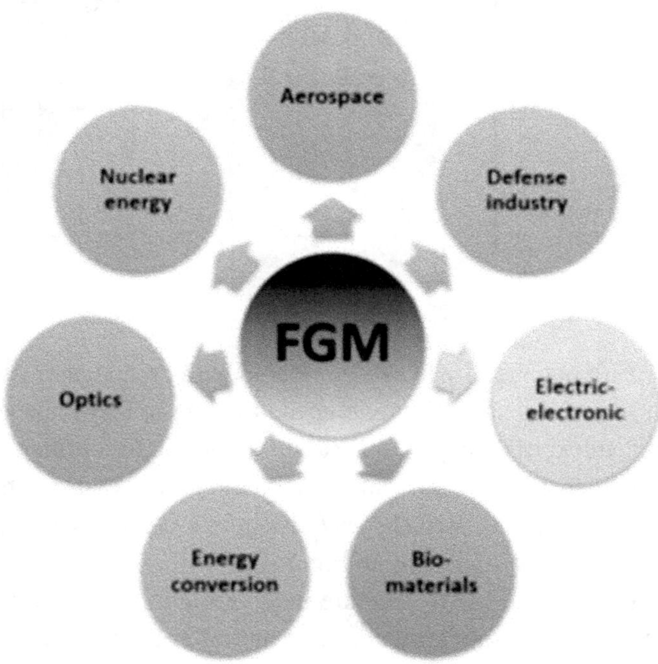

Depending on these applications, considering the characteristics such as abrasion, corrosion, chemical stability, electrical and thermal conductivity, providing different properties in different layers of a single material structure makes FGMs indispensable (Edwin, Anand, & Prasanna, 2017; El-Galy, Saleh, & Ahmed, 2019; A. Gupta & Talha, 2015) . FGMs are often structurally divided into two types. In the first kind FGMs, the composition in the material structure changes step by step, while the composition in the second variant shows a continuous change. In this context, FGMs can be produced from different layers and using different materials in each layer. On the other hand, FGMs with different grain sizes can be produced in the same structure by obtaining materials with different grain sizes in each region (lower-middle-upper) (Chi Zhang et al., 2019). This type of FGMs is shown schematically in Figure 2.

Figure 2. Various gradients on the FGMs; (a) interrupted and (b) uninterrupted gradient, (c, d, e) laminated FGMs or gradual variation in composition, (f, g, h) fractional variation in grain size of FGMs (Chi Zhang et al., 2019)

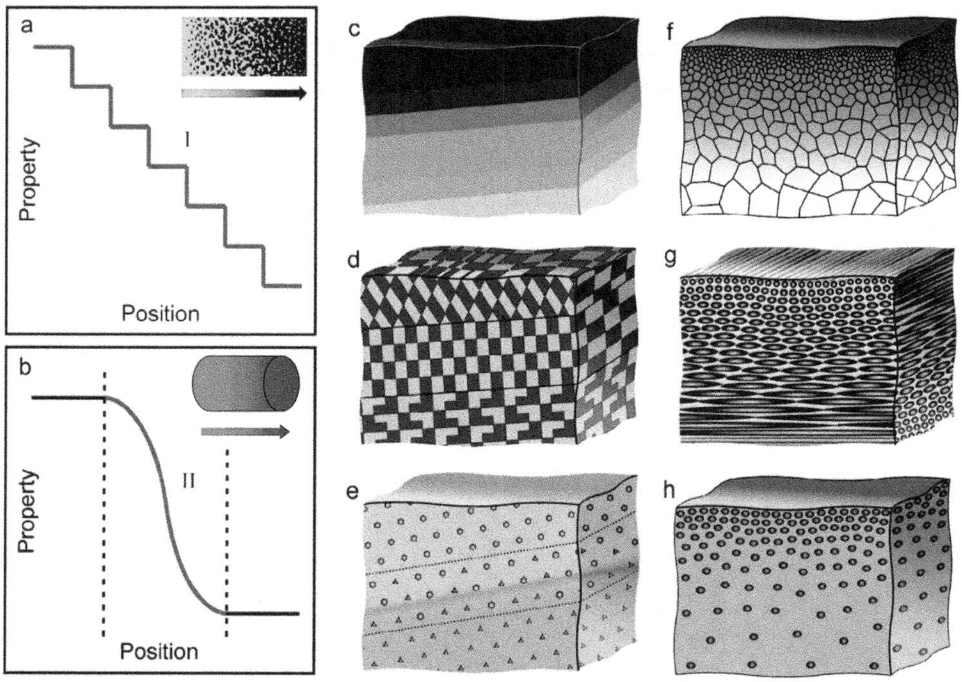

The most important result obtained by FGMs compared with composite materials; while composite materials have a certain material feature throughout the thickness, FGMs have a constantly changing material property along the thickness. They produce a continuous or discontinuous change in their chemical composition and/or microstructures (grain size, thickness, porosity, strength, hardness, thermal and electrical conductivity, etc.) in the specified shapes / distances (Akavci, 2014). This difference between composite material and FGM is given visually in Figure 3.

For example, a recently preferred FGM structure in the production of electrical contact materials is given in Figure 4. As can be seen, using pure copper material in the bottom layer is considered as the region with the highest electrical conductivity, harder layers are obtained by increasing the content of additives as the upper layers are increased. Thus, the advantage of using FGM as an electrical contact material has been seen in the hard layer, which is exposed to electrical contact, and with different properties in a single structure with higher electrical conductivity (Mirazimi, Abachi, & Purazrang, 2016). In addition, for example, as given in Figure 5, the desired high hardness on the surface and the low hardness of the porous structure desired in the core can be achieved through the production of FGM within the same material structure (Afzal et al., 2012).

Moreover, FGMs which are composed of composite layers can meet the demands such as wear, thermal, corrosion and ballistic resistance with superior mechanical properties are widely produced (Ferreira et al., 2011; Übeyli et al., 2014).

Figure 3. Comparison of the gradient in a conventional composite material and the FGM structure (Akavci, 2014)

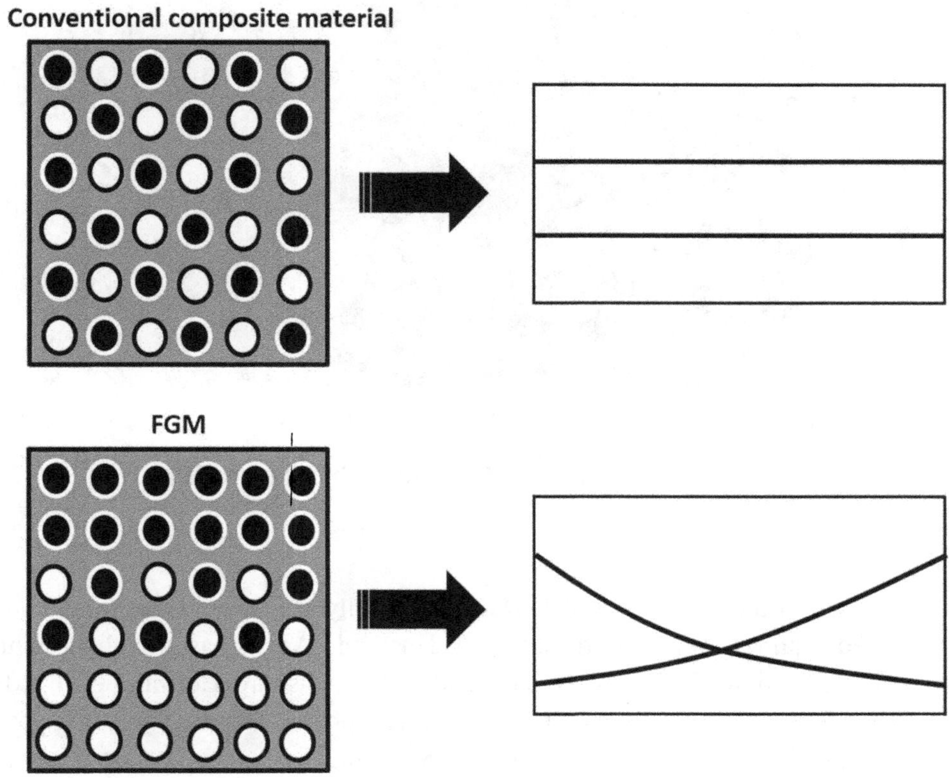

Figure 4. An FGM structure example (Mirazimi et al., 2016)

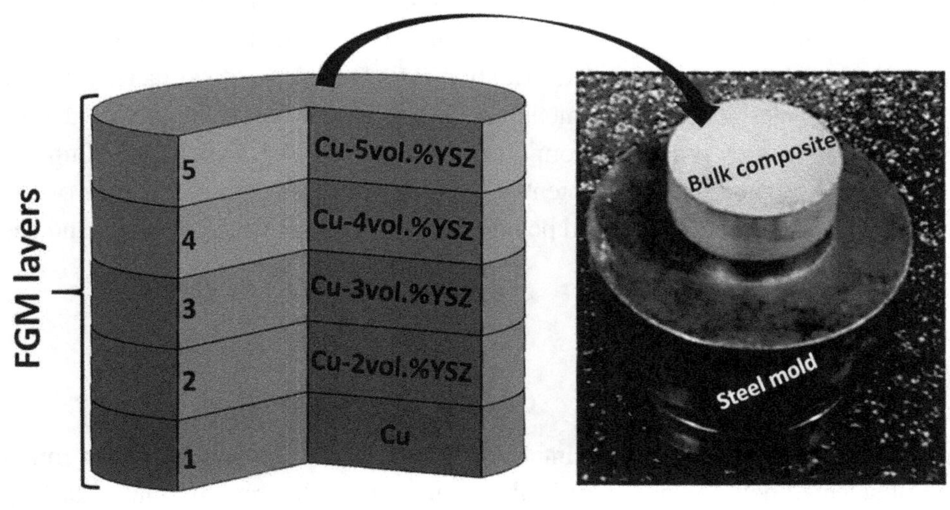

Figure 5. Hardness variation along the layers of an FGM (Afzal et al., 2012)

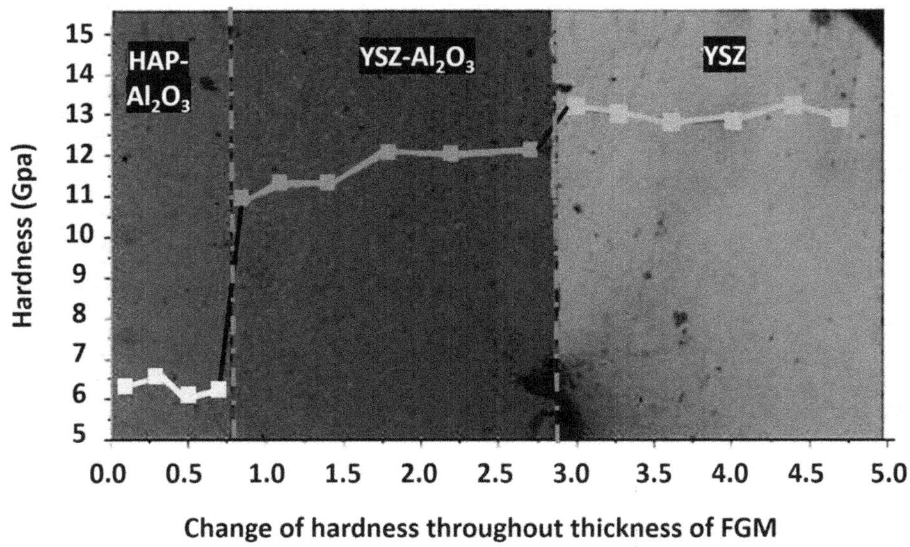

The methods used in the fabrication of FGM are generally classified under three main headings. These are gas-based, liquid phase and solid phase-based methods. As gas-based methods; while chemical vapor deposition / infiltration (CVD / CVI) (Choy, 2003), thermal spray (Gonzalez et al., 2016), surface reaction processes (Somiya, 2013) are mentioned, for liquid phase basis methods; it is possible to talk about spin casting (Arsha et al., 2015; Rahimipour & Sobhani, 2013), strip casting (Hadj et al., 2018), laser deposition (Shi et al., 2019; Wang, & Tao, 2016), electrophoretic deposition (Askari et al., 2012) and ED methods (Liu et al., 2019). Solid phase-based methods such as spark plasma sintering (Kessel et al., 2010), powder metallurgy (F. Chen et al., 2020) and mechanical alloying and/or milling methods (Meng et al., 2015) have important advantages compared to other fabrication method of FGM materials.

Powder metallurgy, which is the most effective method in the fabrication of FGM, is an easy and practical method that provides strong inter-layer bonding. As known; the particle properties such as morphology, particle size, apparent density and oxidation resistance is a critical factor that directly affects the properties of bulk materials in powder metallurgy. Therefore, by changing the powder properties, the bulk material properties to be produced from these powders can also be changed (Gupta et al., 2020). In this regard, the ED method has great potential to significantly alter the properties of metal powders and with this method, the functional graded powders (FGPs) can be produced easily and cheaply (Jafari et al., 2013; Luo et al., 2010).

BACKGROUND

Powder metallurgy method is an important FGM production technique based on mixing, stacking and pressing of selected powder materials. In this technique, high-strength FGMs can be produced by changing factors such as pressing pressure, sintering time and temperature (Mahamood et al., 2012). The hot pressing technique, which is based on the principle of performing sintering and pressing operations under certain temperatures simultaneously, compared to the sintering of bulk materials with raw density obtained by pressing powders under certain pressure, minimizes the negative effects in terms of chemical stability and microstructure (Jin et al., 2005). A typical powder metallurgy workflow and pressing system for an exemplary FGM are schematically given in Figure 6 (Kunimine et al., 2016). Today, traditionally produced FGMs cannot be said to be very cost-effective. However, when it is out of the conventional, the applications where powder metallurgy and sintering methods are used together, make the FGM production done with new non-traditional methods rather than traditional methods. Such production methods ensure that the production of FGM is economical, but it plays an effective role in determining the appropriate material properties by changing production parameters quickly. As a result, the production of materials with a long service life, high wear, corrosion, resistance and electrical properties can be realized at affordable costs (Gasik, 2010).

Figure 6. Powder metallurgy process steps and FGM structure obtained by pressing different powder layers (Kunimine et al., 2016)

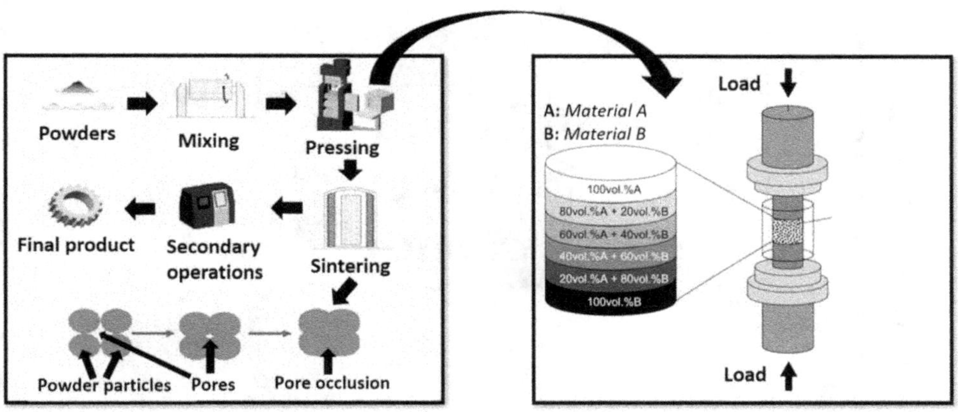

Another method that is frequently used for pressing powders is hot pressing. In this method, the powders are sintered and compacted, as a result of simultaneous application of thermal energy and high pressure at a small deformation rate at sufficiently high temperature. Powders previously pre-pressed or powders fed to a mold made of loosely hot work tool steel can be compressed by pressing them in a vacuum furnace (for oxidation control) providing high temperature up to 2400 ^{0}C. In this way, sintering can be carried out under lower pressure at lower temperatures and shorter times (Samal et al., 2013). Figure 7 shows a typical hot-pressing system.

Figure 7. A schematic hot-pressing system

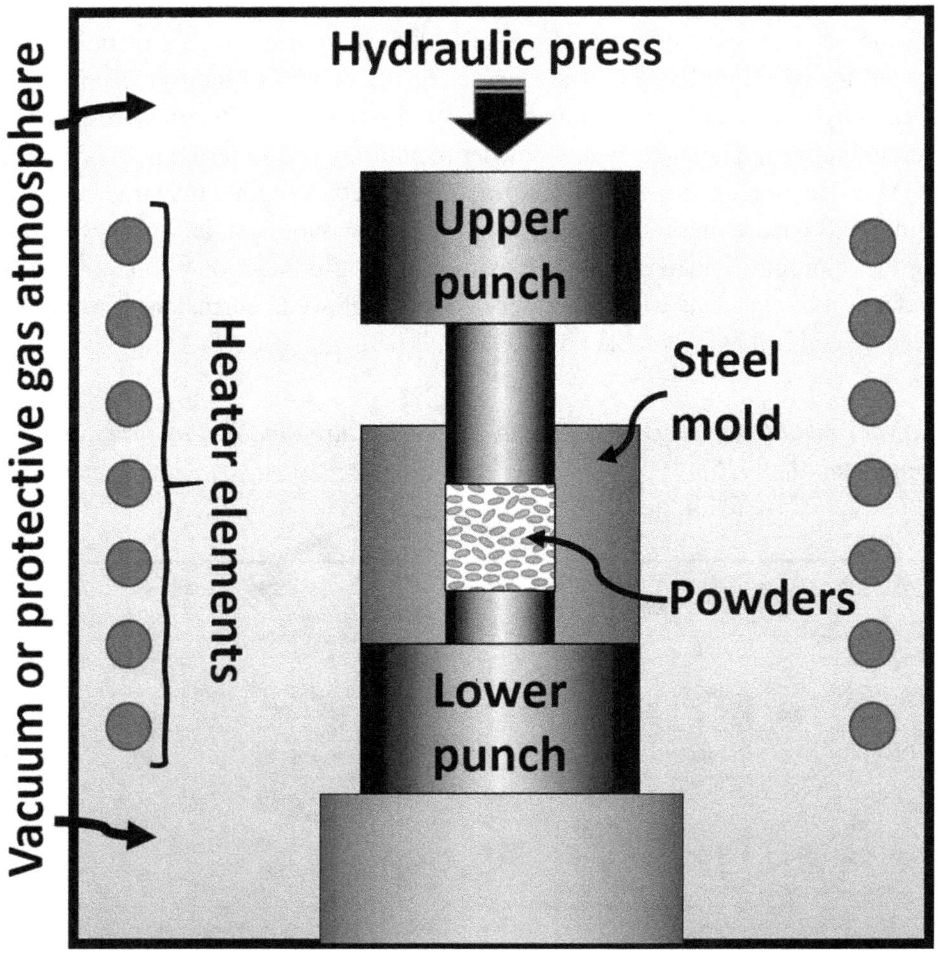

With this regard, advantages such as no loss of scrap, being a suitable method for mass production, alternative in the production of composite material against the difficulty of mixing with matrix by casting method, the proportional control of matrix-reinforcing materials can be easily adjusted as desired, and the production and sintering of layered materials especially for FGMs, powder metallurgy with hot-pressing method is frequently used for FGM production method by researchers (Erdemir et al., 2015; Yusong et al., 2013).

Nassef et al. (2017) compared the material properties obtained by pressing and then sintering Fe-Al alloy powders with the material properties obtained by hot pressing of the same powders under the same conditions in their study. They first pressed Fe-Al alloy powders under approximately 900 MPa pressure and then sintered it under 800 °C for 1 h. The SEM image of the microstructure obtained after this procedure is given in Figure 8a. On the other hand, the same material powders were kept under a temperature of 800 °C for 1 h in a steel mold and pressed under a pressure of about 445 MPa meaning hot-pressing. The microstructure image as a result of this process can be seen in Figure 8b. Although pores are almost completely lost in the internal structure obtained as a result of hot pressing, the presence of highly porous regions in the internal structure is clearly seen as a result of pressing and sintering operations separately.

Figure 8. The microstructure SEM images of Fe-Al alloys obtained by (a)pressing followed by sintering and (b) hot-pressing only (Nassef et al., 2017)

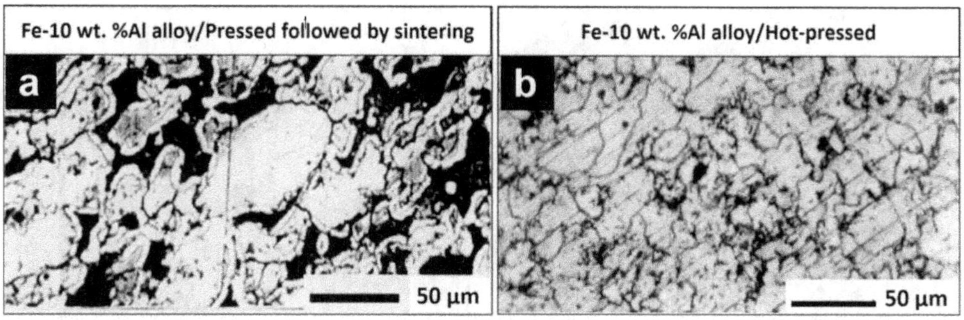

The most important step in the preparation of FGMs obtained by the powder metallurgy and hot-pressing method is the powder preparation stage. Powder particle size, shape and composition play a major role in obtaining FGM. It has been emphasized by researchers that powder characteristics play a critical role on the physical and mechanical properties of the final bulk material obtained after pressing and sintering (Brackx et al., 2015; Slotwinski et al., 2014). The physical properties of the powder particles, such as particle size and shape, play a critical role in the agglomeration of the powders and facing difficulties in mixing. Therefore, the tendency of nano-size powders to agglomerate should always be taken into consideration and mixing should be done under suitable conditions (Barbosa-Canovas et al., 2005). Also, as the powder size decreases, the contact angle between the powder grains increases, so the friction between the powder grains increases and the powder flow slows down. In other words, small sized particles frequently have cohesion among them resulting in low tapped density because of huge surface area causes the more friction and lower flowability. This negatively affects the compactness behavior of pressing powders. This phenomenon results in the lower density of the bulk materials produced from the smaller sized powders (Landillon, Cassan, Morel, & Cuq, 2008).

Although spherical shaped powders are generally used in the powder metallurgy method, the use of powders with irregular and flake-like morphology is frequently encountered. The friction force between spherical morphology powders is very low, their flowability is high and their pressing ability is higher than other shaped powders. In contrast, despite the low flowability in the irregularly shaped or flake-like powder structure, powder grains are always present that can settle in voids that may form among the powders (Narasimhan & Amuda, 2017). Flake powder metallurgy technique, which is a new technique especially in the production of powder metallurgy and particle reinforced composites, is an excellent example to this subject. In this technique, it is desirable to select the matrix material, especially flake, when the powder integrity is ensured in the mold, it can be easily provided that the additives are distributed among the matrix powder particles and that they are homogeneous. There are many studies on this subject in the literature, which have quite impressive results (Fan et al., 2014; Guo et al. 2014; Jiang et al., 2012a; Kai et al., 2013a). It is inevitable that homogeneous particle distribution in the matrix is superior to spherical matrix structures (Figure 9) and as a result of this, more dense materials are obtained and therefore, materials with high physical and mechanical properties are obtained by powder metallurgy method (Jiang et al., 2012b; Kai et al., 2013b).

Figure 9. Uniform distribution of additive particles in the flake matrix (Fan et al., 2014)

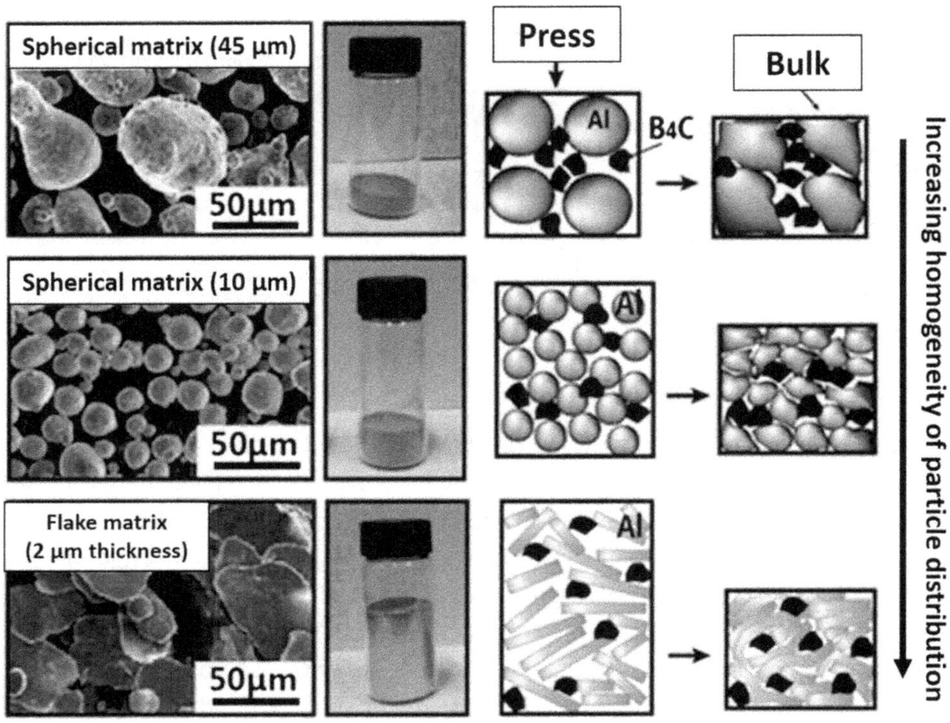

One of the most critical factors affecting the material properties of FGMs prepared the powder metallurgy method is the powder composition. By using powders with different compositions in each layer of FGMs consisting of different layers, structures with desired properties can be created in the desired layer in the same material structure. For example, B_4C reinforced Al-matrix composite can be composed of many layers in order to eliminate thermal stresses. Increasing additives from the lower layers to the upper layers, the B_4C ratio ensures high hardness and thermal conductivity in the areas where the material will be exposed to thermal stress (Camurlu, 2019; Pandey et al., 2016). Moreover, hard ceramic particle reinforced aluminum matrix composites with lightness properties preferred especially in aviation technologies are frequently used. However, the FGM structure consisting of wear-resistant layers developed from bottom to top in the material structure in the parts exposed to wear is an excellent choice for these applications (Bhattacharyya et al., 2008; Padmavathi & Ramakrishnan, 2014; Surya & Prasanthi, 2018).

In addition to using composite powders in different compositions in each layer of layered FGM materials, FGMs can also be produced by using functional graded powder particles (FGPs). The problem of inhomogeneous dispersion of reinforcement particles within the matrix in the traditional composite materials can be minimized with the production of FGPs. In addition, structures with different properties can be created in each FGP grain produced as a core-shell and integration of these FGPs with powder metallurgy can yield advanced featured material (Laima et al., 2016). Moreover, during the production of some ceramic-based nanoparticle reinforced metal matrix composites by casting method, the poor wettability of the reinforcing ceramic phase with the metal matrix and clumping problem can be eliminated to the reinforcing materials by core-shell structure. Since the problem of oxidation of particles, which is another problem, in the melting metal will cause instability in the internal structure, it is also advantageous to form the core-shell structure by coating the powder particles prone to oxidation with metal (Elsharkawi & Esawi, 2018; Uysal et al., 2013).

The ED method was introduced in 1950. With this method, metallic films are formed on conductive, semiconductor or non-conductive surfaces without applying any external current. Metallization with palladium was applied before coating for metal absorption on non-conductive surfaces. Whatever metal is desired to be reduced to the surfaces, solutions containing metal and reducing agents are used. For more than 50 years, this technique is used in applications where wear and corrosion resistance are important, especially in the field of electricity (W. H. Lin & Chang, 1998). However, studies on the properties of FGPs obtained with this technique by making them compact with powder metallurgy are quite limited.

FGPs obtained by the production of metal coated metal and ceramic particles can be studied in the production of FGM. Structural properties of metals and ceramics of different properties can be combined with FGP production in a single structure to produce new and improved materials. In this context, the production of metal-coated metal and ceramic FGPs can be performed by methods such as ball milling, liquid-state methods of sol-gel, electroless deposition (ED) and precipitation deposition, and vapor deposition of PVD. ED method is frequently used as the most effective and fastest method among these coating techniques. In this method, metal ions are reduced from the aqueous solutions containing metal on a substrate, whether conductive or not. Surface metallization occurs as a result of autocatalytic chemical reactions without applying any current. Even on irregularly shaped surfaces, metal shell with homogeneous thickness can be easily formed. It is widely used by researchers because it is very fast, simple and low cost compared to other methods mentioned (Beygi, Sajjadi, & Zebarjad, 2012; Beygi, Vafaeenezhad, & Sajjadi, 2012; L. Luo et al., 2010)

As mentioned, the production of metal-metal and ceramic-metal FGPs obtained by the ED method is quite common in the literature. However, the studies on the characterization of bulk materials produced by the powder metallurgy method, which are provided with different material properties, are quite limited. Therefore, in this study, the production of FGPs by ED, the powder properties affecting FGP production and the properties of powder metallurgy produced from these FGPs are discussed. The use of FGP based materials produced with powder metallurgy can contribute to the literature in many areas thanks to its improved electrical, biocompatibility, hardness, abrasion and corrosion resistance compared to traditional composites.

MAIN FOCUS OF THE CHAPTER

In the light of the information gained in the background of the advantages of the production of FGMs, it is inevitable to obtain various features in a single material structure by pressing the bulk of the FGPs and the bulk materials obtained from these FGPs. Therefore, the main focus of this study is the production of metal-metal and metal-ceramic FGPs and bulk materials by powder metallurgy from such FGPs.

Figure 10. ED mechanism of copper-silver core-shells (Hai et al., 2006)

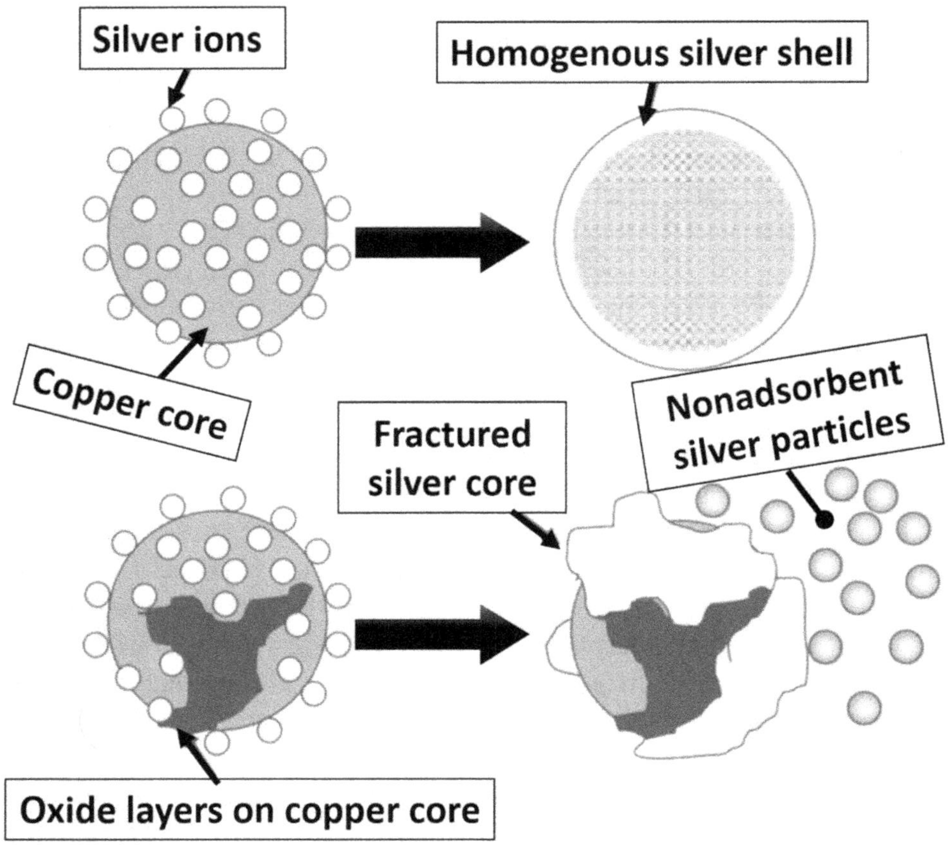

Table 1. Metals that can be coated on metal surfaces by ED method (Djokić, 2005)

Pure metals	Compounding with Ni and Co	Some alloys
Cu, Ag, Ni, Cr, Co, Au, Pd, Pt, In, Sn, Ru, Rh, Bi, Cd	P, B, W, Fe, Mn, Zn, V, Mo, Re, Ti	AuAg, CuNi, AuSn, CuCd, PbSn, AuIn, PdZnP, CuCo, CuAu, PdNiP, PdCoP

Fabrication of Metal-metal FGPs

ED process is a chemical reduction process defined as the catalytic reduction of the metal ion onto the substrate using reducing agents in the aqueous solution, which is the metal source of metal salts. Since this method is a coating technique with metal, alloy or a composite deposit on a substrate material with active conductivity, many researchers have studied the applicability and usefulness of ED of metals such as Ni, Co, Cu, Ag, Au and Pd on different surfaces (Pancrecious et al., 2018). The key point to be considered during the fabrication of metal-metal core-shell particles with the ED technique is that the substrate powder or substrate material surfaces to be coated do not contain an oxide layer. When it is desired to coat the surface of a particle that is not purified from the oxide layer by ED, the formation of uncoated regions is observed, as shown in Figure 10 (Hai et al., 2006).

The surface cleaning or de-oxidation process, which is the first step in the metal-metal core-shell production process with the ED process, is a step that cannot be ignored for the quality of the final coating layer. For this, ammonium hydroxide and ammonium sulphate solutions are generally used. Metal particles with oxide film on the surface should be kept in solution to remove oxide film. For example, in the process of coating copper particles with poor oxidation resistance and containing oxide film on the surface by ED, the following reaction takes place during the cleaning of ammonium hydroxide and ammonium sulphate solution (Choi c Lee, 2019; Güler et al., 2019; Xu et al., 2003).

$$\mathbf{Cu_2O} + 2NH_4OH + (NH_4)_2SO_4 = [Cu\,(NH_3)_2]_2(SO_4) + 3H_2O$$

$$\mathbf{CuO} + 2NH_4OH + (NH_4)_2SO_4 = Cu\,(NH_3)_4\,SO_4 + 3H_2O$$

$$Cu\,(OH)_2 + 2NH_4OH + (NH_4)_2SO_4 = \mathbf{Cu}\,(NH_3)_4SO_4 + 4H_2O$$

All metals and metal alloys that can be electrolytically coated on metallic surfaces can be coated on autocatalytically active surfaces by ED method. However, it is necessary to properly prepare the ED solutions under the influence of factors such as pH, coating time, surface activity and temperature. The list of elements that can be coated on metallic surfaces with ED is given in Table 1 (Djokić, 2005).

Table 2. ED bath compositions for some common metals

Metal	Metal salts	Reducing agent	Complexing agent
Cu (Deckert, 1995)	Copper sulfate, chloride or nitrate	Formaldehyde, dimethylamine borane, borohydride, hypophosphite, hydrazine	Tartrate salts, alkanol amines, EDTA (ethylenediamine tetraacetic acid)
Ag (Chang et al., 2016; Li et al., 2015; Zhang, 2016)	Silver cyanide, silver nitrate	Sodium-potassium tartrate	Sodium thiosulphate, sodium sulfite, EDTA
Ni (Brunner & Picalek, 2012; Picalek et al., 2016)	Nickel chloride or nickel sulphate	Sodium hypophosphite	Glycine-citrate, sodium citrate, lactic acid, malic acid
Au (Toben et al., 2002)	Gold cynanide, potassium gold cyanide	Sodium hypophosphite, phosphonic acid	Potassium cyanide, hydrazine hydrate,
Co (Sudagar et al., 2017)	Cobalt chloride, cobalt sulfate	Dimethylaminebo rane and/or phosphoric acid	Citric acid, triethanolamine, trisaminoathane, glycine
Pd (Ozaki et al., 2003)	Tetraamminepalladium(II) nitrate, palladium chloride	Hydrazine, formaldehyde	Disodium EDTA, ethylenediamine
Cr (Benaben, 2010)	Chromium chloride, chromium fluoride	Sodium hypophosphite	Sodium citrate, potassium bioxalate
Pt (Brachmann et al., 2019; Mizuhashi et al., 2015)	Hexachloroplatinic acid	Sodium borohydride, hydrazine	Platinum hydroxide

Figure 11. Silver coating layer onto the copper core particles with (a) spherical and (b) flake-like morphology obtained by ED method (Güler et al., 2019)

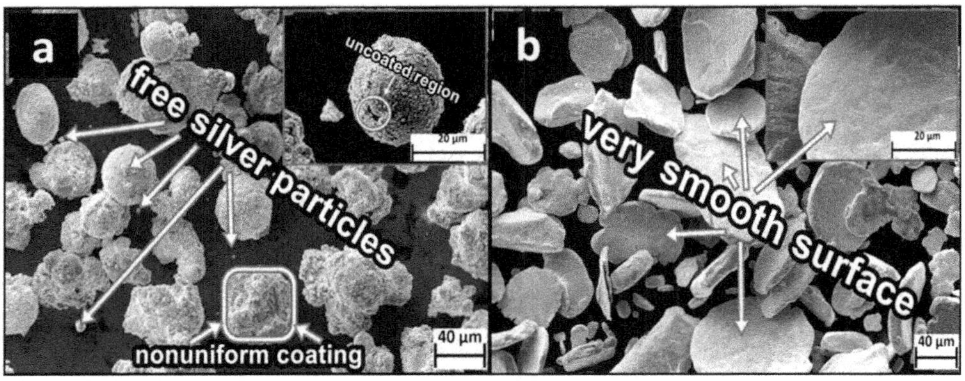

The plating of metals on the surface of other metal with the ED method depends on the salts containing the source of the desired metal, reducing agents and substances used as complexing agents. Metal salts, reducing and complexing agents used in ED of some metals are given in Table 2.

Figure 12. SEM images of the silver adsorbed copper powders with (a) low and (b) high magnification (Peng et al., 2012)

For example, copper-silver core-shell powders obtained by ED are used in many electronic fields, especially conductive adhesives (Sancaktar & Bai, 2011). The excellent electrical conductivity and oxidation resistance of silver has led researchers to work on electroless silver deposition on copper powders with very good electrical conductivity but lower oxidation resistivity (Chen et al., 2018; Hai et al., 2006; Shin et al., 2015; Xu et al., 2003). Some studies in the literature with ED method show that especially the coating layer on spherical surfaces is not as homogeneous as the coating layer on flat surfaces. Therefore, it should be ensured that powder metals have flake or flake-like morphology for the homogeneity of the metal layers that are desired to be obtained on powder metals by the ED method. This has a direct relationship with the specific surface area (SSA) and surface roughness. The ability of flat powder surfaces (flake) to absorb other metal ions during ED method compared to spherical surfaces is quite high (Ghosh & Paria, 2012; Güler et al., 2019; Shin et al., 2015). The difference between the shell coating layer obtained by the ED method using powders with spherical morphology as the core and the shell coating layer obtained from the powders using flake-like morphology is clearly observed in the SEM images of Figure 11. As can be seen in Figure 11a, spherical core morphology caused the nonhomogeneous coating layer with some uncoated regions and free silver particles remaining as unadsorbed on the core particles. Figure 11b shows the core particles in same composition with that of Fig 11a with flake-like morphology. As can be understood, smooth and homogenous coating layer adsorbed onto the core particles without free silver particles among the powders.

Figure 13. Ni coated elemental Ti powders with (a) low and (b) high magnification (Ramaseshan et al., 2001)

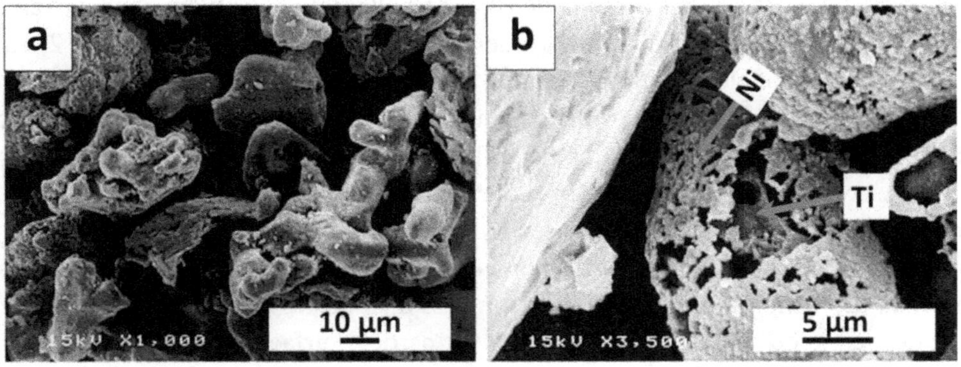

Another critical factor in obtaining metal-metal core-shell powders by the ED method is the particle size. This factor actually has a direct relationship with the surface area. The smaller the powder size, the greater the surface area and the area where the reduced ions can be adsorbed will decrease compared to the coarser powders. Besides, spherical powders have huger surface area than that of flake powders. It is because of the huge specific surface area value of the flake powders due to their high diameter to thickness aspect ratio. This high aspect ratio value can easily increase the chemical reactivity of the powders. Therefore, reducing of the metals with ED on the flake particles with higher specific surface area is easier than that on the spherical powders with higher surface area (S. Kumar, 2015). On the other hand, it will be inevitable to encounter the powders remaining uncoated by reducing the size of the powder, causing the powders to agglomerate in the ED bath (Xu et al., 2009; Zhang et al., 2012). Figure 12 shows the SEM images of the silver coated copper core powders fabricated by ED method. The copper core powders with the size less than 2 μm in this process. As can be seen in Figure 12a-b, very rough surface is observed by silver shell on the copper powders. This phenomenon means that copper powders agglomerated during the ED process and silver layers adsorbed on the agglomerated copper core powders (Peng et al., 2012).

One of the biggest problems in the powder metallurgy process is that a homogeneous distribution of different particles during sintering after pressing. Consequently, pressing and sintering of the powder particles with suitable coatings by techniques such as ED before starting the powder metallurgy stages directly affects the homogeneous bulk materials (Ramaseshan et al., 2001). Moreover, the problem of low oxidation resistance of intermetallic phases that will occur during the fabrication of materials as a result of high temperature process such as sintering or hot pressing (Jia et al., 2019; Nouri et al., 2019). However, very low oxidation resistance of metallic particles such as copper and titanium can be improved by coated with high oxidation resistance metals such as nickel. With this regard, a SEM image example of Ni coated titanium powders obtained by ED method were given in Figure 13 (Ramaseshan et al., 2001).

Fabrication of Metal-ceramic FGPs

Ceramic-metal core-shell FGPs describes the composite powders with different properties owing to the core's and shell's features in same powder structure. For example, with high electrical, magnetic and chemical properties in the same structure can provide to obtain the advanced powders by electroless deposited ceramic powders with metals. Moreover, the weak wettability of metals with ceramic powders for composite materials, metal-ceramic FGPs can be alternative to obtain good wettability and the structure with good distributed particles (Alirezaei et al., 2013; Hajizamani et al., 2012; Kilicarslan et al., 2012). The difficulty of electrolytically coating the powder components by applying current and the lack of homogeneous layer thickness of the coatings obtained by the electrolysis method encouraged metal-ceramic FGPs to be obtained by ED method. Accordingly, in the ED method, a metal layer can be formed on the ceramic powder surfaces by reducing the metal from metal salts containing metal ions by reducing agents as described in previous section of production of metal-metal FGPs (Djokić, 2005; Kang et al., 2012). However, the process of obtaining FGPs by metal coating on ceramic particles by electroless metal deposition involves a different process than that of metal particles. For metallic coatings that are desired to be applied on metallic powder particle surfaces with ED, no activation process is required, while the ceramic powder particles used as core material should be activated and metallized for metal coatings. In Figure 14, the process steps for metal coating by using ED on ceramic powder particles are given schematically.

Figure 14. ED steps of ceramic cores/particles

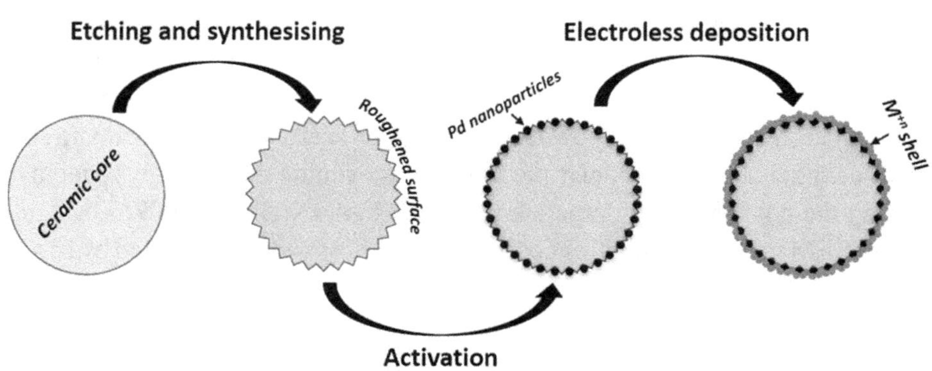

Figure 15. The SEM images of (a) SiC, (b) Ni coated SiC FGPs and (c) these FGPs reinforced Al matrix composite microstructure (Leon & Drew, 2000)

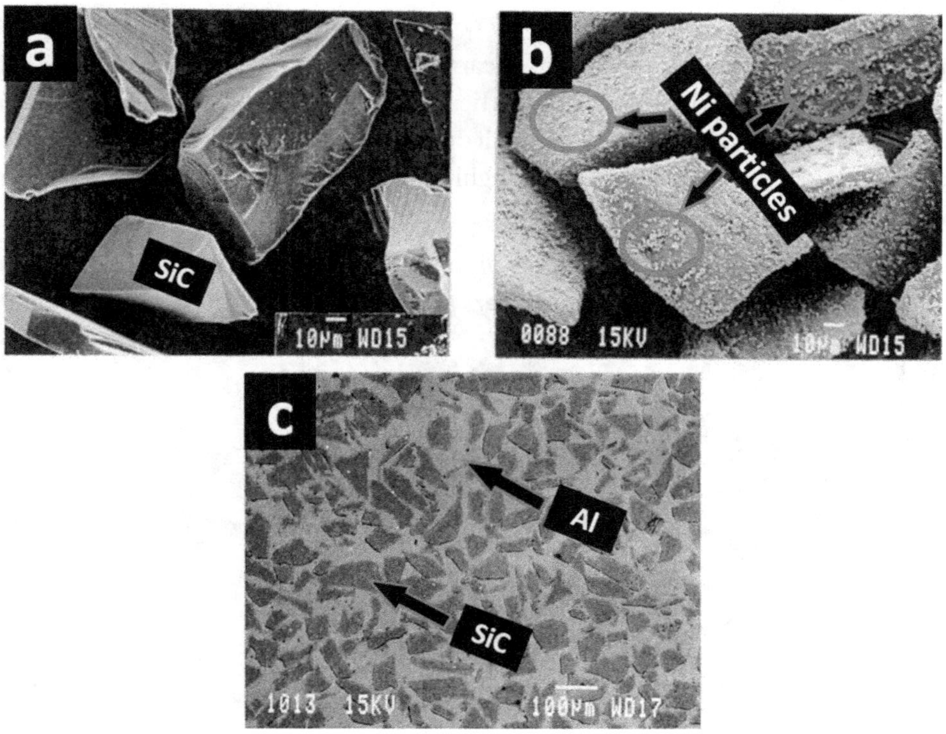

For metal coating on ceramic particles with ED, the core particles used as ceramics need to be metalized. In other words, the ceramic particle surface must be autocatalytically active in order to create the shell coating metal layer desired to be formed on the ceramic particle in ED applications (Pancrecious et al., 2018). FGPs with ceramic core that are intended to be produced by the ED method must first be purified from impurities that may be present on their surface. Although the surface cleaning process carried out in this way may seem like a simple process, it plays a critical role for the final ED coating layer. Any contamination on the core ceramic particle can cause metal ions to adhere to this contamination during the ED process. As a result, the formation of free metal ions/particles can be seen without being attached to the core ceramic surface (Muench et al., 2014). Due to the mentioned advantages regarding surface cleaning, ceramic core surfaces must be cleaned primarily in alkaline solutions because the alkaline solutions are very good cleaner. In addition, the use of alkaline solutions such as sodium hydroxide solution for cleaning ceramic particles removes the impurities on the ceramic particle due to its hydroxide content and makes the surface ready for activation by etching. The micro-rough pits that occur on the surfaces after the surface cleaning process enable the synthesis process and the activation process to be carried out perfectly. Synthesis is the process of placing Sn^{+2} ions in the pits of the surface for ceramic core particles that will be coated with ED. By taking the ceramic particles into the solution containing $SnCl_2$ and acid stabilizers such as HCl and trifluoroacetic acid, Sn^{+2} monolayers are attached to the etching pits formed on the ceramic particles and make the ceramic particles ready for activation. The activation process applied after the synthesis process is the process of metallizing the ceramic particles for ED. Ceramic particles containing Sn^{+2} ions in surface roughness are usually taken into $PdCl_2$ + HCl solution for this process and Pd^{+2} metal ions are reduced on Sn^{+2} ions. Ceramic surfaces whose surface is reduced to Pd^{+2} ions are metalized by activation and become ready for ED. Ceramic-metal core-shell particles, which are desired to be obtained by ED of metals, are taken into the metal baths mentioned under the previous heading and ED processes are carried out (Kim et al., 2013; Wei et al., 2014) as shown in Figure 15. New materials with desired properties can be produced by coating many ceramic-based particles such as Al_2O_3, SiC, B_4C, ZrO_2, Si_3N_4 with different metals because of their improved properties such as wear, corrosion, microhardness etc. (Araghi & Paydar, 2010; Mazaheri & Allahkaram, 2012).

Figure 16. The microstructure of bulk samples produced from (a) copper coated aluminum (Zhang et al., 2019) and (b) nickel coated aluminum powders (Hanyaloglu et al., 2001) by ED method

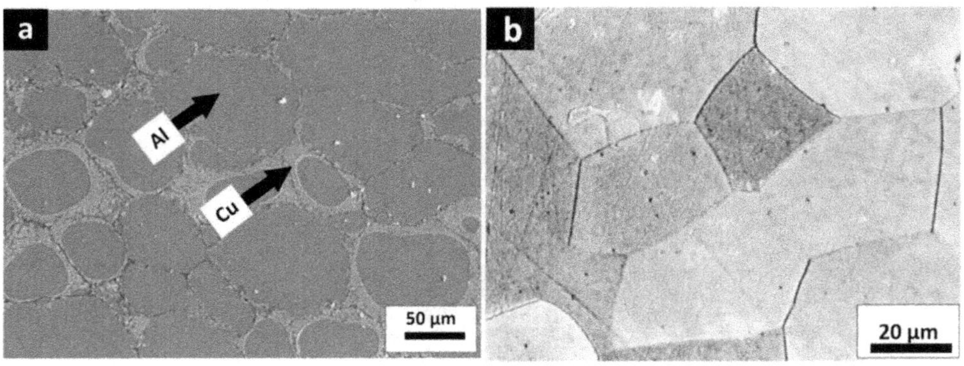

Ceramic-metal core-shell FGPs are mostly used to increase the wetting properties of ceramic reinforcing materials in the metal matrix, during the production of metal matrix composites by the stir casting method. Since the wettability of the ceramic reinforcement is poor with metals, the fact that the ceramic particles are agglomerated in the metal matrix directly affects the physical and mechanical properties of the final composites directly. Therefore, the addition of ceramic reinforcing particles into the matrix after being coated with suitable metals in the matrix minimizes the problem of agglomeration (Elsharkawi & Esawi, 2018; H. Wang et al., 2011). As can be seen in Figure 15, when the SiC particles is coated with Ni by ED method and these FGPs used as reinforcement materials in Al matrix, a uniform microstructure of composite can be obtained because of improved wettability of SiC particles because of its Ni shell on the core. This phenomenon provided the homogenous distribution of SiC particles in Al matrix as seen in Figure 15c.

SOLUTIONS AND RECOMMENDATIONS

FGPs that are produced as functional (layered) with the ED method bring many advantages by being bulk into powder metallurgy method. Layers with different composition around each powder grain provide a homogeneous distribution in the microstructure, greatly improving the final product properties (Hanyaloglu et al., 2001; Loto, 2016; Zhang et al., 2019). Figure 16 exhibits the SEM images of the microstructures obtained from electroless copper coated aluminum powders (Figure 16a) and electroless nickel coated aluminum powders (Figure 16b) with powder metallurgy and sintering processes. As can be understood, copper regions were homogenously distributed among the grain boundaries and in Figure 16b, Ni_3Al alloy created from nickel coated aluminum elemental powders without porosity, concentratedly. This can be contributed that improved properties of the bulk materials can be produced from electroless deposited metal powders (FGPs) owing to distribution of the coating layers onto the substrate metal cores (Hanyaloglu et al., 2001; Zhang et al., 2019)

Figure 17. The electrical conductivity bridges created by copper-silver FGPs in the microstructure of an electrical contact material

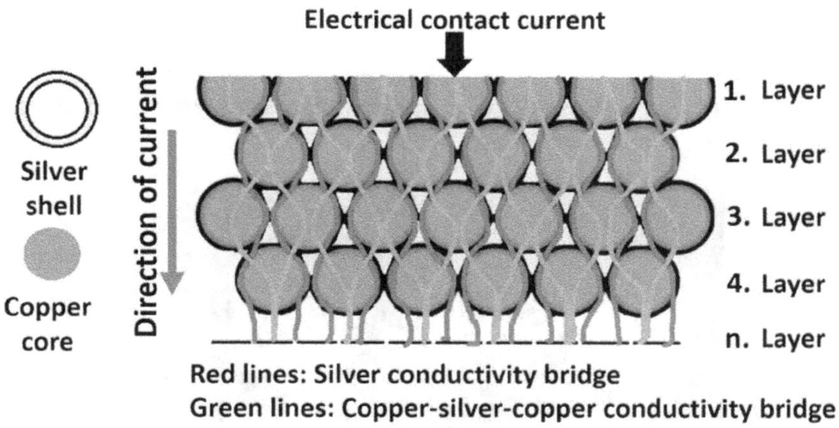

Figure 18. The SEM image of the microstructure of bulk material obtained from copper-silver core-shell FGPs

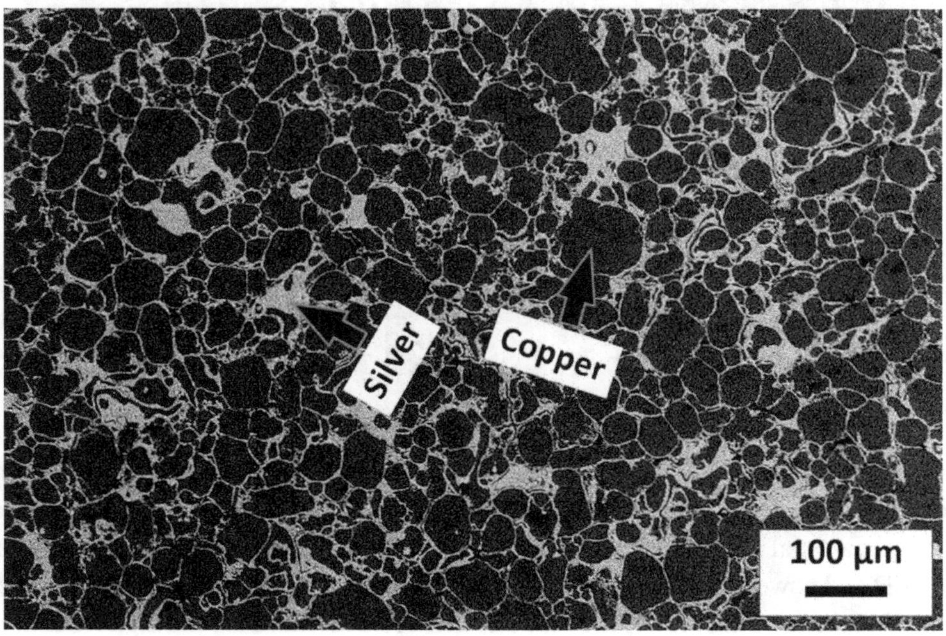

Figure 19. The SEM images of (a) Cu-Ag-Ni FGPs and the microstructure of bulk samples with Cu-Ag-Ni FGPs

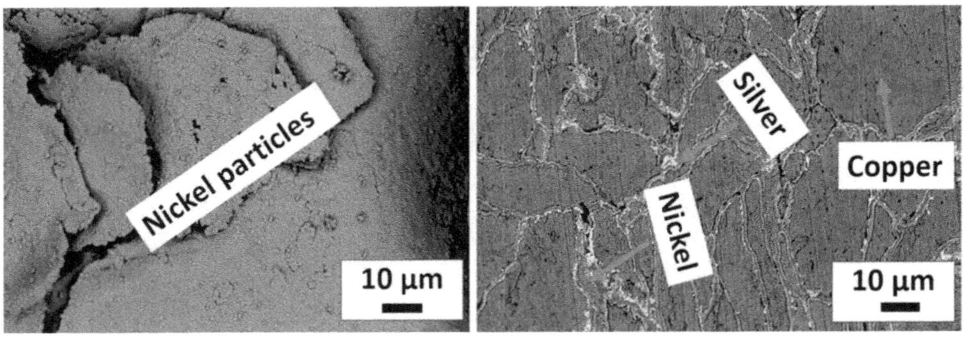

Figure 20. SEM images of (a) Co deposited SiC FGPs, the microstructure of Al matrix (b) undeposited and (c) Co deposited SiC reinforced composites (Noori et al., 2016)

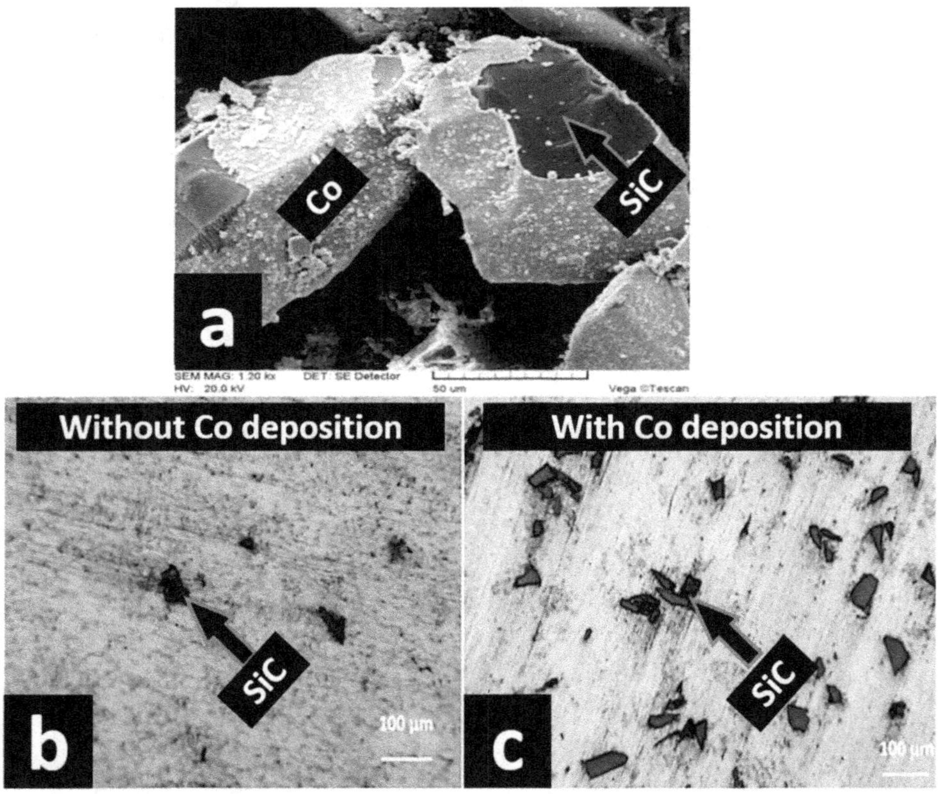

Silver is the metal with the best electrical and thermal conductivity makes it indispensable to use silver in such electrical and energy applications as electrical contact materials (Guo et al., 2019; Oluwalowo et al., 2019; Pompanon et al., 2019). In this context, materials obtained by silver coating on copper FGPs fabricated by ED and powder metallurgy followed by hot pressing method can be used for these applications. Considering the working principle of the electrical contact material, the high oxidation resistance and thermal conductivity of the material under the influence of temperature contributes to its resistance to arc erosion. The oxidation problem of copper at high temperatures causes arc formation during the operation of the electrical contact material (Güler et al., 2019). With the production of copper-silver FGPs, a small amount of silver layer surrounding copper powders will both ensure high conductivity and eliminate the oxidation problem of copper at high temperatures. The transmission of the current during the operation of the contact materials obtained from such FGPs is schematically given in Figure 17. And Figure 18 shows the microstructure of a bulk material produced from silver coated copper FGPs. There are many studies in the literature on the production of hard contact reinforced electrical contact materials with silver and copper matrix (Cui et al., 2019; Y. Wang et al., 2020; Yang et al., 2019; Zhao et al., 2020). However, the use of silver matrix composites is not in demand because silver is expensive while copper matrix composites shorten the service life of these materials due to the oxidation of copper at high temperatures. Therefore, FGMs can be used, in which silver-plated copper FGPs are used as matrices, as reinforcement of hard particles, and also in which the electrical contact regions are used with harder compositions with higher conductivity towards the substrates.

Figure 21. Microstructures of the composites including (a) copper coated graphite FGPs, (b) double copper and nickel (4.5 vol. %Ni) coated graphite FGPs, (c) those of 7 vol. %Ni FGPs and (d) 10 vol. %Ni FGPs (Yuan et al., 2019)

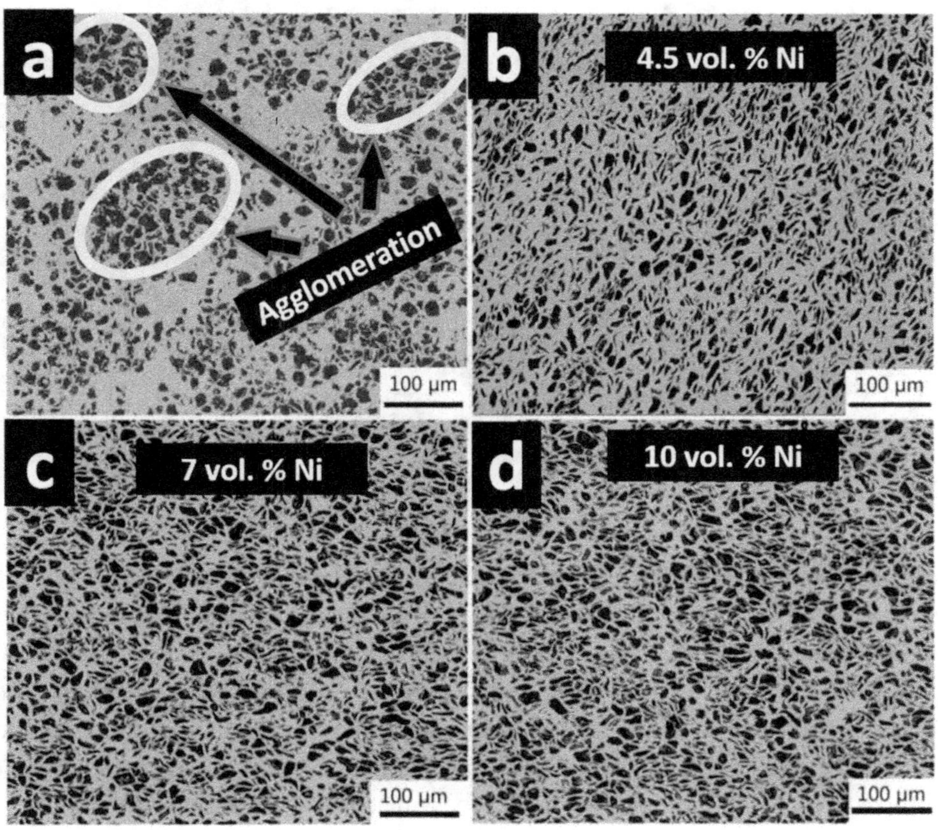

In addition, by using silver-based hard coatings made of copper materials for electrical contact materials, oxidation resistance in contact materials can be increased and wear resistance can be strengthened by hard coating (Romanov et al., 2019). However, the copper layer used as a base after wear during contact operation causes oxidation when exposed to air and when exposed to temperature, and contact performance is badly affected. For this reason, the use of FGPs in such structures may be a factor that improves the wear performance of the material, compared to other conventional composites, due to the different composition metals surrounding each grain of powders (Figure 19a).

Figure 22. Uncoated WC particles with (a) low and (b) high magnification and cobalt coated WC FGPs particles with (c) low and (d) high magnification (Tong et al., 2019)

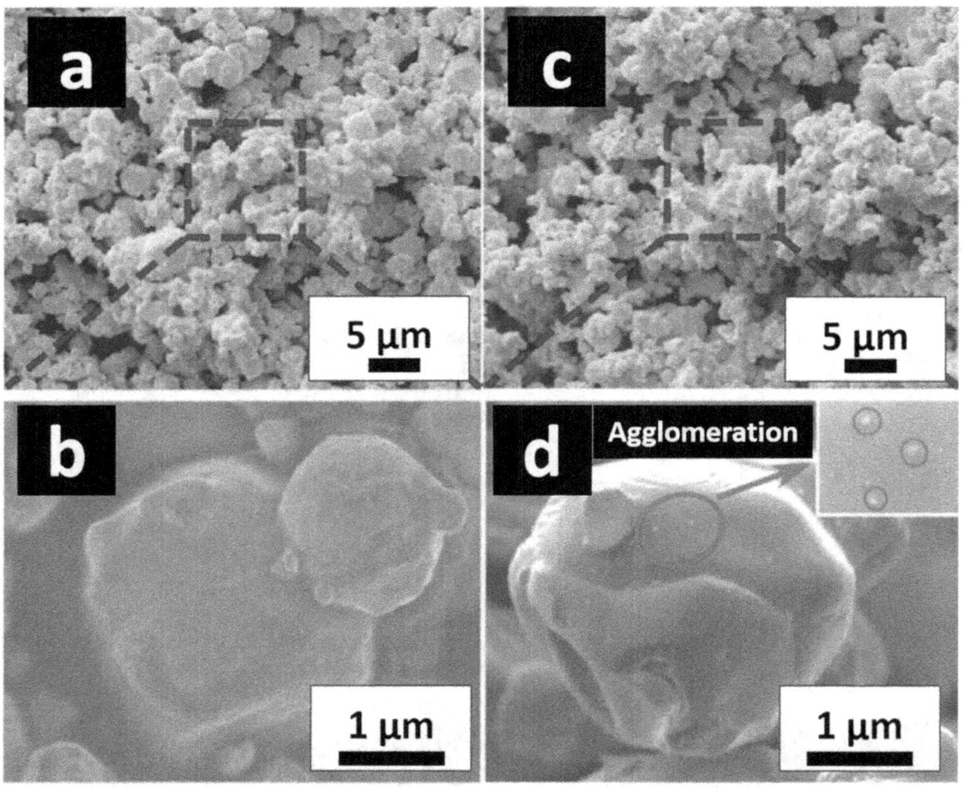

By increasing the number of layers on the metal core, material properties can be further improved. For example, 3-layer FGPs can be produced by ED on copper powders used as cores and then silver layer onto the copper powders and hard layers such as nickel on silver layer. The materials obtained by making these powders compact by powder metallurgy can be low cost (copper content), excellent electrical conductivity and oxidation resistant (silver content) and high wear resistance (nickel etc. content on the surface). By adjusting the thickness of the layer such as nickel, the decrease in electrical conductivity can be minimized and at the same time, high hardness materials can be obtained. In case the materials produced from such FGPs are subject to wear, the hard layer found as a shell around each grain will wear out, and even after the wear of an entire grain, the nickel shell layer that will come from the bottom can improve the wear resistance due to its high hardness. In this way, self-renewing materials can be obtained. Figure 19 shows the SEM images of 3-layer FGPs coated with silver on the copper core and then nickel-plated on the silver layer by the ED method.

Figure 23. Figurative representation of the production of the bulk samples including W-Cu and Cu-Ag FGPs (Yuan et al., 2019)

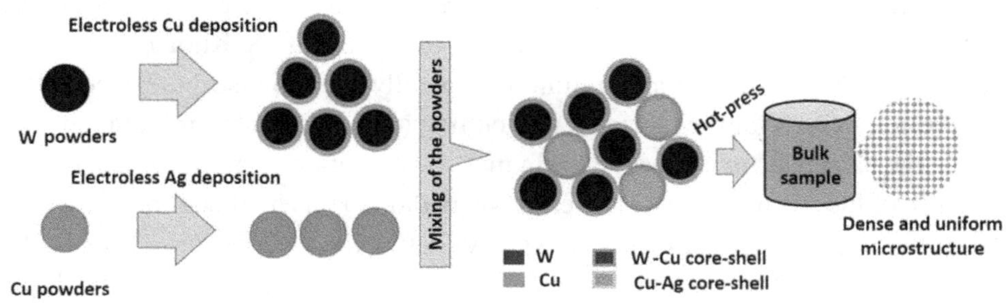

Figure 20 shows the SiC-Co core-shell FGPs (Figure 20a), the microstructure of Al matrix composites with the pure SiC reinforcement ceramic particles (Figure 20b) and with the Co deposited SiC reinforcement FGPs in Al matrix. It is clear that the homogenous reinforcement distribution in the Al matrix can be obtained with using the metal coated ceramic particles because of the good wettability of ceramic particles between the phase boundaries of the Al matrix and reinforcement ceramics (Noori et al., 2016).

Figure 24. Microstructure of composites produced from (a) Cu coated W FGPs and (b) Cu coated W FGPs mixed with Ag coated Cu FGPs (Yuan et al., 2019)

Figure 21 shows the effect of Ni content on the microstructure of the composites including copper coated graphite FGPs. As seen, when the copper coated graphite FGPs are compacted, the graphite agglomeration is existed in some regions resulted with weak physical and mechanical properties of the composites. When a nickel layer is created between the graphite and copper layers, on the other words, when the graphite powders is deposited with electroless nickel first and then copper shell is formed on the nickel layer, according to nickel content the graphite particles is distributed uniformly in the microstructure as seen in Figure 21 b-c-d (Yuan et al., 2019). This situation contributes higher hardness, compressive strength and relative density of composites.

On the other hand, some ceramic-based particles, such as WC, are often used because they provide properties such as high hardness and wear resistance. However, when these particles are compacted, the compacted material shows high brittleness. Therefore, a soft metal phase needs to be incorporated into the WC, a softer phase that melts at the sintering temperatures of the WC. Metals such as nickel, iron and cobalt are often used as binders in such cases (Herd et al., 2018; Kumar et al., 2011; Liu et al., 2018; Moustafa et al., 2011). Mechanical milling is generally used to distribute such soft phases (Co) in hard matrix particles (WC) for the fabrication of cutting tools. However, problems such as the risk of impurities due to the balls used in mechanical grinding and oxidation of metal particles during grinding encouraged the use of the ED method as an alternative method to this method. For instance, compacting the WC-cobalt core-shell FGPs (Figure 22) obtained by coating the WC powders (Tong et al., 2019) with the cobalt using the ED method, can form a bridge connecting each grain of the cobalt phase located between the WC grains. Thus, obtaining materials with very low porosity, high mechanical and physical properties can easily be achieved.

Figure 25. FGPs of (a) copper coated carbon nanotubes and the (b) microstructure of composite produced from these FGPs (Daoush, 2008)

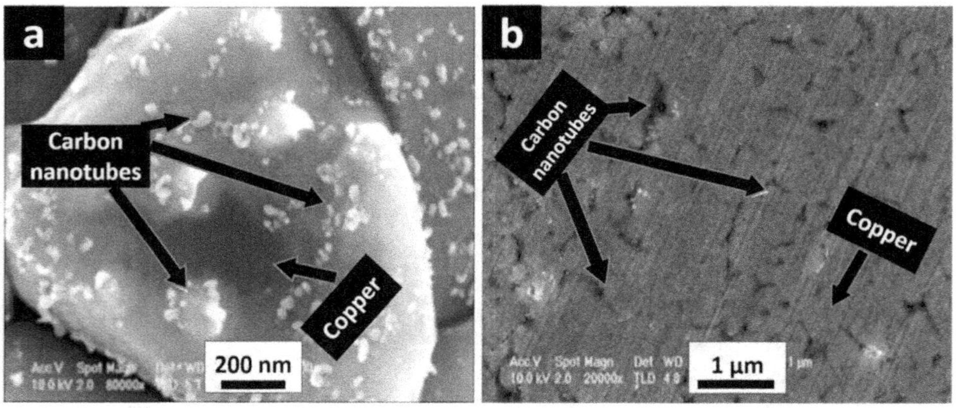

The production of composites obtained by making metal-metal and ceramic-metal core-shell FGPs compact by powder metallurgy method has also been developed recently. It has been demonstrated that the FGPs obtained by the ED method clearly improve the properties of the final microstructure. For example, the high hardness of W powders encourages the use of bulk materials obtained by pressed these powders in many areas. In areas where electrical conductivity and high hardness are desired, W-Cu composites are produced. However, since the bulk materials obtained from ceramic-metal components with minimum porosity will directly affect high hardness and electrical conductivity, first of all, Cu-coated W FGPs are mixed with Ag-coated Cu FGPs and compact metallurgy has achieved superior hardness and electrically conductive materials. A schematic representation of mentioned example production method and the materials used are given in Figure 23. With this regard, in addition to ceramic-metal core-shell FGPs, the effect of using metal-metal core-shell FGPs in the same structure is clearly seen in Figure 24b (Yuan et al., 2019).

In applications where electrical conductivity plays a critical role, the production of copper matrix and carbon nano-tube reinforced metal matrix composites is frequently encountered. However, the problem of adhesion of the carbon nanotube with the copper matrix causes the final composites to be of low density and therefore of low electrical conductivity (Sundaram et al., 2018). Therefore, it is a great advantage to obtain ceramic-metal FGPs as a result of copper coating of carbon nanotube powders with copper, primarily with ED. High density and electrically conductive materials can be easily obtained by subsequent aggregation of these FGPs by powder metallurgy or spark plasma sintering. The SEM images of copper coated carbon nanotubes and the microstructure of copper-coated carbon nanotubes composite materials obtained by making these FGPs compact is shown in Figure 25 (Daoush, 2008).

As a result, materials with many different properties can be obtained thanks to the bulk structures of metal-metal and / or ceramic-metal core-shell powders by powder metallurgy, thanks to their core structures with different properties. In addition, high electrical conductivity, hardness and wear resistance can be achieved in the same structure with multiple different layers stored on the FGPs by ED. (Figure 19)

FUTURE RESEARCH DIRECTIONS

As explained in the focus of this book chapter, the production of FGPs is commonly conducted by researchers. In the light of literature researches, FGPs produced are often added into matrix materials to obtain composites with advanced properties. The importance of direct bulk agglomeration of FGPs with metal or ceramic layers containing different and multiple electroless coating layers is emphasized in this study. Besides, the effect of the ED layers of FGPs used in these bulk materials on the physical and/or mechanical, wear, and corrosion properties of the materials may be a subject of researches in the future. For example, it has been observed in the literature that copper and silver-based composite materials are frequently used in the production of electrical contact materials. However, in order to increase the hardness and wear resistance of the materials, hard particles are used as a contribution to the copper and silver matrix. Although these hard particles added to the structure increase the hardness and wear resistance of the composites, they cause the electrical conductivity values to decrease significantly (Tian et al., 2006; X. Zhang et al., 2019, Mu et al., 2013). In such cases, thin layers such as nickel and chromium can be formed, for example, with high hardness, wear and corrosion resistance formed on copper or silver metal powders. By adjusting the shell layer thickness with different parameters in the ED method, particles with high hardness and abrasion resistance can be obtained without experiencing much decrease in electrical conductivity. In the structure of the compact produced from these particles, hard layer materials on each base material (silver, copper, etc.) can provide continuous renewal during wear. On the other hand, for the EMI shielding properties of materials, nickel ED treatment is frequently applied on polymer materials in the literature (R. H. Guo, Jiang, Yuen, Ng, & Zheng, 2012; R. H. Guo, Jiang, Yuen, Ng, & Lan, 2013; Moazzenchi & Montazer, 2019). In this case, the nickel layer obtained on the materials is coated only on the surface of the material. however, it is envisaged that applying nickel ED treatment on polymer pellets can then increase the EMI shielding properties of bulk material by pressing these particles. In this way, the nickel layer around each grain will be located on the entire microstructure and surface of the material and can improve the magnetic properties of the material.

CONCLUSION

In this chapter, the authors focused on ED method and its importance as coating method, the fundamentals of functionally graded materials and their importance in material science, powder metallurgy and its importance as fabrication methods in engineering applications. Moreover, fabrication and characterization of functionally graded metal and ceramic powders by ED method were discussed with the literature works. Although there are many studies in the literature on the production of metal-metal and ceramic-metal core-shell powders, studies on the core-shell FGPs by powder metallurgy and hot-pressing method are limited. FGPs fabricated by the ED method are used for the production of advanced materials with different properties in a single structure in many areas such as conductive adhesives, electromagnetic shielding with high electrical conductivity, antibacterial applications with high oxidation resistance and ceramic based powders with high wettability in metal matrix.

After the layered metallic powders are obtained by ED method, bulk materials can be obtained by sintering these powders by making them compact with powder metallurgy. After the layered metallic powders are obtained by ED method, bulk materials can be obtained by sintering these powders by making them compact with powder metallurgy. For example, bulk materials containing FGPs can be produced by powder metallurgy as a result of nickel ED being formed on the silver layer after silver coating of copper powders with ED. In this way, many material properties can be achieved in powder metallurgy products since each powder grain has both high electrical and thermal conductivity and high wear and corrosion resistance. In addition, such materials minimize the wear of the nickel layer on the top layer when exposed to wear, when this layer is worn, the nickel layer on the other grain from below acts as an anti-wear. In this way, materials that are self-renewing can be produced with high abrasion resistance. In addition, the ED layer thickness contained in the layered products obtained with ED can be changed by changing the ED parameters. On the other hand, with the production of ceramic-based FGPs, high matrix compatible powders with the good wettability can be obtained for metal matrix composites. In addition, when it is desired to manufacture ceramic bulk materials with high electrical conductivity, it is possible to coat high electrically conductive metals on ceramic particles. As a result of the metal coated ceramic FGPs obtained by powder metallurgy, the conductive metal formed in the grain boundaries can be produced both ceramic based and high electrical conductivity materials by forming a conductivity bridge throughout the material structure. It is also possible to replicate the procedures in obtaining biocompatible materials.

ACKNOWLEDGMENT

The authors would like to thank Scientific and Technological Research Council of Turkey (TÜBİTAK) for their contribution with the projects numbered 218M208 and 119M398.

REFERENCES

Afzal, M. A. F., Kesarwani, P., Reddy, K. M., Kalmodia, S., Basu, B., & Balani, K. (2012). Functionally graded hydroxyapatite-alumina-zirconia biocomposite: Synergy of toughness and biocompatibility. *Materials Science and Engineering C*, *32*(5), 1164–1173. doi:10.1016/j.msec.2012.03.003

Ahn, W., Taylor, B., Dall'Asén, A. G., & Roper, D. K. (2008). Electroless gold island thin films: Photoluminescence and thermal transformation to nanoparticle ensembles. *Langmuir*, *24*(8), 4174–4184. doi:10.1021/la703064m

Akavci, S. S. (2014). Thermal buckling analysis of functionally graded plates on an elastic foundation according to a hyperbolic shear deformation theory. *Mechanics of Composite Materials*, *50*(2), 197–212. doi:10.1007/s11029-014-9407-1

Ali, H. O., & Christie, I. R. A. (1985). A Review of Electroless Gold Deposition Processes. *Circuit World*, *11*(4), 10–16. doi:10.1108/eb043759

Alirezaei, S., Monir Vaghefi, S. M., Ürgen, M., Saatchi, A., & Kazmanli, K. (2013). Evaluation of structure and mechanical properties of Ni-P-Al2O3 nanocomposite coatings. *Journal of Composite Materials*, *47*(26), 3323–3329. doi:10.1177/0021998312464082

Araghi, A., & Paydar, M. H. (2010). Electroless deposition of Ni-P-B4C composite coating on AZ91D magnesium alloy and investigation on its wear and corrosion resistance. *Materials & Design*, *31*(6), 3095–3099. doi:10.1016/j.matdes.2009.12.042

Arsha, A. G., Jayakumar, E., Rajan, T. P. D., Antony, V., & Pai, B. C. (2015). Design and fabrication of functionally graded in-situ aluminium composites for automotive pistons. *Materials & Design*, *88*, 1201–1209. doi:10.1016/j.matdes.2015.09.099

Askari, E., Mehrali, M., Metselaar, I. H. S. C., Kadri, N. A., & Rahman, M. M. (2012). Fabrication and mechanical properties of Al2O3/SiC/ZrO2 functionally graded material by electrophoretic deposition. *Journal of the Mechanical Behavior of Biomedical Materials*, *12*, 144–150. doi:10.1016/j.jmbbm.2012.02.029

Barbosa-Canovas, G. V., Ortega-Rivas, E., Juliano, P., & Yan, H. (2005). Food Powders: Physical Properties, Processing, and Functionality. In Food Engineering Series (1st ed.). doi:10.100713398-014-0173-7.2

Benaben, P. (2010). An overview of hard chromium plating using trivalent chromium solutions. National Association for Surface Finishing Annual Conference and Trade Show 2010, SUR/FIN 2010, 638–646.

Beygi, H., Sajjadi, S. A., & Zebarjad, S. M. (2012). An optimization analysis on electroless deposition of Al2O3/Cu core-shell nanostructures. *Applied Surface Science*, *261*, 166–173. doi:10.1016/j.apsusc.2012.07.134

Beygi, H., Vafaeenezhad, H., & Sajjadi, S. A. (2012). Modeling the electroless nickel deposition on aluminum nanoparticles. *Applied Surface Science*, *258*(19), 7744–7750. Advance online publication. doi:10.1016/j.apsusc.2012.04.132

Bhattacharyya, M., Kumar, A. N., & Kapuria, S. (2008). Synthesis and characterization of Al/SiC and Ni/Al2O3 functionally graded materials. *Materials Science and Engineering A*, *487*(1–2), 524–535. doi:10.1016/j.msea.2007.10.040

Brachmann, E., Seifert, M., Neumann, N., Alshwawreh, N., Uhlemann, M., Menzel, S., Acker, J., Herold, S., Hoffmann, V., & Gemming, T. (2019). Electroless-Deposited Platinum Antennas for Wireless Surface Acoustic Wave Sensors. *Materials (Basel)*, *12*(7), 1002. doi:10.3390/ma12071002

Brackx, E., Pages, S., Dugne, O., & Podor, R. (2015). Recent analytical developments for powder characterization. *The European Physical Journal. Special Topics*, *224*(9), 1787–1804. doi:10.1140/epjst/e2015-02499-2

Brunner, H., & Picalek, J. (2012). Patent No. EP3026143A1. Retrieved from https://patents.google.com/patent/US20170335462A1/en

Camurlu, H. E. (2019). Al matrix FGM having 5, 15 and 25% B 4 C p reinforced layers. In Machines. Technologies. Materials (Vol. 13). Scientific Technical Union of Mechanical Engineering "Industry 4.0."

Canakci, A., Varol, T., Özkaya, S., & Erdemir, F. (2014). Microstructure and Properties of Al-B4C Functionally Graded Materials Produced by Powder Metallurgy Method. *Universal Journal of Materials Science*, *2*(5), 90–95. doi:10.13189/ujms.2014.020502

Chang, C., Lei, Z., Li, Y., & Wang, Z. (2016). Electrochemical Research of a Stable Electroless Silver Bath. *Journal of the Electrochemical Society*, *163*(3), 121–125. doi:10.1149/2.1101603jes

Chebiam, R. V., Dubin, V. M., & Simka, H. S. (2004). Patent No. US20040096592A1. Retrieved from https://patents.google.com/patent/US20040096592A1/en

Chen, F., Jia, M., She, Y., Wu, Y., Shen, Q., & Zhang, L. (2020). Mechanical behavior of AlN/Mo functionally graded materials with various compositional structures. *Journal of Alloys and Compounds*, *816*, 152512. doi:10.1016/j.jallcom.2019.152512

Chen, P., Zhou, Z., Jiang, W., Luo, W., Yang, J., Zhu, J., & Fan, Y. (2018). Enhancing the thermoelectric performance of filled skutterudite nanocomposites in a wide temperature range via electroless silver plating. *Scripta Materialia*, *146*, 136–141. doi:10.1016/j.scriptamat.2017.11.019

Choi, E. B., & Lee, J. H. (2019). Dewetting behavior of Ag in Ag-coated Cu particle with thick Ag shell. *Applied Surface Science*, *480*, 839–845. doi:10.1016/j.apsusc.2019.02.221

Choy, K. (2003). Chemical vapour deposition of coatings. *Progress in Materials Science*, *48*(2), 57–170. doi:10.1016/S0079-6425(01)00009-3

Cui, R., Han, Y., Zhu, Z., Chen, B., Ding, Y., Zhang, Q., Wang, Q., Ma, G., Pei, F., & Ye, Z. (2019). Investigation of the structure and properties of electrodeposited Cu/graphene composite coatings for the electrical contact materials of an ultrahigh voltage circuit breaker. *Journal of Alloys and Compounds*, *777*, 1159–1167. doi:10.1016/j.jallcom.2018.11.096

Daoush, W. M. (2008). Processing and characterization of CNT/Cu nanocomposites by powder technology. *Powder Metallurgy and Metal Ceramics*, *47*(9–10), 531–537. doi:10.1007/s11106-008-9055-x

Deckert, C. A. (1995). Electroless copper plating a review: Part II. *Plating and Surface Finishing*, *82*(3), 58–64.

Deepa, J. P., Rajan, T. P. D., Pavithran, C., & Pai, B. C. (2012). Formation of Electroless Ni-B Coating on Boron Carbide Particles for Composite Processing. *Materials Science Forum*, *710*, 359–364. www.scientific.net/MSF.710.359 doi:10.4028/www.scientific.net/MSF.710.359

Dikici, B., Tekmen, C., Gavgali, M., & Cocen, U. (2011). The effect of electroless Ni coating of SiC particles on the corrosion behavior of A356 Based Squeeze Cast Composite. Strojniski Vestnik/Journal of Mechanical Engineering, 57(1), 11–20. doi:10.5545v-jme.2010.111

Djokić, S. S. (2005). Electroless Deposition of Metals and Alloys. In Modern Aspects of Electrochemistry (pp. 51–133). doi:10.1007/0-306-47604-5_2

Dorfman, M. R. (2018). Thermal spray coatings. In Handbook of Environmental Degradation Of Materials: Third Edition (pp. 469–488). doi:10.1016/B978-0-323-52472-8.00023-X

Edwin, A., Anand, V., & Prasanna, K. (2017). Sustainable development through functionally graded materials: An overview. *Rasayan Journal of Chemistry*, *10*(1), 149–152. doi:10.7324/RJC.2017.1011578

El-Galy, I. M., Saleh, B. I., & Ahmed, M. H. (2019). Functionally graded materials classifications and development trends from industrial point of view. SN Applied Sciences, 1(11), 1–23. doi:10.1007/s42452-019-1413-4

Elsharkawi, M., & Esawi, A. M. K. (2018). Development of an Electroless Plating Process for Multi-wall Carbon Nanotubes (MWCNTS) to Improve Their Dispersion and Wettability in Molten Aluminum. In Minerals, Metals and Materials Series (pp. 29–39). doi:10.1007/978-3-319-72853-7_3

Erdemir, F., Canakci, A., & Varol, T. (2015). Microstructural characterization and mechanical properties of functionally graded Al2024/SiC composites prepared by powder metallurgy techniques [English Edition]. *Transactions of Nonferrous Metals Society of China*, *25*(11), 3569–3577. doi:10.1016/S1003-6326(15)63996-6

Fan, G., Xu, R., Tan, Z., Zhang, D., & Li, Z. (2014). Development of Flake Powder Metallurgy in Fabricating Metal Matrix Composites: A Review. *Acta Metallurgica Sinica. English Letters*, *27*(5), 806–815. doi:10.1007/s40195-014-0148-x

Ferreira, S. C., Rocha, L. A., Ariza, E., Sequeira, P. D., Watanabe, Y., & Fernandes, J. C. S. (2011). Corrosion behaviour of Al/Al3Ti and Al/Al3Zr functionally graded materials produced by centrifugal solid-particle method: Influence of the intermetallics volume fraction. *Corrosion Science*, *53*(6), 2058–2065. doi:10.1016/j.corsci.2011.02.010

Gao, J., Wu, Y., Liu, L., & Hu, W. (2005). Crystallization behavior of nanometer-sized Al2O3 composite coatings prepared by electroless deposition. *Materials Letters*, *59*(2–3), 391–394. doi:10.1016/j.matlet.2004.09.032

Gasik, M. M. (2010). Functionally Graded Materials: Bulk processing techniques. *International Journal of Materials & Product Technology*, *39*(1/2), 20. Advance online publication. doi:10.1504/IJMPT.2010.034257

Ghosh Chaudhuri, R., & Paria, S. (2012). Core/Shell Nanoparticles: Classes, Properties, Synthesis Mechanisms, Characterization, and Applications. *Chemical Reviews*, *112*(4), 2373–2433. doi:10.1021/cr100449n

Gonzalez, R., Ashrafizadeh, H., Lopera, A., Mertiny, P., & McDonald, A. (2016). A Review of Thermal Spray Metallization of Polymer-Based Structures. *Journal of Thermal Spray Technology*, *25*(5), 897–919. doi:10.1007/s11666-016-0415-7

Grasso, S., Sakka, Y., & Maizza, G. (2009). Electric current activated/assisted sintering (ECAS): a review of patents 1906–2008. Science and Technology of Advanced Materials, 10(5), 053001. doi:10.1088/1468-6996/10/5/053001

Güler, O., Varol, T., Alver, Ü., & Çanakçı, A. (2019). The effect of flake-like morphology on the coating properties of silver coated copper particles fabricated by electroless plating. *Journal of Alloys and Compounds*, *782*, 679–688. doi:10.1016/j.jallcom.2018.12.229

Guo, N., Luan, B., He, F., Li, Z., & Liu, Q. (2014). Influence of flake thickness on the shape and distribution of Al2O3 particles in Al matrix composites fabricated by flake powder metallurgy. *Scripta Materialia*, *78–79*, 1–4. doi:10.1016/j.scriptamat.2013.12.019

Guo, R. H., Jiang, S. Q., Yuen, C. W. M., Ng, M. C. F., & Zheng, G. H. (2012). Influence of nickel ions for electroless Ni-P plating on polyester fabric. *Journal of Coatings Technology and Research*, *9*(2), 171–176. Advance online publication. doi:10.1007/s11998-009-9227-8

Guo, R. H., Jiang, S. X., Yuen, C. W. M., Ng, M. C. F., & Lan, J. W. (2013). Optimization of electroless nickel plating on polyester fabric. *Fibers and Polymers*. Advance online publication. doi:10.100712221-013-0459-y

Guo, W., Zhang, S., Zhu, N., Luo, D., & Wu, P. (2019). Recovery of high purity secondary silver from waste Ag/Cu electrical contacts. *Process Safety and Environmental Protection*, *127*, 197–205. doi:10.1016/j.psep.2019.05.027

Gupta, A., & Talha, M. (2015, November 1). Recent development in modeling and analysis of functionally graded materials and structures. *Progress in Aerospace Sciences*, *79*, 1–14. doi:10.1016/j.paerosci.2015.07.001

Gupta, B. (2017). Few Studies on Biomedical Applications of Functionally Graded Material. Barkha Gupta International Journal of Engineering Technology Science and Research, 4(3), 39–43. Retrieved from www.ijetsr.com

Gupta, R., Tripathy, A., & Chaubey, A. K. (2020). Development of CNT Reinforced Al Matrix FGM for Advanced Applications in Lightweight and High Strength Components. In Applications of Robotics in Industry Using Advanced Mechanisms (pp. 350–356). doi:10.1007/978-3-030-30271-9_32

Hadj Mostefa, A., Merdaci, S., & Mahmoudi, N. (2018). An Overview of Functionally Graded Materials FGM. In Proceedings of the Third International Symposium on Materials and Sustainable Development (pp. 267–278). doi:10.1007/978-3-319-89707-3_30

Hai, H. T., Ahn, J. G., Kim, D. J., Lee, J. R., Chung, H. S., & Kim, C. O. (2006). Developing process for coating copper particles with silver by electroless plating method. *Surface and Coatings Technology*, *201*(6), 3788–3792. doi:10.1016/j.surfcoat.2006.03.025

Hajizamani, M., Alizadeh, A., & Ehsani, N. (2012). Deposition of a Ni3P nano-scale layer on B4C nanoparticles by simple electroless plating in an acidic bath. Applied Nanoscience (Switzerland), 2(4), 417–421. doi:10.1007/s13204-011-0055-7

Hanyaloglu, S. C., Aksakal, B., & McColm, I. J. (2001). Reactive sintering of electroless nickel-plated aluminum powders. *Materials Characterization*, *47*(1), 9–16. doi:10.1016/S1044-5803(01)00146-2

Herd, S., Wood, R. J. K., Wharton, J. A., & Higgs, C. F. III. (2018). Explicit fracture modelling of cemented tungsten carbide (WC-Co) at the mesoscale. *Materials Science and Engineering A*, *712*, 521–530. doi:10.1016/j.msea.2017.11.109

Heuer, S., Lienig, T., Mohr, A., Weber, T., Pintsuk, G., Coenen, J. W., Gormann, F., Theisen, W., & Linsmeier, C. (2019). Ultra-fast sintered functionally graded Fe/W composites for the first wall of future fusion reactors. *Composites. Part B, Engineering*, *164*, 205–214. doi:10.1016/j.compositesb.2018.11.078

Jafari, M., Enayati, M. H., Salehi, M., Nahvi, S. M., & Park, C. G. (2013). Microstructural and mechanical characterizations of a novel HVOF-sprayed WC-Co coating deposited from electroless Ni-P coated WC-12Co powders. *Materials Science and Engineering A*, *578*, 46–53. doi:10.1016/j.msea.2013.04.064

Jha, D. K., Kant, T., & Singh, R. K. (2013). A critical review of recent research on functionally graded plates. *Composite Structures*, *96*, 833–849. doi:10.1016/j.compstruct.2012.09.001

Jia, Q., Li, D., Zhang, Z., Zhang, N., & Zhao, W. (2019). Comparison of oxidation resistance of Al–Ti and Al–Ni intermetallic formed in situ by thermal spraying. Materials Research Express, 6(9), 096408. doi:10.1088/2053-1591/ab2cfc

Jiang, L., Li, Z., Fan, G., Cao, L., & Zhang, D. (2012a). Strong and ductile carbon nanotube/aluminum bulk nanolaminated composites with two-dimensional alignment of carbon nanotubes. *Scripta Materialia*, *66*(6), 331–334. doi:10.1016/j.scriptamat.2011.11.023

Jiang, L., Li, Z., Fan, G., Cao, L., & Zhang, D. (2012b). The use of flake powder metallurgy to produce carbon nanotube (CNT)/aluminum composites with a homogenous CNT distribution. *Carbon*, *50*(5), 1993–1998. doi:10.1016/j.carbon.2011.12.057

Jiang, L., Li, Z., Fan, G., & Zhang, D. (2011). A flake powder metallurgy approach to Al2O3/Al biomimetic nanolaminated composites with enhanced ductility. *Scripta Materialia*, *65*(5), 412–415. doi:10.1016/j.scriptamat.2011.05.022

Jin, G., Takeuchi, M., Honda, S., Nishikawa, T., & Awaji, H. (2005). Properties of multilayered mullite/Mo functionally graded materials fabricated by powder metallurgy processing. *Materials Chemistry and Physics*, *89*(2–3), 238–243. doi:10.1016/j.matchemphys.2004.03.031

Kai, X., Li, Z., Fan, G., Guo, Q., Tan, Z., Zhang, W., & Zhang, D. (2013). Strong and ductile particulate reinforced ultrafine-grained metallic composites fabricated by flake powder metallurgy. *Scripta Materialia*, *68*(8), 555–558. doi:10.1016/j.scriptamat.2012.11.024

Kai, X. Z., Li, Z. Q., Fan, G. L., Guo, Q., Xiong, D. B., Zhang, W. L., Su, Y. S., Lu, W. J., Moon, W. J., & Zhang, D. (2013). Enhanced strength and ductility in particulate-reinforced aluminum matrix composites fabricated by flake powder metallurgy. *Materials Science and Engineering A*, *587*, 46–53. doi:10.1016/j.msea.2013.08.042

Kang, C., Lu, H., Yuan, S., Hong, D., Yan, K., & Liang, B. (2012). Superhydrophilicity/superhydrophobicity of nickel micro-arrays fabricated by electroless deposition on an etched porous aluminum template. *Chemical Engineering Journal*, *203*, 1–8. doi:10.1016/j.cej.2012.06.128

Kawasaki, A., & Watanabe, R. (1997). Evaluation of thermomechanical performance for thermal barrier type of sintered functionally graded materials. *Composites. Part B, Engineering*, *28*(1–2), 29–35. doi:10.1016/S1359-8368(96)00017-0

Kessel, H., Hennicke, J., Kirchner, R., & Kessel, T. (2010). Rapid sintering of novel materials by FAST/SPS—Further development to the point of an industrial production process with high cost efficiency. FCT Systeme GmbH. Retrieved from www.fct-systeme.de

Keuler, J. N., Lorenzen, L., Sanderson, R. N., & Linkov, V. (1997). Optimizing Palladium Conversion in Electroless Palladium Plating of Alumina Membranes. *Plating and Surface Finishing*, *84*(8), 34–40.

Kilicarslan, A., Toptan, F., & Kerti, I. (2012). Electroless nickel-phosphorus coating on boron carbide particles. *Materials Letters*, *76*, 11–14. doi:10.1016/j.matlet.2012.02.037

Kim, J. H., Kim, J. D., Park, H. M., Lee, S. B., Yi, J. W., & Jung, S. (2013). A homogeneous Ni-P coating with a unique scalelike structure deposited on VGCNFs by an electroless deposition method. *Applied Surface Science*, *265*, 527–531. doi:10.1016/j.apsusc.2012.11.039

Kobayashi, Y., Salgueiriño-Maceira, V., & Liz-Marzán, L. M. (1630–1633). Salgueiriñ O-Maceira, V., & Liz-Marzá, L. M. (2001). Deposition of Silver Nanoparticles on Silica Spheres by Pretreatment Steps in Electroless Plating. *Chemistry of Materials*, *13*(5), 1630–1633. Advance online publication. doi:10.1021/cm001240g

Kumar, A., Singh, K., & Pandey, O. P. (2011). Sintering behavior of nanostructured WC-Co composite. *Ceramics International*, *37*(4), 1415–1422. doi:10.1016/j.ceramint.2011.01.001

Kumar, S. (2015). *Clean Hydrogen Production Methods* (1st ed.). Springer International Publishing., doi:10.1007/978-3-319-14087-2.

Kumar Chauhan, P., & Khan, S. (2019). Microstructural examination of aluminium-copper functionally graded material developed by powder metallurgy route. Materials Today: Proceedings. doi:10.1016/j.matpr.2019.10.007

Kunimine, T., Sato, H., Miura-Fujiwara, E., & Watanabe, Y. (2016). New Processing Routes for Functionally Graded Materials and Structures through Combinations of Powder Metallurgy and Casting. In F. Ebrahimi (Ed.), Advances in Functionally Graded Materials and Structures. doi:10.5772/62393

Laima, L., Xiaoyue, T., Xiaoyu, D., Zelong, L., Guangnan, L., Xiang, Z., Xiaoyong, Z., & Yucheng, W. (2016). Fabrication of W/Cu Graded Heat-sink Materials by Electroless Plating and Powder Metallurgy. *Rare Metal Materials and Engineering*, *45*(8), 1983–1987. doi:10.1016/S1875-5372(16)30158-8

Lajevardi, S. A., Shahrabi, T., & Szpunar, J. A. (2017). Tribological Properties of Functionally Graded Ni-Al 2 O 3 Nanocomposite Coating. *Journal of the Electrochemical Society*, *164*(6), D275–D281. doi:10.1149/2.0731706jes

Landillon, V., Cassan, D., Morel, M. H., & Cuq, B. (2008). Flowability, cohesive, and granulation properties of wheat powders. *Journal of Food Engineering*, *86*(2), 178–193. doi:10.1016/j.jfoodeng.2007.09.022

León, C., & Drew, R. A. (2002). The influence of nickel coating on the wettability of aluminum on ceramics. *Composites. Part A, Applied Science and Manufacturing*, *33*(10), 1429–1432. doi:10.1016/S1359-835X(02)00161-6

Leon, C. A., & Drew, R. A. L. (2000). Preparation of nickel-coated powders as precursors to reinforce MMCs. *Journal of Materials Science*, *35*(19), 4763–4768. doi:10.1023/A:1004860326071

León-Patiño, C. A., Aguilar-Reyes, E. A., Bedolla-Becerril, E., Bedolla-Jacuinde, A., & Méndez-Díaz, S. (2013). Dry sliding wear of gradient Al-Ni/SiC composites. *Wear*, *301*(1–2), 688–694. doi:10.1016/j.wear.2012.11.052

Li, Y. (2015). Process of electroless silver coated Al-Mg alloy powder. Xiyou Jinshu Cailiao Yu Gongcheng. *Rare Metal Materials and Engineering*, *44*(11), 2633–2636. doi:10.1016/S1875-5372(16)60008-5

Li, Y., Liu, R., Zhang, J., Luo, G., Shen, Q., & Zhang, L. (2019). Fabrication and microstructure of W-Cu composites prepared from Ag-coated Cu powders by electroless plating. *Surface and Coatings Technology*, *361*, 302–307. doi:10.1016/j.surfcoat.2019.01.030

Li, Y., Wang, R., Qi, F., & Wang, C. (2008). Preparation, characterization and microwave absorption properties of electroless Ni-Co-P-coated SiC powder. *Applied Surface Science*, *254*(15), 4708–4715. doi:10.1016/j.apsusc.2008.01.076

Lin, W. H., & Chang, H. F. (1998). Effect of chelating agents on the structure of electroless copper coating on alumina powder. *Surface and Coatings Technology*, *107*(1), 48–54. Advance online publication. doi:10.1016/S0257-8972(98)00553-2

Lin, Y.-S., & Chiu, S.-S. (2004). Effects of oxidation and particle shape on critical volume fractions of silver-coated copper powders in conductive adhesives for microelectronic applications. *Polymer Engineering and Science*, *44*(11), 2075–2082. doi:10.1002/pen.20212

Lin, Y.-S., & Chiu, S.-S. (2008). Electrical Properties of Copper-Filled Electrically Conductive Adhesives and Pressure-Dependent Conduction Behavior of Copper Particles. *Journal of Adhesion Science and Technology*, *22*(14), 1673–1697. doi:10.1163/156856108X320537

Liu, X., Song, X., Wang, H., Liu, X., Tang, F., & Lu, H. (2018). Complexions in WC-Co cemented carbides. *Acta Materialia*, *149*, 164–178. doi:10.1016/j.actamat.2018.02.018

Liu, X. J., Sun, W. C., Dong, Y. R., Ma, M., Zhang, Y. G., Yang, F., & Ruan, Y. Y. (2019). Microstructure and corrosion resistance of Ni-P gradient coating/stannate conversion film on magnesium alloy. *Journal of Materials Research*, *34*(6), 1064–1072. doi:10.1557/jmr.2019.21

Loto, C. A. (2016). Electroless Nickel Plating – A Review. Silicon, 8(2), 177–186. doi:10.1007/s12633-015-9367-7

Luo, H., Sui, Y., Qi, J., Meng, Q., Wei, F., & He, Y. (2017). Copper matrix composites enhanced by silver/reduced graphene oxide hybrids. *Materials Letters*, *196*, 354–357. doi:10.1016/j.matlet.2017.03.084

Luo, L., Yu, J., Luo, J., & Li, J. (2010). Preparation and characterization of Ni-coated Cr3C2 powder by room temperature ultrasonic-assisted electroless plating. *Ceramics International*, *36*(6), 1989–1992. doi:10.1016/j.ceramint.2010.03.003

Luo, Q. G., Dai, Y., Liu, S. L., Yu, K., Shen, Q., & Zhang, L. M. (2014). Fabrication and Properties of W-Cu Functionally Graded Material by Tape-Casting. *Key Engineering Materials*, *616*, 66–71. doi:10.4028/www.scientific.net/KEM.616.66

Mahamood, R. M., Akinlabi, E. T., Shukla, M., & Pityana, S. (2012). Functionally graded material: An overview. In S. I. Ao, L. Gelman, D.W.L Hukins, A. Hunter, A. M. Korsunsky (Ed.), Proceedings of the World Congress on Engineering (pp. 1593-1597). London, UK: Imperial College London.

Mallory, G. O., & Hajdu, J. B. (1990). Electroless plating: fundamentals and applications (W. Andrew, Ed.). American Electroplaters and Surface Finishers Society.

Mazaheri, H., & Allahkaram, S. R. (2012). Deposition, characterization and electrochemical evaluation of Ni-P-nano diamond composite coatings. *Applied Surface Science*, *258*(10), 4574–4580. doi:10.1016/j.apsusc.2012.01.031

Meng, Y., Zhang, J., Duan, C., Chen, C., Feng, X., & Shen, Y. (2015). Microstructures and properties of W–Cu functionally graded composite coatings on copper substrate via high-energy mechanical alloying method. *Advanced Powder Technology*, *26*(2), 392–400. doi:10.1016/j.apt.2014.11.009

Mikli, V., Kaerdi, H., Kulu, P., & Besterci, M. (2001). Characterization of powder particle morphology. Proceedings of the Estonian Academy of Sciences: Engineering, 22–34.

Mirazimi, J., Abachi, P., & Purazrang, K. (2016). Spark plasma sintering of ultrafine YSZ reinforced Cu matrix functionally graded composite. *Acta Metallurgica Sinica. English Letters*, *29*(12), 1169–1176. doi:10.1007/s40195-016-0512-0

Mizuhashi, S., Cordonier, C. E. J., Honma, H., & Takai, O. (2015). Stabilization of an Electroless Platinum Plating Bath Using S-Bearing Additives. *Journal of the Electrochemical Society*, *162*(10), D497–D502. doi:10.1149/2.0081510jes

Moazzenchi, B., & Montazer, M. (2019). Click electroless plating of nickel nanoparticles on polyester fabric: Electrical conductivity, magnetic and EMI shielding properties. *Colloids and Surfaces. A, Physicochemical and Engineering Aspects*, *571*, 110–124. Advance online publication. doi:10.1016/j.colsurfa.2019.03.065

Moustafa, S. F., Abdel-Hamid, Z., Baheig, O. G., & Hussien, A. (2011). Synthesis of WC hard materials using coated powders. *Advanced Powder Technology*, *22*(5), 596–601. doi:10.1016/j.apt.2010.08.011

Mu, Z., Geng, H.-R., Li, M.-M., Nie, G.-L., & Leng, J.-F. (2013). Effects of Y2O3 on the property of copper based contact materials. *Composites. Part B, Engineering*, *52*, 51–55. doi:10.1016/j.compositesb.2013.02.036

Muench, F., Eils, A., Toimil-Molares, M. E., Hossain, U. H., Radetinac, A., Stegmann, C., Kunz, U., Lauterbach, S., Kleebe, H.-J., & Ensinger, W. (2014). Polymer activation by reducing agent absorption as a flexible tool for the creation of metal films and nanostructures by electroless plating. *Surface and Coatings Technology*, *242*, 100–108. doi:10.1016/j.surfcoat.2014.01.024

Müller, E., Drašar, Č., Schilz, J., & Kaysser, W. A. (2003). Functionally graded materials for sensor and energy applications. *Materials Science and Engineering A*, *362*(1–2), 17–39. doi:10.1016/S0921-5093(03)00581-1

Narasimhan, K. S., & Amuda, M. O. H. (2017). Powder Characterization. In Reference Module in Materials Science and Materials Engineering. doi:10.1016/B978-0-12-803581-8.10141-9

Nassef, A., El-Garaihy, W., & El-Hadek, M. (2017). Characteristics of Cold and Hot Pressed Iron Aluminum Powder Metallurgical Alloys. *Metals*, *7*(5), 170. doi:10.3390/met7050170

Nemat-Alla, M. M., Ata, M. H., Bayoumi, M. R., & Khair-Eldeen, W. (2011). Powder Metallurgical Fabrication and Microstructural Investigations of Aluminum/Steel Functionally Graded Material. *Materials Sciences and Applications*, *02*(12), 1708–1718. doi:10.4236/msa.2011.212228

Nienow, A. W., Harnby, N., & Edwards, M. F. (1992). Introduction to mixing problems. In Mixing in the Process Industries (pp. 1–24). doi:10.1016/B978-075063760-2/50022-6

Noori, H., Mousavian, R. T., Khosroshahi, R. A., Brabazon, D., & Damadi, S. R. (2016). Effect of SiC particle morphology on Co–P electroless coating characteristics. *Surface Engineering*, *32*(5), 391–396. doi:10.1179/1743294415Y.0000000035

Nouri, S., Sahmani, S., Asayesh, M., & Aghdam, M. M. (2019). Study on the oxidation resistance of γ -TiAl intermetallic alloy coated via different diffusion coating processes. Materials Research Express, 6(10), 106522. doi:10.1088/2053-1591/ab36d5

Oluwalowo, A., Nguyen, N., Zhang, S., Park, J. G., & Liang, R. (2019). Electrical and thermal conductivity improvement of carbon nanotube and silver composites. *Carbon*, *146*, 224–231. doi:10.1016/j.carbon.2019.01.073

Ozaki, T., Zhang, Y., Komaki, M., & Nishimura, C. (2003). Preparation of palladium-coated V and V?15Ni membranes for hydrogen purification by electroless plating technique. *International Journal of Hydrogen Energy*, *28*(3), 297–302. doi:10.1016/S0360-3199(02)00065-4

Padmavathi, K. R., & Ramakrishnan, R. (2014). Tribological Behaviour of Aluminium Hybrid Metal Matrix Composite. *Procedia Engineering*, *97*, 660–667. doi:10.1016/j.proeng.2014.12.295

Pancrecious, J. K., Ulaeto, S. B., Ramya, R., Rajan, T. P. D., & Pai, B. C. (2018). Metallic composite coatings by electroless technique–a critical review. *International Materials Reviews*, *63*(8), 488–512. doi:10.1080/09506608.2018.1506692

Pandey, V. K., Patel, B. P., & Guruprasad, S. (2016). Mechanical properties of Al/Al 2 O 3 and Al/B 4 C composites. *Advanced Materials Research*, *5*(4), 263–277. doi:10.12989/amr.2016.5.4.263

Pázmán, J., Mádai, V., Tóth, J., & Gácsi, Z. (2010). Investigation of the electroless nickel plated SiC particles in metal matrix composites. *Powder Metallurgy Progress*, *10*(2), 102–109.

Peng, Y., Yang, C., Chen, K., Popuri, S. R., Lee, C.-H., & Tang, B.-S. (2012). Study on synthesis of ultrafine Cu–Ag core–shell powders with high electrical conductivity. *Applied Surface Science*, *263*, 38–44. doi:10.1016/j.apsusc.2012.08.066

Picalek, J., Holger, B., Akhtar, S., Bejan, I., & Pinnau, J. (2016). Patent No. 15/520,558. Washington, DC: US Patent Office.

Pompanon, F., Laporte, J., Fouvry, S., & Alquier, O. (2019). Normal force and displacement amplitude influences on silver-plated electrical contacts subjected to fretting wear: A basic friction energy – contact compliance formulation. *Wear*, *426–427*, 652–661. doi:10.1016/j.wear.2018.12.010

Rahimipour, M. R., & Sobhani, M. (2013). Evaluation of centrifugal casting process parameters for in situ fabricated functionally gradient Fe-TiC composite. *Metallurgical and Materials Transactions. B, Process Metallurgy and Materials Processing Science*, *44*(5), 1120–1123. doi:10.1007/s11663-013-9903-z

Ramaseshan, R., Seshadri, S., & Nair, N. (2001). Electroless nickel-phosphorus coating on Ti and Al elemental powders. *Scripta Materialia*, *45*(2), 183–189. doi:10.1016/S1359-6462(01)01013-2

Romanov, D., Moskovskii, S., Konovalov, S., Sosnin, K., Gromov, V., & Ivanov, Y. (2019). Improvement of copper alloy properties in electro-explosive spraying of ZnO-Ag coatings resistant to electrical erosion. *Journal of Materials Research and Technology*, *8*(6), 5515–5523. doi:10.1016/j.jmrt.2019.09.019

Ruigang, W., Wei, P., Jian, C., Mengning, J., & Minghao, F. (2002). Fabrication and characterization of machinable Si3N4/h-BN functionally graded materials. *Materials Research Bulletin*, *37*(7), 1269–1277. doi:10.1016/S0025-5408(02)00773-0

Samal, C. P., Parihar, J. S., & Chaira, D. (2013). The effect of milling and sintering techniques on mechanical properties of Cu-graphite metal matrix composite prepared by powder metallurgy route. *Journal of Alloys and Compounds*, *569*, 95–101. doi:10.1016/j.jallcom.2013.03.122

Sancaktar, E., & Bai, L. (2011). Electrically Conductive Epoxy Adhesives. *Polymers*, *3*(1), 427–466. doi:10.3390/polym3010427

Shacham-Diamand, Y., Inberg, A., Sverdlov, Y., Bogush, V., Croitoru, N., Moscovich, H., & Freeman, A. (2003). Electroless processes for micro- and nanoelectronics. *Electrochimica Acta*, *48*(20–22), 2987–2996. doi:10.1016/S0013-4686(03)00364-5

Sharma, S. B., Agarwala, R. C., Agarwala, V., & Ray, S. (2002). Dry sliding wear and friction behavior of Ni-P-ZrO2-Al2O3 composite electroless coatings on aluminum. *Materials and Manufacturing Processes*, *17*(5), 637–649. doi:10.1081/AMP-120016088

Shi, J. M., Zhang, L. X., Liu, H., Sun, Z., & Feng, J. C. (2019). Reliable brazing of SiBCN ceramic and TC4 alloy using AgCuTi filler with the assist of laser melting deposited FGM layers. *Journal of Materials Science*, *54*(4), 2766–2778. doi:10.1007/s10853-018-3025-6

Shin, J., Kim, H., Song, K. H., & Choe, J. (2015). Synthesis of Silver-coated Copper Particles with Thermal Oxidation Stability for a Solar Cell Conductive Paste. *Chemistry Letters*, *44*(9), 1223–1225. doi:10.1246/cl.150424

Slotwinski, J. A., Garboczi, E. J., Stutzman, P. E., Ferraris, C. F., Watson, S. S., & Peltz, M. A. (2014). Characterization of Metal Powders Used for Additive Manufacturing. Journal of Research of the National Institute of Standards and Technology, 119(), 460. doi:10.6028/jres.119.018

So, K. P., Lee, I. H., Duong, D. L., Kim, T. H., Lim, S. C., An, K. H., & Lee, Y. H. (2011). Improving the wettability of aluminum on carbon nanotubes. *Acta Materialia*, *59*(9), 3313–3320. doi:10.1016/j.actamat.2011.01.061

Somiya, S. (2013). *Handbook of Advanced Ceramics: Materials* (2nd ed.). Applications, Processing, and Properties., doi:10.1016/C2010-0-66261-4

Sudagar, J., Tamilarasan, R., Sanjith, U., Rajendran, R., & Kumar, R. (2017). Electroless Deposition of Nanolayered Metallic Coatings. In Nanoscaled Films and Layers. doi:10.5772/intechopen.68220

Sundaram, R. M., Sekiguchi, A., Sekiya, M., Yamada, T., & Hata, K. (2018). Copper/carbon nanotube composites: Research trends and outlook. *Royal Society Open Science*, *5*(11), 180814. doi:10.1098/rsos.180814

Surya, M. S., & Prasanthi, G. (2018). Tribological Behaviour of Aluminum Silicon Carbide Functionally Graded Material. Tribology in Industry, 40(2), 247–253. doi:10.24874/ti.2018.40.02.08

Tian, B., Liu, P., Song, K., Li, Y., Liu, Y., Ren, F., & Su, J. (2006). Microstructure and properties at elevated temperature of a nano-Al2O3 particles dispersion-strengthened copper base composite. *Materials Science and Engineering A*, *435–436*, 705–710. doi:10.1016/j.msea.2006.07.129

Toben, M. P., Martin, J. L., Ohta, Y., Takizawa, Y., & Enomoto, H. (2002). Patent No. US6383269B1.

Tong, J., Zhang, J., Wang, Y., Min, F., Wang, X., Zhang, H., & Ma, J. (2019). Preparation of Co-plated WC powders by a non-precious-Co-activation triggered electroless plating strategy. *Advanced Powder Technology*, *30*(10), 2311–2319. doi:10.1016/j.apt.2019.07.012

Travessa, D. N., Silva, M. J., & Cardoso, K. R. (2017). Niobium Carbide-Reinforced Al Matrix Composites Produced by High-Energy Ball Milling. *Metallurgical and Materials Transactions. B, Process Metallurgy and Materials Processing Science*, *48*(3), 1754–1762. doi:10.1007/s11663-017-0959-z

Tripathy, A., Gupta, R., Sarangi, S. K., & Chaubey, A. K. (2020). Design and Fabrication of Aluminium/Alumina Ultra-fine Composite and Functionally Graded Material Using Powder Metallurgy Route. In M. Shunmugam (Ed.), Advances in Unconventional Machining and Composites. doi:10.1007/978-981-32-9471-4_62

Übeyli, M., Balci, E., Sarikan, B., Öztaş, M. K., Camuşcu, N., Yildirim, R. O., & Keleş, Ö. (2014). The ballistic performance of SiC-AA7075 functionally graded composite produced by powder metallurgy. *Materials & Design*, *56*, 31–36. doi:10.1016/j.matdes.2013.10.092

Uysal, M., Karslioğlu, R., Alp, A., & Akbulut, H. (2013). The preparation of core-shell Al2O3/Ni composite powders by electroless plating. *Ceramics International*, *39*(5), 5485–5493. doi:10.1016/j.ceramint.2012.12.060

Van Der Biest, O., Joos, E., Vleugels, J., & Baufeld, B. (2006). Electrophoretic deposition of zirconia layers for thermal barrier coatings. *Journal of Materials Science*, *41*(24), 8086–8092. doi:10.1007/s10853-006-0635-1

Wang, F., Arai, S., & Endo, M. (2005). The preparation of multi-walled carbon nanotubes with a Ni-P coating by an electroless deposition process. *Carbon*, *43*(8), 1716–1721. doi:10.1016/j.carbon.2005.02.015

Wang, H., Jia, J., Song, H., Hu, X., Sun, H., & Yang, D. (2011). The preparation of Cu-coated Al2O3 composite powders by electroless plating. *Ceramics International*, *37*(7), 2181–2184. doi:10.1016/j.ceramint.2011.03.013

Wang, J., Li, L., & Tao, W. (2016). Crack initiation and propagation behavior of WC particles reinforced Fe-based metal matrix composite produced by laser melting deposition. *Optics & Laser Technology*, *82*, 170–182. doi:10.1016/j.optlastec.2016.03.008

Wang, Y., Gao, Y., Li, Y., Li, M., Sun, L., Zhai, W., & Li, K. (2020). Research on synergistic lubrication effect of silver modified Cu–Ni-graphite composite. *Wear*, *444–445*, 203140. doi:10.1016/j.wear.2019.203140

Wei, X., Roper, K., Roper, D. K., & Martin, R. E. (2014). Tin Sensitization for Electroless Plating Review Fundamental chromatography View project Optoplasmonic Heating with Gold Nanoparticle Arrays View project Tin Sensitization for Electroless Plating Review. *Journal of the Electrochemical Society*, *161*(5), 235–242. doi:10.1149/2.047405jes

Wessel, J. K. (2004). Handbook of Advanced Materials. In J. K. Wessel (Ed.), Handbook of Advanced Materials. doi:10.1002/0471465186

Xu, R., Zhou, K., & Hu, M. (2009). Preparation of core-shell Cu-Ag bimetallic powder via electroless coating. Journal of Wuhan University of Technology-Mater. *Science Editor*, *24*(4), 637–639. doi:10.100711595-009-4637-0

Xu, X., Luo, X., Zhuang, H., Li, W., & Zhang, B. (2003). Electroless silver coating on fine copper powder and its effects on oxidation resistance. *Materials Letters*, *57*(24–25), 3987–3991. doi:10.1016/S0167-577X(03)00252-0

Yang, R., Liu, S., Chen, J., Cui, H., Liu, M., Zhu, F., Yang, Y., Xie, M., Sun, X., & Li, X. (2019). Porous Y2O3 fiber-reinforced silver composite exhibiting enhanced mechanical and electrical properties. *Ceramics International*, *45*(2), 1881–1886. doi:10.1016/j.ceramint.2018.10.079

Yin, H. M., Sun, L. Z., & Paulino, G. H. (2004). Micromechanics-based elastic model for functionally graded materials with particle interactions. *Acta Materialia*, *52*(12), 3535–3543. doi:10.1016/j.actamat.2004.04.007

Yuan, Y., Gan, X., Lai, Y., Zhao, Q., & Zhou, K. (2019). Microstructure and properties of graphite/copper composites fabricated with Cu-Ni double-layer coated graphite powders. *Composite Interfaces*, •••, 1–15. doi:10.1080/09276440.2019.1644051

Yusong, P., Qianqian, S., Chengling, P., & Jing, W. (2013). Prediction of mechanical properties of multilayer gradient hydroxyapatite reinforced poly(vinyl alcohol) gel biomaterial. *Journal of Biomedical Materials Research. Part B, Applied Biomaterials*, *101*(5), 729–735. doi:10.1002/jbm.b.32875

Zhang, B. (2016). Green Electroless Plating. In B. Zhang (Ed.), *Amorphous and Nano Alloys Electroless Depositions* (pp. 383–409). Elsevier., doi:10.1016/B978-0-12-802685-4.00007-8.

Zhang, C., Chen, F., Huang, Z., Jia, M., Chen, G., Ye, Y., Lin, Y., Liu, W., Chen, B., Shen, Q., Zhang, L., & Lavernia, E. J. (2019). Additive manufacturing of functionally graded materials: A review. *Materials Science and Engineering A*, *764*, 138209. doi:10.1016/j.msea.2019.138209

Zhang, C., Kong, L., Li, H., & Wu, Y. (2019). Electroless Cu plating on the microstructure and hardness of Al–Cu powder metallurgy alloy. Materials Research Express, 6(8), 086577. doi:10.1088/2053-1591/ab1e1a

Zhang, N., Khan, T., Guo, H., Shi, S., Zhong, W., & Zhang, W. (2019). Functionally Graded Materials: An Overview of Stability, Buckling, and Free Vibration Analysis. *Advances in Materials Science and Engineering*, ●●●, 1–18. doi:10.1155/2019/1354150

Zhang, N., Liu, S., & Xu, Y.-J. (2012). Recent progress on metal core@semiconductor shell nanocomposites as a promising type of photocatalyst. *Nanoscale*, *4*(7), 2227. doi:10.1039/c2nr00009a

Zhang, X., Zhang, Y., Tian, B., An, J., Zhao, Z., Volinsky, A. A., Liu, Y., & Song, K. (2019). Arc erosion behavior of the Al2O3-Cu/(W, Cr) electrical contacts. *Composites. Part B, Engineering*, *160*, 110–118. doi:10.1016/j.compositesb.2018.10.040

Zhao, H., Feng, Y., Zhou, Z., Qian, G., Zhang, J., Huang, X., & Zhang, X. (2020). Effect of electrical current density, apparent contact pressure, and sliding velocity on the electrical sliding wear behavior of Cu–Ti3AlC2 composites. *Wear*, *444–445*, 203156. doi:10.1016/j.wear.2019.203156

Zou, G., Cao, M., Lin, H., Jin, H., Kang, Y., & Chen, Y. (2006). Nickel layer deposition on SiC nanoparticles by simple electroless plating and its dielectric behaviors. *Powder Technology*, *168*(2), 84–88. doi:10.1016/j.powtec.2006.07.002

Chapter 8
Electroless Coating on Non-Conductive Materials:
A Review

Anupam Jana
Haldia Institute of Technology, India

Supriyo Roy
https://orcid.org/0000-0001-8731-0068
Haldia Institute of Technology, India

Goutam Kumar Bose
https://orcid.org/0000-0002-4347-3508
Haldia Institute of Technology, India

Sourav Sarkar
Haldia Institute of Technology, India

ABSTRACT

This chapter attempts to make a review of electroless metal deposition over various non-conducting substrates like for its application in the field of medical research, electrical and electronics units, household aesthetics, automobile and textile industries. Electroless coating of metals over conducting substrates have been developed, critically reviewed, and proven its worth by showing excellent desired properties over the years. This review aims to discuss the techniques that have been applied by the researchers to overcome the difficulties of coating on these materials, their influence in their physical and mechanical properties, and their prospects of use in the industries. With the discussion of the underlying coating fundamentals and its historical backgrounds, the emphasis was put into the coating deposition with sensitizations and activations of various substrates, electroless baths, and the characteristically changed properties of the materials observed in the analysis.

DOI: 10.4018/978-1-7998-4870-7.ch008

INTRODUCTION:

Electroless deposition is a uniform coating method of metallic layer on the substrate surface through chemical reduction. In this technique metals like, nickel, copper, silver and gold can be deposited on substrate surface without the use of electrical energy. The idea of coating metallic nickel, on mostly conducting materials from aqueous bath with the help of sodium hypophosphite came into existence, when Wurtz revealed it in 1844. Then the era of electroless nickel coating started with the rediscovery work of Brenner & Riddell (1947).It was also taken into account that Mallory and Hazdu in 1990 were successful in finding the reducing action of sodium hypophosphite which stood as a stepping stone to successful electroless nickel coatings over conductive materials in the future. Sharma, Agarwala and Agarwala (2006) investigated thoroughly upon the techniques of the coating that helped researchers to find interesting facts too. Extensive studies over nickel-phosphorus and nickel-boron coatings over metallic substrates were made with the use of their respective reducing agents such as sodium hypophosphite and sodium borohydride. With the aim to develop excellent wear, friction and corrosion resistant hard electroless nickel coating, their micro mechanical properties were studied by numerous researchers (Balaraju, Narayanan & Seshadri, 2006).The idea of incorporating a second phase particle into the electroless nickel bath dates back in the year 1960 when Oderkelen (1972) tried to impose a structure of intermediate layer which contained fine divided particles distributed within the metallic matrix. Likewise, the introduction of alumina particles in the nickel-phosphorus matrix was done by Metzer (1973).

Electroless coating on metals, which are fortunately conductive materials, have been being the scope of work for researchers till now. In this field of research, scientists got ample opportunity to study the techniques with dedicated involvements with their aim to get conductive materials with superior physicochemical surface properties. An electroless plating bath in general consists of source of metal ions, reducing and complexing agents, Stabilizers, Wetting and buffering agents. This bath is generally kept at suitable operating temperature. In previous works of scientists, involving electroless coating /plating of conductive materials (metals mostly) included metals like Copper, Nickel, Gold, Palladium, Silver, etc. were deposited over a metallic substrate. Researchers came a long way from single metal deposition to incorporation of phase, 1, 2, 3 element forming electroless alloy, alloy composites and nano-composites of conducting materials. Due to extensive use in industries, the electroless alloy composites of Nickel-Phosphorus or Nickel-Boron gained popularity and the percentage of its use being near to singularity or hundred. Thus, often synonymously the word "electroless" started to relate nickel/phosphorus alloys. From Sudagar, Lian & Sha (2013) one can get a critical review of the electroless nickel/Phosphorus and nickel/Boron alloys.

The applications of Electroless coated materials (conducting/metals) have been really wide considering the aspect of its uses in industries. They found the way in the sector of aerospace, aviation, electrical, electronics, textile, automotive and many more (Agarwala & Agarwala, 2003). The choice of electroless alloy/alloy composites in these industries were based on certain mechanical and physicochemical properties which showed highlighted predominance viz. (a) its characteristics of being uniform with respect to deposition over substrate as well as its physical and chemical properties, (b)Excellent corrosion and wear resistance, (c)The ability to get welded and soldered easily, (d)The structure possessing high hardness with microcrystalline deposits, and (e)even showing magnetic properties with resilience. Likewise, different alloys with economical electroless technique were coated for the desired properties, but among all, nickel singly turned out to be widely accepted and applied in the industries. Several electroless coated alloy of nickel became notably famous with the incorporation of third and fourth phase materials in the

Ni-P or Ni-B matrix like Ni-P-TiO_2,Ni-P-Al_2O_3,Ni-P-SiC,Ni-P-MoS_2,Ni-P-W,Ni-P-PTFE,Ni-P-Cu-Zn, Ni-B-Mo, Ni-B-Cu and many more (Gutzeit, 1958; Sahoo, 2011; Balaraju, 2010; Sudagar, 2013; Hajdu, 2000; Mahallawy, 2008; Kodoma, 1989; Oguocha, 2010; Islam,2011; Choudhury, 2009).

Modern trends of material science have focused the interest of scientists and researchers towards non-metallic materials. From the automobile industry to bio-medical field, in every step of life the conventional metallic components are being replaced by light weight but hard non metallic materials. In almost every cases these non metallic materials are non conductive also. It is not possible to provide a metallic coating or deposition to these non conductive materials by conventional electroplating process. Thus, surface engineers and scientists have faced that challenge and been able to deposit metallic coatings on these non metallic substrates by screen printing, sputtering and electroless deposition. Among these methods, being the least expensive, electroless method has gained much acceptability in the industry.

Non conducting materials are used extensively in automotive industries, home fixtures, electronics and bio-medical implants. An automobile manufacturing company uses non conducting materials in most of the aspect of the vehicle these days, say from interior to exterior for the purpose of reduction in weight as well as to add an aesthetic value. In the meantime care is taken to maintain a no compromise position with respect to the strength of the body. A similar scenario is seen for electronic circuit board and integrated chips now a days. In these cases, technology demands a light weight, strong but existence of conductivity. Sometimes materials required with self conflict characteristics having thermal barrier properties with electrical conductivity. Thus in view of all these respect metallic coating is essential on non conductive materials.

This chapter aims to make a review as to why there had been the necessity of coating non conductive materials and that too by the technique of electroless coating method. Emphasis will be given to the technique of using electroless coating adopted by the researchers for metallic deposition on non conducting materials like ceramics (Alumina, Titenia, Zirconia, Yttria stabilized Zirconia, Yttria treated Alumina, Zirconia treated alumina), plastics, Silicon or glass.

BACKGROUND

Material Science till now has been successful in pin pointing a few non-conductive materials for the metallic coating process overcoming the challenges. Graphite, Silicon carbide, Titanium Oxide, Alumina, Zirconia, Yttria stabilized Zirconia, Yttria treated Alumina, Zirconia treated alumina are some of the names. These are known as difficult substrates when it comes to the word plating/coating. A number of technologies are available for plating/coating such substrates like conversion plating, electrochemical plating, Hybrid coating, organic coatings, vapour phase process but among all, electroless technique proved to be the promising one in almost every respect.

In this context researchers have focused their interest towards two metals, Nickel and Copper. The importance of Nickel can be understood by the synonymously used electroless coating for electroless nickel alloy/alloy composite coatings. The uniform physical properties in the deposition over the substrate, a crystallized structure after annealing gives excellent hardness, corrosion resistance and wear resistance, along with other notable mechanical properties makes them the first choice coating matrix (Krishnan, 2006). Sudagar, Lian, & Sha (2013) covered a typical physical and mechanical properties of pure nickel, Ni-P/B coatings. Another similar material of importance is copper(Ghosh, 2019).One of the large scales extracted and very first extracted metals on earth. It has remained all time useful for its good electrical as well as thermal conduction properties. It is relatively soft when compared with other relative metals. Today, copper has found its place in fabrication of electrical and electronic components say transformers, motors, switches, fabrication of Air conditioning units, moulds, valves, clocks and many more. So the uses of such metals like nickel and copper as a coating over non conducting materials are gaining interest of researchers now days.

Cheng, Gan, Chen, Lai, Xiong & Zhou, (2019) fabricated copper/nickel-iron-coated graphite composites by electroless plating and SPS method. Pantograph slider is an electrical contact part which requires good thermal and electrical conductivities, excellent self-lubricating performance, wear resistance, arc ablative resistance, light weight and low cost at the same time it also needs high mechanical strength so that it can withstand vibration and is not damaged due to the impact and should be light in weight. It should also possess excellent self lubricating capabilities along with good wear resistance. In the above mentioned work, the focus was given on improving the interfacial bonding between graphite and copper by coating graphite powders, which are in spherical form with nickel-iron alloy and simultaneously incorporating copper in it by electroless coating technique.

Silicon is another non conducting substance available extensively used in the electronic industry and forms the heart of integrated circuits (ICs) or so called silicon chips. With the base of this silicon substructure, researchers founded a way to use Copper upon layers of Titanium seeds. In their workAith al,Yenamandra,Gunasekaran,Coane,&Varahramyan,(2006)described a way of depositing Cu on titanium seed layer which has been deposited over Silicon. Titanium served as a adhesion layer as well as a barrier layer to Cu diffusion which prevents copper to migrate into the silicon substructure, which is the way it is actually needed in integrated circuits if copper is to be used as a conducting material.

Apart from graphite and silicon another non conducting material which became quite a use in modern days is Ceramics. The use of ceramic, be it in a powder form or in reinforced structure are used in households to various industries. Therefore, the superiority of copper/nickel was there to be imparted to ceramics and it was achieved by coating ceramic powders with metals. A good number of researchers (Sharma, Agarwala and Agarwala 2006; Zheng, 2017) worked on development of metallic coatings on ceramic powder by electroless technique. They were able to come to a conclusion with evidence, that using the electroless technique, coating of metal over ceramic powders is feasible and efficient. The surface morphology of the electroless copper coating over ceramic powders showed a globular nature. The coating was found to be uniform not only over the plane surfaces but also inside of the porous region of the ceramic particles. A mechanical bonding was established in the globules inside micro-sized pores with the evidence of cauliflower structure by the coating. The coating of Cu upon ceramic powders by electroless technique is not only a process of chemical bonding but also a mechanical bonding. With these results it was evident that the scope of coating Cu (a metal) over ceramic (a non-conducting material) is promising with notable desired physicochemical properties.

The considerable potential for the electroless coated ceramic nano-powders can be understood by the eagerness of the researchers to incorporate superior metals over it. As such, one of the desired ceramic materials used in thermal barrier coatings is Yttria Stabilized Zirconia or YSZ. The YSZ becomes a material of choice due to its low thermal conductivity, high thermal shock resistance, therefore where the primary criterion is non thermal conduction. For the manufacture of solid fuel cell anodes and to be used as elevated temperatures, the application of electroless nickel coated YSZ powders gave a result of expectations. Alavi, Aghajani, &Rasooli, (2019;Alhamadi,2018) worked on the electroless nickel coated YSZ by investigating and developing a single step surface activation methods. But in their work they were able to propose two different methods to do it.

The inertness of the ceramics used in biotic environment (orbio inert ceramics) has found its picture 20 years ahead of now or more. When the application of materials associated with medical fields /applications, like being used in artificial implants as a femoral head, acetabular cups, for Hip replacement, dental implants, bone fillers, tissue engineering and many more, the need for tough and strong biologically inert substances is met by ceramics. J.Chevalier,L.Gremillard of Universite' De Lyon (2009)described an overview of application of ceramics for medical devices. Though the paper does not give a direct review for the electroless coating of the ceramic material, but surely gives an idea of its tremendous importance in the noble service to kind with its sensitive applications to medical field.

With the advent of Graphite, silicon, ceramic and one of its finer form Zirconia, one should not forget Plastic. Plastic with its good strength-to-weight ratio, mold-ability, lower density and excellent surface finish quality found its application in various types of industries in various forms as well. Two of the very famous industries of application are automotive and home appliances and two of the popular known forms are PET(Polyethylene Terephthalate) and PTFE(Polytetrafluoroethylene)(Ramalho,2005;Reid,2000). The addition of electroless nickel and copper metal matrix are demanded popular works of achievements with purposes from strengths to shields (i.e., the strengthened microstructure and shielding of any form of conduction)(Kutnahoersky,2009). Coating on plastics had been emerged widely due to use in the manufacturing of automobile parts, electromagnetic shielding applications, printed circuit boards and many more(Cui,2012).One of the applications in the automotive industry is the production of the reflective surfaces in case of exterior mirrors, body panels, fenders, etc. The coated plastics can mostly be molded to accuracy with no tolerance required without even incurring damages. But with the need for better physical and mechanical properties in the electroless coated plastic, scientist stuck to the trend of plating them with conductive metal matrix which even led to the development of variations in the coating techniques through the process of sensitizing and activating the substrate. Selvam,2010 used a surface activation technique for deposition onto Acrylonitrile-butadiene-styrene plastics was done by employing biopolymer chitosan which in turn gave enhanced adhesive strength between the plating layer and the substrate. This method was different from the conventional sensitizing method. To develop electroless coated plastics as electromagnetic interference shielded composites, the factors like the resistivity and anti-oxidation capacity of the conductive filler materials used became important. Hence,a new aspect of developing a special purpose coating was found. The shielding effectiveness of the Electromagnetic Interference (EMI)of single or double layer electroless metal (say Nickel –phosphorus in pure form or in forms of Ni-NiP,Ni-P-Cu,Ni-Cu) coated reinforced carbon fiber ABS was investigated. The result showed that electroless copper coated reinforced carbon fibre/ABS plastic showed low EMI which was promising.

Further, for the production of selective metal patterns on PET sheets for good electrical and mechanical properties, Garcia and others,(Li,2018;Narcus,1947;Tsai,2007;Nathan,2012 ;Jain,1993) developed a simple 3-D amino induced electroless plating process. Researchers viewed such process as an excellent alternative method for the fabrication of large area plastic electronic devices. Charbonnier, Geopfert and Romand (2003) was able to describe with his work that how noble metal complexes can be adsorbed over the surfaces of polymers (like PET) for modification. Such surfaces when in contact with the electroless bath will reduce the noble metal complex and form clusters for electroless depositions. The work of direct patterning of metallic layers onto insulating substrates like plastics or polymers now has great potential in fabrication of micro to nano devices. Wang,2015 in recent studies was able to demonstrate such deposition of Nickel patterns over these non-conducting surfaces.

Ceramic particles reinforced metal matrix composites or CPRMMCs or simply PRMMCs, are widely used in large scale today in many fields of aerospace, electronic and new energy generation due to its excellent wear resistance, hardness, superior strength and low pricing/costing(Leon,2000).These are ceramic particles coated with metals bearing ceramic cores and metallic coatings over it. One such material is metal coated ZTA. ZTA as it stands, Zirconia treated Alumina or ZrO_2 toughened Al_2O_3.To improve the interface between Zirconia toughened Alumina particles and the metal matrix and nickel by electroless method was deposited over the surface of ZTA particles (Fan, Wang, Yang, Chen, Hong, Zhang, & Ren, 2018). Researchers performed studies of the mechanism behind this particular nickel coating and noted certain effects of parameters like the solution pH, loading capacity of the ZTA particles and the nickel deposition temperature. Conclusive evidence of nickel being plated successfully over the ZTA particles without any noticeable defects by electroless technique was achieved after the characterization.

Copper has been deposited over tungsten powders by electroless method. Wang, Xu, Srinivaskannan, Kopalla, Han, Xia (2018) worked on electroless copper plating of tungsten powders and the preparation of WCu20 composites by microwave sintering. The results of the work were noticeable inferences drawn by performing studies of the prepared WCu20 composites under the variation of controlled parameters like temperature and pH showing changes in micro-properties. Apart from these previous research works upon electroless coating over certain non-conductive materials, it is time to draw attention to some fundamentals of the EL plating/coating technique.

ELECTROLESS DEPOSITION

Sensitization and Activation

Deposition of metallic coatings on non-conductive surfaces by electroless technique is highly influenced by sensitization and activation procedure of the substrates. The sole reason behind the procedure is to develop a built-in-potential in the solution bath for the deposition to occur. To apply electroless nickel, copper or any other noble metal on the surface of conductive, non-conductive or semi-conductive materials, the surface should be activated by a catalyst to start the coating reaction. The way of activating a non-metallic surface or non-conductive surface such as ceramic or plastic is a multi-step process that usually consists of sensitization of the substrate by Sn^{2+} ions followed by an activation step by Pd^{2+} ions (Alavi, 2019; Wen, 2000; Izatt, 1967). On contrary to the two step procedure, Shipley and co-workers proposed a single step of activation method for such surface of non-conducting substrates by the use of colloidal solutions of catalytic metals like platinum, palladium, Gold (Shipley,2001;Boily,2005). For the purpose of coating non-metallic powders with nickel or copper using electroless technique, facing a difficulty in particle recovery is evident in each step. The complexity increases with the decrement in the size of the particles and therefore as such the no of working steps should be reduced as much as possible and Shipley's work was one among it. In this work, stannous chloride used as a reducing agent to reduce catalyst metal ions of palladium is used as a catalyst. But later, Pd-Sn catalysts owing to its particulate nature was declared unsuitable for treating submicron core particles of ceramic by Garg and the coworkers who was working on the sensitization theory. They introduced a single step pre-treatment process (sensitization and activation) with palladium catalyst and sodium hypophosphite as a reducing agent (Garg, 1993). However, some typical examples of sensitization and activation of different materials are described in the following sections.

The work of Palaniappa, Veerababu, Balasubramanium (2007) on electroless nickel-phosphorus plating on graphite powders is a target oriented one as discussed above along with a unique way of activation which opens a direct aspect to the electroless deposition. Here, the activation was carried out by heating the graphite particles in muffle furnace at a temperature of 380°C for nearly 1 hour. It is evident that the sensitization and activation step for graphite particles differs from any other conventional research of electroless coating over graphite. The result was a more uniform nickel-phosphorus deposit on graphite particles. This technique for activation which removed any adsorbed gases from the surface of the graphite powders rendering it a highly active surface. On the other way, graphite powder was electrolessly coated with nickel-iron alloy and copper in sequence by Cheng, Gan, Chen, Lai, Xiong, Zhou with reference to the manufacture of a pantograph slider (Chen, 2017, 2019; Yang, 2011; Xu, 2014; Wang, 2017; Ma, 2018). A short discussion about their work has been done previously in this chapter. The pollutants were removed by immersing the graphite powders into sodium hydroxide for 1 hour. As the powder gets degreased, it was put into HNO_3 for another 1 hour. The sensitization was done in the solution of 15 g/l $SnCl_2.2H_2O$ and 20 ml/l HCl at nearly 30°C for 30 minutes followed by the activation in 0.02 g/l $PdCl_2$ and 20 ml/l of HCl at 50°C for another half an hour. An observation with tiny amount of Palladium atoms deposited over the quasi-spherical surface of graphite was evident after activation.

Al-Si oxide ceramic powder was coated with copper through electroless technique by Rahul Sharma, R.C.Agarwala, V.Agarwala (sharma, 2006). Sensitization of the ceramic powder to deposit copper over the dielectric surfaces, the Al-Si oxide was immersed in the solution of $SnCl_2$ and $PdCl_2$ for around five minutes and at a temperature of 25°C. The bath was kept at pH of 7.0 for both sensitization and the activation. Similar methodology was used by Lei Fan, Wang, Chen, Hong Zhang and Ren (2018) to deposit nickel on ZTA (ZrO_2 toughened Alumina). The sensitization was done using aqueous solution of 15 g/l $SnCl_2.2H_2O$ and 50 ml/l concentrated HCl for 30 minutes followed by the activation in another aqueous solution of 0.25g/l $PdCl_2$ and 10 ml/l of concentrated HCl for 20 minutes. In other way, single step activation method proposed by Garg and coworkers(Garg,1993) for activating YSZ (Yttria stabilized Zirconia) nanopowders. Two variations of single step methods for activating YSZ were implemented. The first method used $PdCl_2$ dissolved in distilled water and Hydrochloric acid with a pH of 1.5. The YSZ nanopowders were gradually added in the solution and stirred for about thirty minutes. The specified amount of sodium hypophosphite was added to the solution after stirring and the activated YSZ particles were recovered from the solution by sedimentation process. The second method, proposed by Alavi, Hossein and Rasooli (Alavi,2019) almost used the same techniques except for a few changes such as, the activation was carried out without the reduction of Pd^{2+} ion. With respect to the first method here after 30 minutes of stirring the solution, addition of 1M NaOH solution was facilitated to increase the pH of the solution to a desired value. The stirring gets continued for another half an hour after which the particles were recovered via sedimentation. With these methods a fact needs to be considered that the activation of nanopowder took place in room temperature.

Electroless Bath

Electroless process is a chemical reduction method to deposit metals on large verity of substrates. This chemical bath consists of basic elements viz. source ofmetal ions,reducing agent, complexing agent, stabilizer, buffering agent and surfactants/wetting agents, which should be kept under certain experimental parameters like pH and temperature for successful deposition. The mechanism of electroless deposition is the reduction of metallic ions to substrate under the action of reducing agent, which are simply nothing but electron donors. The surface of the substrate needs sensitization and activation for a potential to get built up in the ionic medium. The metallic deposition occurs under the influence of this built-in potential difference. Thus without the use of any external current, this is an autocatalytic process, which accelerates the electrochemical reaction allowing the chemical reduction of metal ions or rather saying the oxidation of the reducing agents. The functions of electroless bath constituents are listed in Table 1.

An Electroless bath may be acidic or alkaline depending on the requirement of the type of metal to be deposited, the incorporated hard or soft material to be deposited along with it and the substrate upon which it is to be deposited. An Electroless Ni-P acidic bath provides nearly 3-5% P (low), 6-9%P (medium), 10-14%P (high) phosphorus content in its Ni-P EL alloy coating whereas in alkaline bath it shows constant stability in the phosphorus content. Needless to say, an EL bath becomes acidic or alkaline by the reaction of reducing agents, complexing agents, stabilizers, surfactants and therefore the pH of the solution bath too can be controlled by the use of additives.

The use of an acidic electroless bath almost becomes exclusive when there is a need for thick coated layers of metals over the surface of substrate. The obtained coatings show uniformity, are relatively stable and of good quality. This is taken as an advantage a hot acid bath solution has over an alkaline solution. The low phosphorus EL nickel coatings have excellent wear and corrosion resistance. In addition to it the medium phosphorus content coating shows well enough abrasion resistance to be economically accepted for application in industries. The high P content coating shows good ductility and resistance to corrosion against mechanical stresses and chlorides.

An alkaline EL bath proves advantageous while depositing nickel coatings over one of the non-conducting material. But the downside it carries along with it is the tendency of instability at temperatures above 90°C.Being very much temperature sensitive, the alkaline bath found less application in industries for preparing thick coatings. The sudden decomposition of the warm alkaline bath creates instability, results an increased operating cost which industries tend to avoid. In addition to these other limitation is the lower corrosion resistance of the coating prepared in the alkaline bath. But researchers found a way of using stabilizers to effect for producing a good deposit rate and superior adhesion deposits such as use of organic additives like Sodium Benzene Sulfonate.

Table 1. Fundamental aspects of electroless bath (Mallory and Hazdu, 1990)

Basic constituents	Function	Chemicals used by researchers
Source of Metal ions	These are basically the salts of metals, soluble in water.	Nickel based coatings: · Nickel sulphate · Nickel chloride Copper based coatings: · Copper sulphate · Copper chloride
Reducing agent	This chemical component has a vital role in autocatalytic deposition of metals. This chemical agent produces sufficient electrons in electroless solution, to reduce the metallic ions on substrate. However, selection of reducing agent depends on the type of metal to be deposited and the desired properties of coating.	Nickel based coatings: · Sodium hypophosphite · Sodium borohydride · Dimethyleamine borane · Hydrazine Copper based coatings: · Formaldehyde · Hypophosphite · Borohydride · Hydrazine · Glyoxylic acid
Complexing agent	Complexing agents are added to prevent the rapid or sudden decomposition of the solution bath, these reagents are important in controlling the reaction. These are mostly organic acids, or the salts of these acids. It works by controlling the amount of available/ free electrons available for the reaction. The electroless solution bath gets buffered by the action of these complexing agents and slows down the precipitate formation, say the precipitation of nickel phosphite in EL Ni-P coating. Even the evolvement of hydrogen can be reduced by the use of hydroxides, carbonates, ammonia and thus these can be used as complexing agents too. Also it is evident that a complexing agent impacts the quality of the metal deposition over the substrate thus affecting the mechanical property like internal stress.	Nickel based coatings: · Citric, lactic · Glycolic · Proprionic acids · Sodium citrate · Succinic acid Copper based coatings: · Cyanides · Chlorates · Iodates · Molybdates · Seignette salt · EDTA salt
Buffering agent	Maintaining pH is one critical factor among all major factors to control the rate of deposition in an electroless coating. Hydrogen ions produced during plating causes the pH of the solution to decrease and hence the use of buffer is important in this aspect. In a Ni-P EL deposit, the addition of buffers controls the phosphorus content of the deposit which becomes an important factor for nickel deposition.	For both Nickel and Copper based bath · NaOH
Stabilizer	Stabilizers are chemical agents that are deliberately added in the electroless plating baths to prevent the homogeneous reaction that triggers the spontaneous decomposition of the entire plating operation. An electroless nickel bath is highly sensitive to impurities and therefore it becomes crucial to maintain the stability ad that can be achieved by the use of stabilizers in proper amount in the solution bath.	Nickel based coatings: · Thiourea · Lead acetate · Heavy metal salts Copper based coatings: · Potassium ferricyanide · Sodium thiosulfate
Surfactants	In an electroless bath, surfactants are used as a wetting agent for the improvement of the wettability of the surface to be coated. During electroless reaction, hydrogen evolved, which might affect the surface properties and a chemical additive is required which will be able to counter it. This can be achieved by the use of surfactants which has an influence over the adsorption mechanism followed. It becomes even more important at the time of incorporation of certain soft particles in the EL coating like polytetrafluoroethylene (PTFE), molybdenum disulphide, Graphite etc.	Nickel based coatings: · SDS · CTAB · DTAB Copper based coatings: · complex organic phosphate esters · alkyl phenoxy polyethoxy phosphates

PROPERTIES OF ELECTROLESS COATINGS

On Ceramics

YSZ (Yttria stabilized Zirconia) nanoparticles were coated with nickel with single step activation method rather than applying a two or multistep conventional pretreatments by (Alavi,2019; Li,2001; Wen,2000; Baba,2009; Pratihar,2004; Lopez,2007; Izzatt,1967; Rodriguez,2012). The atomic absorption spectroscopy (AAS) results showed that there was an increase in the amount of deposited Palladium upon the YSZ nanoparticles with the rise of $PdCl_2$ concentration in the bath. But it was seemed that the increase in the quantity of the catalyst hasn't had any considerable effect on the deposited nickel's amount in an equal period of time and the weight of the deposited nickel was almost same. But at the same time it was also observed that the electroless reaction took shorter period of time to complete with high concentration of catalyst particles which was obvious. Likewise, another result was drawn from the AAS estimation on the YSZ particles activated via the two step activation method. It was observed that initially with the raise of bath pH, the amount of Palladium increased but with pH value more than 3 the Pd concentration almost remained unchanged. Palladium Hydrolysis in aqueous solution demonstrated Pd^{2+} ions to be weekly acidic and near neutral at room temperature. The transmission electron microscopy (TEM) and electron diffraction pattern showed a uniform layer of EL coated nickel over the surface of the YSZ with a thickness of around 5 to 10mm. The Differential thermal analysis (DTA) and Thermal Galvanic analysis (TGA) were able to provide the thermal stability test result. Simultaneous thermal analysis (STA) results showed Ni-YSZ powders are thermally stable up to 1200 °C. The XRD patterns and magnetization curves of the coated YSZ with used different concentration bath of nickel salt were analyzed and it was evident that the deposited nickel ion varied in direct proportion with the concentration of the source in the solution bath.

Another ceramic particle ZTA, which showed high toughness, high hardness, outstanding wear resisting capability with economic at the same time were coated with nickel on the surface by electroless method (Fan,2018; Zheng,2017; Naga,2017; Han,2015; Nastic,2015; Chen,2017). The results of XRD showed evidence of nickel coated ZTA particles with nickel peaks present in it and the test sample exhibited a typical pattern for the ZrO_2 and Al_2O_3 as well. The SEM images showed the result of interfacial nucleation and growth of nickel. It was evident that nickel atoms got uniformly distributed over the surface of the ZTA particles with a grain size of nearly 2-5 μm. An observation on the loading capacity of the ZTA particles in the solution was seen which provides an inverse relation with the amount of nickel to be deposited over ZTA. It was found out that the optimal loading limited from 15-20 g/l of ZTA for an optimum nickel deposition along with an optimum operating temperature of 85°C.In the manufacturing of ZTA reinforced iron matrix composite the electroless coated Ni-ZTA plays an important part too where the coated ZTA particles gets embedded in the iron matrix. Ni gets diffused into the Fe matrix while sintering and at the elevated temperature where the diffusion of nickel in the iron matrix took place with the formation of solid solution, and therefore an improved interfacial bond between iron matrix and ZTA were observed.

To understand the mechanism of bonding of copper on Al-Si oxides ceramic powder, Sharma, Agarwala and Agarwala (2006) used electroless method. It was observed that EL copper coating over the activated ceramic (Al-Si) substrate grew layer by layer inside and outside of the surface pores. It was suggested that the adherence of the electroless coating to the substrate is due to mechanical bonding. It refers to the formation of nano sized globules in the coating that entirely filled the surface and the pores. Some other investigators suggested the cause of bonding to be chemical reactions at the interface which did its job well due to the pretreatment provided to the substrate. But with standing evidences, both chemical and mechanical bondings were given importance. It was observed that the coated powder changed its color after undergoing heat treatment from a brighter one (bright Maroon color) to a darker one. Comparing the XRD peak heights of the coated and uncoated phase in the pattern, it was quite understandable that the EL technique was efficient and feasible. The weight difference showed a 60% increase in weight after coating.

On Graphite Powder

The microstructure and surface morphology of the nickel-iron coated graphite composite was studied by (Narayanan,2004; Kong,2002; Pallaniappa,2007; Moustafa,2002; Xu,2014; Wang,2017; Chu,2010) using SEM, BFTEM, SAED and XRD and closely compared with the final copper/nickel-iron coated composite. Analysis and calculations were able to give results those certain mechanical properties like bending and compressive strength, hardness, relative density of copper/nickel-iron coated graphite composites is much higher than the copper/graphite composite. Researchers related that the nickel-iron coated graphite composite can make an attempt to have a higher mechanical property as such of the copper/nickel-iron coated graphite composite but would lag with the Electrical property. Such enhancement in the mechanical property of the final composite is due to the continuously and evenly distributed electroless copper plating on the surface of the nickel-iron coated graphite which led to the complete Cu/Cu contact during sintering. The high electrical resistivity of the composite is a dominating effect of the higher electrical resistance nature of pure nickel and pure iron and hence the composite possesses an enhanced electrical property. Wear behavior of coated specimen were evaluated against pure copper using a block-on –ring tester. It was observed under 40N normal load at 1200 rpm the co-efficient of friction of copper-nickel-iron/graphite composite were higher than that of copper/graphite composite at the initial stage but it became opposite after sliding for about 35 minutes. The average co-efficient of these two different composites were very close due to the same graphite content present in both the composites. But however, the wear rate after hand calculation is found out much lower in case of copper/nickel-iron coated graphite composite than of the copper/graphite composites.

With an aim to investigate the deposition of Ni-P on graphite surface, the non-conventional activation technique applied by Palaniappa, M.Babu, G.V and Balasubramanium (2007). Scanning Electron Microscopy or SEM was used to study the surface morphology of the coated graphite powders. The typical cauliflower type structure was identified as a deposited nickel structure upon graphite surface at higher magnifications. The particle sizes of graphite were in the range of 50-150 μm and that of nickel were in the range of 15-20 nm. The EDS analysis was performed on the EL coated graphite powders to find out the content in the deposits too which depicted peaks or C, Ni and P. It was observed that there was an initial increase in the plating rate but the rate was not linear throughout as it reduced and reached a plateau nearly after fifteen minutes. It was obvious that increase in the loading of graphite powder in the bath lead to decrease in the deposited nickel content as the effective surface area to be coated gets increased which demands to be coated with the same amount of nickel and therefore it was balanced accordingly with the weight percentage of graphite and nickel content in the EL bath. It was noticed that irrespective of the change in powder sizes, the variation in the nickel content deposition over the powder is negligible or near to zero. Rohatgi, Pai (Ashassi-Sorkhabi, 2004; Narayanan, 2004) observed similarities as such in case of copper plating on graphite powders. But on contrary, a higher powder load results in a lower nickel % in the powder as analyzed by chemical wet method. The amount of phosphorus estimated was only 10% of the amount of nickel coated over graphite

On Other Non-conductive Materials

Several researchers have deposited metallic coatings on substrates such as glass, resins, polymers, papers etc. and investigated the properties in details. The difficulty or ease with which the desired coating characteristics were achieved depended on a few facts of which the impact of surface substrate is one. No matter what the reducing agent is or the chemical composition of the electroless bath, the surface of the substrate plays an important role to determine the morphology of the deposited film or coating. Even the nucleation of the depositing metal ions depends on the catalytic properties of the substrate surface (Shukla,2009;Das,2017).Mostly observed that EL Cu deposits have polycrystalline strong orientation which is a preference in the electronic industry to attain low electrical resistivity and electro-migration reliability (Aithal, 2006; Jain, 1993; Chin, 1998) .Irrespective of the additives used in the EL bath, the effect found on the polycrystalline structure is negligible or none except for sulphur and polyethylene glycol with which a strong texture is obtainable (Aithal, 2006; Ghosh, 2019; Nakahara, 1977).The crystalline structure of EL Cu deposits with diameter ranging from 150 to 430 angstrom was obtained on Alumina (Al2O3) due to interaction between Cu and the chelating agents(Kondo, 2014; Ghosh, 2019; Kutnahorsky, 2009).Observed that the surface sensitization and activation procedures did not have any effect of the structure of the substrate. Even the change in the reducing agents has no influence over the crystalline structures (Aithal,2006).

ELECTROLESS COATING OF TWO EXPENSIVE METALS

Electroless Silver Coatings

Mandich, Krulik, Singh (US patent:5322553) worked on numerous experiments to develop succesful EL Ag coating on numerous substrates. This system gave an idea of the use of a redox system comprised of thiosulfate-sulfite-sulfate with no other reducing agents needed.The system said to have eliminated the need for reducing agent such as formaldehyde, reducing sugars, hydrazine or borohydrides. The solution bath of this system is said as free from ammonia or cyanide ions and yet had a higher plating rate and solution stability than the other electroless silver bath. The base of their experiments very much matches with the scope of this chapter as the substrate used was epoxy glass laminate printed circuit board material. It is of utter importance that the epoxy test boards were precoated with thin layers of medium Ni-P before coating them with silver. Rest, the sensitization and the activation followed using Palladium chloride and HCl in the same way even for Ag plating that is seen for Ni-P pre-coating.

Chen Liu, Lei Geng, YiFan Yu, Y.Zhang, B.Zhao, S,Zhang, Q Zhao were able to demostrate for the first time that the incorporation of Ti in nano-particle form into the silver matrix promoted a significant Ag ion release as well as an increase in the electron donor surface energy which stands effective in enhancing the antibacterial effect of the Ag-TiO2 coatings.(Zhao,2007).The choice of electroless technique for the preparation of Ag-TiO2 nanocomposites was again a result to choose an economic and effective option.

Electroless Gold Coating

Another expensive metal, gold was deposited through electroless technique onyarn, commonly known to all in the form of textiles. Yarn is generally non-conductive material made from plastics, animal fibers and plant material. Even silk in pure form is a non-conducting material unless coated or co-woven with copper threads. Though an electroless coating on any yarn contradicts with the fact that the process is expensive/uneconomical (Uddin,2010; Corti,2004; Goodman,2002) unlike any other electroless coating, still it finds its place of being advantageous in metallizing a difficult non conducting substrate as such. Yarns have been coated with noble metals such as silver, copper nickel for their own purpose of activity they needed to perform. Talking about this particular type of material being electroless coated with gold, researchers in textile industries aimed to use it as a sensing and measuring device of body parameters, the reason being the gold-coated wearable would be highly conductive to acquire data, skin friendly and stable (Uddin,2010; Kim,2006; Schwarz-Cordoen,2010; Westbroek,2007). With its inertness to biocompatibility, the EL coated Gold-Yarn has an excellent resistant to corrosion near to that of pure gold. Schwarz, Hakuzimana, Kaczynska, Banaszczyk, westbroek, McAdams, Moody, Chronis, Priniotakis, Mey, Tseles and Langenhove (Schwarz,2010; Ge,2003; Oh-Hong,1999; Green,2003) developed electroless gold coating on para-aramid yarns modified with polypyrole and copper. The SEM images of the coated fiber revealed smoother gold surface with increase in pH of the coating solution. Though at pH value of 7.0, larger grain size of gold with thicker deposit were obtained, still it was observed that at pH 9.0, the film contained closely packed crystals with very uniform distribution. The test result of the electrochemical impedance spectroscopy was quite significant about the fact that the gold coated para-aramid yarn aims to get integrated into fire-fighter garments to form ECG electrodes. These fabric electrodes must withstand the built-up sweat as well as ensure good electrical contact with the underlying hydrating skin. Tests of tensile strength were conducted on pure uncoated para-aramid yarn and on the electroless gold coated PPY-Cu yarn to see if the coating had any influence on the strength of the yarn. It was however statistically observed that the coating did not affect the mechanical strength of the yarn but also indicated that it reduced the yarn stiffness and its resistance to extension. It is an important fact to know that Au gets deposited over PPY-Cu by replacements which is when the material is immersed in the gold plating solution. First an ion exchange occurs between Cu and the yarn surface and the gold ions in the solution. Copper, being less noble than gold dissolves and enters the solution and in turn gold ions gets reduced and gets deposited in pure form over the yarn material. The layer of gold would start to grow unless a dense layer is obtained. At this point theoretically copper cannot be dissolved any longer and the ion exchange/replacement of ions process stops completely leaving a gold surface over the PPY para-aramid yarn.

CLOSURE

Electroless coating on conductive or metallic substrates have mostly shown properties like excellent resistance to corrosion and friction wear, high toughness with better micro-properties, fantastic electrical conductivities and magnetism which found its place in numerous industrial applications and reached its importance. But in certain sensitive applications like in an electronic industry where the substrate needed to be coated and non-conductive, required brainstorming and with experiments performed in the past and present, electroless technique became one of economical and viable idea. A non-conductive to be coated is generally considered as a difficult substrate due to the problems faced with the adherence of the metals to it, but recently scientist are able to do so with the help of updated pretreatment techniques and material specific reducing agents.

With the inculcation of new technologies it has been possible to electroless coat material like steel with nickel or copper but the gap between the technical development and industrial applications needs to be filled to gain benefits of the electroless technique in coating a non conductive substrate. Technologies should be established for large commercial acceptance of these research works discussed in the chapter, provided they should show maximum efficiency on these kinds of versatile materials with lesser bath wastage and inclination towards green plating technology.

REFERENCES

Agarwala, R. C., & Agarwala, V. (2003). Electroless alloy/composite coatings: A review. *Sadhana*, *28*(3-4), 475–493. doi:10.1007/BF02706445

Ahmadi, M., & Aghajani, H. (2018). Structural characterization of YSZ/Al2O3 nanostructured composite coating fabricated by electrophoretic deposition and reaction bonding. *Ceramics International*, *44*(6), 5988–5995. doi:10.1016/j.ceramint.2017.12.185

Aithal, R. K., Yenamandra, S., Gunasekaran, R. A., Coane, P., & Varahramyan, K. (2006). Electroless copper deposition on silicon with titanium seed layer. *Materials Chemistry and Physics*, *98*(1), 95–102. doi:10.1016/j.matchemphys.2005.08.079

Alavi, B., Aghajani, H., & Rasooli, A. (2019). An investigation on electroless nickel coating on yttria stabilized zirconia nanoparticles via single step surface activation methods. *Thin Solid Films*, *669*, 514–519. doi:10.1016/j.tsf.2018.11.053

Andricacos, P. C. (1999). Copperon-chipinterconnections. *The Electrochemical Society Interface*, *8*(1), 6.

Ashassi-Sorkhabi, H., & Rafizadeh, S. H. (2004). Effect of coating time and heat treatment on structures and corrosion characteristics of electroless Ni–P alloy deposits. *Surface and Coatings Technology*, *176*(3), 318–326. doi:10.1016/S0257-8972(03)00746-1

Baba, N. B., Waugh, W., & Davidson, A. M. (2009). Manufacture of electroless nickel/YSZ composite coatings. *Proceedings of World Academy of Science, Engineering and Technology (WASET)*, 715-20.

Balaraju, J. N., Narayanan, T. S., & Seshadri, S. K. (2006). Structure and phase transformation behaviour of electroless Ni–P composite coatings. *Materials Research Bulletin*, *41*(4), 847–860. doi:10.1016/j.materresbull.2005.09.024

Balaraju, J. N., & Rajam, K. S. (2010). Electroless ternary Ni–W–P alloys containing micron size Al2O3 particles. *Surface and Coatings Technology*, *205*(2), 575–581. doi:10.1016/j.surfcoat.2010.07.047

Behrangi, S., & Aghajani, H. (2018). Nano3YSZ electrophoretic deposition from acetylacetone+ ethanol solvent on the surface of AZ91 magnesium alloy. *Micro & Nano Letters*, *13*(5), 611–616. doi:10.1049/mnl.2017.0623

Boily, J. F., & Seward, T. M. (2005). Palladium (II) chloride complexation: Spectrophotometric investigation in aqueous solutions from 5 to 125° C and theoretical insight into Pd-Cl and Pd-OH2 interactions. *Geochimica et Cosmochimica Acta*, *69*(15), 3773–3789. doi:10.1016/j.gca.2005.03.015

Cao, X. Q., Vassen, R., & Stöver, D. (2004). Ceramic materials for thermal barrier coatings. *JournaloftheEuropeanCeramicSociety*, *24*(1), 1–10.

Charbonnier, M., Goepfert, Y., Romand, M., & Leonard, D. (2004). Electroless plating of glass and silicon substrates through surface pretreatments involving plasma-polymerization and grafting processes. *The Journal of Adhesion*, *80*(12), 1103–1130. doi:10.1080/00218460490884132

Chen, J., Ren, S., He, X., & Qu, X. (2017). Properties and microstructure of nickel-coated graphite flakes/copper composites fabricated by spark plasma sintering. *Carbon*, *121*, 25–34. doi:10.1016/j.carbon.2017.05.082

Chen, J., Xie, Z., Zeng, W., & Wu, W. (2017). Toughening mechanisms of ZTA ceramics at cryogenic temperature (77 K). *Ceramics International*, *43*(5), 3970–3974. doi:10.1016/j.ceramint.2016.11.072

Chen, P. Y., Wong, H. C., Hu, S. F., Huang, C. Y., & Liu, R. S. (2011). Functional electroless gold Ohmic contacts in light emitting diodes. *Applied Physics Letters*, *99*(6), 063511. doi:10.1063/1.3624831

Chen, Y., Cao, M., Xu, Q., & Zhu, J. (2003). Electroless nickel plating on silicon carbide nanoparticles. *Surface and Coatings Technology*, *172*(1), 90–94. doi:10.1016/S0257-8972(03)00320-7

Cheng, J., Gan, X., Chen, S., Lai, Y., Xiong, H., & Zhou, K. (2019). Properties and microstructure of copper/nickel-iron-coated graphite composites prepared by electroless plating and spark plasma sintering. *Powder Technology*, *343*, 705–713. doi:10.1016/j.powtec.2018.11.057

Chevalier, J., & Gremillard, L. (2009). Ceramics for medical applications: A picture for the next 20 years. *Journal of the European Ceramic Society*, *29*(7), 1245–1255. doi:10.1016/j.jeurceramsoc.2008.08.025

Chin, B., Ding, P., Sun, B., Chiang, T., Angelo, D., Hashim, I., & Chen, F. (1998). Barrier and seed layers for damascene copper metallization. *Solid State Technology*, *41*(7), 141–145.

Choudhury, B. S., Sen, R. S., Oraon, B., & Majumdar, G. (2009). Statistical study of nickel and phosphorus contents in electroless Ni–P coatings. *Surface Engineering*, *25*(5), 410–414. doi:10.1179/174329407X239081

Chu, K., Liu, Z., Jia, C., Chen, H., Liang, X., Gao, W., Tian, W., & Guo, H. (2010). Thermal conductivity of SPS consolidated Cu/diamond composites with Cr-coated diamond particles. *Journal of Alloys and Compounds*, *490*(1-2), 453–458. doi:10.1016/j.jallcom.2009.10.040

Clarke, D. R., & Levi, C. G. (2003). Materials design for the next generation thermal barrier coatings. *Annual Review of Materials Research*, *33*(1), 383–417.

Corti, C. W., & Holliday, R. J. (2004). Commercial aspects of gold applications: From materials science to chemical science. *Gold Bulletin*, *37*(1-2), 20–26. doi:10.1007/BF03215513

Cui, X., Hutt, D. A., & Conway, P. P. (2012). Evolution of microstructure and electrical conductivity of electroless copper deposits on a glass substrate. *Thin Solid Films*, *520*(19), 6095–6099. doi:10.1016/j.tsf.2012.05.068

Das, S., Weerasiri, L. D., & Yang, W. (2017). Influence of surface tension on bubble nucleation, formation and onset of sliding. *Colloids and Surfaces. A, Physicochemical and Engineering Aspects*, *516*, 23–31. doi:10.1016/j.colsurfa.2016.12.010

Dennis, J. K., & Such'Nickel, T. E. (1993). Chromium Plating (3rd ed.). Woodhead Publishing Ltd.

Dini, J. W., Beat, T. G., Cowden, W. C., Ryan, L. E., & Hewitt, W. B. (1992, June). Plating under reduced pressure. In Proceedings of the 79th AESF Annual Technical Conference. Atlanta: American Electroplaters and Surface Finishers Society.

El Mahallawy, N., Bakkar, A., Shoeib, M., Palkowski, H., & Neubert, V. (2008). Electroless Ni–P coating of different magnesium alloys. *Surface and Coatings Technology*, *202*(21), 5151–5157. doi:10.1016/j.surfcoat.2008.05.037

Fan, L., Wang, Q., Yang, P., Chen, H., Hong, H., Zhang, W., & Ren, J. (2018). Preparation of nickel coating on ZTA particles by electroless plating. *Ceramics International*, *44*(10), 11013–11021. doi:10.1016/j.ceramint.2018.03.055

Garcia, M., Martinez, H. V., & Ortiz, A. (2010). Compoforging of Al-Si metal matrix composites reinforced with beta-SiC: An alternative technique. *Open Industrial & Manufacturing Engineering Journal*, *3*, 1–6.

Garg, A. K., & De Jonghe, L. C. (1993). Metal-coated colloidal particles. *Journal of Materials Science*, *28*(13), 3427–3432. doi:10.1007/BF01159817

Ge, J., Turunen, M. P. K., & Kivilahti, J. K. (2003). Surface modification of a liquid-crystalline polymer for copper metallization. *Journal of Polymer Science. Part B, Polymer Physics*, *41*(6), 623–636. doi:10.1002/polb.10407

Ghosh, S. (2019). Electroless copper deposition: A critical review. *Thin Solid Films*, *669*, 641–658. doi:10.1016/j.tsf.2018.11.016

Goodman, P. (2002). Current and future uses of gold in electronics. *Gold Bulletin*, *35*(1), 21–26. doi:10.1007/BF03214833

Green, T. A., Liew, M. J., & Roy, S. (2003). Electrodeposition of gold from a thiosulfate-sulfite bath for microelectronic applications. *Journal of the Electrochemical Society*, *150*(3), C104–C110. doi:10.1149/1.1541006

Gutzeit, G., Talmey, P., & Lee, W. G. (1958). Chemical nickel plating processes and baths therefor. *US Patent, 2.*

Hajdu, J., & Zabrocky, S. (2000). The future of electroless nickel. *Metal Finishing*, *98*(5), 42–46. doi:10.1016/S0026-0576(00)81730-4

Han, X., Liang, Z., Wang, W., Chen, J., Xue, C., & Zhao, H. (2015). Characterization and synthesis of ZTA nanopowders and ceramics by rotating packed bed (RPB). *Ceramics International*, *41*(3), 3568–3573. doi:10.1016/j.ceramint.2014.10.184

Islam, M., & Shehbaz, T. (2011). Effect of synthesis conditions and post-deposition treatments on composition and structural morphology of medium-phosphorus electroless Ni–P films. *Surface and Coatings Technology*, *205*(19), 4397–4400. doi:10.1016/j.surfcoat.2011.03.042

Izatt, R. M., Eatough, D., & Christensen, J. J. (1967). A study of Pd 2+(aq) hydrolysis. Hydrolysis constants and the standard potential for the Pd, Pd 2+ couple. *Journal of the Chemical Society A: Inorganic, Physical, Theoretical*, 1301-1304.

Jain, A., Kodas, T. T., Jairath, R., & Hampden-Smith, M. J. (1993). Selective and blanket copper chemical vapor deposition for ultra-large-scale integration. *Journal of Vacuum Science & Technology. B, Microelectronics and Nanometer Structures: Processing, Measurement, and Phenomena: An Official Journal of the American Vacuum Society*, *11*(6), 2107–2113.

Kim, B., Koncar, V., & Dufour, C. (2006). Polyaniline-coated PET conductive yarns: Study of electrical, mechanical, and electro-mechanical properties. *Journal of Applied Polymer Science*, *101*(3), 1252–1256. doi:10.1002/app.22799

Kodama, S. H. U. N. I. C. H. I., Horiuchi, M. A. S. A. O., Kuroda, K., & Kunii, T. (1989,April).Ultra-blacknickel-phosphorus alloy optical absorber. In *6th IEEE Conference Record., Instrumentation and Measurement Technology Conference* (pp.495-497) .IEEE. 10.1109/IMTC.1989.36909

Kondo, K., Akolkar, R. N., Barkey, D. P., & Yokoi, M. (Eds.). (2014). *Copper electrodeposition for nanofabrication of electronics devices* (Vol. 14). Springer. doi:10.1007/978-1-4614-9176-7

Kong, F. Z., Zhang, X. B., Xiong, W. Q., Liu, F., Huang, W. Z., Sun, Y. L., Tu, J. P., & Chen, X. W. (2002). Continuous Ni-layer on multiwall carbon nanotubes by an electroless plating method. *Surface and Coatings Technology*, *155*(1), 33–36. doi:10.1016/S0257-8972(02)00032-4

Kozlov, A. S., Palanisamy, T., & Narasimhan, D. (2002). *U.S. Patent No. 6,387,542*. Washington, DC: U.S. Patent and Trademark Office.

Krishnan, K. H., John, S., Srinivasan, K. N., Praveen, J., Ganesan, M., & Kavimani, P. M. (2006). An overall aspect of electroless Ni-P depositions—A review article. *Metallurgical and Materials Transactions. A, Physical Metallurgy and Materials Science*, *37*(6), 1917–1926. doi:10.100711661-006-0134-7

Kundu, S., Das, S. K., & Sahoo, P. (2014). Properties of electroless nickel at elevated temperature-a review. *Procedia Engineering*, *97*(0), 1698–1706. doi:10.1016/j.proeng.2014.12.321

Kutnahorsky, M. R. (2009). *Electroless copper deposition: A sustainable approach* (Doctoral dissertation).

Leon, C. A., & Drew, R. A. L. (2000). Preparation of nickel-coated powders as precursors to reinforce MMCs. *Journal of Materials Science*, *35*(19), 4763–4768. doi:10.1023/A:1004860326071

Li, G. J., Huang, X. X., & Guo, J. K. (2001). Fabrication of Ni-coated Al2O3 powders by the heterogeneous precipitation method. *Materials Research Bulletin*, *36*(7-8), 1307–1315. doi:10.1016/S0025-5408(01)00628-6

López-Honorato, E., Dessoliers, M., Shapiro, I. P., Wang, X., & Xiao, P. (2012). Improvements to the sintering of yttria-stabilized zirconia by the addition of Ni. *Ceramics International*, *38*(8), 6777–6782. doi:10.1016/j.ceramint.2012.05.073

Ma, S., Xu, E., Zhu, Z., Liu, Q., Yu, S., Liu, J., Zhong, H., & Jiang, Y. (2018). Mechanical and wear performances of aluminum/sintered-carbon composites produced by pressure infiltration for pantograph sliders. *Powder Technology*, *326*, 54–61. doi:10.1016/j.powtec.2017.12.027

Mallory, G. O., & Hajdu, J. B. (Eds.). (1990). *Electroless plating: fundamentals and applications*. Cambridge University Press.

Marie,O.J.(1966). *U.S.PatentNo.3,282,810*. Washington, DC: U.S. Patent and Trademark Office.

Metzger, W., Schmidt, H., Ott, R., & Pappe, G. (1973). *U.S. Patent No. 3,753,667*. Washington, DC: U.S. Patent and Trademark Office.

Moustafa, S. F., El-Badry, S. A., Sanad, A. M., & Kieback, B. (2002). Friction and wear of copper–graphite composites made with Cu-coated and uncoated graphite powders. *Wear*, *253*(7-8), 699–710. doi:10.1016/S0043-1648(02)00038-8

Naga, S. M., Awaad, M., Bondioli, F., Fino, P., & Hassan, A. M. (2017). Thermal diffusivity of ZTA composites with different YSZ quantity. *Journal of Alloys and Compounds*, *695*, 1859–1862. doi:10.1016/j.jallcom.2016.11.019

Nakahara, S. (1977). Microporosity induced by nucleation and growth processes in crystalline and non-crystalline films. *Thin Solid Films*, *45*(3), 421–432. doi:10.1016/0040-6090(77)90229-2

Narayanan, T. S., & Seshadri, S. K. (2004). Formation and characterization of borohydride reduced electroless nickel deposits. *Journal of Alloys and Compounds*, *365*(1-2), 197–205. doi:10.1016/S0925-8388(03)00680-7

Narcus, H. (1947). Practical copper reduction on nonconductors. *Metal Finishing*, *45*(9), 64–67.

Nastic, A., Merati, A., Bielawski, M., Bolduc, M., Fakolujo, O., & Nganbe, M. (2015). Instrumented and Vickers indentation for the characterization of stiffness, hardness and toughness of zirconia toughened Al2O3 and SiC armor. *Journal of Materials Science and Technology*, *31*(8), 773–783. doi:10.1016/j.jmst.2015.06.005

Nathan, A., Ahnood, A., Cole, M. T., Lee, S., Suzuki, Y., Hiralal, P., ... Haque, S. (2012). Flexible electronics: the next ubiquitous platform. *Proceedings of the IEEE, 100*, 1486-1517.

Oguocha, I. N. A., Taheri, R., Yannacopoulos, S., Uju, W. A., Sammynaiken, R., Wettig, S., & Hu, Y. F. (2010). Temperature effects on the chemical composition of nickel–phosphorus alloy thin films. *Thin Solid Films*, *518*(8), 2045–2049. doi:10.1016/j.tsf.2009.09.007

Oh, K. W., Hong, K. H., & Kim, S. H. (1999). Electrically conductive textiles by in situ polymerization of aniline. *Journal of Applied Polymer Science*, *74*(8), 2094–2101. doi:10.1002/(SICI)1097-4628(19991121)74:8<2094::AID-APP26>3.0.CO;2-9

Oita, M., Matsuoka, M., & Iwakura, C. (1997). Deposition rate and morphology of electroless copper film from solutions containing 2, 2¢-dipyridyl. *Electrochimica Acta*, *42*(9), 1435–1440. doi:10.1016/S0013-4686(96)00367-2

Palaniappa, M., Babu, G. V., & Balasubramanian, K. (2007). Electroless nickel–phosphorus plating on graphite powder. *Materials Science and Engineering A*, *471*(1-2), 165–168. doi:10.1016/j.msea.2007.03.004

Pang, J. N., Jiang, S. W., Lin, H., & Wang, Z. Q. (2016). Significance of sensitization process in electroless deposition of Ni on nanosized Al2O3 powders. *Ceramics International*, *42*(3), 4491–4497. doi:10.1016/j.ceramint.2015.11.137

Paul, T. (1956). *U.S. Patent No. 2,766,138*. Washington, DC: U.S. Patent and Trademark Office.

Paunovic, M. (1968). Electrochemical aspects of electroless deposition of metals. *Plating (East Orange, N.J.)*, *55*, 1161–1167.

Pauw, L. (1958). A method of measuring specific resistivity and Hall effect of discs of arbitrary shape. *Philips Research Reports*, *13*(2), 1–9.

Pratihar, S. K., Sharma, A. D., Basu, R. N., & Maiti, H. S. (2004). Preparation of nickel coated YSZ powder for application as an anode for solid oxide fuel cells. *Journal of Power Sources*, *129*(2), 138–142. doi:10.1016/j.jpowsour.2003.11.023

Ramalho, A., & Miranda, J. C. (2005). Friction and wear of electrolessNiP and NiP+ PTFE coatings. *Wear*, *259*(7-12), 828–834. doi:10.1016/j.wear.2005.02.052

Reid, J., Mayer, S., Broadhent, E., Klawuhn, E., & Ashtiani, K. (2000). Factors influencing damascene feature fill using copper PVD and electroplating. *Solid State Technology*, *43*(7), 86–86.

Rodriguez-Suarez, T., Bartolomé, J. F., & Moya, J. S. (2012). Mechanical and tribological properties of ceramic/metal composites: A review of phenomena spanning from the nanometer to the micrometer length scale. *Journal of the European Ceramic Society*, *32*(15), 3887–3898. doi:10.1016/j.jeurceramsoc.2012.06.026

S. (2009). Electroless deposition of nickel and copper on titanium substrates: Characterization and application. *Journal of Alloys and Compounds, 472*(1-2), 595-601.

Safranek, W. H., Dahle, F. B., & Faust, C. L. (1948). Electroforming—A Literature Review and a NewApplication. *Plating (East Orange, N.J.)*, *35*, 39–49.

Sahoo, P., & Das, S. K. (2011). Tribology of electroless nickel coatings–a review. *Materials & Design*, *32*(4), 1760–1775. doi:10.1016/j.matdes.2010.11.013

Saxena, V., Rani, R. U., & Sharma, A. K. (2006). Studies on ultra high solar absorber black electroless nickel coatings on aluminum alloys for space application. *Surface and Coatings Technology*, *201*(3-4), 855–862.

Schwarz, A., Cardoen, J., Westbroek, P., Van Langenhove, L., Bruneel, E., Van Driessche, I., & Hakuzimana, J. (2010). Steps Towards a Textile-Based Transistor: Development of the Gate and Insulating Layer. *Textile Research Journal*, *80*(16), 1738–1746. doi:10.1177/0040517510365948

Schwarz, A., Hakuzimana, J., Kaczynska, A., Banaszczyk, J., Westbroek, P., McAdams, E., ... Tseles, D. (2010). Gold coated para-aramid yarns through electroless deposition. *Surface and Coatings Technology*, *204*(9-10), 1412–1418. doi:10.1016/j.surfcoat.2009.09.038

Seeböck, R., Esrom, H., Charbonnier, M., Romand, M., & Kogelschatz, U. (2001). Surface modification of polyimide using dielectric barrier discharge treatment. *Surface and Coatings Technology*, *142*, 455–459.

Selvam, M. (2010). Electroless silver deposition on ABS plastic using Co (II) as reducing agent. *Transactions of the IMF*, *88*(4), 198-203. 10.1179/002029610X12694310988964

Shacham-Diamand, Y., Dubin, V., & Angyal, M. (1995). Electroless copper deposition for ULSI. *Thin Solid Films*, *262*(1-2), 93–103. doi:10.1016/0040-6090(95)05836-2

Sharma, R., Agarwala, R. C., & Agarwala, V. (2006). Development of copper coatings on ceramic powder by electroless technique. *Applied Surface Science*, *252*(24), 8487–8493. doi:10.1016/j.apsusc.2005.11.085

Shijitha, T., Baiju, K. V., Shukla, S., Patil, K., & Warrier, K. G. K. (2009). Novelelectrolessprocessforcoppercoatingofflyashusingtitania/ultraviolet-radiation/metalcatalyst-system. *Applied Surface Science*, *255*(13-14), 6696–6704. doi:10.1016/j.apsusc.2009.02.066

Shukla, S., & Koelling, K. W. (2009). Classical nucleation theory applied to homogeneous bubble nucleation in the continuous microcellular foaming of the polystyrene– CO2 system. *Industrial & Engineering Chemistry Research*, *48*(16), 7603–7615. doi:10.1021/ie8011243

Sudagar, J., Lian, J., & Sha, W. (2013). Electroless nickel, alloy, composite and nano coatings–A critical review. *Journal of Alloys and Compounds*, *571*, 183–204. doi:10.1016/j.jallcom.2013.03.107

Toozandehjani, M. (2017). *Synthesis And Characterization of Carbon Nanotubes and Nano Alumina-Reinforced Aluminum Matrix Nanocomposites*. Academic Press.

Trademark Odekerken, J. M. (1972). *U.S. Patent No. 3,644,183*. Washington, DC: U.S. Patent and Office.

Tsai, T. K., Wu, S. S., Liu, W. L., Hsieh, S. H., & Chen, W. J. (2007). ElectrolessCoWP as a diffusion barrier between electroless copper and silicon. *Journal of Electronic Materials*, *36*(11), 1408–1414. doi:10.100711664-007-0223-4

Uddin, A. J. (2010). Coatings for technical textile yarns. In *Technical Textile Yarns* (pp. 140–184). Woodhead Publishing. doi:10.1533/9781845699475.1.140

Wang, G., Jong, Y.W., Balakumar, S., Seah, C.H., & Hara, T. (2004). Nanotechnology and Microelectronics: Materials, Processing, Measurement, and Phenomena. *Journal of Vacuum Science & Technology B*.

Wang, Y., Gao, Y., Li, Y., Zhang, C., Sun, L., & Zhai, W. (2017). Research on nickel modified graphite/Cu composites interface. *Surface and Coatings Technology*, *328*, 70–79. doi:10.1016/j.surfcoat.2017.08.036

Wen, G., Guo, Z. X., & Davies, C. K. L. (2000). Microstructural characterisation of electroless-nickel coatings on zirconia powder. *Scripta Materialia*, *43*(4), 307–311. doi:10.1016/S1359-6462(00)00409-7

Westbroek, P., Schwarz, A., Hakuzimana, J., Gasana, E., Van Langenhove, L., & Tseles, D. (2007). Corrosion resistant textiles electrodes obtained through electroless coating of fibers with gold. In eRA-2 Conference, Athens, GA.

Whang, T. J., Hsieh, M. T., & Kao, Y. C. (2010). Studies of single-step electrode position of CuInSe2 thin films with sodium citrate as a complexing agent. *Applied Surface Science*, *257*(5), 1457–1462. doi:10.1016/j.apsusc.2010.08.072

Wojcik, H., Lehninger, D., Neumann, V., & Bartha, J. W. (2011, September). *Characterization of barrier and seed layer integrity for copper interconnects. In 2011 Semiconductor Conference Dresden*. IEEE.

Wurtz, A. (1844). Sur LHydrure De. *Ann. Chim. Phys*, *11*, 250–251.

Xu, X., Cui, Z. D., Zhu, S. L., Liang, Y. Q., & Yang, X. J. (2014). Preparation of nickel-coated graphite by electroless plating under mechanical or ultrasonic agitation. *Surface and Coatings Technology*, *240*, 425–431. doi:10.1016/j.surfcoat.2013.12.070

Yang, L., Ran, L., & Yi, M. (2011). Carbon fiber knitted fabric reinforced copper composite for sliding contact material. *Materials & Design*, *32*(4), 2365–2369. doi:10.1016/j.matdes.2010.11.063

Zhang, B., Wenyuan, W., Shaohua, Y., Shiwei, L., Yao, L., Xue, B., & Ganfeng, T. (2010). Process optimization of electroless copper plating and its influence on electrochemical properties of AB5-type hydrogen storage alloy. *Journal of Rare Earths*, *28*(6), 922–926.

Zhao, Q., Wang, C., Liu, Y., & Wang, S. (2007). Bacterial adhesion on the metal-polymer composite coatings. *International Journal of Adhesion and Adhesives*, *27*(2), 85–91. doi:10.1016/j.ijadhadh.2006.01.001

Zheng, X., Tan, J., Zhang, Q., Wang, M., & Meng, L. (2017). Effect of laser surface texturing depth on the adhesion of electroless plated nickel coating on alumina. *Surface and Coatings Technology*, *311*, 151–156. doi:10.1016/j.surfcoat.2017.01.002

Chapter 9
Tribological and Micro-Structural Characterization of Ni-Cu-P-W Coatings

Bal Mukund Mishra
Haldia Institute of Technology, India

Supriyo Roy
https://orcid.org/0000-0001-8731-0068
Haldia Institute of Technology, India

Goutam Kumar Bose
https://orcid.org/0000-0002-4347-3508
Haldia Institute of Technology, India

ABSTRACT

Ni-Cu-P-W coating was deposited by electroless method on mild steel substrate to study the crystallization and tribological behavior at different annealing temperatures. Energy dispersive x-ray (EDX) analysis, scanning electron microscopy (SEM), x-ray diffraction (XRD), and differential scanning calorimeter (DSC) were used to study the composition, surface morphology, phase behavior, and thermal behavior of the coating, respectively. Tribological study was conducted using Pin-on-Disc tribotester. EDX analysis confirms the presence of Ni, Cu, P, and W in the deposit. SEM image shows the surface is dense, smooth, and without any observable nodule. Some of the samples were heat treated to 300°C, 500°C, and 700°C for 1 hour to observe the crystallographic change by XRD. One sharp crystalline peak of Ni (111) is present in all condition, but the intensity increases rapidly with the heat treatment temperature. The phase transition temperature of this quaternary coating analyzed by DSC was 431.8°C.

DOI: 10.4018/978-1-7998-4870-7.ch009

INTRODUCTION

Tungsten and copper already have gained much attention to material scientists and researchers as alloying elements due to their excellent material properties. Tungsten has high hardness, high melting point, high density and good corrosion resistance, on the other hand, copper possess high corrosion resistance, good conductivity, ductility etc. Thus for last two decades, researchers have deposited these elements as alloy along with Ni-P binary electroless coating to explore their properties and also to correlate the structure of alloy with their properties(Sudagar, Lian, and Sha 2013; Agarwal and Agarwala 2003; Ghosh,2019). Several researchers have studied the thermal stability and correlation between crystalline phase and hardness of Ni-W-P ternary alloy coatings (Tsai et al. 2001; Zhou et al. 2019; Luoet al. 2018). It was reported, that crystallization temperature of ternary Ni-W-P deposit is higher than binary Ni-P deposit due to elemental diffusion of W which has high melting point. The hardness of ternary coatings is higher than binary coatings in both as-plated and heat treated condition due to formation of Ni-W solid solution. The hardness of Ni-W-P coatings increased sharply within the temperature range 350°C to 400°C. Wu et al. (2003) studied surface characteristics of as deposited and heat treated ternary Ni-W-P coatings. Nodular surface was observed in as-plated condition of electrolessly deposited ternary alloy with an average roughness of 10 nm. After heat treatment to 400°C and 450°C the nodules started disappearing and smoother surface was generated. XRD and DSC results confirmed the precipitation of Ni_3P phase after heating 300°C for long time (Tien and Duh 2004; Tien, Duh, and Chen 2004a; Tien, Duh, and Chen 2004b; Wu, Tien, Chen, and Duh 2004). At higher temperature all the metastable phases (Ni_5P_2, $Ni_{12}P_5$) transform in to stable Ni_3P phase. Roy and Sahoo studied the friction, wear and corrosion behavior of ternary Ni-P-W coatings at different coating conditions and heat treatment temperature (Roy and Sahoo 2012; Roy and Sahoo 2013). It was revealed that, incorporation of tungsten in to binary Ni-P matrix enhanced the wear and corrosion properties up to 500°C heat treatment temperature due to formation of stable Ni_3P phase (Palaniappa and Seshadri 2008). On the other hand, crystallization characteristics of electroless ternary Ni-Cu-P alloy were studied by several researchers (Yu, Luo, and Wang, 2001; Valova et al. 2005; Yu et al. 2002). It was reported that during heat treatment process low P with high Cu alloy first transform to metastable phase of Ni_5P_2, $Ni_{12}P_5$ then at higher temperature it is converted to Ni_3P phase. As like Ni-W-P, Ni-Cu-P alloy also has higher crystallization temperature compared to Ni-P binary alloy. Incorporation of copper in to Ni-P deposit significantly improves the corrosion resistance (Liu et al. 2010; Liu and Zhao, 2004; Meng, M., Leech, A., & Le, H. 2019; Chen et al. 2019a; Chen et al. 2019b). Concentration of copper ion, pH of the electroless bath and bath temperature has great influence on the corrosion properties of the coating. However, detailed study on the concentration of electroless bath constituents and heat treatment temperature on the tribological and corrosion characteristics of Ni-P-Cu ternary coatings were investigated (Roy and Sahoo, 2014). The comparative study between Ni-P, Ni-W-P and Ni-Cu-P revealed that, in respect to tribological behavior ternary Ni-W-P is better than binary Ni-P and with respect to corrosion characteristics ternary Ni-Cu-P is better that binary Ni-P alloy (Sahoo and Roy, 2017). Thus, to have a coating with good tribological as well as anti-corrosive property electroless quaternary Ni-Cu-W-P can be deposited. There are very few papers reported about the deposition of quaternary Ni-W-Cu-P from sulphate or chloride-based baths and their corrosion properties (Balaraju and Rajam, 2005; Balaraju et al. 2006; Balaraju, Anandan and Rajam, 2006). But, dependency of bath constituents and heat treatment temperature on the tribological and corrosion behavior of this quaternary coating is still untouched. The structural characterization and phase transformation behavior within a wide range of heat treatment temperature also needs to explore. This research work is a preliminary step towards this aim.

Thus the present work aims to deposit quaternary Ni-Cu-P-W coatings by electroless method using sulphate and chloride mixed bath. The surface morphology and chemical composition of the deposited alloy was studied by scanning electron microscopy (SEM) and energy dispersive x-ray analysis (EDX). Within a wide range of heat treatment temperature (up to 700°C), the phase transition behavior was studied by x-ray diffraction analysis (XRD). The crystallographic transition and calorimetric study was carried out by differential scanning calorimeter (DSC). Tribological tests (wear and friction) were carried in both dry and lubricated conditions.

EXPERIMENTAL DETAILS

Sample Preparation

Cylindrical mild steel samples (φ6mm × 30mm) were used as substrate for deposition of quaternary Ni-Cu-P-W alloy by electroless method. The samples were wiped first, to make it free from foreign particles and corrosion products. In next step, samples were mechanically polished with SiC papers of fine grits to make the surface smooth. After polishing, samples were rinsed with deionized water and cleaned by dipping in to acetone bath for 5 min at room temperature. To degrease, the samples were dipped in to 50% HCL solution and rinsed with deionized water. Thoroughly cleaned samples were activated in an aqueous solution of 0.25g/L palladium chloride and 10ml/L concentrated hydrochloric acid (HCl) for 1 min.

Electroless Deposition

After completing the pre-treatment and activation, the samples were placed in to electroless bath for the deposition of Ni-Cu-P-W quaternary alloy. For each sample separate fresh bath was used. The electroless bath constituents and operating condition for coating are given in Table 1. In this study both sulphate and chloride based compound of nickel were used simultaneously in the electroless bath as nickel source. Thus, a mixed sulphate-chloride based bath was used instead of either sulphate or chloride based bath, used by previous researcher (Balaraju and Rajam,2005; Balaraju, Jahan, Anandan and Rajam,2006; Balaraju, Anandan and Rajam,2006). Bath temperature was monitored continuously by a digital thermometer to maintain in the range of 90±2 °C. After 3 hours deposition the samples were taken out and again submerged in to a fresh bath for another 3 hours to increase the deposition thickness, which is essential for tribological testing. After complete deposition samples were taken out of bath and rinsed with deionized water and dried in air. Some samples were heat treated in a bench top muffle furnace at 300°C, 500°C and 700°C respectively for 1 hour and cooled within the furnace.

Table 1. Chemical composition and operating condition for electroless Ni-Cu-P-W coating

Bath constituents	Concentration
Nickel Sulphate ($NiSO_4$)	10 g/l
Nickel Chloride ($NiCl_2$)	10 g/l
Sodium Hypophosphite ($NaPO_2H_2$)	20 g/l
Sodium Citrate ($Na_3C_6H_5O_7$)	35 g/l
Ammonium Chloride (NH_4Cl)	12 g/l
Sodium Tungstate (Na_2WO_4)	30 g/l
Copper Chloride ($CuCl_2$)	0.3 g/l
Lactic Acid ($C_3H_6O_3$)	5 g/l
Operating conditions	
pH	8
Bath temperature	90±2 °C
Bath volume	200 ml
Deposition time	3 hrs

Tribological Study

A pin-on-disc wear and friction monitoring apparatus (DUCOM, TR-20LE) was used to conduct the dry and lubricated tribological tests. As-deposited and heat treated samples were held against an EN-31 disc of higher hardness than the coatings to ensure the wear of coated surface only. During all the tests, the load, sliding velocity and sliding distance were kept constant at 10 N, 0.432 m/s and 259 m respectively. For lubricated tests, commercially available SAE 10 W 30 grade lubricant was used. All the tests were carried out following ASTM G99 standard. Wear depth and friction force were directly measured by attached sensors with the apparatus. To ensure the stability and accuracy of experimental data each experiment was repeated three times.

Structural and Calorimetric Study

After deposition the samples were taken out and washed with deionized water. Energy dispersive x-ray analysis (EDX) (EDAX Corporation) was carried out for the confirmation of presence of all the alloying elements. Microstructure of the coated specimen was studied by scanning electron microscopic (SEM) (JEOL, LSM 6360) image. X-ray diffraction analyses (XRD) of each heat treated specimen were carried out with (PANalytical X-PERT PRO) powder diffractometer to study the phase transition and crystallographic characteristics of the quaternary coatings. The calorimetric study during the thermal processing was studied by differential scanning calorimeter (DSC).

RESULTS

Tribological Characterization

Figure 1(a) and (b) shows the wear behaviour of as-deposited (WHT) and heat treated (HT) samples at dry and lubricated environment respectively. The load, speed and sliding distance kept constant for all the experiments. It is evident from figure 1(a), at dry environment the wear depth decreases rapidly with the increase of heat treatment temperature up to 500°C, but beyond this temperature, wear depth increased. Similar trend of wear depth can be observed in case of lubricated environment in figure 1(b), but within the range of temperature from 500°C to 700°C, the increase of wear depth is not much compared to dry environment. Thus in both the cases, 500°C heat treated coating showed minimum wear. In figure 1(c) and (d), variations of coefficient of friction (COF) are shown at dry and lubricated environment respectively. It is evident from the experimental data, that in dry environment, COF is lesser for 300°C HT than as-deposited coating, but increased with increase in heat treatment temperature. Whereas, in lubricated environment, COF for 300°C HT sample is higher than as-deposited sample and in next higher heat treatment temperature (500°C) it reduced down but again increased for 700°C heat treated sample. Thus lubrication and heat treatment temperature plays a vital role in controlling coefficient of friction of these coatings.

Structural Characterization

Chemical compositions of as-plated quaternary coatings were analyzed by energy dispersive x-ray analysis (EDX). Results revealed that, the coating consisted of 81.08 wt% - 83.64 wt% of Ni, 5.81 wt% - 7.08 wt% Cu, 5.23 wt% - 6.45 wt% P and 4.22 wt% - 6.41 wt% W. The spectrum obtained from the analysis is shown in Figure 2. Surface morphology of the coating was studied by scanning electron microscope (SEM). Micrographs obtained from the analysis are shown in Figure 3. Figure 3 (a) shows the deposit is dense, apparently smooth and without any porosity. Highly magnified image at Figure 3 (b) also confirms those characteristics mentioned above. This finding is quite consistent with the result obtained by Balaraju et al. (2005) (Balaraju and Rajam,2005). It was reported that, incorporation of copper in to the ternary Ni-P-W alloy coatings produces smooth microstructure with disappearance of nodularity at surface.

Crystallization Behavior

Diffraction patterns of as-plated and heat treated quaternary Ni-Cu-P-W coatings are shown in Figure 4. Figure 4(a) shows the x-ray diffraction pattern of as-plated quaternary coating. It consists of a single crystalline peak of fcc Ni (111) at 44.72° (2θ) within a mixture of amorphous and nano crystalline structure. This type of crystalline structure appears in as-plated condition is due to presence of less than 7 wt.% of phosphorous in the deposited alloy (Balaraju and Rajam,2005; Balaraju, Jahan, Anandan and Rajam,2006). Figure 4(b) depicts sharper peak with continuous heating at higher temperature. No other new phases have been identified at this stage. More sharp and narrow peak has been identified by heating at 500°C. At this condition the major peak also shifts a little to lower angle (44.24°). This can be explained by diffusion of tungsten and copper in to nickel matrix causing lattice expansion (Tien and Duh 2004; Yu, Sun, Luo, Wang, and Wu, 2002). At this heat treated condition small crystalline peaks of

Figure 1a. Variation of wear and COF with sliding distance: Dry wear

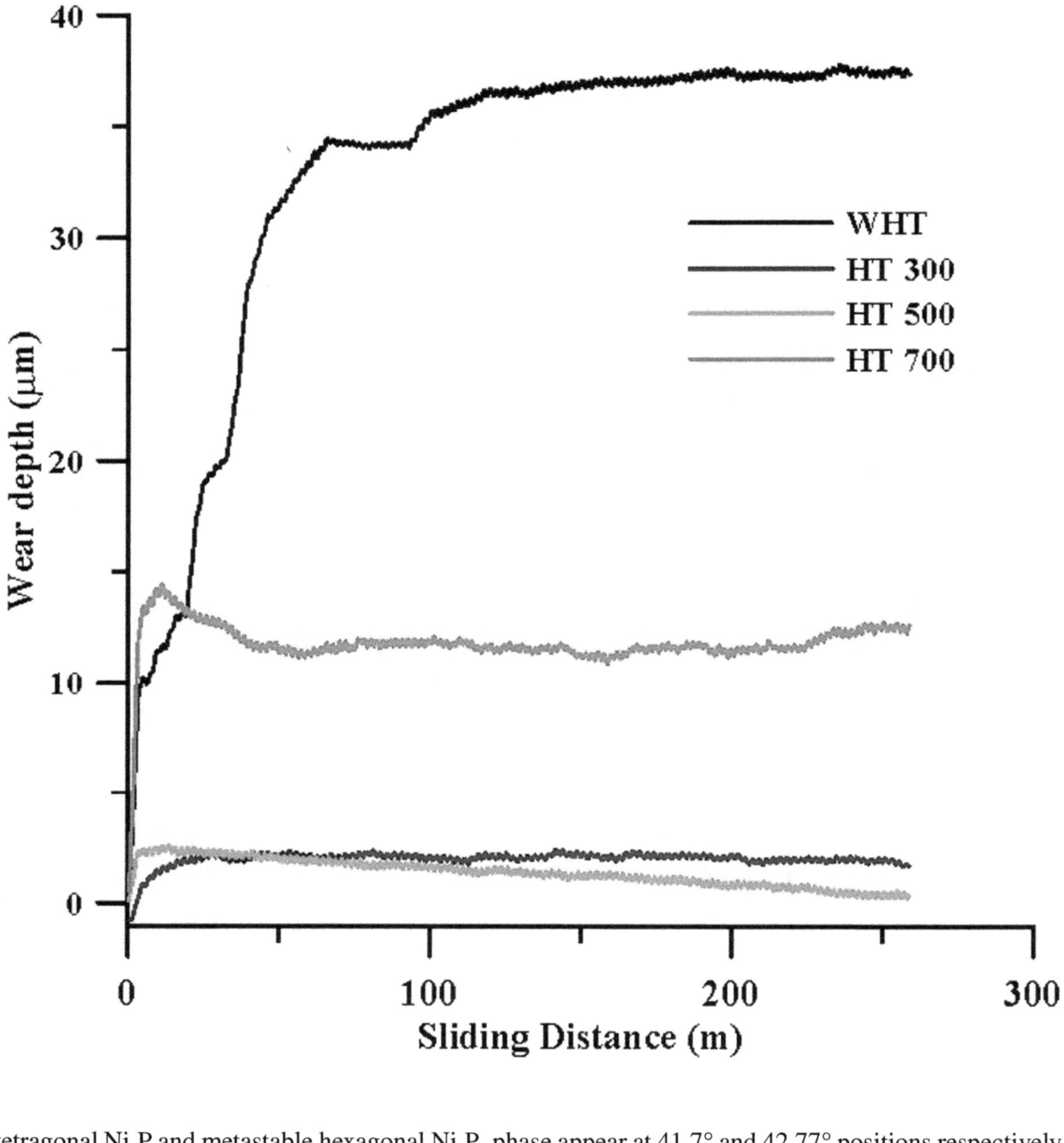

tetragonal Ni_3P and metastable hexagonal Ni_5P_2 phase appear at 41.7° and 42.77° positions respectively (Figure 5(a)). Further increase in heat treatment temperature to 700°C causes increase of intensity of all crystalline peaks, shown in Figure 4(d) and 5(b). It is clear from the enlarged view of XRD plot of 700°C heat treated condition in Figure 5(b), that not only increase in intensity but, the metastable Ni_5P_2 phase transforms in to stable Ni_3P phase with the generation of some more Ni_3P crystalline phase in the alloy. As in this deposited alloy, no other tungsten phase or copper phosphide phase have been identified, thus, the Ni (111) peak is suggested as Ni-W overlapped with Ni-Cu phase at higher temperature due to dissolving of tungsten and copper. The grain size of major crystalline peak at different heat treatment

Figure 1b. Variation of wear and COF with sliding distance: Lubricated wear

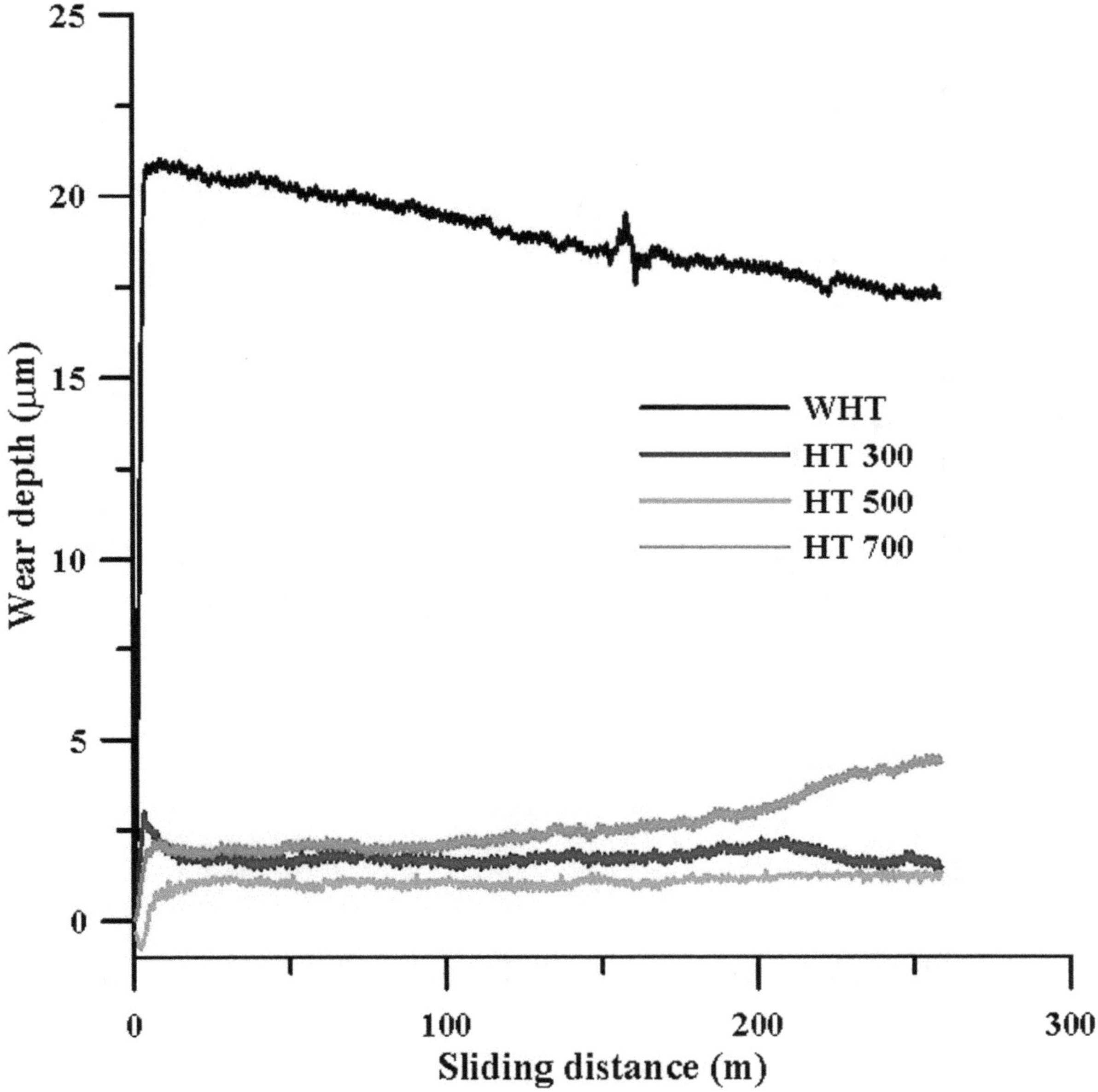

temperatures was calculated using Scherrer equation. The full width at half maximum (FWHM) and grain size corresponding to peak position is given in Table 2 and explained in Figure 6. It indicates that, grain growth rate is smaller at lower temperature but increases rapidly at higher temperature. Heating at lower temperature the grain size increased from 9.94 nm to 12.23 nm (increased by 2.29 nm). At second higher heating temperature the grain size reached 34.00 nm (increased by 21.77 nm) and at 700°C it reached 66.79 nm (increased by 32.79 nm). This crystalline structure formation is greatly influenced by the presence of tungsten and copper atom in fcc phase of nickel within the coating (Tsai, Wu, Chen, Peng, Duh, and Tsai,2001; Lu and Zangari,2002).

Figure 1c. Variation of wear and COF with sliding distance: Dry COF

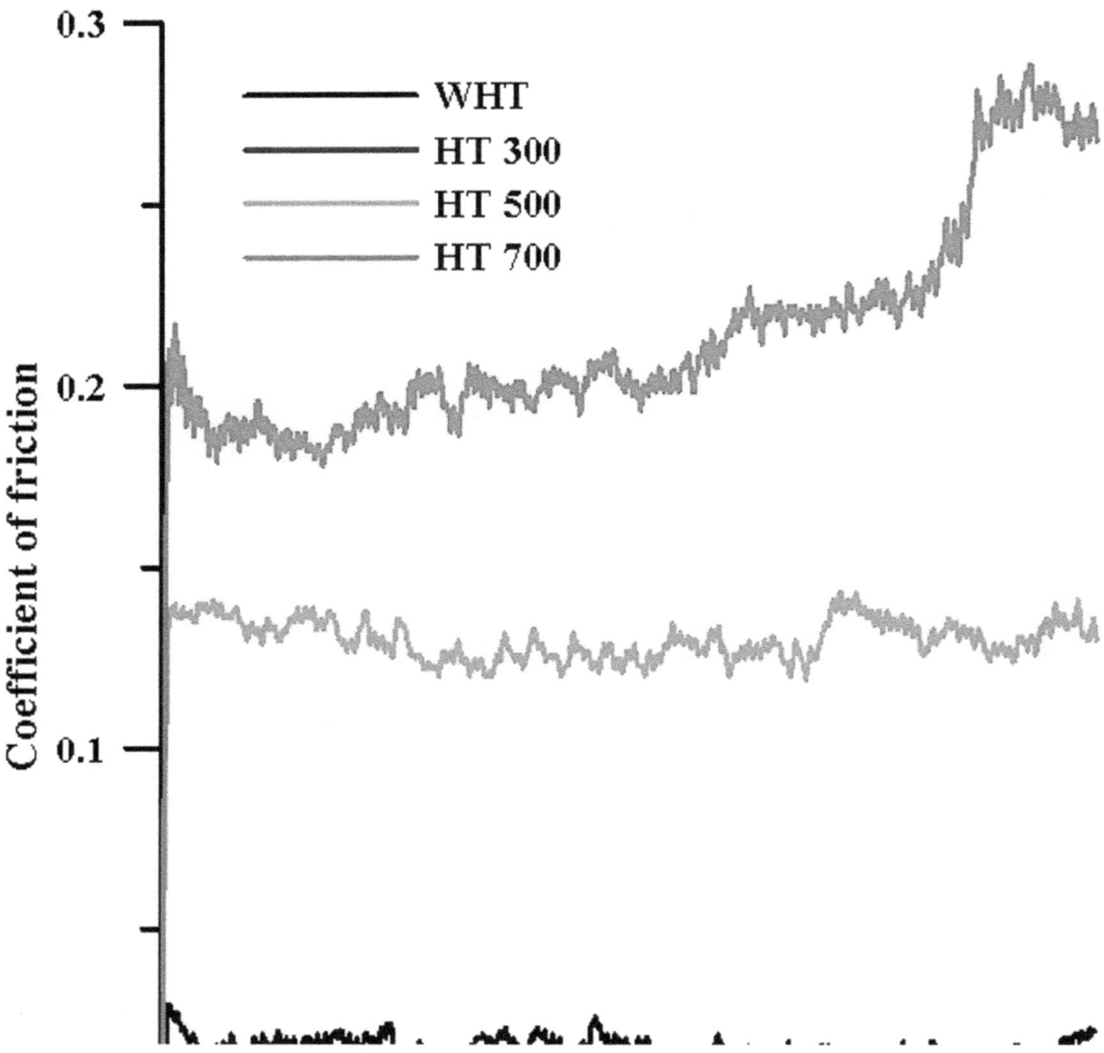

Calorimetric Study

The DSC plot for Ni-Cu-P-W quaternary alloy coating consists of 81.9 wt% Ni-6.87 wt.% Cu-5.63 wt.% P-6.41 wt.% W with heating rate 10°C/min is shown in Figure 7. Upper range of temperature was set to 550°C due to aluminium trey that had been used for holding sample during testing. The thermogram depicts a broad exothermic peak within the range of 316.6°C to 366.4°C. It can be clearly observed from XRD plots at Figure 4(b) and 4(c) that after 300°C peak sharpness corresponding to Ni (111) rapidly increased. Figure 4 also revealed, a sharp increase in grain size of Ni (111) phase after 300°C. Thus it can be concluded that after 300°C grain growth of Ni-W and Ni-Cu occurs and these phases coexist at 44.24° (2θ) (Figure 4(c)). Another big exothermic peak is identified at 431.8°C. The peak starts from

402.7°C and completes at 479.5°C. This peak corresponds to the formation of new crystalline phase of Ni_3P within the deposited alloy along with significant grain growth of crystalline phases. It is also confirmed from the Figure 4(c). It depicts the existence of Ni_3P phase at 500°C heat treated condition, which is not present at 300°C heat treated condition in Figure 4(b).

Figure 1d. Variation of wear and COF with sliding distance: Lubricated COF

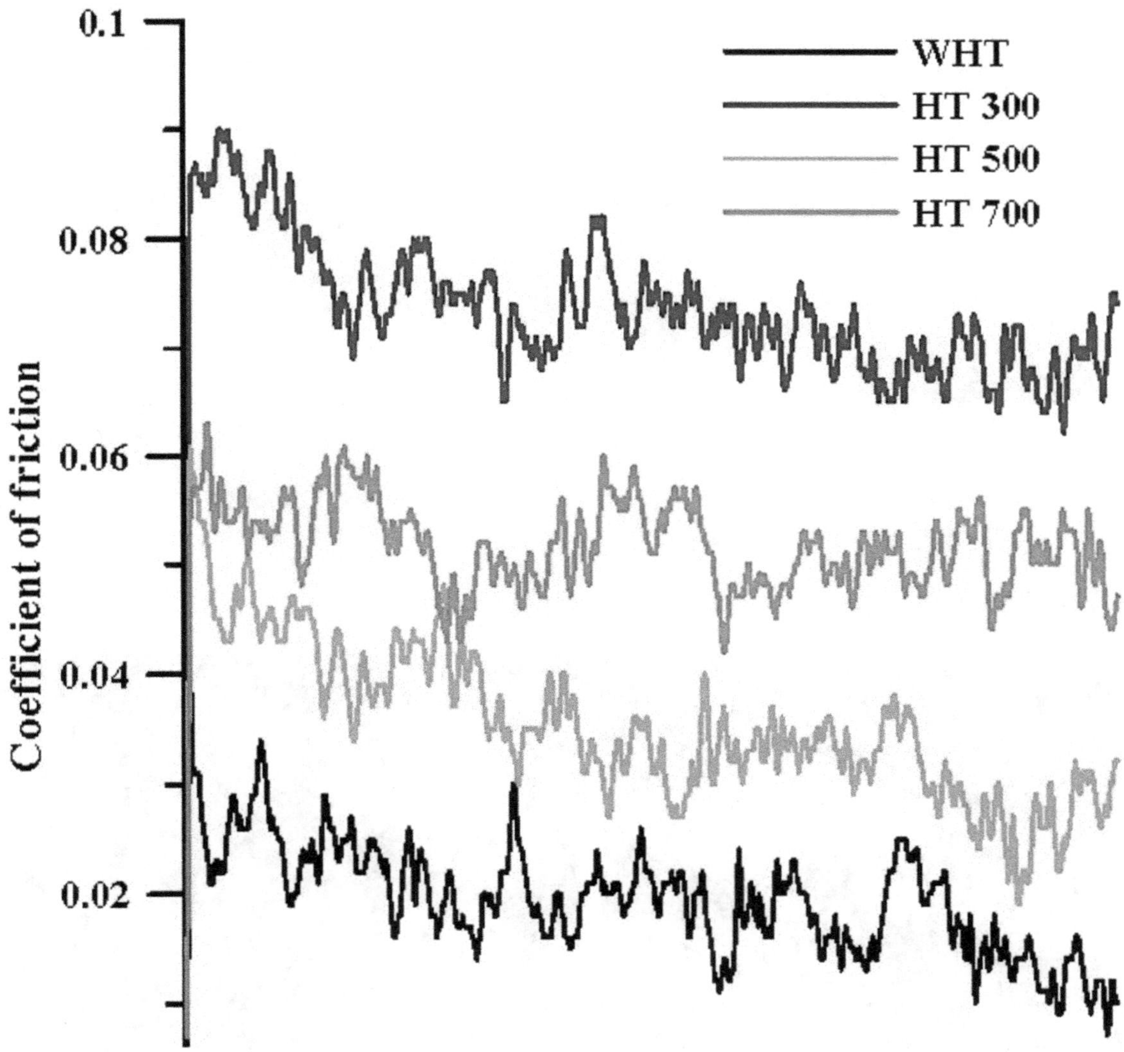

Figure 2. EDX spectrum of quaternary Ni-Cu-P-W coatings (as deposited)

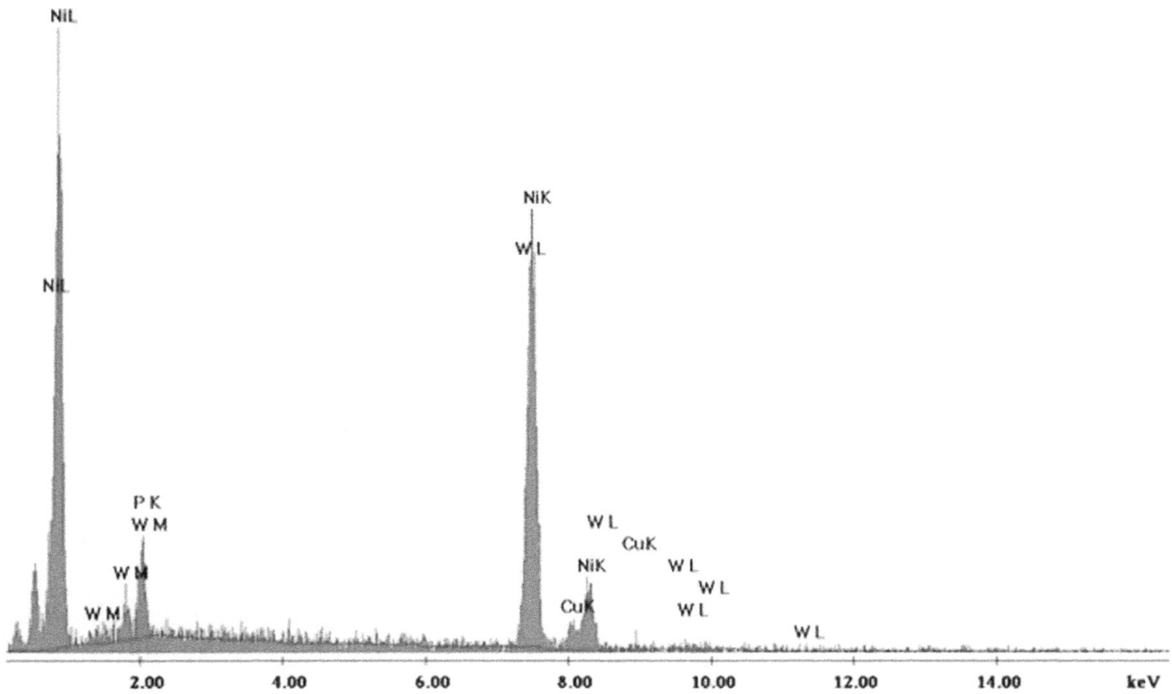

Figure 3. SEM image of as deposited quaternary Ni-Cu-P-W coatings

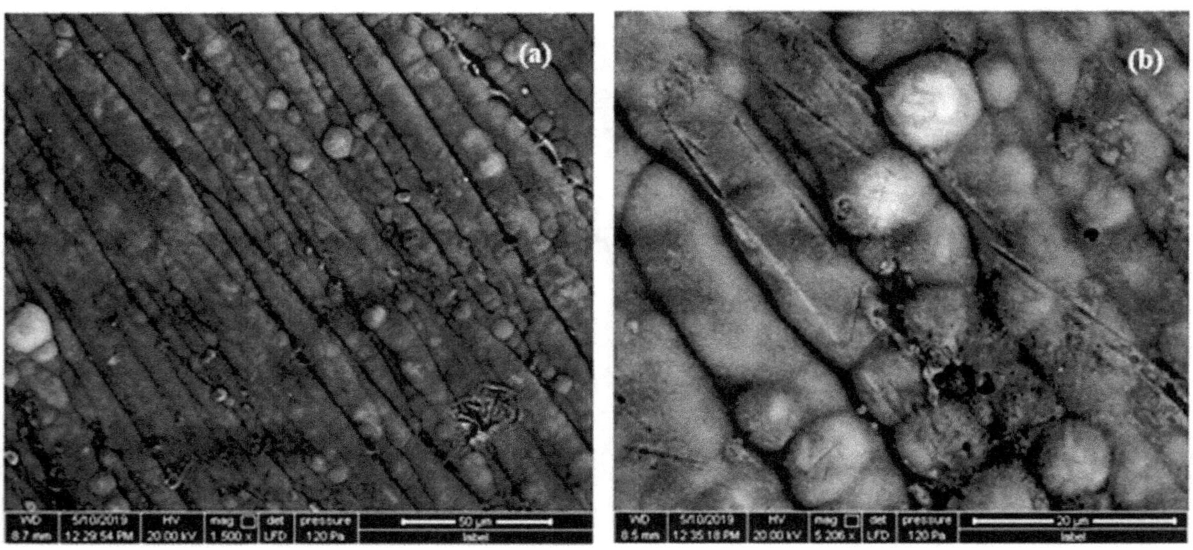

Figure 4. X-ray diffraction pattern of quaternary Ni-Cu-P-W coatings (a) as-plated, (b) 300°C, (c) 500°C, (d) 700°C

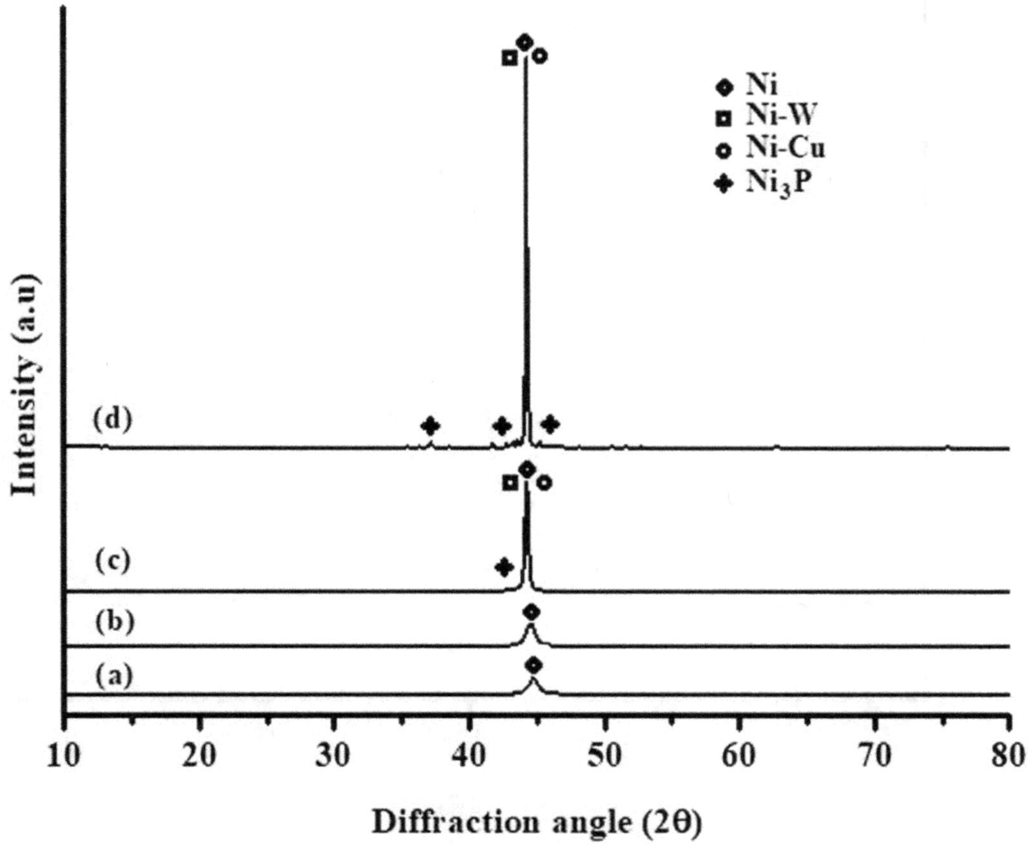

DISCUSSION

It is an established fact that, wear is very much related to hardness and in turn, depends on crystalline structure of the surface. In this work also, an attempt has been made to correlate the wear characteristics with crystalline structure at different heat treatment temperature. Balaraju and Rajam (2005) in their work showed that near 400°C the Ni_3P phase starts to precipitate in this quaternary alloy and achieved highest hardness. Beyond this temperature grain coarsening starts and hardness falls down at 600°C. This result has a consistency with present investigation. In this investigation, the wear depth is gradually decreases with heat treatment temperature up to 500°C and then increases. The crystallographic study also revealed that near 500°C, Ni_3P phase precipitation starts and beyond that temperature grains become coarse nodular. Thus it can be concluded that due to formation of Ni_3P phase within temperature range 400°C to 500°C the hardness increase to its maximum value thus the wear depth is minimum and beyond that heating temperature due to grain coarsening effect the hardness decreases and more wear takes place. The wear debris, which are getting entrapped within the work piece and rotating disc, result increasing friction force at constant external load at higher temperature.

Figure 5a. Enlarged view of XRD plot 500°C

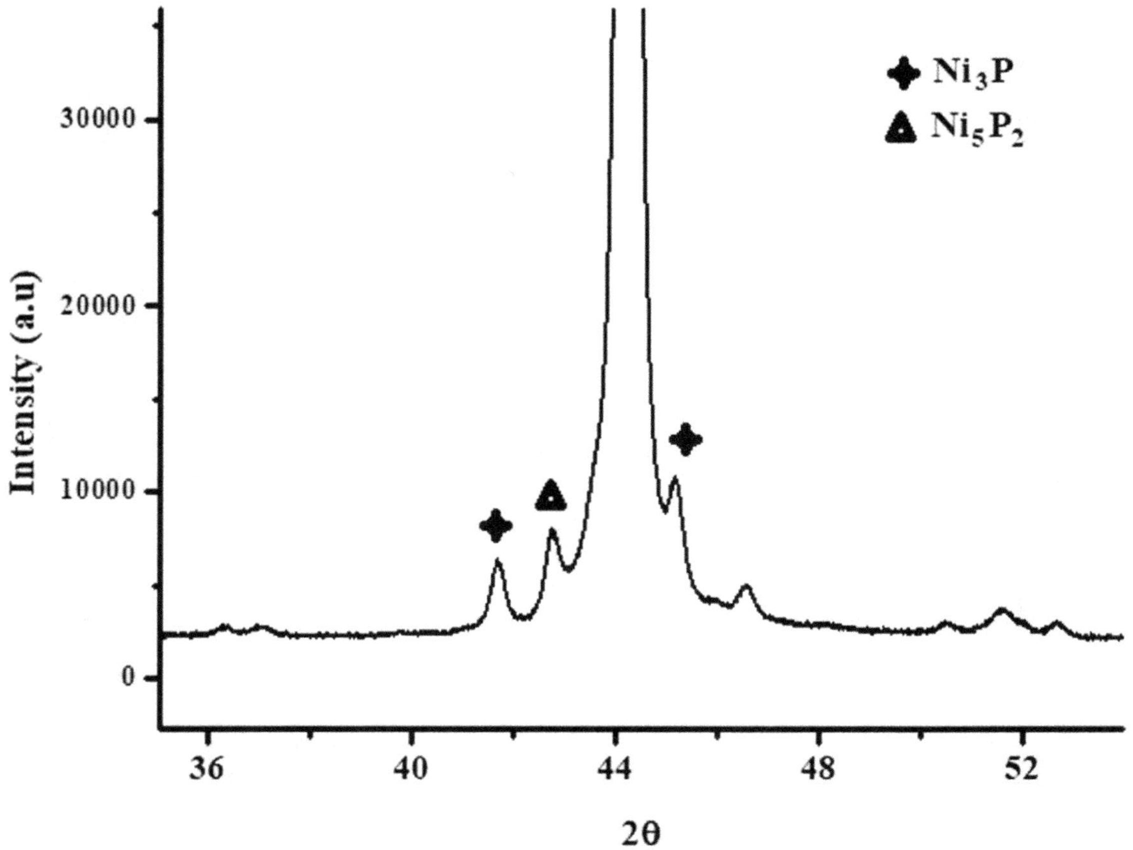

CONCLUSION

Electroless Ni-Cu-P-W coating was deposited from sulphate-chloride mixed bath. EDX confirms the presence of all the four elements within the deposit. XRD result of as-plated quaternary coating showed a crystalline peak of fcc Ni (111) which become sharper and narrower with the increase of heat treatment temperature. A broad exothermic peak in DSC plot was observed within the temperature range of 316.6°C to 366.4°C due to diffusion of Cu and W in to the nickel matrix and forming Ni-Cu and Ni-W phase. Due to this diffusion, rapid grain growth occurred after 300°C. A sharp exothermic peak at 431.8°C was also observed in DSC plot due to formation of new crystalline phase of tetragonal Ni_3P phase. The appearance of new Ni_3P phase in 500°C XRD plot confirms this phenomenon, which was absent in 300°C XRD plot. After heating at 700°C for 1 hour, intensity of all crystalline peaks increased and all metalstable phases converted to stable Ni_3P phase. Tribological study reveals that wear depth is minimum at 500°C and beyond that heating temperature; it starts to increase due to grain coarsening effect.

Figure 5b. Enlarged view of XRD plot 700°C

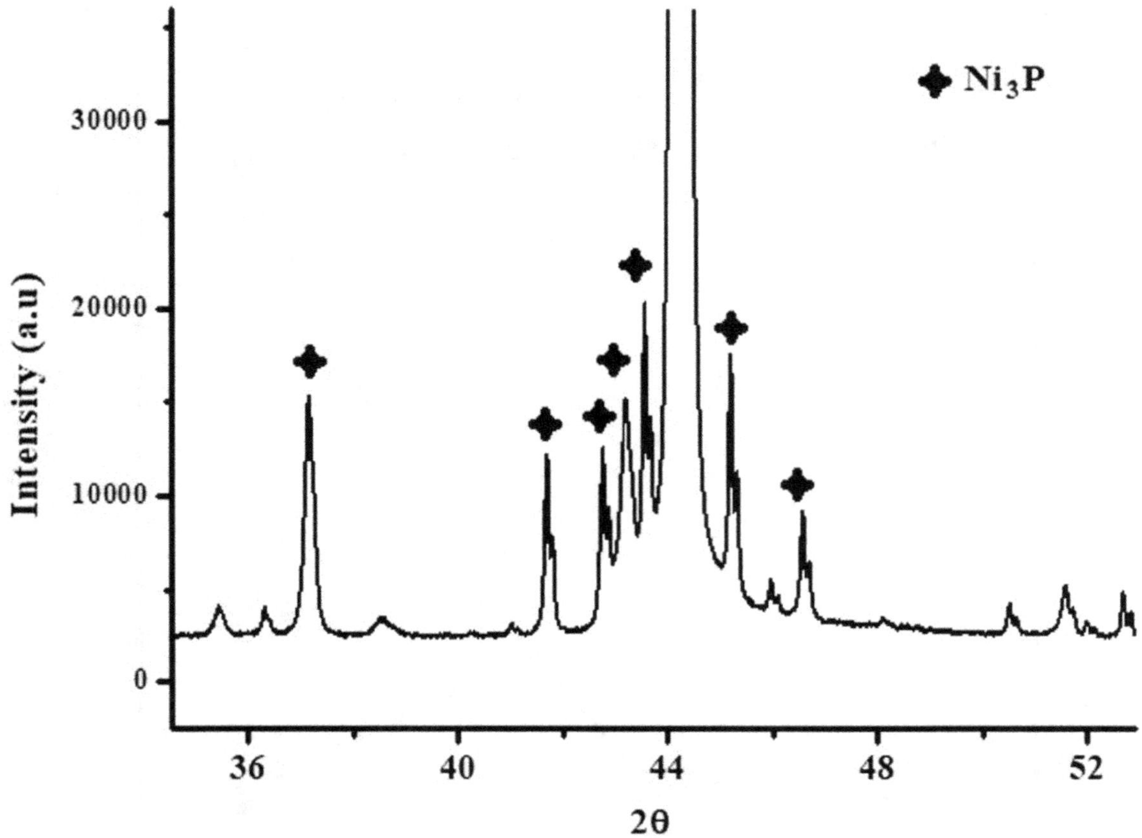

Table 2. FWHM and grain size of electroless Ni-Cu-P-W coating at different heat treated condition

Heat treatment temperature (°C)	Peak position 2θ (°)	FWHM (2θ (°))	Grain size (nm)
As plated	44.72	0.86403	9.94
300	44.52	0.70187	12.23
500	44.24	0.25218	34.00
700	44.25	0.12839	66.79

ACKNOWLEDGMENT

Authors are grateful for financial support provided by CSIR-HRDG, India, vide sanction letter no.: 22(0804)/19/EMR-II dated 25/07/2019, to carry out this work.

Figure 6. Variation of grain size and FWHM with temperature

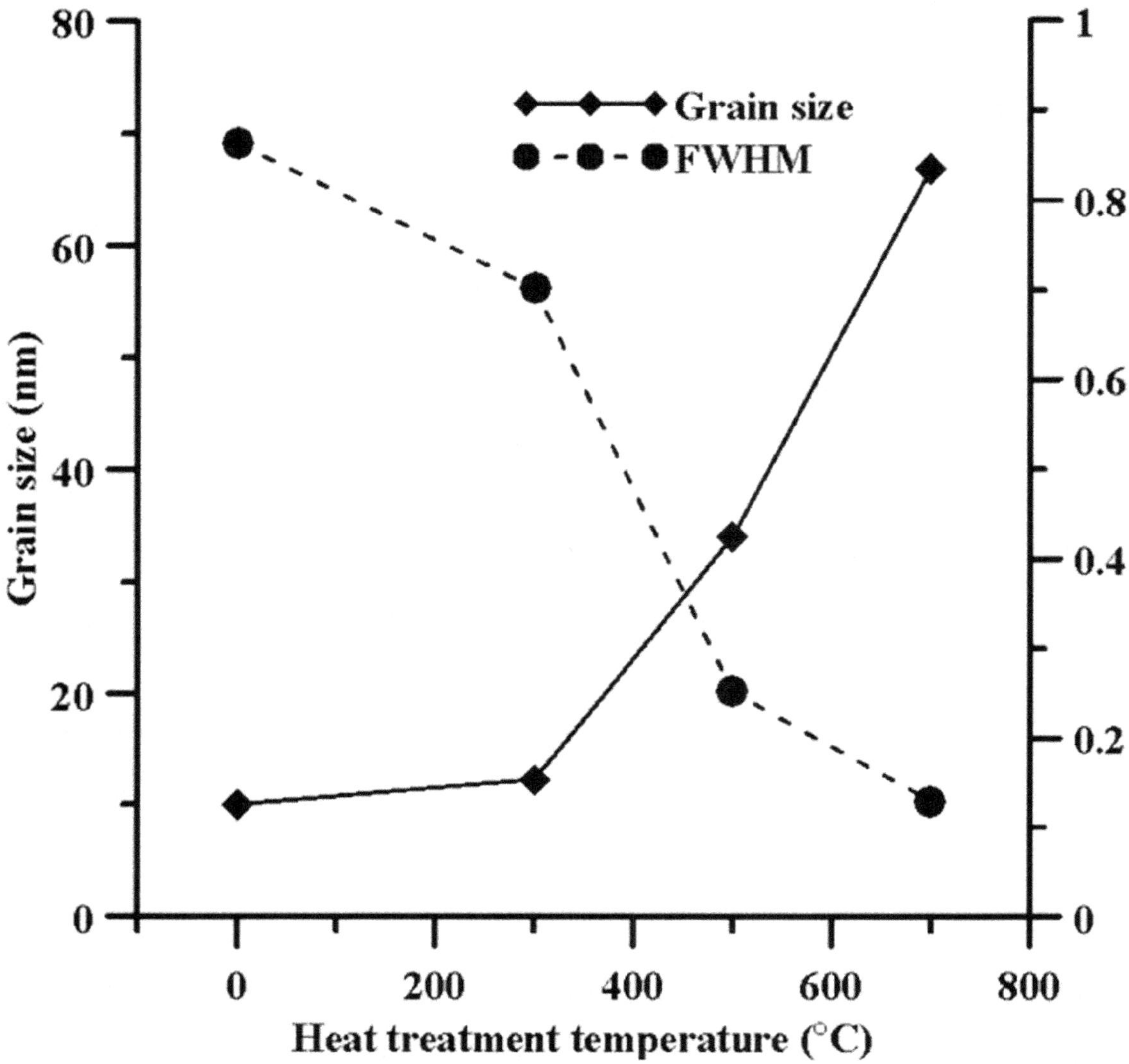

Figure 7. DSC plot for Ni-Cu-P-W quaternary alloy

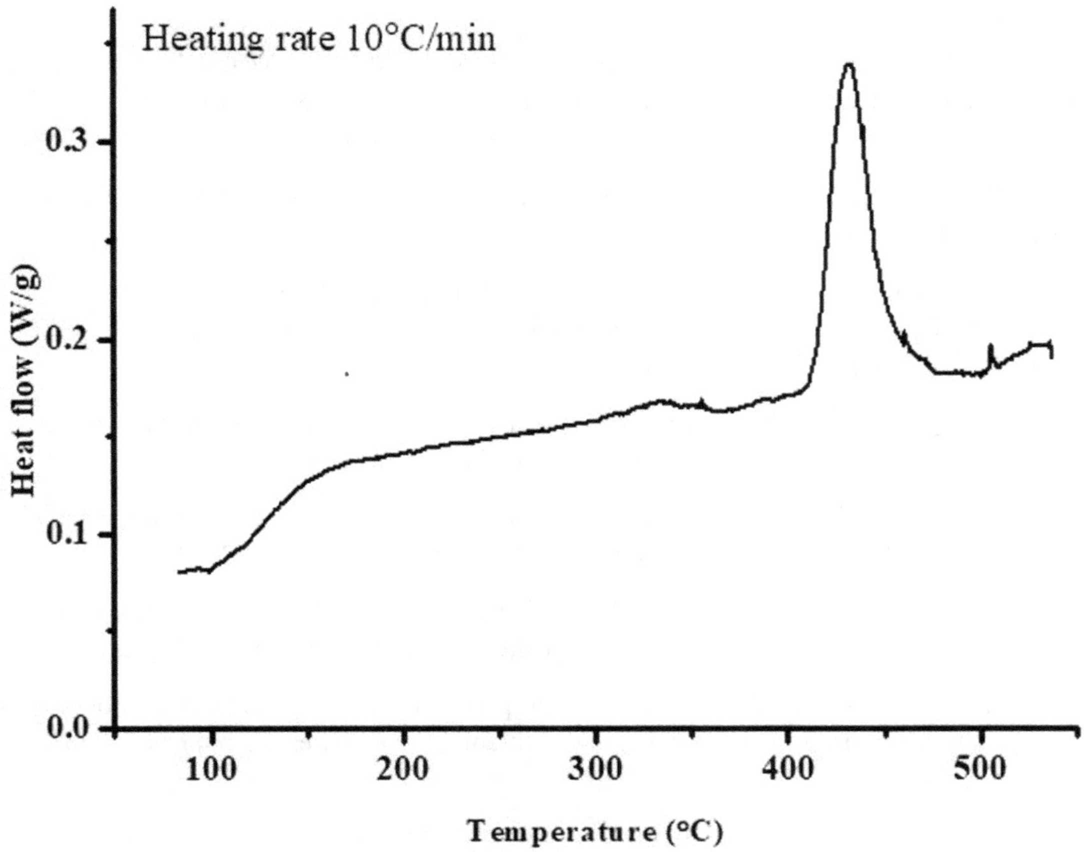

REFERENCES

Agarwala, R. C., & Agarwala, V. (2003). Electroless alloy/composite coatings: A review. *Sadhana*, *28*(3-4), 475–493. doi:10.1007/BF02706445

Balaraju, J. N., Anandan, C., & Rajam, K. S. (2006). Influence of codeposition of copper on the structure and morphology of electroless Ni–W–P alloys from sulphate-and chloride-based baths. *Surface and Coatings Technology*, *200*(12-13), 3675–3681. doi:10.1016/j.surfcoat.2004.09.010

Balaraju, J. N., Jahan, S. M., Anandan, C., & Rajam, K. S. (2006). Studies on electroless Ni–W–P and Ni–W–Cu–P alloy coatings using chloride-based bath. *Surface and Coatings Technology*, *200*(16-17), 4885–4890. doi:10.1016/j.surfcoat.2005.04.053

Balaraju, J. N., & Rajam, K. S. (2005). Electroless deposition of Ni–Cu–P, Ni–W–P and Ni–W–Cu–P alloys. *Surface and Coatings Technology*, *195*(2-3), 154–161. doi:10.1016/j.surfcoat.2004.07.068

Chen, J., Zhao, G., Matsuda, K., & Zou, Y. (2019b). Microstructure evolution and corrosion resistance of Ni–Cu–P amorphous coating during crystallization process. *Applied Surface Science*, *484*, 835–844. doi:10.1016/j.apsusc.2019.04.142

Chen, J., Zhao, G., Zhang, Y., Duan, S., Matsuda, K., & Zou, Y. (2019a). Metastable phase evolution and nanoindentation behavior of amorphous Ni–Cu–P coating during heat treatment process. *Journal of Alloys and Compounds*, *805*, 597–608. doi:10.1016/j.jallcom.2019.07.068

Ghosh, S. (2019). Electroless copper deposition: A critical review. *Thin Solid Films*, *669*, 641–658. doi:10.1016/j.tsf.2018.11.016

Liu, G., Yang, L., Wang, L., Wang, S., Chongyang, L., & Wang, J. (2010). Corrosion behavior of electroless deposited Ni–Cu–P coating in flue gas condensate. *Surface and Coatings Technology*, *204*(21-22), 3382–3386. doi:10.1016/j.surfcoat.2010.03.056

Liu, Y., & Zhao, Q. (2004). Study of electroless Ni–Cu–P coatings and their anti-corrosion properties. *Applied Surface Science*, *228*(1-4), 57–62. doi:10.1016/j.apsusc.2003.12.031

Lu, G., & Zangari, G. (2002). Corrosion resistance of ternary Ni- P based alloys in sulfuric acid solutions. *Electrochimica Acta*, *47*(18), 2969–2979. doi:10.1016/S0013-4686(02)00198-6

Luo, H., Leitch, M., Zeng, H., & Luo, J. L. (2018). Characterization of microstructure and properties of electroless duplex Ni-WP/Ni-P nano-ZrO2 composite coating. *Materials Today Physics, 4*, 36-42.

Meng, M., Leech, A., & Le, H. (2019). Mechanical properties and tribological behaviour of electroless Ni–P–Cu coatings on corrosion-resistant alloys under ultrahigh contact stress with sprayed nanoparticles. *Tribology International*, *139*, 59–66. doi:10.1016/j.triboint.2019.06.031

Palaniappa, M., & Seshadri, S. K. (2008). Friction and wear behavior of electroless Ni–P and Ni–W–P alloy coatings. *Wear*, *265*(5-6), 735–740. doi:10.1016/j.wear.2008.01.002

Roy, S., & Sahoo, P. (2012). Corrosion study of electroless Ni-PW coatings using electrochemical impedance spectroscopy. *Portugaliae Electrochimica Acta*, *30*(3), 203–220. doi:10.4152/pea.201203203

Roy, S., & Sahoo, P. (2013). Tribological performance optimization of electroless Ni-PW coating using weighted principal component analysis. *Tribology in Industry*, *35*(4), 297–307.

Roy, S., & Sahoo, P. (2014). Parametric optimization of corrosion and wear of electroless Ni-P-Cu coating using grey relational coefficient coupled with weighted principal component analysis. *International Journal of Mechanical and Materials Engineering*, *9*(1), 10. doi:10.118640712-014-0010-y

Sahoo, P., & Roy, S. (2017). Tribological behavior of electroless Ni-P, Ni-PW and Ni-P-Cu coatings: A comparison. *International Journal of Surface Engineering and Interdisciplinary Materials Science*, *5*(1), 1–15. doi:10.4018/IJSEIMS.2017010101

Sudagar, J., Lian, J., & Sha, W. (2013). Electroless nickel, alloy, composite and nano coatings–A critical review. *Journal of Alloys and Compounds*, *571*, 183–204. doi:10.1016/j.jallcom.2013.03.107

Tien, S. K., & Duh, J. G. (2004). Thermal reliability of electroless Ni–P–W coating during the aging treatment. *Thin Solid Films*, *469*, 268–273. doi:10.1016/j.tsf.2004.08.179

Tien, S. K., Duh, J. G., & Chen, Y. I. (2004). Structure, thermal stability and mechanical properties of electroless Ni–P–W alloy coatings during cycle test. *Surface and Coatings Technology*, *177*, 532–536. doi:10.1016/j.surfcoat.2003.08.032

Tien, S. K., Duh, J. G., & Chen, Y. I. (2004). The influence of thermal treatment on the microstructure and hardness in electroless Ni–P–W deposit. *Thin Solid Films*, *469*, 333–338. doi:10.1016/j.tsf.2004.08.146

Tsai, Y. Y., Wu, F. B., Chen, Y. I., Peng, P. J., Duh, J. G., & Tsai, S. Y. (2001). Thermal stability and mechanical properties of Ni–W–P electroless deposits. *Surface and Coatings Technology*, *146*, 502–507. doi:10.1016/S0257-8972(01)01462-1

Valova, E., Dille, J., Armyanov, S., Georgieva, J., Tatchev, D., Marinov, M., Delplancke, J.-L., Steenhaut, O., & Hubin, A. (2005). Interface between electroless amorphous Ni–Cu–P coatings and Al substrate. *Surface and Coatings Technology*, *190*(2-3), 336–344. doi:10.1016/j.surfcoat.2004.03.050

Wu, F. B., Tien, S. K., Chen, W. Y., & Duh, J. G. (2004). Microstructure evaluation and strengthening mechanism of Ni–P–W alloy coatings. *Surface and Coatings Technology*, *177*, 312–316. doi:10.1016/j.surfcoat.2003.09.010

Wu, F. B., Tien, S. K., Duh, J. G., & Wang, J. H. (2003). Surface characteristics of electroless and sputtered Ni–P–W alloy coatings. *Surface and Coatings Technology*, *166*(1), 60–66. doi:10.1016/S0257-8972(02)00725-9

Yu, H. S., Luo, S. F., & Wang, Y. R. (2001). A comparative study on the crystallization behavior of electroless Ni- P and Ni- Cu- P deposits. *Surface and Coatings Technology*, *148*(2-3), 143–148. doi:10.1016/S0257-8972(01)01345-7

Yu, H. S., Sun, X., Luo, S. F., Wang, Y. R., & Wu, Z. Q. (2002). Multifractal spectra of atomic force microscope images of amorphous electroless Ni–Cu–P alloy. *Applied Surface Science*, *191*(1-4), 123–127. doi:10.1016/S0169-4332(02)00170-8

Zhou, P., Cai, W., Yang, Y., Li, X., Zhang, T., & Wang, F. (2019). Effect of ultrasonic agitation during the activation process on the microstructure and corrosion resistance of electroless Ni-WP coatings on AZ91D magnesium alloy. *Surface and Coatings Technology*, *374*, 103–115. doi:10.1016/j.surfcoat.2019.05.080

Chapter 10
Role of Alloying Elements on Powder Metallurgy Steels and Spectroscopic Applications on Them

Sefa Celik
Istanbul University, Turkey

Nuray Bekoz Ullen
https://orcid.org/0000-0003-2705-2559
Istanbul University-Cerrahpasa, Turkey

Sevim Akyuz
Istanbul Kultur University, Turkey

Aysen E. Ozel
Istanbul University, Turkey

ABSTRACT

The powder metallurgy (PM) technique is suitable for mass production and is a well-established process for the low production costs of net-shaped products close to long series. The properties of the iron-based PM parts, such as strength, hardness, magnetic properties, impact, wear, and corrosion resistance, can be improved with adding various alloying elements. The desired surface performance can be achieved with various surface coating technologies. Recently, various coating techniques have been developed as discussed in the chapter. The alloying elements have a significant effect on the coating quality of the final product. Surface coating can be analyzed by examining the surface-coated powder metallurgy (PM) using infrared and Raman spectroscopy technique. This contribution focuses on the role of alloying elements on properties and coating technologies of powder metallurgy steels and Fourier transform infrared (FT-IR) and Raman spectroscopic applications on them.

DOI: 10.4018/978-1-7998-4870-7.ch010

1. THE IMPORTANCE OF POWDER METALLURGY IN INDUSTRY

Powder metallurgy (PM) process is a near-net-shape production process that combines the properties of shape-making technology for powder compression with the improving of ultimate material and design properties (mechanical and physical) along with the subsequent densification or consolidation processes (e.g., sintering).

PM methods produce complicated components in a simple process, gas atomization, due to rapid solidification, easily and safely allows microstructural modifications such as grain refinement, reduced segregation, and increased solubility. Low cost and high quality in mass production, wide alloying probability and low-density part production are other advantages of PM technology (Angelo and Subramanian, 2008; Chawla et al., 1999; Lee et al., 1998). Nowadays lots of PM-products are used in various industries; these are automotive and off-highway vehicle applications, machinery, computers and fabric sector equipment, home appliances and biomedical device manufacturing sectors. PM products have high surface quality and their properties such as strength and wear resistance may be improved by applying heat treatment. PM methods ensure part-to-part reproducibility and are appropriated to medium-to-high volume manufacturing. If required, controlled microporosity may be supplied for self-lubrication or filtration. The advantage of the method is that the PM alloy can be exactly designed for environmental agents. Then, it may be applied in low amount to form a metallurgical bond with a low-cost substrate that will withstand application conditions (Samal, 2015; Lee et al., 1998). The desired surface performance can be achieved with various surface engineering technologies that can be applied to PM components. In recent times, many numbers of surface coating techniques have been developed such as organic coatings, ceramic coatings, metallic coatings, laser-based techniques, thermal spray coatings, chemical vapour deposition (CVD), physical vapour deposition (PVD) coatings, and other methods of surface modification, which help to improve the tribological properties of the PM materials (Boing et al., 2020; El-Eskandarany, 2015; Donghai et al., 2017; Lee at al., 1998).

PM technology combines different fields, such as polymer and chemical engineering, physical metallurgy and surface sciences. Advances in the field of PM in the last decade are the result of a method of a multi-disciplinary approach (Narasimhan, 1996; Klar and Samal, 2007; Dangsheng, 2002). An extensive range of engineering materials are available, and through proper material and method selection, the necessary microstructure may be created in the material with the addition of variety alloying elements. Alloying additives such as nickel, copper, molybdenum, graphite or carbon are added to the base iron powder to provide the desired performance characteristic of a PM product. A few alloying techniques are available to create PM-products from low alloy powders (Samal and Newkirk, 2015; Salgado et al., 2001; Klar and Samal, 2007). In this chapter; alloying techniques of PM steel products and the effects of alloying elements on PM steel parts, and their surface treatment are discussed. In recent years, surface the coating technologies used in coating of PM-products have been mentioned. Moreover Fourier Transform Infrared (FT-IR) and Raman spectroscopic studies on the alloy composition of PM steels will be discussed.

2. ALLOYING METHODS IN POWDER METALLURGY

The microstructure of PM materials, the size, morphology and distribution of the pores in it, directly affect the mechanical properties of the product. Alloy elements are added to increase material performance characteristics. For example, it is very significant to add alloying elements to develop the mechanical properties of pure iron because it is essential for the production of components such as strength, hardness, ductility, impact resistance and fatigue strength. At the same time, the alloying technique has an important influence on the microstructure and porosity of the final product (Narasimhan, 2001; Dincer, 2014; Skoglund et al., 2013). Carbon is normally mixed into the iron powder in the form of graphite, but different metallic alloy elements are often admixed by various techniques (Narasimhan, 1996; Lee et al., 1998; Narasimhan, 2001). The methods used in the alloying of ferrous PM materials can be classified as follows; (Samal and Newkirk, 2015; James and West, 1998; Yu et al., 2004; Lee et al., 1998).

a) Admixed alloys
b) Partially alloys (diffusion alloys)
c) Pre-alloys
d) Hybrid alloys (mix of diffusion-alloyed powder with pre-alloyed powder)

The schematic representation of the above-mentioned different alloy methods is illustrated schematically in Figure 1.

Figure 1. Schematic representation of alloying methods (Lee et al., 1998; Yu et al., 2004).

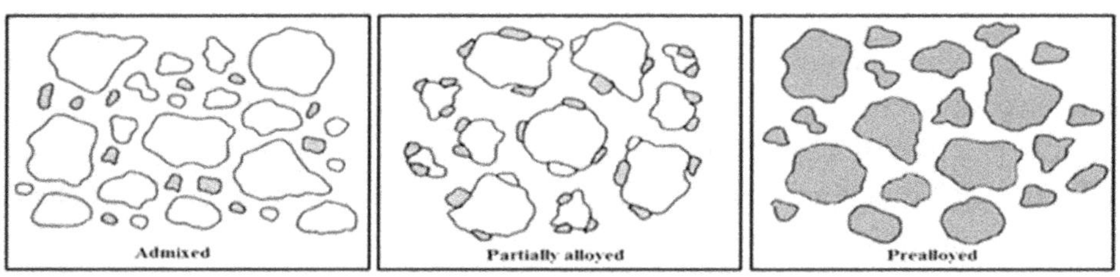

Admixed Powders

Alloy additions may be mixed into iron powder in elemental form or as ferroalloys such as Fe_3P, FeCr or FeMn. This alloy method is the most common and cheapest method used in generally ferrous powder metallurgy. The additions are not alloyed with the iron powder when the mix is pressed, so, the admixed materials maintain most of the compressibility of the base iron powder. The diffusion of the alloying elements in iron at sintering temperature during sintering limits the degree of alloying. Therefore, the observed microstructures may be microstructurally and chemically heterogeneous. The density needed to obtain the requested mechanical properties usually determines which type of powder to use. Water atomized iron powders or directly reduced sponge iron may be used as the base powder for admixed structures (Yu et al., 2004; Lee et al., 1998; Bergman, 2011; El-Eskandarany, 2015). The benefits of admixed processing with binders can be listed as follows: (Lee et al., 1998; Yu et al., 2004; Angelo and Subramanian, 2008).

- Provides reduction of variability in sintered properties
- Opportunities for new alloy improvement,
- Developed (or consistent) flow rate and good die filling while maintaining same green strength,
- Potential for obtaining enhanced PM part weight and densities control without the require,
- Potential for developed efficiency,
- Use of fine particle additives.

Partially Alloyed (Diffusion Alloyed) Powders

The demand for making high ductility, high strength PM parts by only pressing and sintering has been influential in the development of various composite types of partially pre-alloyed steel powder containing molybdenum, nickel and copper. In partially alloyed powders, also known as diffusion alloyed powders; alloy additions are metallurgically bonded to either a pre-alloy powder base or an elemental iron powder. Since the compressibility of the base powder is not desired to be significantly reduced, the diffusion alloying of the additions is controlled to ensure sufficient bonding to the base powder while limiting the diffusion degree of the additives. Products made from partially alloyed powders have a complicated as sintered microstructure occurring of regions of martensite, fine unresolved pearlite, bainite, nickel-rich ferrite including blocky carbides, in addition, scattered nickel-rich areas of retained austenite. This complicated microstructure ensures a combination of better tensile ductility and tensile strength in addition to a material with great impact energy. Diffusion alloyed materials may be heat treated to developed for such as apparent hardness, strength, and wear resistance. Diffusion alloyed powders have also enabled the removal of double sintering-pressing and sizing because of the high hardness achieved in the sintered condition (Lee et al., 1998; Bergman, 2011; El-Eskandarany, 2015; Samal and Newkirk, 2015). The benefits of diffusion-alloyed steel powders are (Lee et al., 1998; Yu et al., 2004; Angelo and Subramanian, 2008):

- Enhanced green strength,
- Uniform distribution of alloying elements,
- Protection of high compressibility of the base iron,
- Less alloy separation and dusting risk when transporting the powder mixture,
- Excellent stability of this composite type base powder,
- Good sinterability.

Pre-alloyed Powders

Pre-alloyed low alloy steel powders are beneficial in the production of PM components manufactured for use in high performance applications. The amount and type of the alloy used may be controlled the hardenability of the pre-alloyed materials. For example, nickel, manganese and molybdenum compositions are added to the molten iron during the steel production process to produce a chemically consistent powder. During melting and atomization, applications are followed to ensure the production of powders that do not contain non-metallic inclusion and suitable for the most demanding requirements. The reactivity of carbon in graphite form is excellent with pre-alloyed powders, and it provides close control of the combined carbon content during sintering. These powders are utilized in a wide variety of applications. Changing performance requirements are met with a combination of alloy selection, carbon additions and processing conditions. The compressibility of admixed or partially alloyed powders is generally higher than that of pre-alloyed powders. Generally, pre-alloyed powders are often used for parts that need heat treatment. The amount and type of the alloy used may be controlled the hardenability of the pre-alloyed materials. For example, the compressibility of pre-alloyed powders using only molybdenum as the main alloy material is higher than Nickel-Molybdenum pre-alloyed powders. Products made from pre-alloyed powders have a uniform microstructure and apparent hardness. Carbides formed in the as-sintered condition are not in lamellar in nature. The characteristic carbide morphology may be defined in the as-sintered condition with pre-alloyed materials (Yu et al., 2004; Lee et al., 1998; Bergman, 2011; El-Eskandarany, 2015; Samal and Newkirk, 2015; Angelo and Subramanian, 2008).

Hybrid Alloy Powders

Hybrid alloys are formed by adding elemental and ferroalloy additions to partial or pre-alloyed base powders to obtain the desired chemical composition. When added to nickel or copper pre-alloy powders, it can be highly compacted with molybdenum, which is the addition of the main alloy. It has become widespread with the use of pre-alloyed powders. It is particularly suitable for the addition of high-carbon ferroalloys to these pre-alloyed powders to produce chromium-molybdenum, manganese-molybdenum and chromium-molybdenum-manganese steels. Addition of elemental Ni to partially alloyed powders may be to provide greater impact properties. The compressibility and hardenability of hybrid alloy materials are mainly related to the properties of the base powder used. The general characteristics of hybrid materials are that it has heterogeneous microstructure and apparent hardness. Generally, higher sintering temperatures and longer sintering times caused to more diffusion of alloy additions and the formation of more homogeneous microstructures (Lee et al., 1998; Bergman, 2011; El-Eskandarany, 2015; Samal and Newkirk, 2015).

3. THE EFFECT OF ALLOYING ELEMENTS ON THE PROPERTIES OF POWDER METALLURGY PARTS

The addition of alloying elements (pre-alloyed, admixed, or diffusion-alloyed) is an important aspect of the design stage as the part of PM component manufacturing and has a marked effect on the mechanical properties of the final product. Alloyed PM steels are a particular group of PM materials, having various composition of alloying elements (Cr, Mo, Ni, Mn, V, Cu, Ti and W). Each alloy elements specifically affects the properties of the PM steel and additionally plays an essential role in heat treatment behavior. Strong bonding of alloy additions offers an economic advantage in terms of less green scrap, higher alloy efficiency and an easier maintenance environment. The amount of respirable dust in the immediate surrounding of the compaction press is decreased in order of magnitude and plant purity is significantly developed. Variety alloying system such as Fe-C, Fe-P, Fe-Cu, Fe-Mo-C, Fe-Cu-C, Fe-Ni-C, Fe-Cr-Mo-C, Fe-Ni-Cu-Mo -C, Fe-Mo-Cu-C have definition of ferrous materials (Lee et al., 1998; Angelo and Subramanian, 2008; Shanmugasundaram and Chandramouli, 2009; Yu et al., 1998).

Carbon is the main element for PM steel and steel. Because of the increasing in temperature of the martensitic transformation, the carbon enhances the tensile strength and hardness but reduces the elongation. This is valid for wrought steel as well as for PM steels. In the components containing more than 0.8% carbon concentration, residual austenite formation takes place, which causes a decrease in tensile strength. Carbon dissolves in austenite during heating. Subsequently, the transformation of austenite to ferrite-carbide aggregates results in changing strength levels after cooling from the austenitizing temperature. The porosity of sintered PM steels affects the properties that may be seen after heat treatment. As mentioned above, the properties depend on the processing conditions (Yu et al., 2004; Lee et al., 1998; Danninger and Gierl, 2008).

Nickel is added to enhance the strength, impact and abrasion resistance, and fatigue strength of steels. Shrinkage may occur during the sintering. Nickel steels are the most widespread high strength and heat treated materials. Usually 1-4% Ni alloy is added in PM steels. The addition of nickel does not enhance the sizes, but causes shrinkage and forms a solid solution with a limited solubility in iron. The impact of nickel on tensile strength is lower than copper. Because the solubility of nickel to iron is quite low, but it increases the hardening ability of the steel. Thus, it provides additional benefit to the parts to be heat treated to obtain enhanced mechanical properties (Lee et al., 1998; Danninger and Gierl, 2008; Angelo and Subramanian, 2008; Shanmugasundaram and Chandramouli, 2009).

Copper is added to steel for enhance strength, wear resistance, and apparent hardness. It is usually added in combination with other alloying elements. The addition of copper and graphite leads to growth in sintered compact. Copper is mostly added to iron because it melts and quickly dissolves, thereby significantly enhancing the strength of the iron. Copper melts (T_m=1083þC) at the sintering temperature normally used (1120þC). Liquid copper penetrates the spaces between the grain boundaries and the powder particles within single powder particles. After copper melts, it leaks along the grain boundaries of the structure during sintering, which explains the swelling behavior. For this reason it is difficult to separate grain boundaries from particle to particle contacts. Copper steels can be heat treated to enhance strength, wear resistance and apparent hardness (Angelo and Subramanian, 2008; Mocarski et al., 1996; Angel et al., 2014).

Chromium creates unsteady carbides in steel. But chromium is inexpensive and widespread. It enhances corrosion resistance, particularly to intercrystalline corrosion. Additionally, it can delay transformation throughout tempering. Cr is the most widely used alloying element in heat treated structural steel (Bergman, 2011; Wang, 2005; Angelo and Subramanian, 2008; Engström et al., 1992).

Molybdenum is generally used for nitriding, because the nitrides it forms strengthens the steel. Mo enhances creep resistance and hardenability. The prevalence area of austenite is narrowed with the addition of Mo. The steel containing 3% Mo has a ferritic structure. Molybdenum has a high affinity for carbon and creates carbides, for example Mo_6C. Molybdenum is an effective hardening agent and develops densification because it is a ferrite stabilizer (Mocarski et al., 1996; Bergman, 2011; Sułowski and Matusiewicz, 2013; Dincer, 2014).

Manganese provides enhanced strength and hardness in combination with molybdenum. Manganese is easily reachable and cheap, PM steels with this element have similar properties to Ni steels. On the other side, manganese has a high affinity for oxygen and causes problems during manufacture, therefore this element is added to powder mixtures as ferroalloy. Since manganese easily forms oxide, it is necessary to carefully select the sintering atmosphere. With the addition of sulfur, the workability of sintered iron and its alloys is developed. Also manganese sulfide particles have been added to iron to provide similar benefit. Recently days, the sulfur in the liquid melt is dissolved in the melt before being atomized (pre-alloyed sulfur) to create manganese sulfide (MnS) with carefully controlled amounts of manganese. If it was pre-alloyed, as a result of the addition of these particles of MnS causes as a gray phase in the pores or a MnS phase to form inside the iron particles (Wang, 1994; Lawley and Murphy, 2003; Šalak and Selecká, 2012; El-Eskandarany, 2015).

Silicon has a positive effect on the magnetic properties of iron. Silicon reduces the saturation induction and enhances the resistivity. Liquid phase occurs during the sintering (Bergman, 2011; Danninger and Gierl, 2008).

Phosphorus has a positive effect on the soft magnetic features when it is added to iron. The phosphorus element stabilizes the alpha phase, which is more open, and creates the liquid phase at 1050°C. The sintering occurs at the mostly used temperature of 1120 °C. Phosphorus causes to enhance the resistivity and saturation induction and reduces the coercive area. Phosphorus forms a liquid phase, like copper, under the normal sintering conditions. However, the difference in microstructure is important. In contrast to the copper in low concentrations, it forms shrinkage because of the formation of phosphorus sintering in the ferrite phase. The diffusion rate of phosphorus is not as fast as a carbon, but it is fast enough to provide a good dispersion in the material at normal sintering times and temperatures (Bergman, 2011; Dobrzanski et al., 2007; Šalak et al., 2005).

The combination of alloying elements significantly affects the properties of steels. There are many studies in the literature about the effect of alloying elements on the microstructure and mechanical properties of PM steels. In the Continuous-Cooling-Transformation (CCT) and Time-Temperature-Transformation (TTT) diagrams, changes on the temperature and composition ratio of a dedicated alloy are observed. In these diagrams, the composition specified for an alloy is determined experimentally. The location and shape of the curves in the diagrams depend on the grain size of the austenite, the amount of alloying elements and according to the time axis. On the other hand, with some exceptions, an increase in the austenite grain size or alloy elements content leads to a reduction in the ratio of austenite-ferrite and austenite-perlite transformations. There are basically two ways alloying elements can decrease the austenite transformation rate: alloying elements can do this by reducing the nucleation and growth rates of bainite, pearlite and ferrite. The transformation of austenite is delayed, probably because of the fact that it is required to wait for the alloying elements to diffuse before the nuclei of new phases may form and grow. It is dependent on molybdenum, nickel and copper to provide the desired hardenability in most alloyed PM steels. As a consequence; alloy compositions in sintered steels for structural parts must be meticulously selected to achieve the required properties (Angelo and Subramanian, 2008; Baran et al., 1999; Kahraman and Gulenc, 2002; Warke et al., 2009; Hu and Tian, 2012; Park et al., 2015).

4. SURFACE TREATMENT OF POWDER METALLURGY PARTS

The majority of the parts produced by the powder metallurgy method are iron-based parts. Corrosion, like all other materials, is a major problem in sintered ferrous parts, and the porous structures of powder metallurgy parts make this problem even more important. PM metal parts in various structures are often subject to corrosion damage due to loss of functionality and size. The reduction of the corrosion resistance of the pores formed in the parts produced by powder metallurgy to environmental effects is explained by the fact that there is more surface area for corrosion and oxidation effects in a porous material. After sintering, many surface treatments are applied to improve the chemical, physical and mechanical properties of the powder metallurgy material and to increase the aesthetics of the product (Baran et al., 1999; Krishna and Patowari, 2013; Tian et al., 2006; Donghai et al., 2007). Steam treatment and phosphatizing processes are classical techniques that have been proven to protect ferrous PM materials from corrosion (Lee et al., 1998; Samal and Newkirk, 2015; Bergman, 2011; Danninger and Gierl, 2008; Konnov et al., 2019; Grimanelis and Eyre, 2007).

Steam treatment is a frequently used to develop the wear properties and to increase the corrosion resistance of ferrous PM parts. On the treatment, all open surfaces (interior and exterior) are coated with Fe_3O_4, which is hard black magnetic iron oxide. Firstly; the materials are purified from oil or grease and then positioned in a forced convection furnace. Steam treating improves the properties of PM materials by: (Samal and Newkirk, 2015; James and West, 1998; Lee et al; 1998; Grimanelis and Eyre, 2007).

- Improving surface integrity
- Improving compressive strength
- Improving wear resistance
- Increasing surface hardness
- Increasing corrosion resistance

Phosphatizing is used to prevent rust by coating sintered iron materials with a thin film of iron phosphates. The materials are immersed in an aqueous solution consisting of primarily phosphoric acid (Chen et al., 2008; Konnov et al., 2019; Angelo and Subramanian, 2008).

Despite the mentioned surface protection processes; in many ways, surface coating treatments is a requirement for most PM metals. Because interconnected porosity in sintered parts causes surfaces to be exposed to the environment. For this reason, sintered parts are more sensitive to corrosion than their cast or machined counterparts. Because of that, the pores in the PM parts should be closed to prevent the coating solutions from being trapped in the pores. Copper, chromium, cadmium, nickel and zinc are widely used for coating. Additionally, electroless nickel plating can be used. Acid copper plating usually gives the best result on PM components because of its perfect throwing power. Nickel is used alone or on a copper, base to provide a bright corrosion-resistant surface and increase wear resistance. Chromium plating is applied to obtain decorative applications or wear-resistant parts. Cadmium plates have faster and better-throwing power than zinc. To plate a PM part with more than 10% porosity, it is important to fill the pores with a resin or to harden the surface so as not to pass the coating salts through the pores of the part. In addition, surface treatments such as ion implementation or anodization may be used to develop the corrosion properties of ferrous based PM parts. Since the powder metallurgy parts contain pores, corrosion resistance is not achieved, such as those obtained with wrought parts. This is probably due to the expansion of gases in the pores of the powder metallurgy parts when the material is heated to 300-400 ° C during the cooking process, which damages the coating. To avoid this, a pore-closing treatment is needed, which can withstand high-temperature heating, good economics and offers perfect dimensional tolerance (Krishna and Patowari, 2013; Tian et al., 2006; Donghai et al., 2007; Liu et al., 1994; Cameron, 2014; Attar, 1996; Tengzelius and Grinder, 2008; Xu et al., 2015).

The coatings act as a layer that protects the powder metal against corrosion. The coating is applied in order to harden the surfaces according to the effect of powder metals used in various areas; and if powder metal is used in an abrasive environment, coating processes are applied to prevent abrasion. All these features can be gained to powder metals during the production phase, but they are preferred because the coating process will provide better satisfaction and will also be more economical. There are two processes to improve existing materials by reducing corrosion: by adding alloy materials or using surface coating methods. Surface coatings are an applicable way to extend the life of metal components by increasing the corrosion resistance of metallic materials. In order to provide better tribological properties to the surface, a suitable coating strategy must be selected (Gill and Kumar, 2019; Boing et al., 2020). The coating techniques vary for different applications. So, the classification of the coating techniques can be very confused. In recent times, numerous surface coating techniques have been developed to achieve the desired mechanical, physical and chemical properties, by means of thin and hard layers on PM structure, and these techniques are generally described below: (Angelo and Subramanian, 2008; Samal and Newkirk, 2015; Sundararajan and Sivakumar, 2005; Šolic et al.,2019; Gill and Kumar, 2019; Mahade et al.,2019; Ganesan et al.,2019; Boing et al., 2020; Cabezon at al., 2020; Konnov et al.,, 2019, Goldschmidt and Streitberger, 2003).

- **The thermal spray coating process** is commonly applied to protect PM alloys. The thermal spray coating is by far the most versatile modern surfacing method with regard to economics, range of materials, and scope of applications. Thermal spraying reduces wear and corrosion and greatly prolongs part service life by allowing the use of high-performance coating material over a low-cost base metal. Among the commercially available thermal spray coating techniques, detonation spray (DS), high-velocity oxy-fuel (HVOF) powder spray, plasma arc (PA) powder spray, oxyfuel powder (OFP) spray and electric arc wire (EAW) spray are the best choices to get hard, dense and consequently wear-resistant coatings are desired. The purpose of the coating process can be categorized as wear resistance, electrical conductivity or resistivity, oxidation resistance, corrosion resistance, thermal barriers restoration of dimension (Kahraman and Gulenc, 2002; Mahade et al.; 2019; Sundararajan et al., 2005).
- **The physical vapour deposition (PVD) coatings** have become a widespread practice nowadays. Multilayer coatings exhibit increased fracture resistance, adhesion and improved wear resistance. Chemical inertness and low friction coefficient can be realized by combining distinctive properties of various monolayers in a multilayer coating. High wear resistance can be achieved by thin and hard coatings particularly with PVD / Especially with PVD coating, thin and hard coatings with high wear resistance can be achieved (Goldschmidt and Streitberger, 2003; Boing et al., 2020; Šolic et al., 2019).
- **The chemical vapour deposition (CVD) coatings** are widely used in industry because of their versatility for depositing a very large variety of elements and compounds covering a wide range of materials from amorphous deposits to epitaxial layers having a high degree of perfection and purity. Chemical vapour deposition can be defined as a process in which the gaseous chemical reactants are transported to the reaction chamber, activated thermally (conventional CVD) or by other means (plasma-assisted CVD or laser-induced CVD) in the vicinity of the substrate, and made to react to form a solid deposit on the substrate surface. It is possible to deposit films of uniform thickness and low porosity even on substrates of complicated shape with this process. If a metalorganic or organic-metallic precursor species is used in a CVD process, the process is designated MOCVD or OMCVD. A large percentage of hard metals cutting tool inserts and sintered high-speed steel tools are now coated using CVD and PVD. The lower temperature PVD process also allows steels to be given a wear-resistant layer of titanium carbide, titanium nitride, alumina or a combination of these materials (Goldschmidt and Streitberger, 2003; Boing et al., 2020; Angelo and Subramanian, 2008).
- **Plasma-assisted chemical vapour deposition (PACVD),** known as plasma-enhanced CVD, is an important technique used for depositing films of cutting tools. The area of application for these types of coated materials is expanding with the development and application of ultra-fine powder grain sizes, as well as the advancements in technology. One method of coating hard metals is

PACVD technique rarely used for coating hard metals with ultrafine grain size (Boing et al., 2020; Angelo and Subramanian, 2008).

- In recent times, **laser-based techniques** such as laser cladding and laser deposition have made headway in the area of surface modification. These techniques can quickly melt the metal surface and then the metal solidifies rapidly, which helps to improve the surface properties of the materials. Laser deposition techniques have attracted great interest in the synthesis of semiconducting and insulating thin films in recent years. Additionally, these techniques have been successfully applied to the deposition of high critical temperature superconducting films. In the laser ablation technique, the material is vaporized and ejected from the surface of a target as it is irradiated by a laser beam. Films are formed by condensing material ablated from the target onto a solid substrate (Angelo and Subramanian, 2008; Mirzaali et al., 2019; Goldschmidt and Streitberger, 2003).
- **Ion implantation** is one of a number of surface modification processes that are emerging as an economical and viable process for improving the near-surface tribological properties of engineering components. The microstructural and chemical changes caused by ion implantation can significantly affect the surface performance of PM components (Angelo and Subramanian, 2008; Kolawole et al., 2019).

Apart from these coating techniques; there are various surface modifying techniques such as plasma spraying, biomimetic mineralization, electrophoretic deposition and sol–gel method (Boing et al., 2020; Cabezon at al., 2020; Konnov et al.,, 2019, Chen et al., 2008; Goldschmidt and Streitberger, 2003).

5. THE EFFECT OF ALLOYING ELEMENTS ON THE SURFACE COATING

PM structural parts can be covered with many coating techniques as mentioned in the previous section. Copper, nickel, cadmium, zinc and coating can be made for PM parts. However, porous parts must be filled with resin before coating so as to avoid penetration and entrapment of coating solutions in the pores (Attar, 1996; Angelo, 2008; Cameron, 2014). The coating properties of PM parts are dependent not only on the nature and distribution of phases present in the coating but also on a host of other properties like materials microstructure (largely determined by powder composition), porosity, the nature of residual stress and its magnitude within the coating and finally the coating-substrate adhesion (Link, 1998; Dangsheng, 2002; El-Eskandarany, 2015; Samal and Newkirk, 2015; Danninger and Gierl, 2008; Wang, 2005; Shanmugasundaram and Chandramouli, 2009; Mocarski et al., 1996; Wang, 1994; Angel et al., 2014; Engström et al., 1992; Mitchell et al., 2005; Hu and Tian, 2012; Park et al., 2015; Xu et al., 2015). The factors affecting the coating processes largely depend on understanding the properties of the material that changes with the alloying elements. There are many scientific studies on PM in the literature due to the wide usage area of PM-products and their superior properties. However, there are few studies on the effect of alloying elements of PM parts on the coating. In some of these studies, the following information is given about the effect of alloying elements on the coating. (Link, 1998; Dangsheng, 2002; El-Eskandarany, 2015; Samal and Newkirk, 2015; Danninger and Gierl, 2008; Wang, 2005; Shanmugasundaram and Chandramouli, 2009; Mocarski et al., 1996; Wang, 1994; Angel et al., 2014; Engström et al., 1992; Mitchell et al., 2005; Hu and Tian, 2012; Park et al., 2015; Xu et al., 2015).

- Copper increases the performance of the coating due to closing the pores with liquid phase sintering.
- Carbon has no significant effect on the coating appearance and thickness. Increasing C ratio causes an increase in coating hardness value. However, in PM steels with high carbon content, when the product is taken into the acid bath for surface cleaning, it can capture hydrogen and cause hydrogen brittleness.
- The high manganese content in PM steels can form a thick, low-impact coating. This creates a negative situation both in terms of aesthetics and coating quality.
- Silicon content in PM steels has a negative effect on thickness and appearance. If the Si content is high, a thick and matt coating is obtained.
- The high phosphorus content in PM steels forms a coating with a high tendency to break.

An ideal surface coating is directly related to the content of the alloying elements in the material. Residual stress formed by the microstructure that changes depending on the alloy element is among the most important factors affecting coating integrity. Besides the composition and properties of the PM part, the working conditions and design of the spray and curing equipment are the key factors affecting coating quality and process performance. This brief survey of the relationship between PM material coating technology and alloy elements will enable researchers interested in the subject to better to understand the coating process.

Techniques like Energy Dispersive Spectroscopy (EDS) and X-ray diffraction (XRD) are used to analyze the change in composition of alloying elements and phases of the coated PM material surface. Surface coating can be analyzed by examining surface coated powder metallurgy (PM) with Raman and FT-IR spectrometry.

6. FT-IR AND RAMAN SPECTROSCOPIC APPLICATIONS ON THEM

FT-IR spectroscopy is one of the important techniques in which the vibrations and structures of molecules are defined (Christy et al., 2001; Smith, 2011; Ciurczak and Igne, 2015; Ordonez and Prieto, 2012; Stuart, 2004). Once the molecule is irradiated by infrared light, which includes all IR wavelengths, the molecules absorb characteristic wavelengths for themselfs, which equal to the difference in their vibrational energy levels. Using IR spectroscopy, the recognition of the molecular structure is performed by the identification of functional groups and chemical bonds. In the industry loss of time is very important and it is desirable to check the product quality quickly. It is important that FT-IR spectroscopy performs this analysis in a few minutes (Celik et al., 2020).

Raman scattering is inelastic scattering of the laser light by the molecules. The energy of scattered photon is increased (Stokes Raman scattering) or reduced (anti-Stokes Raman scattering) during the scattering. The energy difference between the incident electromagnetic radiation (laser light) and the scattered light, gives information about the vibrational energy levels of the molecule and therefore the vibrational frequencies. The vibrational frequencies are characteristic for the molecular structure.(Jacob et al., 2018).

In many studies it was shown that specific characterization of all kind of materials, amorphous or crystalline, can be determined by means of Raman spectroscopy. This is possible, even though the crystal or polycrystal insulators that are composed of several layers of material, or though they are metals (Cantarero, 2015).

Numerous defects and damages can occur in crystals during myco / nano-processing and crystal growth. These have a major effect on the mechanical, optical or electronic efficiency and sustainability of the machined workpiece, such as microstructural alteration, residual stress, phase transition, etc., Such defects and damages can be determined by means of Raman spectroscopy. Raman spectroscopy is significantly sensitive to the microstructural changes, impurities, residual stress, is a good technique for evaluating depth and dimension of these defects or damages It is an easy-to-use, non-destructive method of analysis. Raman spectra provides structural fingerprints and for this reason it is used for molecular recognition and verification within a wide range of scientific disciplines (Xu et al., 2018). Raman spectroscopy also allows detection of complex structural changes, sample mapping and depth scanning. It has important applications in material sciences including semiconductor industry, nanotechnology, solid state physics, synthetic polymer materials and has importance for identifiying all kinds of materials. Raman and FT-IR spectroscopy are known as complementary techniques. Both methods, which have difference in its principles, are good tools for material identification (Vaskova, 2011).

The studies of FT- IR and Raman spectroscopic applications in this field are given below:

In the study performed by Bach et al. in 2019, group theory in a combination with DFT+ U simulation was applied to crystalline materials to predict Raman spectra of anatase and rutile phases of TiO_2. Simulated Raman spectra were calculated and compared to the experimental spectrum and the coatings have been observed to have polycrystalline structures, contain anatase and rutile phases (Bach et al., 2019).

Micro-Raman spectroscopic analysis of different copper sulfate species was performed in a study conducted by Hayez et al (2004). Products such as antlerite, brochantite, posnjakite, langite and chalcanthite, which are likely to occur as a result of atmospheric corrosion of copper alloys, have been investigated. They developed a database of the characteristic Raman spectra of alloys (Hayez et al., 2004).

The interfaces of two diamond/metal composites with high thermal conductivity were analyzed by Raman spectroscopy (Makino et al., 2016).

In 1989, Yang et al., applied Raman spectroscopy to characterize the microscopic bond structure of a-SiGe: H alloys. The correlation between the degree of clustering of Ge-Ge and the transport properties of alloys is shown. Experimental results have been discussed due to an improvement in the quality of the alloy materials (Yang et al., 1989).

In a work performed by Rath *et al.* in 2003, the Raman spectrum of bulk and $Si_{1-x}Ge_x$ epitaxial alloys was investigated to create a metrology for evaluating alloy compositions from the first order Si-Si, Ge-Ge, and Si-Ge optical phonon Raman modes (Rath *et al.*, 2003).

In 2019, Hashemi and his team also created an effective method for simulating the Raman spectra of alloys, comparing them with density-functional theory calculation, and applying to several bidimensional (2D) transition metal dichalcogenides (Hashemi et al., 2019).

In 2018, Rana et al., recorded the Raman spectrum of the Ge-S-Ga chalcogenide glassy alloy and analyzed it for diverse compounds compositions to investigate both high-frequency molecular-like modes and low-frequency acoustic zones. The addition of Ga has been observed to cause the formation of Ga-S bonds responsible for the formation of the tetrahedral device in the glassy network. Besides that, there are some homopolar S-S and Ge-Ge bonds that are characteristic of glasses for the basic GeS_4 structural tetrahedra groups. The addition of Ga also leads to replacement of S - S bonds with Ga - Ga bonds and has been found to shift to a slightly lower wavelength side in the Raman spectrum (Rana et al., 2018).

In a research conducted by Lara et al. in 2017, the phase transition of Gd^{3+}-doped $BaTiO_3$ ceramics was investigated by means of Raman spectroscopy. The Raman peaks, characteristic tetragonal ferroelectric phase of $BaTiO_3$, were determined as 205 cm^{-1} {E(TO +LO), A_1(LO)}, 265 cm^{-1} {A_1(TO)}, 304 cm^{-1} {B_1, E(TO +LO)}, 513 cm^{-1} {A_1(TO), E(TO)}, and 717 cm^{-1} {A_1(LO), E(LO)}. Raman peak, around 304 cm^{-1}, showed a decrease in intensity by increasing Gd^{3+}content, as a result of the decrease of tetragonality of $BaTiO_3$ ceramics (Lara et al., 2017).

In a report by Goff et al., on Raman spectroscopic investigation of crevice corrosion in Beryllium-containing nickel-chrome dental alloys, a simple design to model and classify products of corrosion likely to occur, using Raman spectroscopy, was suggested (Goff et al., 1998).

The role of molybdenum in the passivation and corrosion of nickel alloys has been investigated by Raman spectroscopy (Delichere et al., 1988).

In the work done by Reis and his team, 304L alloy stainless steel powder and multi-walled carbon nanotubes were mixed under the ambient atmosphere with ball milling and for a long grinding time ranging from 0 to 120 min. The morphological and structural changes related to the crystallinity of the samples due to doping process were investigated by the Lorentzian deconvolution from the Raman spectrum. In this study they provided Raman spectroscopic signatures for produced several distinct composites. It was shown that Raman spectra of nanocomposites based on SS 304L alloy metal nanoparticles and MWCNTs present multiple splits in the D, G and 2D-bands region as fingerprints (Reis et al., 2018).

Wu and his team's research in 2020 examined the tribological properties of the Ni_3Si alloy at high temperatures. The effect of adding Ti was investigated by means of Raman spectroscopy. The results showed that the coefficient of friction decreased as the temperature goes up, and the wear rate ranged marginally from 25 to 400 °C (Wu et al., 2020).

Raman scattering experiments were performed in a bulk $Si_{1-x}Ge_x$ single crystal with compositional gradient, which can be used as an X-ray monochromator for focusing and diffraction, to determine both residual strain and alloy composition. Both the residual strain and alloy composition are evaluated by analyzing strain effect on RS in conjunction with a strain model developed for bulk mixed crystals. It is found that the alloy composition evaluated from the RS results is in good agreement with that measured by the standard chemical analysis (Islam et al., 2004).

In the study of Kim at al., Raman spectroscopy was used when examining the effects of chromium and manganese as alloying elements on the corrosion resistance of carbon steel (Kim et al., 2018).

Nouicer et al. studied Ti_6Al_4V alloy by Raman spectroscopy in order to characterize titanium alloy surface in a biological medium (Nouicer et al., 2017). The effect of a biological medium simulated body fluid, on surface modification of titanium alloy Ti_6Al_4V, was investigated and depending on the Raman spectroscopy it was found that titanium dioxide consisted of rutile phase.

In a study conducted in 2017, tribofilms of carbon steel were analyzed by Raman spectroscopy and the technique proved to be effective in direct observation of films with lubricating oil in ambient air (Miyajima et al., 2017).

In 2011, Fernandes et al. was reported that zinc phosphate coatings modified with manganese and nickel were isomorphic with hopeite mineral (hydrated zinc phosphate) which was difficult to identify using X-ray diffraction, for this reason the phosphate coatings were determined using FTIR (Fernandes et al., 2011)

In the study conducted in 1988 (Jasinski et al., 1988), it was stated that by means of FT-IR spectroscopy, iron oxide and iron carbonate corrosion scales could be easily determined and the spectra were related to the chemical history of the corrosion process. In this study, it was found that the IR absorption bands of most of the 11 crystalline iron oxides and iron carbonates were sufficiently separated from each other. However, it was also reported that vacuum exposure of some of these films, or preparation of KBr pellets made band shifts depending on the special oxide composition (Jasinski et al., 1988).

In a study conducted in 1992, oxide films formed on stainless steel SUS430 under different conditions were studied using infrared reflection spectroscopy, using a FT-IR spectrometer equipped with a polarized reflectance attachment and a wire grid polarizer. Comparison of the FTIR spectra of oxidized samples in a humid atmosphere and oxidized samples in a dry atmosphere showed the importance of water vapor at 873 and 973 K. The abnormal rapid oxidation at 873 K in a humid atmosphere with low oxygen pressure was characterized by a rapid increase in the intensity of the *ca.* 550 cm^{-1} IR band, depending on the increasing oxidation time (Honda et al., 1992).

Swiech et al. (Swięch et al., 2018) analyzed the corrosion products, that formed on the surfaces of Fe and Cu metal after being left in solution containing chloride, by means of FT-IR absorption and Raman spectroscopy. Depending on the spectral analysis, it was found that the main component types of corrosion products for Fe metal sample were lepidocrocite and paratacamide for Cu surface. The findings indicated that applying vibrational spectroscopic methods is a precise tool for recognizing and evaluating the products of corrosion.

In the 2012 report, comprehensive study was conducted on improving the corrosion properties of AISI D3 steel surfaces coated with multi-layer $(CrN/AlN)_n$ framework, which is accumulated by spraying with magnetron over various periods. For practical effects, the properties of CrN and AIN single layers accumulated under the same conditions as multi-layer systems were compared. Coatings were examined by various methods, especially by FT-IR spectroscopy. Depending on FTIR results it was reported that identification of the characteristic bonding in single layer coatings was possible (Cabrera et al., 2012).

Ti-6Al-4V-4B_4C metal matrix composite covering were coated on AISI 1040 steel using magnetron spraying in the 2020 study and the dry slippery wear behavior at room temperature was investigated. FT-IR and some other methods were applied on the investigation of coating morphology. The vibrational frequencies of Ti-6Al-4V-4B_4C coating are determined and the presence of B_4C in the coating confirmed by FT-IR spectroscopy (Prince et al., 2020).

The goal of the 2007 research, conducted by Sheban and her team, was to establish corrosion protection systems for chloride attacking concrete structures. Benzotriazole (BTA) and BTA derivatives were used as the steel corrosion protection product and it has been found that the BTA derivatives in simulated concrete pore solutions have good steel protection. The formation of either mono-or bi-dentate bonds between the triazolic nitrogen ring and the steel surface was shown by the FTIR (Sheban et al., 2007).

The effect of green lubrication on the tribological and mechanical properties of AISI 4140 steel was explored in the study conducted in 2019 and performance of three bio-based lubricants was examined, compared to five of the lubricants, and lubricants were chemically characterized using FT-IR spectroscopy (Sierra et al., 2019).

CONCLUSION

PM products are used in variety applications in many sectors. High surface quality of the products produced by powder metallurgy method and their mechanical and physical properties can be adjusted according to the working conditions are important aspects of the method. Properties of the iron-based PM parts such as strength, hardness, impact resistance, corrosion properties, wear resistance and magnetic properties can be improved with adding various alloying elements. Corrosion is a major problem due to the pore content in the iron-based PM product. Surface treatments can be applied to improve the corrosion properties of these parts. As mentioned in this chapter, besides the classical methods, the latest technology surface coating processes are used to improve the surface properties of PM products. The alloying elements have an effect on the thickness, bonding properties, appearance and brittleness of the coating. With using FT-IR and Raman spectroscopic techniques, the early detection of defects and impurity in the manufacturing process can be obtained quickly, that is very useful for controlling the production process and taking action.

REFERENCES

Anderson, A. (1995). *ASM Metals Handbook* (Vol. 18). Friction, Lubrication and Wear.

Angel, W. D. W., Jurado, L. T., Alcalá, J. F. C., Martínez, E. C., & Cedeño, V. F. V. (2014). Effect of copper on the mechanical properties of alloys formed by powder metallurgy. *Materials & Design*, *58*, 12–18. doi:10.1016/j.matdes.2014.02.002

Angelo, P. C., & Subramanian, R. (2008). *Powder metallurgy: science, technology and applications*. PHI Learning Pvt. Ltd.

Attar, F. (1996). Hardness evaluation of thin ceramic coatings on tool steel. *Surface and Coatings Technology*, *78*(1-3), 78–86. doi:10.1016/0257-8972(94)02395-6

Bach, L. X., Thuan, D. V., Thu, V. T. H., Phan, T. B., Vu, N. S. H., & Nam, N. D. (2019). An investigation on titania multilayer coatings for enhanced corrosion resistance of carbon steel in simulated seawater by sol–gel dip coating. *Journal of Materials Research and Technology*, *8*(6), 6400–6406. doi:10.1016/j.jmrt.2019.09.061

Baran, M. C., Graham, A. H., Davala, A. B., Causton, R. J., & Schade, C. (1999). A superior sinter-hardenable material. *Advances in Powder Metallurgy & Particulate Materials*, *2*(Part 7), 185–203.

Bergman, O. (2011). *Key aspects of sintering powder metallurgy steel prealloyed with chromium and manganese*. Chalmers University of Technology.

Boing, D., Oliveira, A. J. D., & Schroeter, R. B. (2020). Evaluation of wear mechanisms of PVD and CVD coatings deposited on cemented carbide substrates applied to hard turning. *International Journal of Advanced Manufacturing Technology*, *106*(11), 5441–5451. doi:10.100700170-020-05000-x

Cabezon, C. G., Hernandez, C. G., Mendez, M. L. R., Herranz, G., & Pedrosa, F. M. (2020). Role of carbon and nitrogen in the improvement of corrosion resistance of new powder metallurgy Co-Cr-Mo alloys. *Corrosion Reviews*. Advance online publication. doi:10.1515/corrrev-2019-0109

Cabrera, G., Torres, F., Caicedo, J. C., Aperador, W., Amaya, C., & Prieto, P. (2012). Improvement of electrochemical surface properties in steel substrates using a nanostructured CrN/AlN multilayer coating. *Journal of Materials Engineering and Performance*, *21*(1), 128–136. doi:10.100711665-010-9798-7

Cameron, D. (2014). Introduction to Films and Coatings: Technology and Recent Development. In Comprehensive Materials Processing (Volume 4.01, pp. 1-2). Elsevier.

Cantarero, A. (2015). Raman scattering applied to materials science. *Procedia Mater. Sci.*, *9*, 113–122. doi:10.1016/j.mspro.2015.04.014

Capus, J. (2013). PM light alloys gaining applications in automotive sector. *Metal Powder Report*, *68*(6), 12–15. doi:10.1016/S0026-0657(13)70210-8

Celik, S., Albayrak, A. T., Akyuz, S., & Ozel, A. E. (2020). The Role of Multivariant Analysis on the Interpretation of FTIR and Raman Spectra. In Design of Experiments for Chemical, Pharmaceutical, Food, and Industrial Applications (pp. 104-129). IGI Global. doi:10.4018/978-1-7998-1518-1.ch005

Chawla, N., Fillari, G., & Narasimham, K. S. (1999). Recent advances in ferrous powder metallurgy. In Powder Materials: Current Research and Industrial Practices(as held at the 1999 TMS Fall Meeting) (pp. 247-255). Academic Press.

Chen, W. G., Duan, X. M., Dong, T., Gao, L. N., Chen, M. Z., & Ye, H. (2008). Study on Steam Treatment Process for Steel. *Surface Technology*, *37*(6), 57–60.

Christy, A. A., Ozaki, Y., & Gregoriou, V. G. (2001). *Modern Fourier transform infrared spectroscopy*. Elsevier.

Ciurczak, E. W., & Igne, B. (2015). *Pharmaceutical and medical applications of near-infrared spectroscopy*. CRC Press.

Dangsheng, X. (2002). The Effect of Rare Earth Compounds on the Friction and Wear Properties of PM Ni-based Alloy. *PM Technology-China, 20*(82), 67-70.

Danninger, H., & Gierl, C. (2008). New alloying systems for ferrous powder metallurgy precision parts. *Science of Sintering*, *40*(1), 33–46. doi:10.2298/SOS0801033D

Delichere, P., Goff, A. H., & Joiret, S. (1988). Study of thin corrosion films by in situ Raman spectroscopy combined with direct observation of nuclear reactions. *Surface and Interface Analysis*, *12*(7), 419–423. doi:10.1002ia.740120710

Dincer, M. (2014). *Sinter hardening studies of ASTALOY 85 Mo powder with varying copper contents* (Master's thesis). Middle East Technical University.

Dobrzanski, L. A., Brytan, Z., Grande, M. A., & Rosso, M. (2007). Properties of duplex stainless steels made by powder metallurgy. *Archives of Materials Science and Engineering*, *28*(4), 217–223.

Donghai, Y., Chengyong, W., & Fenglin, Z. (2007). Development of Cutting Tool Coating Material. *Tool Engineering*, *41*(6), 25–32.

El-Eskandarany, M. S. (2015). *Mechanical alloying: nanotechnology, materials science and powder metallurgy* (Vol. 2). Elsevier.

Engström, U., Lindberg, C., & Tengzelius, J. (1992). Powders and processes for high performance PM steels. *Powder Metallurgy*, *35*(1), 67–73. doi:10.1179/pom.1992.35.1.67

Fernandes, K. S., Alvarenga, E. D. A., Brandão, P. R. G., & Lins, V. D. F. C. (2011). Infrared-spectroscopy analysis of zinc phosphate and nickel and manganese modified zinc phosphate coatings on electrogalvanized steel. *REM. Revista Escola de Minas*, *64*(1), 45–49. doi:10.1590/S0370-44672011000100005

Ganesan, S., Padmanabhan, S., Hemanandh, J., & Venkatesan, S. P. (2019). Influence of substrate temperature on coated engine piston head using multi-response optimisation techniques. *International Journal of Ambient Energy*, 1-8.

Gill, A. S., & Kumar, S. (2019). Surface characteristics investigation of tool steel machined by powder metallurgy tool in EDA. *Materials Research Express, 6*(12), 1265b1.

Goff, A. H., Joiret, S., & Abourazzouk, D. (1998). Raman investigation of crevice corrosion in nickel-chromium dental alloys containing beryllium. *Electrochimica Acta, 43*(1-2), 53–62. doi:10.1016/S0013-4686(97)00234-X

Goldschmidt, A., & Streitberger, H. J. (2003). *BASF handbook on basics of coating technology*. Primedia.

Grimanelis, D., & Eyre, T. S. (2007). Wear characteristics of a diffusion bonded sintered steel with short term surface treatments. *Wear, 262*(1-2), 93–103. doi:10.1016/j.wear.2006.04.006

Hashemi, A., Krasheninnikov, A. V., Puska, M., & Komsa, H. P. (2019). Efficient method for calculating Raman spectra of solids with impurities and alloys and its application to two-dimensional transition metal dichalcogenides. *Physical Review Materials, 3*(2), 023806. doi:10.1103/PhysRevMaterials.3.023806

Hayez, V., Guillaume, J., Hubin, A., & Terryn, H. (2004). Micro-Raman spectroscopy for the study of corrosion products on copper alloys: Setting up of a reference database and studying works of art. *Journal of Raman Spectroscopy: JRS, 35*(8-9), 732–738. doi:10.1002/jrs.1194

Honda, K., Atake, T., & Saito, Y. (1992). Infrared spectra of oxide films on SUS430 stainless steel formed at temperatures between 473 and 1073 K. *Journal of Materials Research, 7*(1), 80–84. doi:10.1557/JMR.1992.0080

Hu, B., & Tian, G. (2012). Morphological instability of γ'phase in nickel-based powder metallurgy superalloys. *Chin Shu Hsueh Pao, 48*(3), 257–263. doi:10.3724/SP.J.1037.2011.00731

Islam, M. R., Yamada, M., Abrosimov, N. V., Kiyama, M., & Tatsumi, M. (2004). Raman scattering characterization of residual strain and alloy composition in bulk Si1-xGex crystal. *The European Physical Journal Applied Physics, 27*(1-3), 325–328. doi:10.1051/epjap:2004124

Jacob, L. J., & Deigner, H. P. (2018). Nanoparticles and Nanosized Structures in Diagnostics and Therapy. In Precision Medicine (pp. 229-252). Academic Press. doi:10.1016/B978-0-12-805364-5.00010-X

James, W. B., & West, G. T. (2002). *Ferrous powder metallurgy materials. In powder technologies and applications* (Vol. 7). ASM International.

Jasinski, R., & Iob, A. (1988). FTIR measurements of iron oxides on low alloy steel. *Journal of the Electrochemical Society, 135*(3), 551–556. doi:10.1149/1.2095656

Kahraman, N., & Gulenc, B. (2002). Abrasive wear behaviour of powder flame sprayed coatings on steel substrates. *Materials & Design, 23*(8), 721–725. doi:10.1016/S0261-3069(02)00075-4

Kim, B., Kim, S., & Kim, H. (2018). Effects of Alloying Elements (Cr, Mn) on Corrosion Properties of the High-Strength Steel in 3.5% NaCl Solution. *Advances in Materials Science and Engineering, 2018*, 1–13. doi:10.1155/2018/7638274

Klar, E., & Samal, P. K. (2007). *Powder metallurgy stainless steels: processing, microstructures, and properties*. ASM international.

Kolawole, M. Y., Aweda, J. O., Iqbal, F., Ali, A., & Abdulkareem, S. (2019). Mechanical Properties of Powder Metallurgy Processed Biodegradable Zn-Based Alloy for Biomedical Application. *International Journal of Materials and Metallurgical Engineering*, *13*(12), 558–563.

Konnov, A. Y., Demchenko, M. V., Sisanbaev, A. V., Nafikov, M. Z., & Farkhshatov, M. N. (2019). Estimation of the Corrosion Resistance of the Metal Coatings Formed by Powder Metallurgy. *Russian Metallurgy (Metally)*, *2019*(12), 1285–1288. doi:10.1134/S0036029519120097

Krishna, M. E., & Patowari, P. K. (2013). Parametric optimisation of electric discharge coating process with powder metallurgy tools using Taguchi analysis. *Surface Engineering*, *29*(9), 703–711. doi:10.1179/1743294413Y.0000000207

Lara, J. P. H., Labra, M. P., Hernández, F. R. B., Serrano, J. A. R., Dávila, E. O. A., Thangarasu, P., & Ramirez, A. H. (2017). Structural Evolution and Electrical Properties of $BaTiO_3$ Doped with Gd^{3+}. *Materials Research*, *20*(2), 538–542. doi:10.1590/1980-5373-mr-2016-0606

Lawley, A., & Murphy, T. F. (2003). Metallography of powder metallurgy materials. *Materials Characterization*, *51*(5), 315–327. doi:10.1016/j.matchar.2004.01.006

Lee, P. W., Trudel, Y., German, R. M., Ferguson, B. L., Eisen, W. B., Mover, K., . . . Davidson, G. M. (1998). ASM handbook. Powder Metal Technologies and Applications, 7, 1146.

Link, R. (1998). Modern Pre-alloyed PM Steels and Advanced Process Technology. *PM World Congress Microstructure*, 596-606.

Liu, Y. B., Lim, S. C., Lu, L., & Lai, M. O. (1994). Recent development in the fabrication of metal matrix-particulate composites using powder metallurgy techniques. *Journal of Materials Science*, *29*(8), 1999–2007. doi:10.1007/BF01154673

Mahade, S., Narayan, K., Govindarajan, S., Björklund, S., Curry, N., & Joshi, S. (2019). Exploiting suspension plasma spraying to deposit wear-resistant carbide coatings. *Materials (Basel)*, *12*(15), 2344. doi:10.3390/ma12152344 PMID:31344804

Makino, Y., Mizuuchi, K., Agari, Y., & Ito, M. (2016). Characterization of Interfaces in Diamond/Metal Composites by Raman Scattering Spectroscopy. *Journal of the Japan Society of Powder and Powder Metallurgy*, *63*(7), 479–483. doi:10.2497/jjspm.63.479

Mirzaali, M. J., Bobbert, F. S. L., Li, Y., & Zadpoor, A. A. (2019). Additive Manufacturing of Metals Using Powder Bed-Based Technologies. In *Additive Manufacturing* (pp. 94–127). CRC Press. doi:10.1201/9780429466236-4

Mitchell, R. J., Rae, C. M. F., & Tin, S. (2005). Grain boundary transformations during isothermal exposure of powder metallurgy nickel base superalloys for turbine disc applications. *Materials Science and Technology*, *21*(1), 125–132. doi:10.1179/174328405X16261

Miyajima, M., Kitamura, K., & Matsumoto, K. (2017). Characterization of Tribochemical Reactions on Steel Surfaces. *Nippon Steel Sumitomo Met Tech Rep*, *114*, 101–107.

Mocarski, S., Hall, D. W., Chernenkoff, R. A., Yeager, D. A., & McHugh, C. O. (1996). Master alloys to obtain premixed hardenable powder metallurgy steels. *Powder Metallurgy*, *39*(2), 130–137. doi:10.1179/pom.1996.39.2.130

Narasimhan, K. S. (1996). Recent advances in ferrous powder metallurgy. *Advanced Performance Materials*, *3*(1), 7–27. doi:10.1007/BF00136857

Narasimhan, K. S. (2001). Sintering of powder mixtures and the growth of ferrous powder metallurgy. *Materials Chemistry and Physics*, *67*(1-3), 56–65. doi:10.1016/S0254-0584(00)00420-X

Nouicer, E., Benlahreche, F., & Nouicer, A. (2017). Spectroscopic Characterization of Titanium Alloy Surface in a Biological Medium. *Acta Physica Polonica A*, *131*(1), 28–31. doi:10.12693/APhysPolA.131.28

Ordonez, A. A., & Prieto, M. (2012). Technical and methodological aspects of fourier transform infrared spectroscopy in food microbiology research. In *Fourier Transform Infrared Spectroscopy in Food Microbiology* (pp. 1–18). Springer. doi:10.1007/978-1-4614-3813-7_1

Park, J., Lee, S., Kang, S., Jeon, J., Lee, S. H., Kim, H. K., & Choi, H. (2015). Complex effects of alloy composition and porosity on the phase transformations and mechanical properties of powder metallurgy steels. *Powder Technology*, *284*, 459–466. doi:10.1016/j.powtec.2015.07.029

Prince, R. M. R., Selvakumar, N., Arulkirubakaran, D., Sreedharan, C. E. S., Thulasiram, R., & Kumar, R. M. (2020). Surface structural features and wear analysis of a multilayer Ti6Al4V-B_4C thin film coated AISI 1040 steel. *Materials Research Express*, *7*, 1–17.

Rana, A., Singh, B., & Sharma, R. (2018). Raman Study of Ga Doped Ge-S Glassy Alloy. *International Journal of Chemical and Physical Sciences*, *7*, 638–641.

Rath, S., Hsieh, M. L., Etchegoin, P., & Stradling, R. A. (2003). Alloy effects on the Raman spectra of $Si_{1-x}Ge_x$ and calibration protocols for alloy compositions based on polarization measurements. *Semiconductor Science and Technology*, *18*(6), 566–575. doi:10.1088/0268-1242/18/6/330

Reinshagen, J. H., & Witsberger, J. C. (1991). Processing and properties of 400 series PM stainless steels. *Metal Powder Report*, *46*(5), 38–44. doi:10.1016/0026-0657(91)91757-W

Reis, M. A. L. D., Neto, N. M. B., Sousa, M. E. S. D., Araujo, P. T., Simões, S., Vieira, M. F., ... Assuncao, P. D. (2018). Raman spectroscopy fingerprint of stainless steel-MWCNTs nanocomposite processed by ball-milling. *AIP Advances*, *8*(1), 015323. doi:10.1063/1.5018745

Šalak, A., & Selecká, M. (2012). *Manganese in powder metallurgy steels*. Springer Science & Business Media. doi:10.1007/978-1-907343-75-9

Šalak, A., Selecká, M., & Danninger, H. (2005). *Machinability of powder metallurgy steels*. Cambridge International Science Publishing.

Salgado, L., Filho, F. A., Neto, R. L., Rossi, J. L., Colosio, M. A., & Filho, C. H. (2001). P/M high-speed steels for valve seat insert application. *SAE Transactions*, 261–264. doi:10.4271/2001-01-0395

Samal, P., & Newkirk, J. (2015). ASM handbook: Vol. 7. *Powder metallurgy*. ASM International. doi:10.31399/asm.hb.v07.9781627081757

Schade, C. T., Schaberl, J. W., & Lawley, A. (2008). Stainless Steel AISI Grades for PM Applications. *International Journal of Powder Metallurgy*, *44*(3), 57–67.

Shanmugasundaram, D., & Chandramouli, R. (2009). Tensile and impact behaviour of sinter-forged Cr, Ni and Mo alloyed powder metallurgy steels. *Materials & Design*, *30*(9), 3444–3449. doi:10.1016/j.matdes.2009.03.020

Sheban, M., Abu-Dalo, M., Ababneh, A., & Andreescu, S. (2007). Effect of benzotriazole derivatives on the corrosion of steel in simulated concrete pore solutions. *Anti-Corrosion Methods and Materials*, *54*(3), 135–147. doi:10.1108/00035590710748605

Sierra, M. T. H., Sánchez, M. G. B., Báez, J. E., Camacho, L. D. A., Miranda, J. S. G., & Moreno, K. J. (2019). Improvement Effect of Green Lubricants on the Tribological and Mechanical Performance of 4140 Steel. *Applied Sciences (Basel, Switzerland)*, *9*(22), 4896. doi:10.3390/app9224896

Skoglund, P., Litstrom, O., & Flodin, A. (2013). Improvement of Powder Metallurgy Gears for Engines and Transmissions. *SAE International*, *26*, 1–8. doi:10.4271/2013-32-9102

Smith, B. C. (2011). *Fundamentals of Fourier transform infrared spectroscopy*. CRC Press., doi:10.1201/b10777

Šolic, S., Schauperl, Z., & Tropša, V. (2019). Wear behaviour of TiAlN PVD coating on deep cryogenically treated HSS substrate. *Kovove Mater*, *57*, 429–434.

Stuart, B. H. (2004). *Infrared spectroscopy: Fundamentals and applications*. John Wiley & Sons., doi:10.1002/0470011149

Sułowski, M., & Matusiewicz, P. (2013). Comparison of mechanical properties of Mn-Cr-Mo steels sintered in different conditions. *Diffusion and Defect Data, Solid State Data. Part B, Solid State Phenomena*, *197*, 33–40. doi:10.4028/www.scientific.net/SSP.197.33

Sundararajan, G., Sen, D., & Sivakumar, G. (2005). The tribological behaviour of detonation sprayed coatings: The importance of coating process parameters. *Wear*, *258*(1-4), 377–391. doi:10.1016/j.wear.2004.03.022

Swięch, D., Paluszkiewicz, C., Piergies, N., Pięta, E., Borkowska, U. L., & Kwiatek, W. (2018). Identification of Corrosion Products on Fe and Cu Metals using Spectroscopic Methods. *Acta Physica Polonica A*, *133*(2), 286–288. doi:10.12693/APhysPolA.133.286

Tengzelius, J., & Grinder, O. (2008). Powder Metallurgy in Denmark, Finland, and Sweden. *International Journal of Powder Metallurgy*, *50*(2), 57–66.

Tian, Y., Zhang, S. T., Ma, J. H., & Xie, J. G. (2006). Development and application of abradable sealing coating. *Nonferrous Metals*, *1*, 96–99.

Upadhyay, R. K., & Kumaraswamidhas, L. A. (2014). Surface modification by multilayered W/W2N coating. *Surface Engineering*, *30*(7), 475–482. doi:10.1179/1743294414Y.0000000260

Vaskova, H. (2011). A powerful tool for material identification: Raman spectroscopy. *Int. J. Math. Model. Methods Appl. Sci.*, *5*, 1205–1212.

Wang, M. C. (1994). Properties of high density powder forged iron based alloy. *Powder Metallurgy*, *37*(3), 201–205. doi:10.1179/pom.1994.37.3.201

Wang, W. F. (2005). Effect of alloying elements and processing factors on the microstructure and hardness of sintered and induction-hardened Fe–C–Cu alloys. *Materials Science and Engineering A*, *402*(1-2), 92–97. doi:10.1016/j.msea.2005.04.016

Warke, V. S., Sisson, R. D. Jr, & Makhlouf, M. M. (2009). FEA model for predicting the response of powder metallurgy steel components to heat treatment. *Materials Science and Engineering A*, *518*(1-2), 7–15. doi:10.1016/j.msea.2009.04.003

Wu, C., Niu, M., Bao, S., Sun, Y., & Zhang, X. (2020). Friction and Wear Properties of Ni3Si Alloy with Ti Addition at High Temperatures. *Materials (Basel)*, *13*(4), 982. doi:10.3390/ma13040982 PMID:32098307

Xu, Z., He, Z., Song, Y., Fu, X., Rommel, M., Luo, X., Hartmaier, A., Zhang, J., & Fang, F. (2018). Topic review: Application of Raman spectroscopy characterization in micro/nano-machining. *Micromachines*, *9*(7), 361. doi:10.3390/mi9070361 PMID:30424294

Xu, Z., Hodgson, M. A., & Cao, P. (2015). A comparative study of powder metallurgical (PM) and wrought Fe–Mn–Si alloys. *Materials Science and Engineering A*, *630*, 116–124. doi:10.1016/j.msea.2015.02.021

Yang, L., Newton, J., & Fieselmann, B. (1989). Raman Spectroscopy of a. SiGe: H Alloys. *MRS Online Proceedings Library Archive, 149*.

Yu, C. J., Eifert, H. H., Banhart, J., & Baumeister, J. (1998). Metal foaming by a powder metallurgy method: Production, properties and applications. *Materials Research Innovations*, *2*(3), 181–188. doi:10.1007100190050082

Yu, Y., Lindqvist, B., & Tengzelius, J. (2004). Höganäs Chair–an industry/university cooperation. *Powder Metallurgy*, *47*(4), 304–308. doi:10.1179/pom.2004.47.4.304

Chapter 11
Surface Engineering of Materials Through Weld-Based Technologies: An Overview

Magdaline N. Muigai
Dedan Kimathi University of Technology, Kenya

Fredrick M. Mwema
https://orcid.org/0000-0001-6116-5587
Centrer for Nanomaterials and Nanoscience Research (CNSR), Dedan Kimathi University of Technology, Kenya & University of Johannesburg, South Africa

Esther T. Akinlabi
Pan African University for Life and Earth Sciences Institute (PAULESI), Nigeria

Japheth O. Obiko
Jomo Kenyatta University of Agriculture and Technology, Kenya

ABSTRACT

In this chapter, an overview of welding as a technology for surface engineering is explored. According to literature, all types of welding techniques are appropriate for coating applications. However, as a result of process characteristics, some welding processes stand out. The most used welding techniques in the metal coating are arc welding (MIG, TIG, and PAW) and oxyacetylene welding. In the coating of metals using welding techniques, the coatings produced usually have a thickness that ranges between 1 and 6 millimeters. Applications of surface coating have been studied extensively. Such applications include aeronautic industry, sports, transport industries, petroleum and chemical industries, mining, food, and in the electronic industry. Plasma MIG welding is an advanced plasma process that combines the advantages of both MIG and plasma welding. Applications of plasma MIG welding in the surface coating of metals are expected to be explored extensively in the future.

DOI: 10.4018/978-1-7998-4870-7.ch011

BACKGROUND

Surface Engineering refers to a multidisciplinary activity aimed at tailoring the surface properties of engineering components to improve their serviceability and functional properties. In other words, surface engineering is the modification of the surface of a material or a component to achieve excellent properties (Mishra, 2002). Some properties which are improved by surface engineering of metallic materials are wear resistance, corrosion, oxidation resistance, fatigue, toughness, electrical properties, thermal insulation, and aesthetics. Historically, methods of surface engineering have included solvent swelling, mechanical abrasion, solvent wiping to remove contaminants on the surface, application of coatings, and wet chemical etching (Kutz, 2010).

This chapter explores coating as a method of surface engineering and in particular welding as a possible method of surface engineering. There are various metal coating techniques including Physical Vapor Deposition(PVD)coating, Chemical Vapor Deposition(CVD) coating, Micro Arc Oxidation (MAO) coating, Plasma Spray Coating, Thermal Spray coating, Sol-gel coating, Welding coating, and among others (Fotovvati et al., 2019). Welding is used to applying a coating to a metallic substrate through partially or fully molting the coating material. The process is related to the laser cladding technique for additive manufacturing. Since welding is used as an additive manufacturing process, it can be used as a technique for protecting or repairing surfaces. It is due to this motivation is considered as a surface engineering method in this paper. In this case, the coating material and the substrate are melted to create a bond between the two materials. When compared to other types of additive manufacturing/coating techniques, wielding produces coatings/clads with higher adhesion and thickness properties (Gebert & Bouaifi, 2006). It is also possible to coat larger samples compared to other techniques. These types of coatings are appropriate particularly for applications of in areas of high wear conditions. These coatings also exhibit great edge strength. Generally, all types of welding techniques are appropriate for additive manufacturing/coating applications. The most used welding techniques in the metal coating are arc welding and oxyacetylene welding (Holmberg & Matthews, 2009. In Oxyacetylene welding, coating material and substrate are placed close to each other and the touching parts are melted using an Oxyacetylene flame (Hazledine, 2019). Arc welding uses an electric arc to produce heat which is used in the coating of metals. Examples of arc welding techniques include; Plasma arc welding (PAW), Tungsten inert gas welding (TIG), and Metal inert gas welding (MIG).

The coating of metals by welding is used on a small scale as well as in large scale applications. The coatings achieved usually resist surface attacks such as wear, corrosion, or a combination of both wear and corrosion (Gebert & Bouaifi, 2006). To attain the part strength desired, a suitable identification of the substrate material is necessary. Sometimes, the substrate is heat-treated before the coating takes place. Because of the potential of these technologies there has been a lot of interest in their applications in surface engineering. As such, a lot of literature on this subject is available. Applications of a surface coating by welding have been studied extensively. Such applications are in the aeronautic industry, sports, transport industries, petroleum and chemical industries, mining, food, and in the electronic industry. Through surface coating, problems that cause engineering failures such as wear, corrosion, and friction are solved (Kennedy et.al, 2005). Surface coating of metals by welding, is also used in areas experiencing issues of surface refinement to enhance resistance to thermal shock and corrosion, as well as fatigue strength and impact. These methods also repair damaged tools and components. In industries, metal coatings are not applied to nonferrous materials with melting points less than 1100°C. The coating materials normally used are carbides, tungsten, boron, or chromium spread in a matrix of nickel, iron, or cobalt. Coatings used to improve resistance to impact usually contain austenitic structures that harden them. For example, austenitic manganese steel (Holmberg & Matthews, 2009).

The purpose of this chapter is to explore the use of welding techniques in surface engineering of metallic materials. Materials such as steel are used in engineering applications that require excellent properties such as high resistance to wear, oxidation, and corrosion properties. Similarly, engineering components need to be maintained and repaired regularly. All these activities are simply surface engineering.

WELDING METHODS FOR SURFACE ENGINEERING

According to literature, all types of welding techniques are appropriate for coating applications. However, as a result of process characteristics, some welding processes such as oxyacetylene and arc welding stand out. Some of the benchmark qualities are deposition rate, coating quality that can be obtained, and transition of property to the substrate. When coating metallic materials using welding techniques, the coatings achieved usually have a thickness that ranges between 1 and 6mm. The coating thickness that can be achieved is limited as a result of internal stresses that occur as the coating thickness increased. Some of the welding techniques which are commonly used in the coating of metals include MIG welding, TIG welding, PAW welding, and oxyacetylene welding (Gebert & Bouaifi, 2006). These techniques are discussed in this chapter.

MIG Welding

Metal inert gas (MIG) welding was established in the 1950s. A typical MIG welding process is shown in Fig. 1. This process has been traditionally used in melting and joining metals by developing an arc between a given metal and a filler wire which is fed continuously. Today MIG is used to apply coatings in metals. In this case, metallic materials are deposited on the surface of the substrate to form a coating. In MIG, two poles produce an electric arc. One pole is from the wire electrode which is continuously consumed while the other one originates from the metallic workpiece. The wire is usually positioned perpendicular to the metallic substrate (Ugla et al, 2019). The electric arc formed heats and melts the coating material together with the workpiece causing them to join (Patil & Waghmare, 2013). An inert gas is used to shield the arc and the weld pool from contaminants such as oxygen, hydrogen, and oxygen found in air (Bhakthavatchalam et al., 2018). The process of coating metals using MIG can be either automatic or semi-automatic.

Figure 1. MIG process (Ugla et al., 2019)

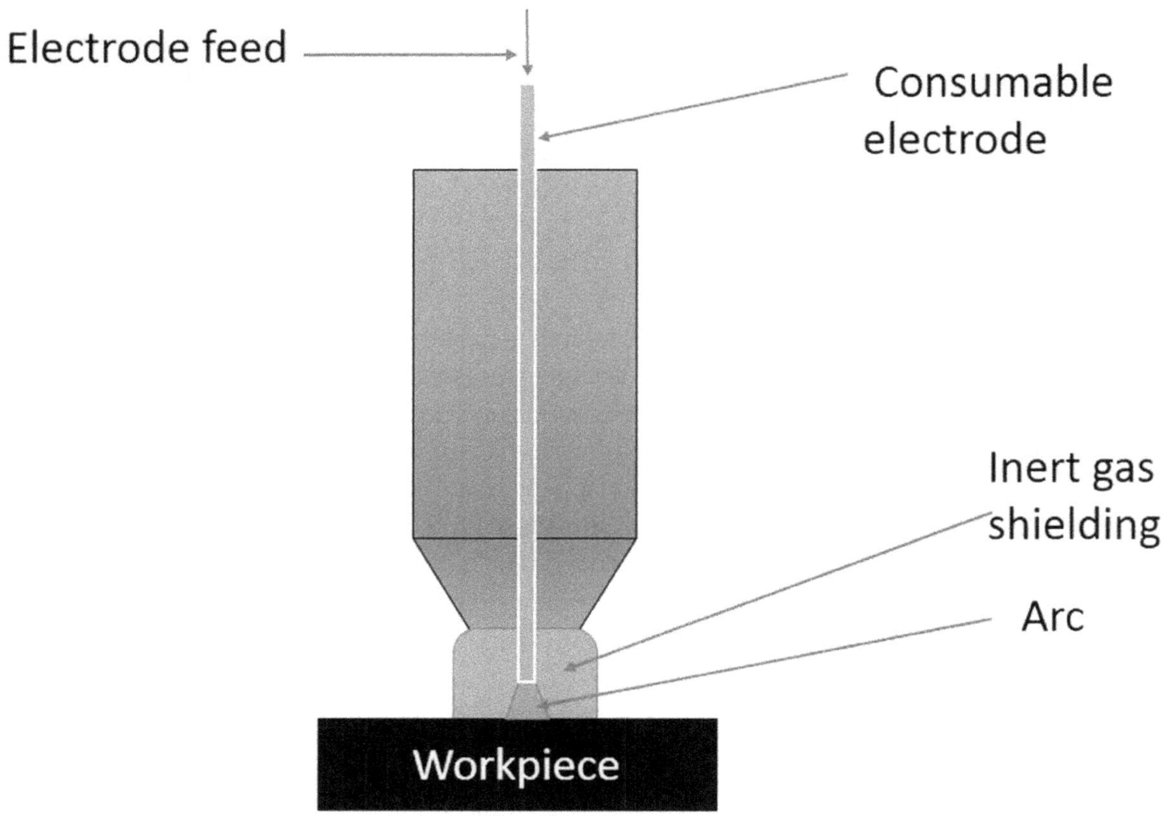

Several parameters affect the process of MIG welding including welding position, gas flow rate, the composition of shielding gas, electrode size, welding speed, welding current, and arc voltage (Patil & Waghmare, 2013). These parameters influence the quality of the coating formed. In MIG welding, the deposition of coatings occurs as stringer beads by forward mechanized feeder or as weave beads by oscillating the welding torch in a transverse motion. Manual operation is also possible; however, the quality of the coating is poor. For this technique, the fusion penetration is high and reaches up to 30. Advancement of the MIG coating technique is the GMAW pulse welding which involves superimposition of a high and low current pulse. This method has been reported to reduce welding spatter, fusion, and enhances detachment of molten material in the form of droplets from the electrode.

TIG Welding

Tungsten inert gas (TIG) welding is also applied in the metal coating process. This type of welding has an arc that burns between a metallic workpiece and a tungsten electrode which is usually not consumable. A typical TIG process is shown in Fig. 2. During the process of deposition, the orientation of the wire feed affects the transfer of material as well as the quality of the deposit. In this process, front feeding, back feeding, and side feeding can be applied (Pan et al., 2018). The different types of feeding during TIG process are illustrated in Fig. 3.

Figure 2. A typical TIG process (Ding et al., 2015)

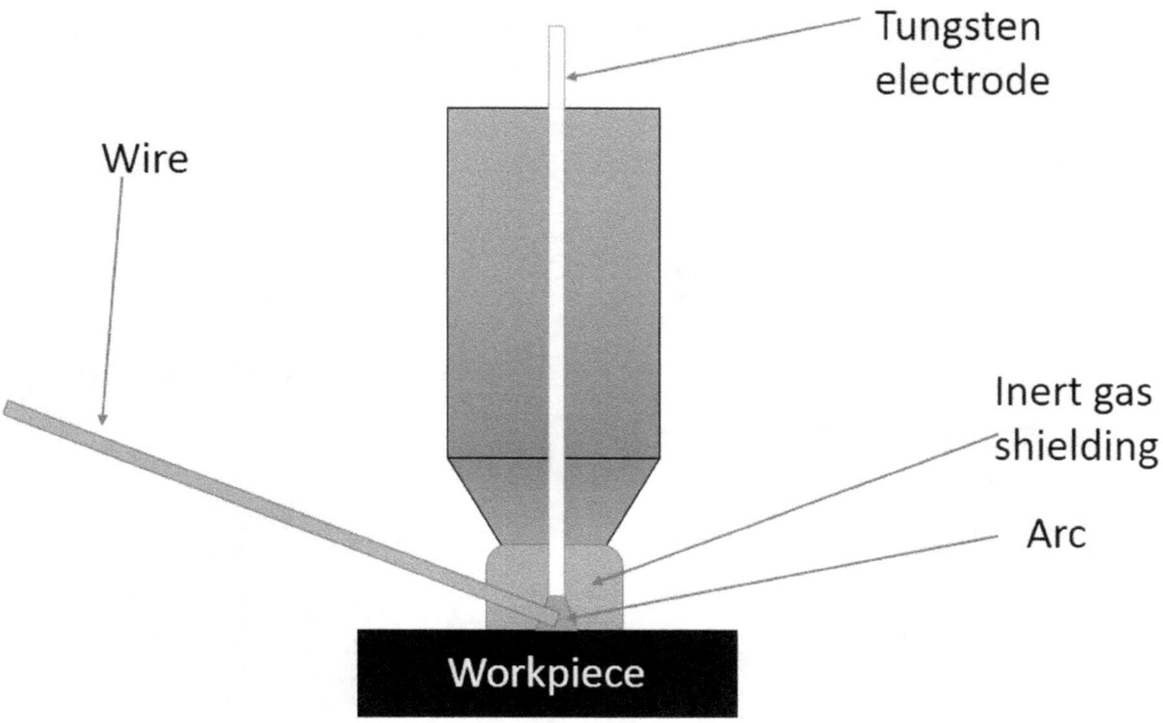

In traditional TIG, the welder delivers the wire feed manually, however, in large scale applications specifically in industries, automated feeding systems are used to increase the rates of deposition as well as the weld repeatability. A wire feeder available commercially is coupled to the TIG power source to adequately automate the process deposition. The feeder resists heat from the wire feed, which helps to minimize thermal shock towards the weld pool when a new wire enters. Since the TIG heat source uses less energy to melt the wire, the rates of deposition and linear feed of the TIG touch are increased (Santangelo et al., 2016). To control this process thoroughly, the welding torch has a compact design, and welders usually observe the molten pool throughout. This process has a low deposition rate which makes mechanization limited to some special applications. In coating light metals, catholic cleaning is used to clean the oxide layer usually found on the surface of light metals.

Various parameters affect the quality of the coatings achieved through TIG welding. These parameters are welding current, welding voltage, inert gases, and welding speed. Very high welding current usually results in splattering and damaging the workpiece. Lower current on the other hand makes the filler wire stick. Welding voltage is usually adjusted or fixed depending on the equipment used for TIG welding. Using a voltage that is too high leads to a huge variable in the quality of coatings obtained (Bhavsar, & Patel, 2016). When an initial voltage is made high, it makes it easy to initiate a welding arc and allows a wider range of distance of working tip.

Inert gases are used in TIG welding coating to shield the weld pool and the arc from interacting with atmospheric oxygen, nitrogen, and hydrogen. The type of inert gas used has effects on the quality of coating achieved. It affects porosity, hardness, brittleness, corrosion resistance, strength, and surface profile of the coating. For TIG welding, Argon or Helium gases are used. Pure argon is usually used when dealing with structural steels, stainless steel, copper, aluminum, titanium, and magnesium. Pure helium is used for copper and aluminum. A mixture of hydrogen and argon is used for nickel alloys and stainless steels (Mohan, 2014). The type of shielding gas used depends on the type of metals used.

Figure 3. Types of the wire feed in TIG (Pan et al., 2018)

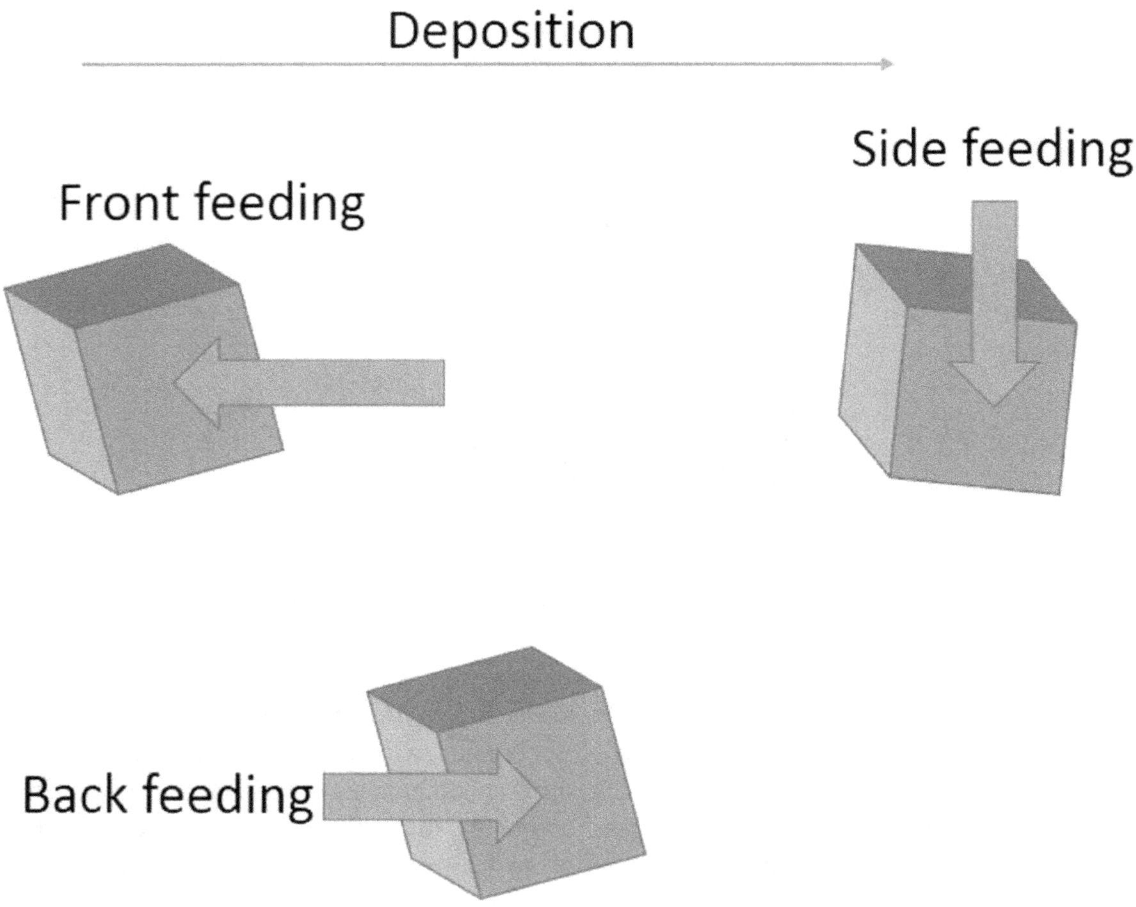

Wending speed also affects the process of TIG welding. An increase in welding speed results in a decrease in power per unit length of the coating. Therefore, less coating reinforcement occurs. Welding speed is interdependent with welding current. Very high welding speed leads to a decrease in wetting action, increases porosity, tendency undercut. The use of a lower welding speed decreases the likelihood of porosity (Bhavsar, & Patel, 2016). It can be seen that welding speed is a fundamental parameter for TIG welding.

PAW Welding

Plasma arc welding (PAW) welding has also been investigated by various researchers. In this case PAW is used as a source of heat to melt the necessary materials and fuse to different sections of the parts under investigations. Plasma arc welding uses a non-consumable electrode made of tungsten for welding purposes. The orientation of wire feed in PAW can be varied and affects the quality of the weld produced (Ding et al., 2015). Therefore, the planning for the process tends to be complicated. The typical features of plasma arc welding (PAW) are illustrated in Fig. 4.

Figure 4. Plasma arc welding (Ding et al., 2015)

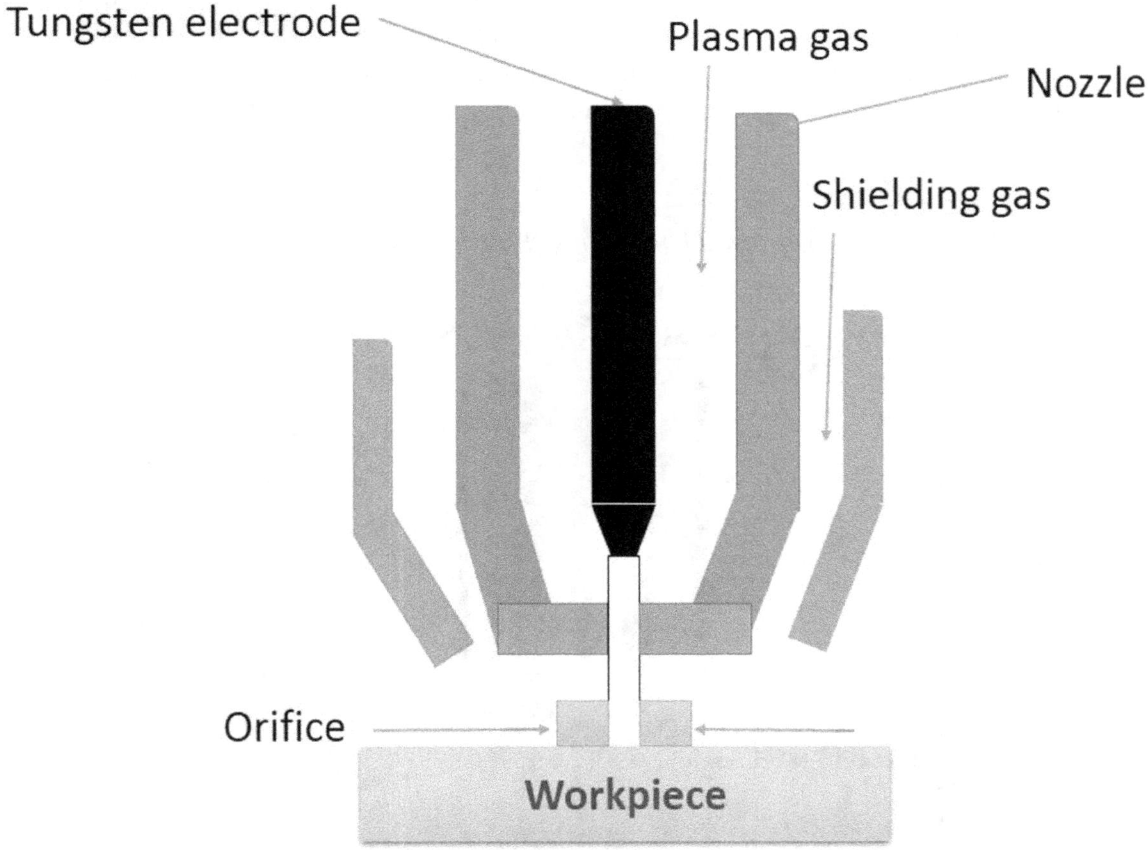

In Plasma arc welding an arc is produced between a cathode (tungsten electrode) and an anode mostly copper within the welding torch. Argon is passed through the orifice between the anode and the cathode, therefore constricting the arc. As a result, energy density is increased, arc stability is increased and contamination is reduced as compared to other welding processes used surface coating (Martina et al., 2012). It is reported that higher speeds of travel are attained using this process.

In Plasma arc welding beads, the heat-affected zone (HAZ) is thin and narrow. As a search, it is easier to control the geometry of the weld bead as a result of the increased flexibility of controlling the fundamental process parameters such as wire feed speed and current. In PAW, it is easier to attain various operating modes such as keyhole plasma, medium current, and microplasma by varying the orifice diameter of the torch, gas flow rate of the plasma, as well as current intensity, it is possible to achieve different operating modes in PAW such as microplasma. Microplasma has the advantage of requiring little welding currents, in ranges of 0.1 and 15A (Rodrigues et al., 2019). This process produces coatings of the highest grades.

Oxyacetylene Torch Welding

Oxyacetylene gas welding uses a mixture of oxygen and acetylene that burns to produce a flame at 35000°C (Holmberg & Matthews, 2009) The flame melts the metallic substrate together with the coating material forming a molten pool. The melted materials stick together to form a coating. This method does not require any specific knowledge or technology. The quality of the coatings produced is dependent on the experience of the welder. Coating metals by use of Oxy-Acetyl welding is easy to learn and the equipment used is cheaper and easy to carry around compared to those equipment used in arc welding (Hazledine, 2019).

Figure 5. Oxy-fuel welding process used in a coating (Holmberg & Matthews, 2009)

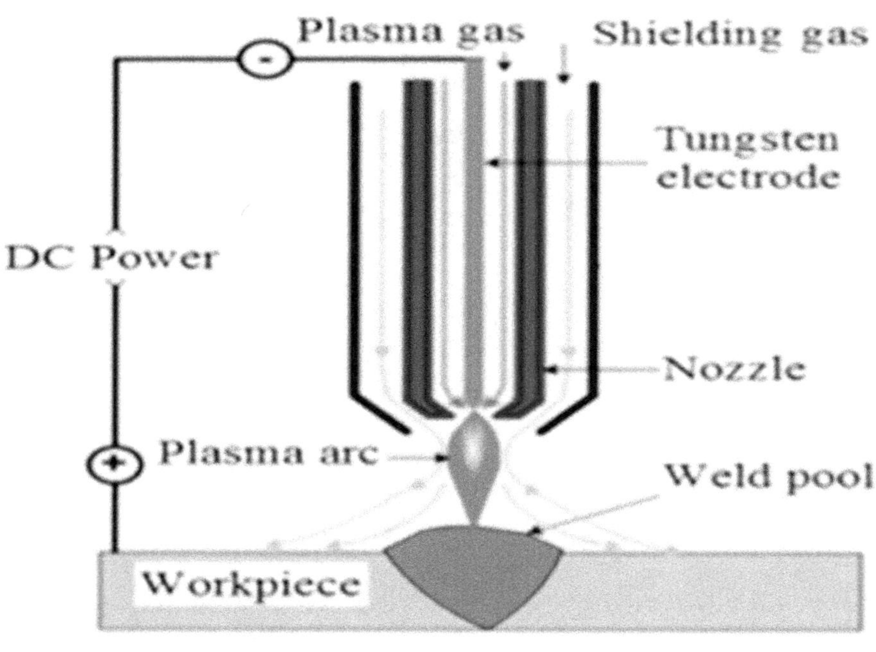

PRACTICAL APPLICATIONS OF WELDING AS COATING TECHNIQUE

Mechanical components and tools wear regularly as a result of corrosion, excess friction, and fatigue. Replacing such components once they are damaged by new ones can be quite expensive. However, by the use of repair welding, these components and tools can be repaired at a lower cost than purchasing new ones. Repair welding has advantages including prolonging the life cycle of a tool while at the same time minimizing downtime caused by replacement delays and saving the money that could have been used to purchase new tools.

Several researchers have studied the application of welding as a coating technique in repairing of engineering components. Tanasković et al., 2018, studied the repair of steel burner pipe. The pipe had been exposed to high temperatures up to 850°C. As a result cracks had occurred on it. The process of repairing the pipe involved removing the worn out part and replacing it with a new material that was joined to the burner using manual arc welding An alloy of stainless steel (steel X12NiCrSi35-16) was used as the base material. The quality of the joint welded was affected by the filler material quality, and the activities performed during welding. Proper selection of the filler material was important to prevent defects such as cracks that do occur on the welded joints as a result of moisture and impurities present. In this process, Castolin 6825 was used as the filler material. Castolin 6825 is an electrode that is used to join duplex stainless steels, nickel alloys, steels with 9% nickel among other materials. It has outstanding resistance to pitting, stress corrosion cracking, intergranular corrosion, and crevice. To repair the damaged burner pipe a current of 70 A and a voltage ranging between 20-22V were used. Short welds were performed on the joint to ensure that the material was not overheated to temperatures above 100°C. After welding, the penetrant method was used to test the weld obtained, and there were no defects detected. Afterward, the pipe was exploited, and no problems occurred. Therefore, the use of welding as a repair technique was justified. Also, the repair of the burner pipe resulted in financial savings. The money that could have been used to purchase anew pipe was saved, also the downtime that could have been experienced while waiting for a new burner pipe to be manufactured was avoided.

In another study, the super TIG welding process was used to carry out butt welding of steel material (9% nickel steel). The electrode used was alloy 625 with a cross-section which was C-type. The samples welded were found to be free of defects such as porosities and cracks. By using an optical microscope, it was found that welding the steel material improved its mechanical properties (Park et al., 2019).

Welding as a coating method is also used to repair worn-out gear shafts. In another study conducted by, Tanasković et al., 2017, eight gear shafts made of steel 42CrMo4, were repaired using welding. These shafts had been exposed to variable loads at high temperatures. As a result, the gears experienced surface fatigue which led to lateral damage of the teeth. Two methods were used to repair the damaged gear shafts; flux core arc welding (FCAW) and manual arc welding. FCAW is a semi-automatic or fully automatic arc welding. For this process, WLDC 3 of diameter 3.2mm wire was used as an electrode. The current used for welding, in this case, was 800A. Two electrodes were used for manual arc welding including, Piva 29/9 R of diameter 3.25 mm which was used to apply an intermediate layers and Piva 430 B R of diameter 5mm which was used to finish the layers. The wire WLDC 3 and the Piva 430 B R were located at the adhesion resistant areas of the shafts, which corresponded partially to the welded layers requirements. The welding current used for manual arc welding was varied from 190 to 230A. The shafts repaired were tampered to minimize residue stresses. In this case, the shafts were heated to a temperature of 620°C a rate of 50°C per hour, and this temperature was maintained for 3hours. The shafts were cooled at a rate of 50°C per hour until a temperature of 150°C was achieved after which they were allowed to continue cooling at room temperature. The hardness of the welded shafts was measured, and found to be 290-310 HB and matched the hardness of the shafts before repair. One shaft repaired using FCAW recorded hardness of 320-340 HB. It is reported that the gear shafts repaired have been performing well without failure. This is a confirmation that repair welding is an excellent technical solution for components repair.

In a study conducted by Sadeq et al., 2020, the repair of worn-out low carbon steel shaft using discontinuous arc welding was investigated. In this study, the base material used was steel st37. The type of electrode used was OK 43.32 of diameter 2 mm and length 350 mm. The welding process was carried out in a parallel and sequential manner at time intervals of 10 to 15 minutes to make sure that the temperature surrounding the welded section did not exceed 1200°C for a given continuous period. The welded shaft was then ground and polished before its microstructure was studied. Other tests carried out were hardness and tensile strength tests. The results revealed that the microstructure of the shaft welded contained ferrite and pearlite batches, indicating that ferrite was not formed inside pearlite grains. Use of discontinuous welding to repair damaged steel shaft increased surface hardness. However, elongation and surface strength of the shafts were not affected.

Milovanović et al., 2017, conducted a study to explore repair of a boiler bottom panel. The damage occurred due to the operation condition in which the boiler was exposed to. For instance, the boiler was subjected to sulphuric acid and low temperature which led to corrosion. As a result, there was thinning of the boiler walls at the bottom panel. The repair process involved replacing the damaged parts through welding. The welding techniques used in this case were gas welding combined with Metal Active Gas (MAG) welding. The electrode used was VAC 60 of radius 1.2 mm. The welding current used was 135-150A with a voltage of 23V. From the metallurgical analysis of the welded parts, the microstructure of the welded surfaces complied with the state required for a boiler material. The boilers repaired complied with the safety requirements for a boiler to operate.

In a second study carried by Tanasković, et al., 2018, a damaged toothed shaft of roller was repaired through welding. The shaft had fractured at the bearing position as a result of continuous exploitation. The base material was made of Steel 42CrMo4 and the filler material was steel 30CrMoV9. Manual arc welding was used to repair the damaged parts. The repaired shaft worked effectively. It is reported that the use of welding to repair the toothed shaft led to significant indirect and direct savings compared to the purchase of a new toothed shaft.

FUTURE PERSPECTIVE

With the advancement in technology welding techniques are now automated to perform additive manufacturing. Additive manufacturing or rapid prototyping is a method of surface engineering metallic materials. It refers to a process that derives information from computer-aided designs and transform them into stereolithography files. In this case, computer drawings are approximated by sliced and triangles containing the information related to every layer that will be printed. (Wong & Hernandez, 2012). Addictive manufacturing can establish complex shapes in three-dimension in a process that involves the deposition of materials layer by layer. This ability has led to a breakthrough in the manufacturing processes of parts (Rodrigues et al., 2019).

Wire arc additive manufacturing (WAAM) processes apply arc welding techniques to melt a wire that is automatically fed. The melted wire solidifies on the substrate material to firm layers. This process is controlled by a computer. Arc welding techniques including MIG), TIG, and PAW are used in additive manufacturing (Santangelo et al., 2016).

Figure 6. MIG based additive manufacturing (Bhakthavatchalam et al., 2018).

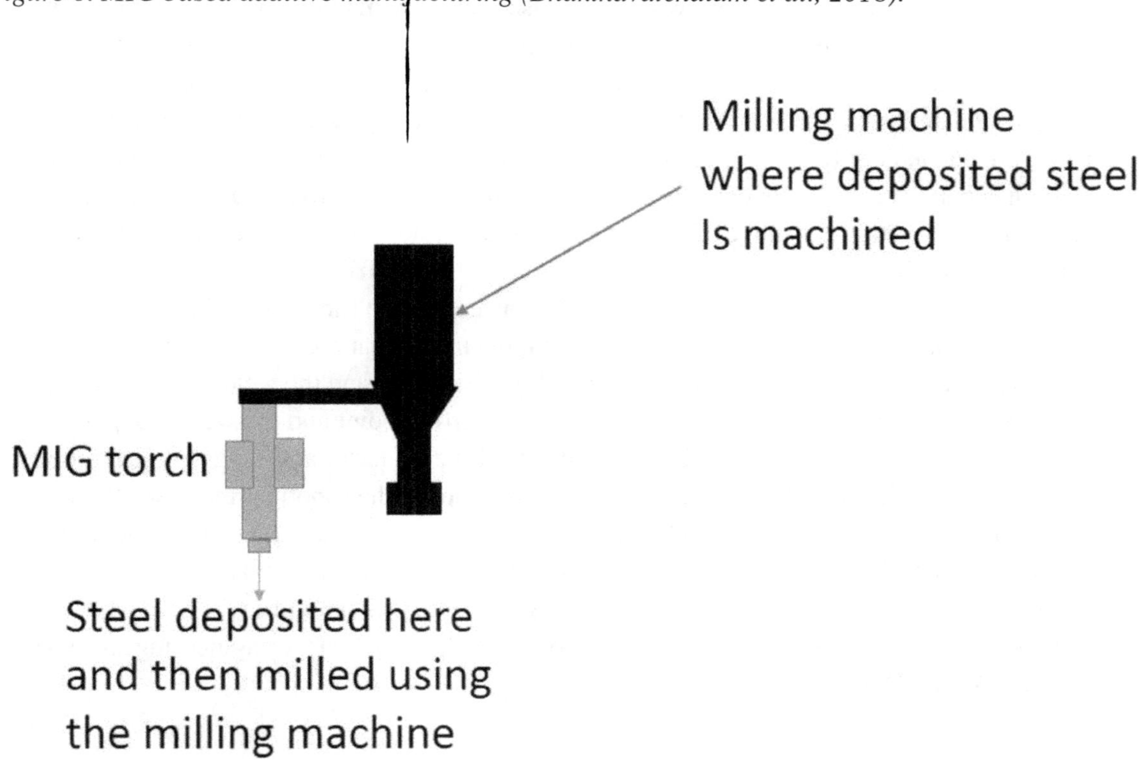

Figure 7. TIG additive manufacturing (Santangelo et al., 2016)

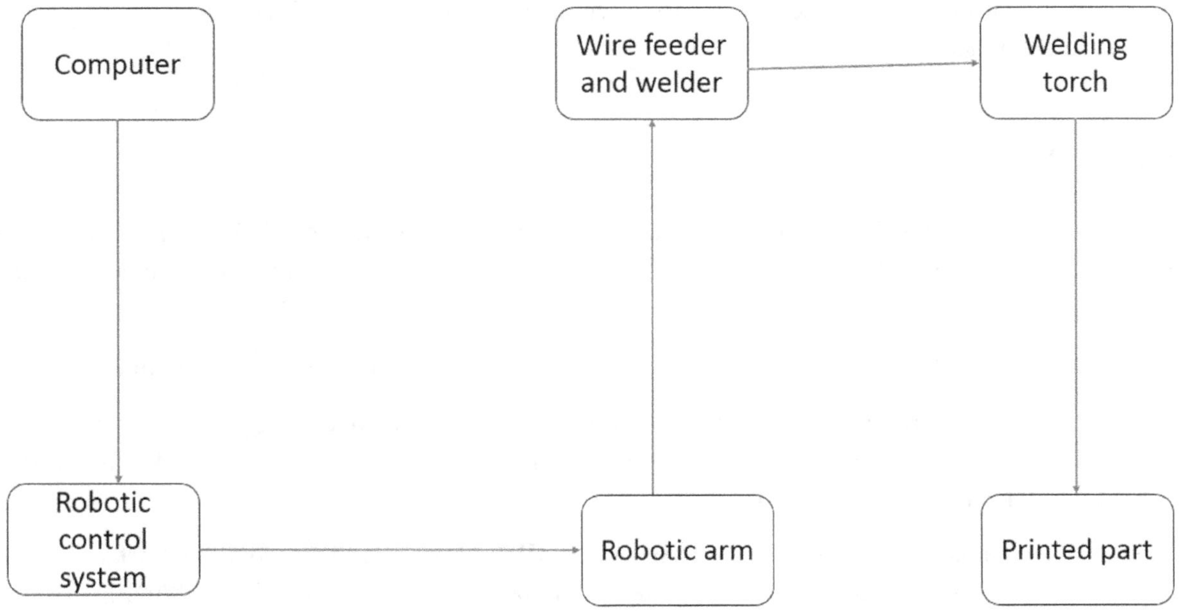

Researchers have paired CNC milling with MIG based additive manufacturing to come up with a complete additive manufacturing system. In this system, a layer of molten wire is deposited across a specified geometric area by use of MIG and then the layer is face milled. The right parameters for this process are selected and optimized before the deposition of material is carried out. Later surface finishing processes are carried out to achieve the final dimensions desired. These processes usually produce parts with high accuracy, acceptable density as well as minimum surface roughness. Using MIG in additive manufacturing has the advantage of ensuring high rates of deposition as well as excellent ease of implementation (Bhakthavatchalam et al., 2018).

In TIG additive manufacturing a computer is used to convert a CAD file to G & M codes. The control program created is later converted into linear and joint movements that are read by a robotic arm. The machine codes program the stop and the start commands of the feeder and the welder. When the robotic arm interpolates the linear movements, it goes to the layer starting point and the welder applies a high frequency start creating an electrical arc between the substrate material and tungsten electrode. The wire feeding occurs via the arc where it liquefies to the weld pool. The robotic arm causes the welding touch to move and the deposition occurs. When deposition of the first layer is done, the robot re-orients the welding torch to another starting position and another layer is created (Santangelo et al., 2016). A schematic illustration of the TIG additive manufacturing process is shown in Fig. 7.

As technology continues to advance more applications of welding in surface engineering are expected to happen (Tanasković, Dordevic, et al., 2018).

CONCLUSION

The purpose of this chapter was to study the use of various welding techniques in surface engineering of metallic materials. It was revealed that two major welding methods are used in surface engineering since they have superior advantages compared to other welding methods. Such techniques are oxyacetylene welding and arc welding. Welding methods are used to repair gears, pipes, and other components. It can be concluded that welding as a method of surface engineering is evolving as it is seen as the emergence of additive manufacturing.

REFERENCES

Bhavsar, A. N., & Patel, V. A. (2016). Influence of process parameters of TIG welding process on mechanical properties of SS304L welded joint. *International Research Journal of Engineering and Technology,* 977-984.

Davis, J. R. (Ed.). (2001). *Surface engineering for corrosion and wear resistance.* ASM international.

Ding, D., Pan, Z., Cuiuri, D., & Li, H. (2015). Wire-feed additive manufacturing of metal components: Technologies, developments and future interests. *International Journal of Advanced Manufacturing Technology, 81*(1-4), 465–481. doi:10.100700170-015-7077-3

Fotovvati, B., Namdari, N., & Dehghanghadikolaei, A. (2019). On Coating Techniques for Surface Protection: A Review. *Journal of Manufacturing and Materials Processing, 3*(1), 28. doi:10.3390/jmmp3010028

Gebert, A., & Bouaifi, B. (2006). Surface Protection by Means of Build-up Welding. *Modern Surface Technology*, 263-296.

Hazledine, F. (2019). Oxy-Acetylene Welding. *Proceedings - Institution of Mechanical Engineers*, *96*(1), 99–189. doi:10.1243/PIME_PROC_1919_096_010_02

Holmberg, K., & Matthews, A. (2009). *Coatings tribology: properties, mechanisms, techniques and applications in surface engineering*. Elsevier.

Kennedy, D., Xue, Y., & Mihaylova, E. (2005). *Current and future applications of surface engineering*. Academic Press.

Kutz, M. (Ed.). (2011). *Applied plastics engineering handbook: processing and materials*. William Andrew.

Martina, F., Mehnen, J., Williams, S. W., Colegrove, P., & Wang, F. (2012). Investigation of the benefits of plasma deposition for the additive layer manufacture of Ti–6Al–4V. *Journal of Materials Processing Technology*, *212*(6), 1377–1386. doi:10.1016/j.jmatprotec.2012.02.002

Milovanović, N., Dordević, B., Tatić, U., Sedmak, S., & Štrbački, S. (2017). Low-temperature corrosion damage and repair of boiler bottom panel tubes. *Structural Integrity and Life*, *17*(2), 125–131.

Mishra, S. K. (2002). *Advanced Techniques for Surface Engineering of Industrial Materials*. http://eprints.nmlindia.org/4215/1/128-140-S_K_Mishra.PDF

Mohan, P. (2014). *Study the effects of welding parameters on TIG welding of Aluminium plate* (Doctoral dissertation).

Pan, Z., Ding, D., Wu, B., Cuiuri, D., Li, H., & Norrish, J. (2018). Arc welding processes for additive manufacturing: a review. In *Transactions on Intelligent Welding Manufacturing* (pp. 3–24). Springer. doi:10.1007/978-981-10-5355-9_1

Park, J. H., Kim, Y. H., Baek, H. J., & Cho, S. M. (2019). A study on process development of super-TIG welding for 9% nickel steel with Alloy 625. *Journal of Manufacturing Processes*, *40*(February), 140–148. doi:10.1016/j.jmapro.2019.03.017

Patil, S. R., & Waghmare, C. A. (2013). Optimization of MIG welding parameters for improving the strength of welded joints. *Int. J. Adv. Engg. Res. Studies, 14*, 16.

Sadeq, B. R., Sahib, B. S., & Alher, M. (2020). Study of the Effects of Welding Process on the Microstructure of Worn Carbon Steel Shaft. *IOP Conference Series. Materials Science and Engineering*, *671*(1), 012161. Advance online publication. doi:10.1088/1757-899X/671/1/012161

Santangelo, M., Silwal, B., & Purdy, A. (2016). *Vibration assisted robotic hot-wire gas tungsten arc welding (GTAW) for additive manufacturing of large metallic parts*. Solid Freeform Fabrication.

Tanasković, D., Dordevic, B., Gajin, M., Arandelović, M., & Gostović, N. (2018, January). Damages of burner pipes due to the working conditions and its repair welding. *Procedia Structural Integrity*, *13*, 404–409. doi:10.1016/j.prostr.2018.12.067

Tanasković, D., Đorđević, B., Sedmak, S., & Mihajlo, A. (2018). The effect of exploitation conditions on the damage of a roller reducer toothed shaft and its repair. *Machine Design*, *10*(4), 157–162. doi:0.24867/MD.10.2018.4.157-162

Tanasković, D., Đorđević, B., Tatić, U., Sedmak, A., & Opačić, M. (2017). Repair Welding of Gear Shafts of Service Rollers at the Železara Smederevo. *8th International Scientific Conference IRMES: Machine Elements and Systems in Energy Sector Development of Power Production Systems,* 365–370.

Ugla, A. A., Khaudair, H. J., & Almusawi, A. R. (2019). Metal Inert Gas Welding-Based-Shaped Metal Deposition in Additive Layered Manufacturing: A Review. *International Journal of Mechanical and Materials Engineering*, *13*(3), 244–257.

Chapter 12
Surface Engineering for Coating: A Smart Technique

Partha Pratim Bag
https://orcid.org/0000-0002-0012-3137
Department of Chemistry, School of Basic Sciences, SRM University, India

Gourisankar Roymahapatra
https://orcid.org/0000-0002-8018-5206
School of Applied Science and Humanities, Haldia Institute of Technology, India

ABSTRACT

Surface engineering includes augmentation of intrinsic properties of the boundary of the component, which isolates the continuum from surroundings known as the surface. Two main purposes of surface engineering encapsulate primarily the hardness of the surface for enhanced wear resistance and also to poise up with inter-surface frictional behavior. Today, there are many different surface engineering techniques available: starting from vacuum to atmospheric pressure, wet to dry, simple to sophisticated, and low-cost to high-cost to obtain the required purposeful distinctiveness of material. Most methods used today are dry and thus environmentally sound. This chapter describes various types of coatings over materials to get an overall idea of the technique keeping prime focus on graduate and undergraduate students.

1. INTRODUCTION:

Surface engineering for coating is a smart technique. Surface engineering is a high-throughput technology encompasses the physical (macro) and chemical (micro) aspects for better electrical, magnetic, electronics, thermal and optical properties. Surface being a membrane plays a crucial role pertinent to permeability factors like diffusion, osmosis, reverse osmosis, absorption and adsorption. Surface coating technology is an interlayer technology which eventually protects the substances without affecting the properties of

DOI: 10.4018/978-1-7998-4870-7.ch012

neighboring layers. To ascertain a good tribological response, surface integrity plays a great role when it is substrated with the concepts of surface engineering. Surfaces which gets germinated owing to the proper directive of surface engineering, deals both the interfaces of the surface with the environment and also with the substrate. It is well-known from long back in different parts of society. The prima facie objective of surface coating is to amend and emphasize more on the surface functions rather to reform the composition of the material of concern. Literature survey reveals the (Pfeifer, 2009) mention of thin film coating technology as adopted by artists and craftsmen more than two thousand years ago without any knowledge about the chemico-physical processes involved in it. Fire gilding, silvering are the longstanding mercury-based processes used to coat surface of less precious substrates with thin layers of gold or silver. Ancient civilization systematically manipulated these metals to create functional and decorative artistic objects (Maria Ingo et al., 2013). Even before, in different ancient civilization, surface coating techniques found in other form such as paints prepared from iron oxide, chalk or charcoal used either with the fingertips or with the tips of soft twigs of brushes. People also used to paint the rock walls of their caves with pictures of their pets, other animals and their family fellows (Gooch, 2002).

Later they started to use natural resins and wax for coating. Artists used lacquers on dried oils to protect their paintings. Interestingly ancient Egyptian scientists developed a fine coating technology which is similar with nanotechnology. Thereafter several theories tried to establish nanotechnology as a re-innovated technology. At the present time, there are several thousand of coating techniques applied to simple systems based on one or two coating steps to sophisticated systems based on multi layers and complicated instruments. Conversely, most are not environment friendly and, in many cases, can't fulfill the demands of the manufacturing industries or of society. This is the driving force to conduct an extensive research and development in coatings science and surface technology.

2. BACKGROUND AND MAIN THEME

From the starting of our civilization, people made many things to fulfill their daily requirements. Those will be stronger and harder if made by hard materials. For specific applications mechanical parts and structures are designed. It is very necessary to choose some extensive material rather than fabrication of these parts. These limitations consist of body materials, mechanical behavior (tension, compression, yield, torsion, fatigue, bending, creep etc.), desired functionality (friction, hydrophobicity, wear resistance etc.), thermal behavior (expansion and thermal conductivity), electrical conductivity, dynamic load bearing (vibrations, high-speed rotation etc.), and corrosion resistance. Some important aspects also to be kept in mind; such as availability, cost, safety, and toxicity of materials. All these factors play an important role to manufacture mechanical parts and structures. Selection of materials is the important key to provide all protection purposes. To protect all the parts, a successful coating technique need to chose, including metals, ceramics, and polymers, that can form a protective layer (DeMasi-Marcin & Gupta, 1994). However, coating processes and material properties may generate complexity to choose the best composition of the deposited layer. Even though coating processes are obliging to afford the profits mentioned above, still they undergo some disadvantages which worsen their trustworthiness. Among these, negative thermal effects, disparaging thermal effects (distortion, crack, delamination, etc.), damaging effects of atmospheric protection (penetration of inclusions and contaminations into the substrate) and coating materials' properties (melting point, availability in different forms of foil, powder, rods, biocompatibility, etc.) are the most vital ones to be considered. (DeMasi-Marcin & Gupta, 1994;

Thakare et al., 2007). Thus, in this point of view a good knowledge of coating process is very crucial. In this chapter we recapitulate different techniques employed for the application of different coating in liquids, gases or solids onto different substrate. Also, we will discuss recent development of coating process for surface engineering.

3. DIFFERENT COATING METHODS

Generally, the purpose of coating processes is to protect the surface of explicit part of a structure when it exposed to environments in different fields. So, it is very important for different industry to tiny bio-medical devices even inside the human body.

3.1 Electro and Electroless Chemical Plating

These chemical processes of plating are most common, cost-effective and simple desirable techniques for applying a metallic coating to a substrate to modify the surface properties of the materials to improve corrosion and wear resistance, electrical conductivity or decorative appearance. In both cases, surface is coated by reduced its metallic form a metal salt in solution (Gu et al., 2006).

3.1.1 Electrochemical Plating

Electroplating is the process of applying a metal coating on another piece of metal (or another conductive surface) through an electrochemical process. The process is quite similar to the electroforming process; both are a form of additive manufacturing, and both work through an electro-deposition process. In electroplating, the deposited metal becomes part of the existing product with the plating/coating. In this process, metal to be plated is taken as anode and metal to surface to be coated is taken as cathode, and both an anode and a cathode are immersed in an electrolytic tub that is composed of a solution of salts. A direct current (DC) of electricity is passed through the solution, resulting the transfer of metal ions from anode onto the cathodic surface, plating the metal onto the item. In this process most challenging is to choose appropriate chemical surface because it needs to prevent oxide formation during the whole process (Gray & Luan, 2002). Generally, the electrons for reduction are supplied from an external source. High reacting materials can quickly form an oxide layer when exposed to air and this layer must be removed prior to plating (Hajdu et al., 1990; Innes, 1974; Luo & Cui, 1998; Sharma et al., 1997). Another important issue is that the quality of the final coating. In the plating bath, non-homogeneous distribution of current density results non-uniform coatings which is again further problem with this technique. It is significantly an expensive process as it needs a large amount of electricity. Now days, in most of the cases electroforming is used in place of electroplating, especially in case of micro material manufacturing (Hernández et al., 2015). In electroforming, as in plating, metal ions are transferred electrochemically through an electrolyte from an anode to a surface where they are deposited as atoms of plated metal. But in electroforming, the surface that is to receive the plated metal, called a mandrel, is conditioned so that the plating does not adhere. Instead, the plated metal, or electroform, is lifted away and retains its as-deposited shape as a discrete component. The process id used to add a glaze to improve corrosion-resistance or to get a smarter product. So rather than the two methods being used independently, they can actually be used in a cooperative manner. A micro product formed by this process has several remarkable characteristics. It can have extremely thin walls (< 1 mm). The thinnest thickness is generally limited only

Figure 1. Electroplating of a metal (M) with copper in a copper sulfate bath

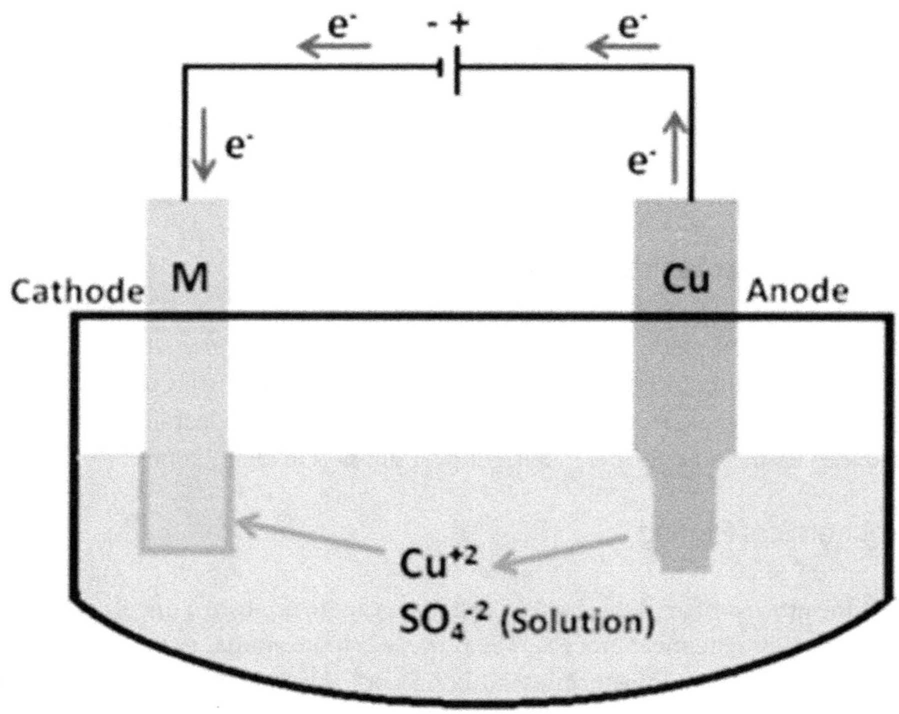

by the fact that the micro product requires a certain amount of durability to avoid being bent or broken by normal handling. Surface cover texture of the mandrel is reproduced with extreme reliability with electroforming process. High surface finish and convoluted detail are easily obtained. Complex contours are produced quite easily and dimensional tolerances can be held to high accuracy (+/- 0.0001 mm). The most interesting aspect of electroforming is the ease of forming complex shapes. The mandrel and electroform bear the same geometric relationship that exists between a mold and a cast metal.

3.1.2 Electroless Chemical Plating

Electroless plating which is also a chemical and autocatalytic plating, involve a non-galvanic plating method. It involves several synchronized reactions in an aqueous solution, without the use of external electrical power (Montazer & Harifi, 2018). It is the plating of metals such as silver, aluminum, copper, nickel, and iron with a uniform coating of metallic layer on the surface of fibers through chemical reduction of metal ions in an aqueous solution and the subsequent deposition of metal without the use of electrical energy. In this process electrons are supplied by a chemical reducing agent in solution or from the material itself. It is quite advantageous than electroplating for complex shapes for anticorrosive coating (Chen et al., 1998). Another advantage is second-phase particle co-deposition during the plating process which improve some properties such as wear resistance, hardness or abrasion, and also good for soldering and brazing applications (Chen et al., 1998; Innes, 1974; Montazer & Harifi, 2018; Sharma et al., 1997). The term electroless plating was first described by Von Liebig in 1835 with the reduction of silver salts by reducing aldehydes. Then the technique was followed by several scientists in coming years. Wurtz, in 1988, used this technique with metallic nickel deposition in an aqueous solution in the existence of hypophosphite (reducing agent) followed by Breteau in 1911. Development in this field not received much sponsorship until World War II. After initiation of electroless nickel deposit, rest of electroless coatings was followed. In 1970, semiconductors and circuit boards were plated with pure gold, at Bell Laboratories (Ali & Christie, 1984). Narcus first reported the electroless copper deposition, and its profitable application was explained by Cahill and Zeblisky et al. (Zeblisky et al., 1963). Baum et al. developed the technique for selective deposition of copper (Cu) by using palladium (Pd) seeds in 1986. Palladium and alloys of palladium been acknowledged as an economic alternative to applications involving gold plating. Electroless cobalt discovered along with nickel deposition by Brenner and Riddell. The commercial application of electroless plating of silver (Ag), aluminum (Al), platinum (Pt), ruthenium (Ru), and rhodium (Rh) were developed. The field of electroless coating chemistry has materialized as one of the promising fields of the surface engineered metal finishing technique in 1984 onwards, when materials like plastics, ceramics, polymers and other non-conducting materials coated with electroless technique before even being subjected to electroplating (Sudagar et al., 2017).

3.2 Conversion Coatings

Conversion coating is a special type used on metal body in which the metal surface is turned into a coating with chemical or electrochemical processes. In addition to anticorrosive property, conversion coating also tenders improved surface hardness. It also can be used as a decorative primer. Some of the best examples of this coating include chromate and phosphate coating as well as black oxide. It is normally applied on alloys like aluminum through chromate conversion or anodizing. The process is widely used due to low-cost involvement. By this process metal substrate is coated by insulating protective barrier in the form of metal oxides such as chromate, vanadate, between the metal surface and the environment which prevent from corrosion.

3.2.1 Chromate/chromate-free Conversion Coating

It is one of the most common coating processes and applied to get better corrosion protection performance of many metals and their alloys. It is widely used due to its self-healing nature of the coating, user friendly application with low cost, the high efficiency and electric conductivity. Additionally, it provides a great under-film corrosion resistance which facilitates further finishing treatment. However, hexavalent chromate is ban for coating processes due to its carcinogenic effect.

There is another method chrome-free conversion coating where non-chromate salts such as cerate, stannate, vanadate, zirconate etc. have been developed. It has strong coating adhesion and can act as a barrier by covalent bonding for coating, limiting (Chen et al., 1998; Hamdy, 2008; Hamdy & Butt, 2007; Hamdy & Farahat, 2010).

3.2.2 Anodizing

Anodizing is in reverse of electroplating. To make a thick oxide layer on the metals and alloys surface, an electrolytic process namely anodizing process is employed to improve corrosion resistance and paint adhesion to the substrate (Mittal, 1995). In anodizing (electrochemical conversion coating), the surface of certain aluminum, magnesium, titanium, and zinc alloys convert to an oxide of the metal. This process involves immersion of a component in an acid solution and passing electricity through the component, with the component being the anode in the electrical circuit with an acid electrolyte and a cathode (Pb is used in general). The temperature of the solution is controlled to give the desired properties. The process is illustrated in Figure-2. This process also includes the mechanical treatment; degreasing, cleaning and pickling; electropolishing; anodizing using AC or DC current; dyeing or post-treatment; and sealing (Mittal, 1995). By this process with a thin barrier layer at the metal–coating interface, a relatively thick layer of a cellular structure also added. Each cell contains a pore the size of which is determined by the. The quality of the anodized film depends on pore size of cell, type of electrolyte and the experimental conditions (Hamdy & Butt, 2007). The anodized film are formed by a brittle ceramic material, so appropriate mechanical properties does not fulfill the requirements of some industrial applications.

Figure 2. Annodizing of metal body (M = Al) in acid electrolyte (H_2SO_4) bath

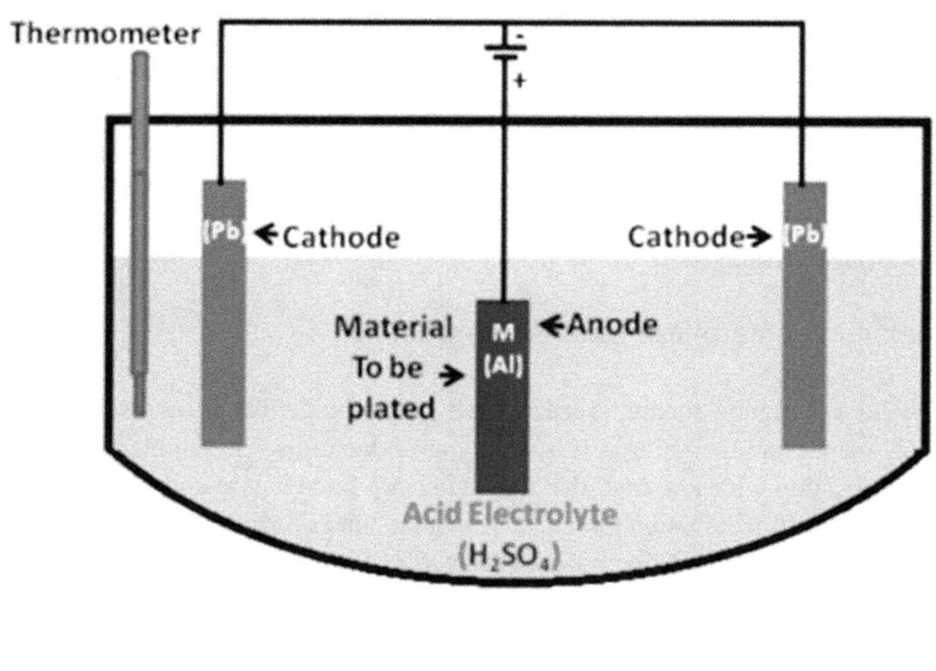

3.3 Chemical and Physical Vapor Deposition (CVD and PVD)

3.3.1 Chemical Vapor Deposition

The most commonly used coating process is chemical vapor deposition (CVD). In 1993, CVD technique was first reported by Endo and his research group. Three years later in 1996, Smalley's group successfully adapted Carbon monoxide-based CVD technique. It is applicable to any metallic or ceramic compound, including elements, metals and their alloys and intermetallic compounds. The CVD processes have been done by the deposition of a gaseous phase onto solid material. The phenomena of this deposition involve a chemical reaction between volatile precursors and hot materials surface to be coated. So the temperature of the material surface is most crucial and can persuade different reactions. Example of different types of CVD are atmospheric pressure chemical vapor deposition, metal-organic chemical vapor deposition, low pressure chemical vapor deposition, laser chemical vapor deposition, photochemical vapor deposition, chemical vapor infiltration etc. (Dai et al., 1996; Endo et al., 1993).The process is illustrated in Figure-3

Figure 3. Schematic diagram of the chemical vapor deposition apparatus.

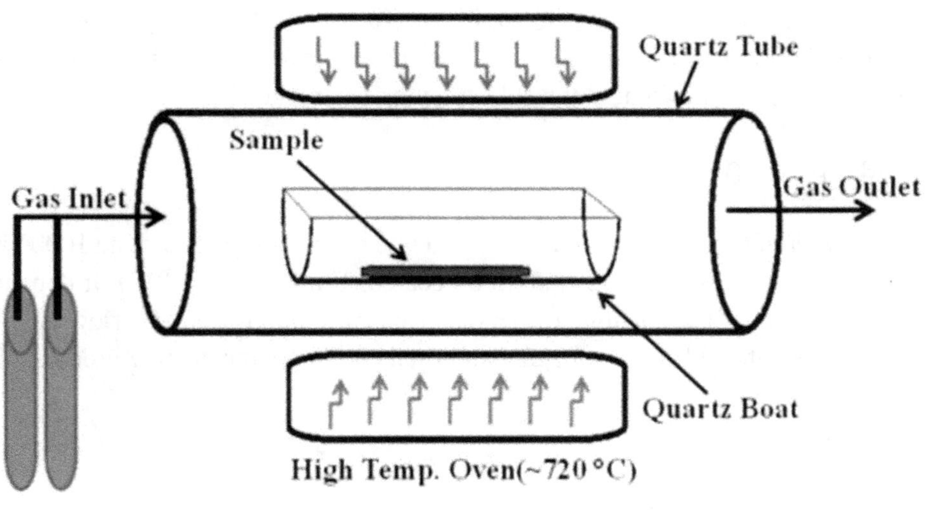

3.3.2 Physical Vapor Deposition

The process is similar to CVD, except the deposited start out in solid form. It is the technique where transfer of material on an atomic level under vacuum conditions. At first, atoms from the surface (which is chosen for coating purpose) evaporate by an electron beam or ions (high energy source) and it coated at the substrate surface. Advantages of PVD are enhancing the properties of original material, environmentally friendly. But main drawback is the expensive process but low output and complicated process.

3.4 Spray Coating

Spray coating process involves molten / soften particles, coated onto a substrate. It is mainly used in industry. Some common spray coating techniques are explained below.

3.4.1 Thermal Spraying

In the thermal spraying process, particles of 1–50 µm are partially melted coating materials are sprayed on the substrate in accelerated high velocities by a flame or an arc. The quality of coating is determined by the oxide content, porosity and adhesion to the substrate. The main advantages of this technique is thick coatings ranging from 15 µm to a few mm. (Kasten et al., 2001). Another modified version of the thermal spray technique is High-velocity oxy-fuel spraying (HVOF). In this technique, a mixture of gaseous fuel such as hydrogen, natural gas, methane, propane, or liquid fuels such as kerosene in presence of oxygen is fed into a combustion chamber for reaction under ignition (Gadow et al., 2010; Paulussen et al., 2005). Cold spraying coating is similar to the HVOF spraying technique, but particles are accelerated to very high speeds by the carrier gas through a nozzle by which solid particles deform plastically and bound mechanically to the substrate. The velocity depends on the properties of the material, powder size and temperature. The advantage of this technique is the possibility to use of soft material for coating and disadvantage is low deposition efficiency (Paulussen et al., 2005). Again, warm spraying is the modification of HVOF where temperature of the combustion gas is lowered by mixing it with nitrogen, which is similar to the principle behind the cold spraying technique.

3.4.2 Plasma Spraying

In this process powder of the coating materials are fed into the plasma jet at around 10000 K, and melted materials and are sprayed over the substrate to be coated (Emran et al., 2018). It depends on several factors such as the nature of the coating powders, plasma gas composition, gas flow rate, energy input, distance between substrate and source of spray and final cooling parameters (Kuroda et al., 2008).

Figure 4. Schematic diagram of Plasma spraying

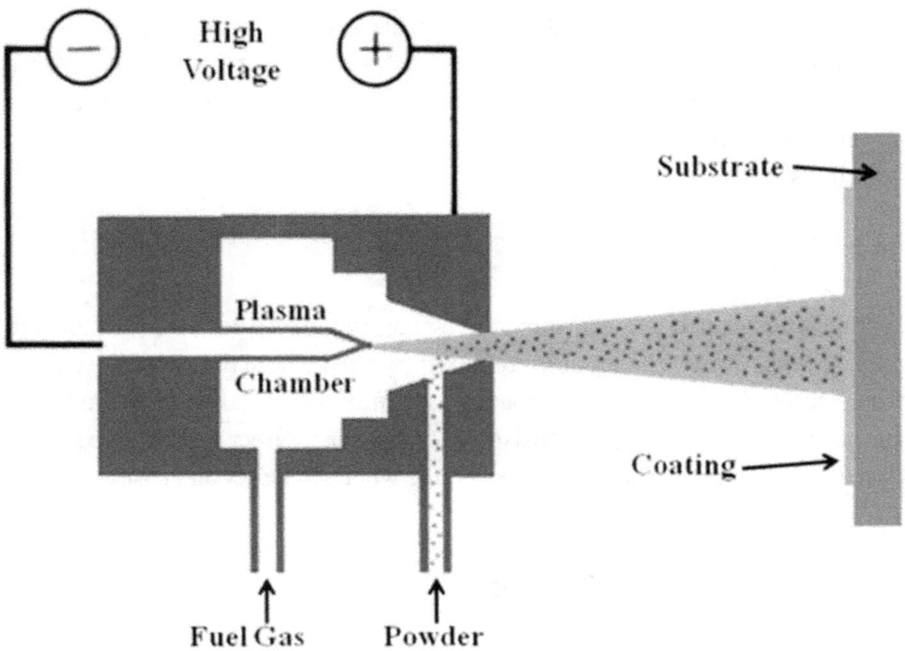

3.4.3 Vacuum Plasma Spraying

It is similar to plasma spray technique except relatively low operating temperatures (40–120°C). So, thermal damage can be avoided. The main advantage is conduction of non-thermally-activated surface reactions, which is not possible in molecular chemistries at atmospheric pressure.

4.5 Other Coating Techniques

Beside these well-known techniques there are several other techniques which are also advantageous as like other. These are discussed below.

4.5.1 Sol–gel Coatings

This is chemical tailorable, environmentally friendly methods which form long lasting coating for many ferrous and nonferrous alloys (Makhlouf & Gajarla, 2019). Poor adhesion performance and lack of high-performance are the main drawback of this coating method. But use of silica, ceria, vanadia and molybdate can improve the coating by making covalent bond. This can make strong coating adhesion and limiting water transport-induced attack on the surface (Hamdy, 2006a; Hamdy, 2006b; Hamdy, 2006c; Hamdy, 2009; Hamdy, 2010; Hamdy & Butt, 2006a; Hamdy & Butt, 2006b; Hamdy & Butt, 2007; Plasma Spraying Process, 2010).

4.5.2 Spin Coating

This process is known for last several decades for thin films coating. In this process, at first a small drop of the coating material is dropped on the centre of a substrate and spun around an axis perpendicular to the coating area, at a controlled high speed. Due to high spin the coating material spreads over substrate surface to form thin film of coating. The thickness and other properties will depend on the nature of the coating and spinning speed (Ching, 2015).

4.5.3 Roll-to-roll Coating

Roll-to-roll coating is performed to a flat substrate by passing it between two (or more) rollers. At first the coating material passes through one or more auxiliary rolls where the gap adjusted between the upper roller and the second roller. After final coating, substrate is then shaped to the final form without affecting the properties of the coating. It consist two techniques; direct and reverse roll coating. In first case, the rotation direction of roll and substrate are same whereas the second case is just opposite (Malik et al., 2009).

4.5.4 Slot/die and slot/extrusion Coating

In this process, the surface of substrate coating is compressed by gravity or external applied pressure through a slot. The process can be termed as 'extrusion' if the composed coating is entirely of solids. The speeds of the extrusion can control the thickness of coating (Malik et al., 2009).

4.5.5 Dip Coating

In this process, the substrate is dipped into a coating material bath of known viscosity. The speed and dipping time are carefully controlled at last the substrate take out from the bath and allowed to drain. The quality and thickness of coating is dependent on the viscosity of the coating and it kept always constant during whole process.

5. NEW LIGHTWEIGHT MATERIALS

Interest in lightweight materials is increasing both in industry and in research circles in course of time, as they can be used to reduce weight and save energy. Light metals such as aluminum (Al) titanium (Ti) have also been generally used to manufacture hard coatings materials such as physically vapor deposited (PVD) TiN or AlN, and chemically vapor deposited (CVD) Al_2O_3 (Korhonen & Harju, 2012). Combination of plasma-aided coating and diffusion processes developed a new term 'duplex treatment', consisting of plasma nitriding and subsequent hard coating. To reduce the degassing and make the cleaning of the surfaces easier inside vacuum pump, another intriguing innovation is TiN coating on aluminum parts. Including many profitable side of light alloys, their surface properties sometimes cause problems. (Korhonen & Harju, 2012) For example, galling or annoying may be a severe problem with titanium parts, and plasma nitriding is an immediate and solution to combat it, though due to adherent oxide scale, plasma nitriding of aluminum is a difficult procedure (Technical Coating International Inc, 2010).

Magnesium alloys are other example lightweight materials to replace heavy alloys. It is now widely used in the automotive and aerospace industries. Main benefit is ensuing reduction of fuel consumption as well as CO_2 emissions. These alloys have an admirable property such as - a high strength-to-weight (S/W) ratio, low density, dimensional stability and castability. In spite of their outstanding mechanical properties, magnesium alloys remain very vulnerable to corrosion, but several coating schemes have been developed to improve the corrosion resistance (Korhonen & Harju, 2012).

5.1 Environmentally Friendly Coatings

Chrome plating or chromium plating), is a technique of electroplating can be decorative, provide corrosion resistance, ease cleaning procedures, or increase surface hardness. Industrial uses of chromium (VI) compounds include chromate pigments in dyes, paints, inks, and plastics. But, the workplace exposure to may cause; lung cancer, irritation or damage to the nose, throat, respiratory tract, irritation or damage to the eyes and skin if hexavalent chromium contacts these organs in high concentrations (Makhlouf, 2011). These health hazard problems influenced the demand for more environmentally friendly coatings. Researcher are in a continual investigation to find out chrome-free, and low in solvent or solvent-free coating materials resulting both properties, in terms of protection and acceptable appearance. Though the coating industry has a bad reputation for causing environmental damage during the manufacture, application and uses of products, but the collective efforts of global regulatory agencies and researchers have significantly improved the environmental paw marks for the industry over the last couple of decades. Significant reductions in hazardous air pollutants, releases of volatile organic compounds (VOCs) and hazardous waste have been realized. Issues of air pollution in China have prompted Chinese government to impose a VOC tax on all coatings that have more than 420 g/L VOC (OSHA, n.d.). Concerning cobalt driers in alkyd paints has impelled a hunt for environmentally alternatives (China Slaps VOC Tax on Coatings, n.d.). Bisphenol-A (BPA) has been banned in coatings for baby food and beverage and food containers or cans. N-methyl pyrrolidone (NMP) solvent has been aloof from coatings to certain automotive customers. Isothiazolone which is used as a preservative in coatings are under scrutiny. Even TiO_2 safety has been called into question (CEPE Guidance on Labelling Decorative Paint with Skin Sensitizing Biocides, n.d.; CEPE Information Bulletin About Cobalt Driers, n.d.; EU Directive Restricting the Use of Nonylphenol Ethoxylate and Nonylphenol Products, n.d.; Lydecker, 2015). Having a wonder for the coating industry; 'what could be the next !!' continued the drive for an alteration in coating chemistry. Environmental concerns have increased the requirement and attention to develop alternative sources for chemical feedstock. Chemistry plays a leading role in this context. The development of environmentally friendly coatings can be grouped into seven general categories: (i) Solvent content reduction; (ii) Elimination of toxic raw materials; (iii) Reducing curing temperature; (iv) Simplifying paint processes; (v) Increasing coating efficiency; (vi) Increasing lifetime (from corrosion, photo-oxidation etc.) of coating-enabled asset; (vii) Coating-enabled energy savings (lightweight substrates, drag reduction, heat dissipation, heat reflection)etc. Many coatings industries have forcefully been developing solutions away from the exemplary petrochemical feedstock. Bio-based technologies are flowering more and more for the coating industry. Resins and additives can be based on renewable resources and formulated into paints with regulations compliances. New binder resins with tailor-made properties and well-defined molecular structures, is a solution for molecular weight distribution to get excellent viscosity of the coating solution. More attention is paid to the calculated design of functionality in polymers to ensure greater uniformity of cross linking during curing, and to optimize film performance (Brock, 2005).

5.2 Self-assembling Molecules

Replacement of organic solvent-based coatings by water-based latex polymer coatings has significantly renovated the coating industry (Brock, 2005). It wiped out a huge environmental and health hazard. Considering the continuing demand for higher efficiency and lower costs, the coating industry has to uphold a sustainability standard (Kamp et al., 2020). To address the standard, polymer latex binders with diverse particle morphologies have been industrialized to improve coating performance. Colloidal self-assembly has been incorporated to help manufacturers to produce low cost best paint (Brock, 2005). Self-assembling molecules can organize themselves in a closely regular arrangement on a metal surface, and then polymerize. This strongly-affixed, flexible coating layer improves coating adhesion, corrosion protection, mechanical and chemical resistance. In this process biochemical gene modified natural raw materials are used paint binders, instead of oils or resins. Natural polymers such as cellulose or chitin can be modified to make them more acceptable in paint binders (Brock, 2005; Jiang et al., 2017; Kamp et al., 2020; Murata, 1996).

5.3 Self-cleaning Coatings

A new fascinating innovation is the self-cleaning coatings which are resistant to dirt, water and oil when applied to a surface. Micro-patterns on surfaces are known to resist water and dirt penetration, and this so-called 'lotus effect' is already used in self-cleaning roof tiles and sanitary ware. Bright surfaces on car bodies, however, do not seem to be a realistic goal: artificial lotus surfaces are not self-renewing, so their sensitivity to cleaning and other mechanical treatments presents a problem. Other surface effects such as shark skin and fur-like films might find outlets in the coating of plastics (EU Directive Restricting the Use of Nonylphenol Ethoxylate and Nonylphenol Products, n.d.). Because of the wide-ranging applications, starting from textile industry (self cleaning clothing) to automobile industry (self-cleaning wind shield glass, car bodies and mirrors), optical industry (cameras, sensors, lenses, telescopes) to aerospace industry (non sticky surfaces), solar modules (self-cleaning solar modules) to exterior paints with self-cleaning properties; this technology arriving a great deal of attention in late 20th century; and now copious research works are going on in this field to produce high proficient and long-lasting self-cleaning coating surfaces with enhanced optical qualities (Thakur, 2019). This technology offers a wide range of benefits, including low maintenance cost, elimination of tiresome manual effort, and above all reduction in cleaning time. This technology is mostly classified into two major categories; hydrophilic and hydrophobic and both related to the action of water. In a hydrophilic self-cleaning coating, the water is spread over the surfaces, which carries away the dirt and other impurities (sheeting), and in hydrophobic process, the water droplets skate-and-roll over the surface thereby cleaning it. Sometimes in hydrophilic coatings, suitable metal oxides are used to chemically destroy the complex-dirt by sunlight-assisted cleaning mechanism (Wang et al., 2002).

Figure 5. Schematic diagram of Self-cleaning coating

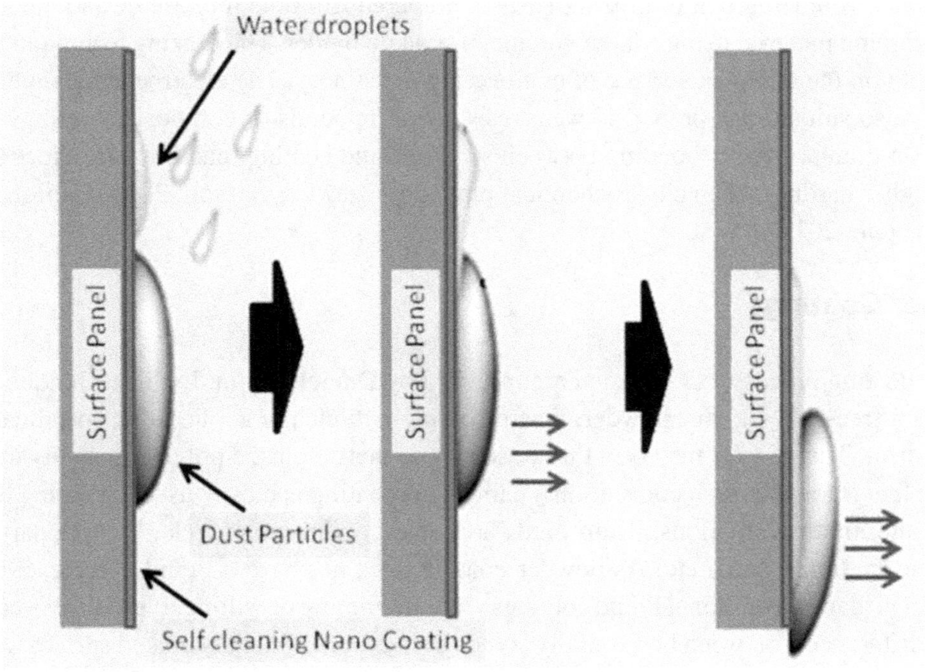

5.4 Composite Coatings

A composite coating is the combination of two or more substances that offer protection against corrosion. This special type of coating is mostly based on epoxy and polyurethane as well as resin. It is produced through a high-tech resin technology according to a particular sequence to meet the requirements of industries. It offers excellent corrosion protection and waterproofing, making it one of the top coating materials used by industries. Composite coating with refractory ceramic particles can result in good wear resistant properties, similar to CVD or PVD coatings. It can be done by four different techniques; (i) by Laser cladding, (ii) by Laser melt injection, (iii) brazing and (iv) coating from aqueous solutions by electrochemical and electroless methods. Laser cladding has been developed for its potentiality of introducing hard particles such as WC, TiC, SiC and Cr_3C_2 etc. as reinforcements. Each of these coposite have very high hardness and high wear resistance. Intensive research on Ni alloys is mode of interest in recent years due to its excellent resistance against abrasion and corrosion at higher temperatures. In Laser Melt Injection (LMI) technology. the surface of the substrate can be melted by laser beam (the depth of the pool is approximately 1 mm), and the solid particles (particle size 10 to 100 μm) can be delivered into this pool by inert gas. In the brazing process, high vacuum or an inert gas atmosphere is used and the whole sample is heated up. In laser technologies there are some difficulties in the preparing of the composite coating. One is the poor wettability between the melted metal and the reinforcing particles and another is the chemical reaction between the two phases. Both the poor wettability and chemical reaction can decrease the mechanical properties of the composite coatings. These problems can be overcome by reinforcing particles formation in-situ during the cooling of the melted metal. Because of rapid cooling of the melted part and the non-uniform temperature in laser method, mechanical stress may occur in the sample. In order to avoid that, the brazing technique is used where whole sample is heated up during the brazing process. A brazing belt or brazing paste is prepared from the mixture of brazing alloy powder and the reinforcing material using a large volume of organic binder. The brazing technique is applicable for the coating on the complex surface of engineering machinery. The electrochemical and electroless methods are also suitable for preparing wear-resistant compounds or composite coatings on complex surfaces, but it grant a weaker bonding between substrate and coating than brazing. Electroless plating may yield higher hardness than electrochemical plating (Azzawi & Baumli, 2015; Ganesh et al., 2011; Karimzadeh et al., 2019).

5.5 Powder Coatings

The powder coating process was invented around 1945 by Daniel Gustin. In this technique coating that is applied as a free-flowing, dry powder, where coating is typically applied electrostatically and then cured under heat. The powder may be a thermosetting or thermoplastic polymer, usually used to create a hard finish that is tougher than conventional paint. This coating is mostly used in coating of household appliances, aluminium extrusions, drum hardware, automobile, motorcycle, bicycle parts and MDF (medium-density fibre board) etc. As powder coating does not have a liquid vehicle, it can produce thicker coatings than conventional liquid coatings without running or wilting. It produces very negligible appearance differences; between horizontally coated surface and vertically coated surface. As no carrier fluid evaporates away, the process emits hardly any VOC. Curing time is significantly faster in powder coating technique than liquid coating.

6 TRENDS IN NANOCOATINGS

The present scientific research and technology highly concern to the nanotechnology due to easy synthetic methods, materials with unique physical, (Ünal et al., 2019) chemical and mechanical properties. In the surface coatings technology, nano size effects can significantly enhanced coating properties by tailor-made coatings (Zhang, Sun, Fu et al, 2003). The assembly of hierarchical systems in nanostructures may determine the strength of coating area. The design models for nanocoatings are nanocomposite coatings, nanoscale multilayer coatings, super lattice coatings, etc. (Baer et al., 2003).

6.1 Micro- and Nanocapsule-based Coatings

Micro- and nanocapsules are of emerging interest to the research community for their extensive applications in many research fields such as molecular biology, medical imaging, electronic materials etc. Moreover, they have also been increasingly used as fillers, coatings, capsule agents, etc. Generally nanocapsules/containers were synthesized by polymerization in suspension using charged polystyrene nanospheres and then its lattices were coated by a layer of inorganic oxide(s) using the sol–gel method. The final products were burn in air to form polystyrene latex (Zhang, Sun, Fu et al, 2003). Nevertheless a thorough research is required to improve the quality of coating thickness. In this regards, researchers started to make self-repairing coatings by encapsulation of a catalyst into spheres (diameter <100 µm) or an inhibitor or a healing agent into similar sized microcapsules. These microcapsules are acting as coating material (Multiprotect Newsletter, 2009). During corrosion or scratching, inhibitor contents leaking out by breaking the capsules at the damaged region. At this time healing agent exposed to air and reacts with the environment to form a protective oxide leading the repair of damage.

6.2 Nanocomposite Coatings

These types of coating compose several techniques, among which reactive magnetron sputtering is most commonly used of a nanocrystalline phase and an amorphous phase. It offers superior hardness to the substrate. Thus a thorough study and extensive experiments carried out to optimize the hardness for industrial applications. The research considers both hardness and roughness of substrate with respect to corrosion behavior. It involves prevention of dislocations in the nanocrystalline phase; blocking of the grain-boundary and natures; lattice parameter; crystal size; and effects of impurities and intermediate phases (Brock, 2005).

7 ADVANCED POLYMERS AND FILLERS

A continuous research for the developments for coatings using advanced polymers represents a future trend in coating technology.

7.1 Hyperbranched Polymers

Hyperbranched polymers can be prepared by controlled radical polymerization with low melt viscosities. This type of coating is highly applicable to a high solid content and for powder coatings with low film thickness and low temperature. It produces solvent-free binders with narrow melting ranges (Multiprotect Newsletter, 2009; Zhang, Yu, Chen et al, 2003).

7.2 Organic–Inorganic hybrid Polymers

The development of future coating systems is highly interested on hybrid polymers which are the combinations of organic polymers and inorganic silicates to improve the overall coating qualities. The combination of inorganic networks with organic polymers may improve the stability and scratch resistance with elasticity. The sol–gel method has been successfully used for the synthesis of hybrid polymers (Multiprotect Newsletter, 2009). An improved translucent quality of polymers was prepared and reported by Brock, where he construct a block and comb polymers with intramolecular incompatibility which was then phase separated in the nano field (Murata, 1996). Further developments in phase separation will guide to improved adhesion to the surface without degrading the coating qualities.

7.3 Conductive Polymers

For the improved corrosion resistance coating of non-metallic substrates a newly developed coating material *i.e.*, conductive polymers were introduced. These are mainly constructing polyanilines and polythiophens moiety and synthesized by electrostatic and electrochemical method. The main advantage of this coating is the presence of heat conductive layers, for which highly acceptable to electrically heat surfaces.

7.4 Water soluble Paint

In architectural paint industries, water soluble paint is the strongest competitive, recently-developed future trends due to it is less expensive, lower in odor without toxic waste production. In the coatings and paints industries, water-soluble paints have met with superior resins with unique characteristics.

7.5 Fillers

Even though fillers don't have direct implementation rather negative effect to the coating quality, yet these are extensively used in every coatings industry to reduce costs. Further research on the fillers may reveals with high mechanical strength or improving the coating quality and decorative effects and also improves corrosion resistance. Nowadays with the help of nanotechnology, it is easy to design ultra-thin films by using nanolayers of special fillers or additives which can improve the matrix of the polymer. This technique is used for thermal and electrical insulation, and especially for fire protection (Multiprotect Newsletter, 2009; Zhang, Yu, Chen et al, 2003).

8. DEVELOPMENTS IN COATING PROCESSES

Starting from typical methods, the surface coating engineering enters into a new era, namely nanocoatings. Scientists and researchers are highly interested to nanocoating due to the superior hardness and strength. So, they are trying to improve the coating processing by which nanopowders will be available. But the requirement of sophisticated instruments and multistep processes oppose fast development for large-scale synthesis of nanoparticles.

For thermal spray processes especially plasma spraying or HVOF spraying, the main feedstock materials are nanopowders. In thermal spraying, use of nanopowders which is synthesized from vapor, liquid and solid provide a big advantage of moderate to high rate throughput and the formation of coat target materials having complex shapes, better wear resistance (Murata, 1996; Tjong & Chen, 2004). The research on HVOF is most attractive to the next generation of nanocoatings as it can deposite in dense nanocrystalline ceramic coatings with superior wear properties, with low porosity, high strength at low spraying temperature.

9. FUTURE RESEARCH DIRECTION

The genesis of surface engineered coating lie in ancient times. In ancient Greece and China with the practice of hardening, tempering and the crude forms of case hardening using solid organic materials, the cultivation of surface engineering started. In 20th century, with the introduction of electrical phenomena, the extensive assortment of modern surface technology emerged. This includes laser, plasma, and ion beam techniques to develop the technology day by day. In recent years, there has been a paradigm shift in surface engineering from age-old electroplating to modern techniques like vapor phase deposition, diffusion, thermal spray, plasma spray, laser, ion, electron, microwave, solar beams, pulsed arc, pulsed combustion, spark, friction and induction etc. But still there are some issues to address. The future improvements in coating processes must consider the thinnest layer of coating, complete automated coating, and use of module method to control the color of end product. Novel nano-composites and hybrids for high-temperature lubricating coating applications is an emerging field to develop. Self-assembled nano-phase particle coating, surface-coated pH-responsive membranes, modifications of electrode surfaces, energy efficient windows, surface coated fabrics, coating for building temperature control are the emerging field of future research. Anti-icing and anti-fouling coatings in aerospace sector, improved bulk materials for aircraft leading edges, thermally resistant coatings and advanced composite materials for aircraft pipelines etc. are the time demanded technology to be developed. The construction industry is always seeking improved ways to repel moisture, limit the impact of fouling and preserve dimensional stability, appearance and insulation performance. The wind sector also needs fire-retardant composites and advanced bulk materials, developed through nano-enabled resins, for applications such as lightweight blades. In the solar sector, easy-clean anti-reflective coatings can significantly improve the efficiency of current photovoltaic systems. Another focus is to protect the offshore wind turbine used in marine energy generation, from tidal wave, corrosion and marine bio-fouling. So, the potential application for coatings and surface engineering is vast. If we want to work on it, there is enough scope to further research and to play with this surface engineering for coating techniques.

9. CONCLUSION

Surface engineering is strongly connected with coating. For successful coating deposition on a substrate, several parameters can affect the quality, such as both the deposition and substrate materials, feedstock form (powder, wire, rods, precursors, etc.), and deposition methods. Among all the parameters deposition processes is the most important as it is connected with chemical alteration of materials. The most commonly used techniques and successful methods are deposition means are physical/chemical vapor deposition (PVD/CVD), electrodeposition (*i.e.*, Electro- and electroless chemical plating, electrochemical plating, electroless chemical plating etc.), sol–gel, and different types of thermal spraying processes (i.e., HVOF, plasma, cold, warm, and vacuum plasma spraying).

Surface engineering provides one of the most important means of engineering product differentiation in terms of quality, performance and life-cycle cost. The surface characteristics of engineering materials have a significant effect on the serviceability and life of a component, thus it cannot be neglected in design. Engineering environments are severe. They are normally complex, combining loading with chemical and physical degradation to the component surface. Surface engineering can help deal with these circumstances to improve the service life, and to enhance the overall performance of the components. In a word, Surface Engineering technology provides effective solutions to extreme applications. Surface engineering will provide many applications to the sports industry of the future.

REFERENCES

Ali, H. O., & Christie, I. R. A. (1984). A review of electroless gold deposition processes. *Gold Bulletin*, *17*(4), 118–127. doi:10.1007/BF03214674

Azzawi, A., & Baumli, P. (2015). Methods of composite coating: A review. *Materials Science and Engineering*, *40*(1), 26–32.

Baer, D. R., Burrows, P. E., & El-Azab, A. A. (2003). Enhancing coating functionality using nanoscience and nanotechnology. *Progress in Organic Coatings*, *47*(3-4), 342–356. doi:10.1016/S0300-9440(03)00127-9

Brock, T. (2005). Trends in coatings technology [Europea Coatings]. *Pitture e Vernici*, *81*, 15–24.

CEPE Guidance on Labelling Decorative Paint with Skin Sensitizing Biocides. (n.d.). http://sveff.se/files/2015/04/CEPE-guidance-on-labelling-with-biocides-skin-sensitizers-final-28-Oct-14.pdf

CEPE Information Bulletin About Cobalt Driers. (n.d.). https://www.vslf.ch/download.php?file_id=695&download=true

Chen, J., Bradhurst, D., Dou, S., & Liu, H. (1998). The effect of chemical coating with Ni on the electrode properties of Mg2Ni alloy. *Journal of Alloys and Compounds*, *280*(1-2), 290–293. doi:10.1016/S0925-8388(98)00731-2

China Slaps VOC Tax on Coatings. (n.d.). https://www.paintsquare.com/news/?fuseaction=view&id=12741

Ching, A. Y. (Ed.). (2015). *Handbook of Manufacturing Engineering and Technology*. Springer-Verlag London.

Dai, H., Rinzler, A. G., Nikolaev, P., Thess, A., Colbert, D. T., & Smalley, R. E. (1996). Single-wall nanotubes produced by metal-catalyzed disproportionation of carbon monoxide. *Chemical Physics Letters*, *260*(3-4), 471–475. doi:10.1016/0009-2614(96)00862-

DeMasi-Marcin, J. T., & Gupta, D. K. (1994). Protective coatings in the gas turbine engine. *Surface and Coatings Technology*, *68*, 1–9. doi:10.1016/0257-8972(94)90129-5

Emran, K. M., Ali, S. M., & Al Lehaibi, H. A. (2018). *Green Methods for Corrosion Control*. IntechOpen. doi:10.5772/intechopen.72762

Endo, M., Takeuchi, K., Igarashi, S., Kobori, K., Shiraishi, M., & Kroto, H. W. (1993). 'The production and structure of pyrolytic carbon nanotubes. *Journal of Physics and Chemistry of Solids*, *54*(12), 1841–1848. doi:10.1016/0022-3697(93)90297-5

EU Directive Restricting the Use of Nonylphenol Ethoxylate and Nonylphenol Products. (n.d.). https://www.tfl.com/web/files/Statement_NPE-surfactants.pdf

Gadow, R., Killinger, A., & Stiegler, N. (2010). Hydroxyapatite coatings for biomedical applications deposited by different thermal spray techniques. *Surface and Coatings Technology*, *205*(4), 1157–1164. doi:10.1016/j.surfcoat.2010.03.05

Ganesh, V. A., Raut, H. K., Nair, A. S., & Ramakrishna, S. (2011). A review on self-cleaning coating. *Journal of Materials Chemistry*, *21*(41), 16304–16322. doi:10.1039/c1jm12523k

Gooch, J. W. (2002). *History of Paint and Coatings Materials*. Springer Science and Business Media LLC.

Gray, J. E., & Luan, B. (2002). Protective coatings on magnesium and its alloys— A critical review. *Journal of Alloys and Compounds*, *336*(1–2), 88–113. doi:10.1016/S0925-8388(01)01899-0

Gu, C., Lian, J., He, J., Jiang, Z., & Jiang, Q. (2006). High corrosion-resistance nanocrystalline Ni coating on AZ91D magnesium alloy. *Surface and Coatings Technology*, *200*(18-19), 5413–5418. doi:10.1016/j.surfcoat.2005.07.001

Hajdu, J., Yarkosky, E., Cacciatore, P., & Suplicki, M. (1990). Electroless nickel processes for memory disks. In *Proceedings of the Fourth International Symposium on Magnetic Materials, Processes and Devices*. Pennington, NJ: Electrochemical Society.

Hamdy, A. S. (2006a). Corrosion protection of aluminum composites by silicate/cerate conversion coating. *Surface and Coatings Technology*, *200*(12–13), 3786–3792. doi:10.1016/j.surfcoat.2005.03.012

Hamdy, A. S. (2006b). Advanced nano-particles anti-corrosion ceria based sol gel coatings for aluminum alloys. *Materials Letters*, *60*(21–22), 2633–2637. doi:10.1016/j.matlet.2006.01.049

Hamdy, A. S. (2006c). A clean low cost anti-corrosion molybdate based nano-particles coating for aluminum alloys. *Progress in Organic Coatings*, *56*(2–3), 146–150. doi:10.1016/j.porgcoat.2006.03.002

Hamdy, A. S. (2008). The effect of surface modification and stannate concentration on the corrosion protection performance of magnesium alloys. *Surface and Coatings Technology*, *203*(3-4), 240–249. doi:10.1016/j.surfcoat.2008.08.070

Hamdy, A. S. (2009). The correlation between electrochemical impedance spectroscopy and other polarization techniques for the corrosion evaluation of coated and bare metals in aqueous solutions. In Corrosion Protection: Processes, Management and Technologies. New York: Nova Science Publishers.

Hamdy, A. S. (2010). *High Performance Coatings for Automotive and Aerospace Industries*. Nova Science Publishers.

Hamdy, A. S., & Butt, D. (2007). *Eco-friendly conversion coatings for automotive and aerospace materials*. In European Coatings Conference New Concepts for Anti-Corrosive Coatings, Berlin, Germany.

Hamdy, A. S., & Butt, D. P. (2006a). Environmentally compliant silica conversion coatings prepared by sol–gel method for aluminum alloys. *Surface and Coatings Technology*, *201*(1–2), 401–407. doi:10.1016/j.surfcoat.2005.11.142

Hamdy, A. S., & Butt, D. P. (2006b). Corrosion protection performance of nanoparticles thin-films containing vanadium ions formed on aluminum alloys. *Anti-Corrosion Methods and Materials*, *53*(4), 240–245. doi:10.1108/00035590610678947

Hamdy, A. S., & Butt, D. P. (2007). Novel anti-corrosion nano-sized vanadia-based thin films prepared by sol–gel method for aluminum alloys. *Mater. Proc.Technol.*, *181*(1–3), 76–80. doi:10.1016/j.jmatprotec.2006.03.042

Hamdy, A. S., & Farahat, M. (2010). Chrome-free zirconia-based protective coatings for magnesium alloys. *Surface and Coatings Technology*, *204*(16-17), 2834–2840. doi:10.1016/j.surfcoat.2010.02.063

Hamdy, A. S., Shoeib, M., & Butt, D. P. (2009). A novel approach in designing environmentally compliant sol–gel based ceramic coatings and nanocomposite coatings for industrial applications. In A. Malik & R. J. Rawat (Eds.), *New Nanotechniques* (pp. 649–659). Nova Science Publishers.

Hernández, P., Campos, D., Socorro, P., Benítez, A., Ortega, F., Díaz, N., & Marrero, M. D. (2015). Electroforming Applied to Manufacturing of Microcomponents. *Procedia Engineering*, *132*, 655–662. doi:10.1016/j.proeng.2015.12.544

Innes, W. (1974). Electroplating and electroless plating on magnesium and magnesium alloys. Wiley-Interscience.

Jiang, S., Van Dyk, A., Maurice, A., Bohling, J., Fasano, D., & Brownell, S. (2017). Design colloidal particle morphology and self-assembly for coating applications. *Chemical Society Reviews*, *46*(12), 3792–3807. doi:10.1039/C6CS00807K PMID:28470250

Kamp, M., de Nijs, B., van der Linden, M. N., de Feijter, I., Lefferts, M. J., Aloi, A., Griffiths, J., Baumberg, J. J., Voets, I. K., & van Blaaderen, A. (2020). Multivalent Patchy Colloids for Quantitative 3D Self-Assembly Studies. *Langmuir*, *36*(9), 2403–2418. doi:10.1021/acs.langmuir.9b03863 PMID:32097015

Karimzadeh, A., Aliofkhazraei, M., & Walsh, F. C. (2019). A review of electrodeposited Ni-Co alloy and composite coatings: Microstructure, properties and applications. *Surface and Coatings Technology*, *372*, 463–498. doi:10.1016/j.surfcoat.2019.04.079

Kasten, L. S., Grant, J. T., Grebasch, N., Voevodin, N., Arnold, F. E., & Donley, M. S. (2001). An XPS study of cerium dopants in sol–gel coatings for aluminum 2024-T3. *Surface and Coatings Technology*, *140*(1), 11–15. doi:10.1016/S0257-8972(01)01004-0

Korhonen, A., & Harju, E. (2012). Surface Engineering with Light Alloys - Hard Coatings, Thin Films, and Plasma Nitriding. *Journal of Materials Engineering and Performance*, *9*(3), 302–305. doi:10.1361/105994900770345953

Kuroda, S., Kawakita, J., Watanabe, M., & Katanoda, H. (2008). Warm spraying – a novel coating process based on high-velocity impact of solid particles. *Science and Technology of Advanced Materials*, *9*(3), 17. doi:10.1088/1468-6996/9/3/033002 PMID:27877996

Luo, J., & Cui, N. (1998). Effects of microencapsulation on the electrode behavior of Mg2Ni-based hydrogen storage alloy in alkaline solution. *Journal of Alloys and Compounds*, *264*(1-2), 299–305. doi:10.1016/S0925-8388(97)00277-6

Lydecker, R. (2015). *Is Copper Bottom Paint Sinking?* https://www.boatus.com/magazine/2012/february/copper.asp

Makhlouf, A. S. H. (2011). *Current and advanced coating technologies for industrial applications*. Woodhead Publishing Ltd. doi:10.1533/9780857094902.1.3

Makhlouf, A. S. H., & Gajarla, Y. (2019). Advances in smart coatings for magnesium alloys and their applications in industry. Elsevier BV.

Maria Ingo, G., Guida, G., Angelini, E., Di Carlo, G., Mezzi, A., & Padeletti, G. (2013). Ancient Mercury-Based Plating Methods: Combined Use of Surface Analytical Techniques for the Study of Manufacturing Process and Degradation Phenomena. *Accounts of Chemical Research*, *46*(11), 2365–2375. doi:10.1021/ar300232e PMID:23829823

Mittal, C. K. (1995). Merbromin as a colouring agent for anodized surfaces of Al and its alloys. *Trans. Metal Finishers Association of India*, *4*, 227.

Montazer, M., & Harifi, T. (2018). *Nanocoating and lamination. In Nanofinishing of Textile Materials*. Elsevier BV.

Multiprotect Newsletter. (2009). Available from: http://multiprotect.org

Murata, K. (1996). Trends in coatings technology. *J. Macromol. Sci. Part A*, *33*(12), 1837–1841.

OSHA. (n.d.). https://www.osha.gov/Publications/OSHA_FS-3649_Bridge_Painting.pdf

Paulussen, S., Rego, R., Goossens, O., Vangeneugden, D., & Rose, K. (2005). Plasma polymerization of hybrid organic–inorganic monomers in an atmospheric pressure dielectric barrier discharge. *Surface and Coatings Technology*, *200*(1–4), 672–675. doi:10.1016/j.surfcoat.2005.02.134

Pfeifer, M. (2009). Manufacturing Process Considerations. *Materials Enabled Designs*, 115-160.

Plasma Spraying Process. (2010). *Plasma spraying process*. Available at: https://www.zircotec.com/page/plasma-spray_processing/39

Sharma, A. K., Bhojraj, H., Kaila, V. K., & Narayanamurthy, H. (1997). Anodizing and inorganic black coloring of aluminum alloys for space applications. *J Metal Finishing*, *95*(12), 14–20. doi:10.1016/S0026-0576(97)82621-9

Sudagar, J., Tamilarasan, R., Sanjith, U., Rajendran, R., & Kumar, R. (2017). *Electroless Deposition of Nanolayered Metallic Coatings*. Nanoscaled Films and Layers, Laszlo Nanai, Intech Open. doi:10.5772/intechopen.68220

Technical Coating International Inc. (2010). Available at: http://www.tciinc.com/coating.html

Thakare, M., Wharton, J., Wood, R., & Menger, C. (2007). Exposure effects of alkaline drilling fluid on the microscale abrasion–corrosion of WC-based hard metals. *Wear*, *263*(1-6), 125–136. doi:10.1016/j.wear.2006.12.047

Thakur, S. (2019). Sustainable Nanostructural Materials for Shape Memory, Self-Healing, and Self-Cleaning Applications. In Dynamics of Advanced Sustainable Nanomaterials and Their Related Nanocomposites at the Bio-Nano Interface. Elsevier. doi:10.1016/B978-0-12-819142-2.00010-0

Tjong, S. C., & Chen, H. (2004). Nanocrystalline materials and coatings. *Materials Science and Engineering R Reports*, *45*(1-2), 1–88. doi:10.1016/j.mser.2004.07.001

Ünal, E., Yaşar, A., & Karahan, İ. H. (2019). A review of electrodeposited composite coatings with Ni–B alloy matrix. *Mater. Res. Exp.*, *6*(9), 0965b5.

Wang, M., Jiang, M., Ning, F., Chen, D., Shiyong, L., & Duan, H. (2002). Blockcopolymer-free strategy for preparing micelles and hollow spheres: Self-assembly of poly(4-vinylpyridine) and modifi ed polystyrene. *J. Macromolecules*, *35*(15), 5980–5989. doi:10.1021/ma0201330

Zeblisky, R. J., McCormack, J. F., Williamson, J. D., & Shneble, F. W. (1963). *Electroless copper plating*. U.S. Patent 3,095,309.

Zhang, G., Yu, Y., Chen, X., Han, Y., Di, Y., Yang, B., Xiao, F., & Shen, J. (2003). Silica nanobottles templated from functional polymer spheres. *Journal of Colloid and Interface Science*, *223*(2), 467–472. doi:10.1016/S0021-9797(03)00340-0 PMID:12909037

Zhang, S., Sun, D., Fu, Y., & Du, H. (2003). Recent advances of superhard nanocomposite coatings: A review. *Surface and Coatings Technology*, *167*(2-3), 113–119. doi:10.1016/S0257-8972(02)00903-9

Compilation of References

Abdulhussein, A. T., Kannarpady, G. K., Wright, A. B., Ghosh, A., & Biris, A. S. (2016). Current trend in fabrication of complex morphologically tunable superhydrophobic nano scale surfaces. *Applied Surface Science*, *384*, 311–332. doi:10.1016/j.apsusc.2016.04.186

Achanta, S., Drees, D., & Celis, J.-P. (2011). *Nanocoatings for tribological applications*. Woodhead., doi:10.1533/9780857094902.2.355

Advanced, A. M. T. A. G. (n.d.). *Materials Technology, Twin Wire Arc Spraying*. Retrieved from: https://www.amt-ag.com/equipment/spray-processes/twin-wire-arc

Advanced-Coating. (n.d.). Retrieved from: http://www.advanced-coating.com/english/spraying-hvof.htm.

Afzal, M. A. F., Kesarwani, P., Reddy, K. M., Kalmodia, S., Basu, B., & Balani, K. (2012). Functionally graded hydroxyapatite-alumina-zirconia biocomposite: Synergy of toughness and biocompatibility. *Materials Science and Engineering C*, *32*(5), 1164–1173. doi:10.1016/j.msec.2012.03.003

Agarwala, R. C., & Agarwala, V. (2003). Electroless alloy/composite coatings: A review. *Sadhana*, *28*(3-4), 475–493. doi:10.1007/BF02706445

Ahmad, I., & Kan, C. (2016). A Review on Development and Applications of Bio-Inspired Superhydrophobic Textiles. *Materials (Basel)*, *9*(11), 1–34. doi:10.3390/ma9110892 PMID:28774012

Ahmadi, M., & Aghajani, H. (2018). Structural characterization of YSZ/Al2O3 nanostructured composite coating fabricated by electrophoretic deposition and reaction bonding. *Ceramics International*, *44*(6), 5988–5995. doi:10.1016/j.ceramint.2017.12.185

Ahn, W., Taylor, B., Dall'Asén, A. G., & Roper, D. K. (2008). Electroless gold island thin films: Photoluminescence and thermal transformation to nanoparticle ensembles. *Langmuir*, *24*(8), 4174–4184. doi:10.1021/la703064m

Aithal, R. K., Yenamandra, S., Gunasekaran, R. A., Coane, P., & Varahramyan, K. (2006). Electroless copper deposition on silicon with titanium seed layer. *Materials Chemistry and Physics*, *98*(1), 95–102. doi:10.1016/j.matchemphys.2005.08.079

Akavci, S. S. (2014). Thermal buckling analysis of functionally graded plates on an elastic foundation according to a hyperbolic shear deformation theory. *Mechanics of Composite Materials*, *50*(2), 197–212. doi:10.1007/s11029-014-9407-1

Akedo, Nakano, Park, Baba, & Ashida. (2008). The aerosol deposition method. *Synthesial. England.*, *1*, 121-130.

Akhtar-uz Zaman, I. M. (1997). *Development of nanocoating on cutting tools*. Academic Press.

Alavi, B., Aghajani, H., & Rasooli, A. (2019). An investigation on electroless nickel coating on yttria stabilized zirconia nanoparticles via single step surface activation methods. *Thin Solid Films*, *669*, 514–519. doi:10.1016/j.tsf.2018.11.053

Alebeida, O. K., Taoa, Z., & Seedahmed, A. I. (2015). New Approach for Dyeing and UV Protection Properties of Cotton Fabric Using Natural Dye Extracted from Henna Leaves. *Fibres & Textiles in Eastern Europe*, *23*(5), 60–65. doi:10.5604/12303666.1161758

Alhulaifi, Buck, & Arbegast. (2012). Numerical and Experimental Investigation of Cold Spray Gas Dynamic Effects for Polymer Coating. *Journal of Thermal Spray Technology, 21*(5), 852.

Alhulaifi, A. S., Buck, G. A., & Arbegast, W. J. (2012). Numerical and experimental investigation of cold spray gas dynamic effects for polymer coating. *Journal of Thermal Spray Technology*, *21*(5), 852–862. doi:10.100711666-012-9743-4

Ali, H. O., & Christie, I. R. A. (1985). A Review of Electroless Gold Deposition Processes. *Circuit World*, *11*(4), 10–16. doi:10.1108/eb043759

Alirezaei, S., Monir Vaghefi, S. M., Ürgen, M., Saatchi, A., & Kazmanli, K. (2013). Evaluation of structure and mechanical properties of Ni-P-Al2O3 nanocomposite coatings. *Journal of Composite Materials*, *47*(26), 3323–3329. doi:10.1177/0021998312464082

Alkhimov, A., Papyrin, A., Kosarev, V., Nesterovich, N., & Shushpanov, M. (1995). *Method and Device for Coating*. European Patent 0488533B1.

Anand, S. C., Miraftab, M., Rajendran, S., & Kennedy, J. F. (2005). *Medical Textiles and Biomaterials for Healthcare*. Woodhead.

Anderson, A. (1995). *ASM Metals Handbook* (Vol. 18). Friction, Lubrication and Wear.

Andricacos, P. C. (1999). Copperon-chipinterconnections. *The Electrochemical Society Interface*, *8*(1), 6.

Angelo, P. C., & Subramanian, R. (2008). *Powder metallurgy: science, technology and applications*. PHI Learning Pvt. Ltd.

Angel, W. D. W., Jurado, L. T., Alcalá, J. F. C., Martínez, E. C., & Cedeño, V. F. V. (2014). Effect of copper on the mechanical properties of alloys formed by powder metallurgy. *Materials & Design*, *58*, 12–18. doi:10.1016/j.matdes.2014.02.002

Anisur, M., Chakrobarty, P., Easton, D., & Raman, R. (2018). Controlling hydrogen environment and cooling during CVD graphene growth on nickel for improved corrosion resistance. *Carbon*, *127*, 131–140. doi:10.1016/j.carbon.2017.10.079

Apelian, D., Paliwal, M., Smith, R. W., & Schilling, W. F. (1983). Melting and solidification in plasma spray deposition—Phenomenological review. *International Metals Reviews*, *28*(1), 271–294. doi:10.1179/imtr.1983.28.1.271

Araghi, A., & Paydar, M. H. (2010). Electroless deposition of Ni-P-B4C composite coating on AZ91D magnesium alloy and investigation on its wear and corrosion resistance. *Materials & Design*, *31*(6), 3095–3099. doi:10.1016/j.matdes.2009.12.042

Arsha, A. G., Jayakumar, E., Rajan, T. P. D., Antony, V., & Pai, B. C. (2015). Design and fabrication of functionally graded in-situ aluminium composites for automotive pistons. *Materials & Design*, *88*, 1201–1209. doi:10.1016/j.matdes.2015.09.099

Artus, G. R. J., Jung, S., Zimmermann, J., Gautschi, H.-P., Marquardt, K., & Seeger, S. (2006). Silicone Nanofilaments and Their Application as Superhydrophobic Coatings. *Advanced Materials*, *18*(20), 2758–2762. doi:10.1002/adma.200502030

Aruna, S. T. (2017). Thermal Spray Coatings Using Solution Combustion Synthesized Powders. In Concise Encyclopedia of Self-Propagating High-Temperature Synthesis (pp. 382-384). Elsevier. doi:10.1016/B978-0-12-804173-4.00155-1

Ashassi-Sorkhabi, H., & Rafizadeh, S. H. (2004). Effect of coating time and heat treatment on structures and corrosion characteristics of electroless Ni–P alloy deposits. *Surface and Coatings Technology*, *176*(3), 318–326. doi:10.1016/S0257-8972(03)00746-1

Askari, E., Mehrali, M., Metselaar, I. H. S. C., Kadri, N. A., & Rahman, M. M. (2012). Fabrication and mechanical properties of Al2O3/SiC/ZrO2 functionally graded material by electrophoretic deposition. *Journal of the Mechanical Behavior of Biomedical Materials*, *12*, 144–150. doi:10.1016/j.jmbbm.2012.02.029

Assadi, H., Gärtner, F., Klassen, T., & Kreye, H. (2019). Comment on "Adiabatic shear instability is not necessary for adhesion in cold spray". *Scripta Materialia*, *162*, 512–514. doi:10.1016/j.scriptamat.2018.10.036

Assadi, H., Gärtner, F., Stoltenhoff, T., & Kreye, H. (2003). Bonding mechanism in cold gas spraying. *Acta Materialia*, *51*(15), 4379–4394. doi:10.1016/S1359-6454(03)00274-X

Assadi, H., Kreye, H., Gärtner, F., & Klassen, T. (2016). Cold spraying – a materials perspective. *Acta Materialia*, *116*, 382–407. doi:10.1016/j.actamat.2016.06.034

Ates, E. S., & Unalan, H. E. (2012). Zinc oxide nanowire enhanced multifunctional coatings for cotton fabrics. *Thin Solid Films*, *520*(14), 4658–4661. doi:10.1016/j.tsf.2011.10.073

Ates, M. (2016). A review on conducting polymer coatings for corrosion protection. *Journal of Adhesion Science and Technology*, *30*(14), 1510–1536. doi:10.1080/01694243.2016.1150662

Attar, F. (1996). Hardness evaluation of thin ceramic coatings on tool steel. *Surface and Coatings Technology*, *78*(1-3), 78–86. doi:10.1016/0257-8972(94)02395-6

Baba, N. B., Waugh, W., & Davidson, A. M. (2009). Manufacture of electroless nickel/YSZ composite coatings. *Proceedings of World Academy of Science, Engineering and Technology (WASET)*, 715-20.

Bach, L. X., Thuan, D. V., Thu, V. T. H., Phan, T. B., Vu, N. S. H., & Nam, N. D. (2019). An investigation on titania multilayer coatings for enhanced corrosion resistance of carbon steel in simulated seawater by sol–gel dip coating. *Journal of Materials Research and Technology*, *8*(6), 6400–6406. doi:10.1016/j.jmrt.2019.09.061

Backman, D. G., & Williams, J. C. (1992). Advanced Materials for Aircraft Engine Applications. *Science*, *255*(5048), 1082–1087. doi:10.1126cience.255.5048.1082 PMID:17817782

Bae, G., Kumar, S., Yoon, S., Kang, K., Na, H., Kim, H. J., & Lee, C. (2009). Bonding features and associated mechanisms in kinetic sprayed titanium coatings. *Acta Materialia*, *57*(19), 5654–5666. doi:10.1016/j.actamat.2009.07.061

Balani, K., Agarwal, A., Seal, S., & Karthikeyan, J. (2005). Transmission electron microscopy of cold sprayed 1100 aluminum coating. *Scripta Materialia*, *53*(7), 845–850. doi:10.1016/j.scriptamat.2005.06.008

Balaraju, J. N., Anandan, C., & Rajam, K. S. (2006). Influence of codeposition of copper on the structure and morphology of electroless Ni–W–P alloys from sulphate-and chloride-based baths. *Surface and Coatings Technology*, *200*(12-13), 3675–3681. doi:10.1016/j.surfcoat.2004.09.010

Balaraju, J. N., Jahan, S. M., Anandan, C., & Rajam, K. S. (2006). Studies on electroless Ni–W–P and Ni–W–Cu–P alloy coatings using chloride-based bath. *Surface and Coatings Technology*, *200*(16-17), 4885–4890. doi:10.1016/j.surfcoat.2005.04.053

Balaraju, J. N., Narayanan, T. S., & Seshadri, S. K. (2006). Structure and phase transformation behaviour of electroless Ni–P composite coatings. *Materials Research Bulletin*, *41*(4), 847–860. doi:10.1016/j.materresbull.2005.09.024

Balaraju, J. N., & Rajam, K. S. (2005). Electroless deposition of Ni–Cu–P, Ni–W–P and Ni–W–Cu–P alloys. *Surface and Coatings Technology*, *195*(2-3), 154–161. doi:10.1016/j.surfcoat.2004.07.068

Balaraju, J. N., & Rajam, K. S. (2010). Electroless ternary Ni–W–P alloys containing micron size Al2O3 particles. *Surface and Coatings Technology*, *205*(2), 575–581. doi:10.1016/j.surfcoat.2010.07.047

Bao, W., Deng, Z., Zhang, S., Ji, Z., & Zhang, H. (2019) Next-generation composite coating system: Nanocoating. *Frontiers in Materials, 6.*

Baran, M. C., Graham, A. H., Davala, A. B., Causton, R. J., & Schade, C. (1999). A superior sinter-hardenable material. *Advances in Powder Metallurgy & Particulate Materials*, *2*(Part 7), 185–203.

Barbosa-Canovas, G. V., Ortega-Rivas, E., Juliano, P., & Yan, H. (2005). Food Powders: Physical Properties, Processing, and Functionality. In Food Engineering Series (1st ed.). doi:10.100713398-014-0173-7.2

Barmouz, M., & Givi, M. K. B. (2011). Fabrication of in situ Cu/SiC composites using multi-pass friction stir processing: Evaluation of microstructural, porosity, mechanical and electrical behavior. *Composites. Part A, Applied Science and Manufacturing*, *42*(10), 1445–1453. doi:10.1016/j.compositesa.2011.06.010

Behrangi, S., & Aghajani, H. (2018). Nano3YSZ electrophoretic deposition from acetylacetone+ ethanol solvent on the surface of AZ91 magnesium alloy. *Micro & Nano Letters*, *13*(5), 611–616. doi:10.1049/mnl.2017.0623

Beltran, A. M., & Shores, D. A. (1972). Hot Corrosion. In The superalloys (pp. 4289-4296). Academic Press.

Benaben, P. (2010). An overview of hard chromium plating using trivalent chromium solutions. National Association for Surface Finishing Annual Conference and Trade Show 2010, SUR/FIN 2010, 638–646.

Benxxonglobal. (n.d.). Retrieved from: http://www.bexxonglobal.com/what-is-wire-and-powder-flame-spray-.html

Bergman, O. (2011). *Key aspects of sintering powder metallurgy steel prealloyed with chromium and manganese*. Chalmers University of Technology.

Berndt, C. C., Brindley, W., Goland, A. N., Herman, H., Houck, D. L., Jones, K., & Smith, M. (1992). Current problems in plasma spray processing. *Journal of Thermal Spray Technology*, *1*(4), 341–356. doi:10.1007/BF02647162

Betancourt-Dougherty, L., & Smith, R. (1998). Effects of load and sliding speed on the wear behaviour of plasma sprayed TiC NiCrBSi coatings. *Wear*, *217*(1), 147–154. doi:10.1016/S0043-1648(97)00212-3

Beygi, H., Sajjadi, S. A., & Zebarjad, S. M. (2012). An optimization analysis on electroless deposition of Al2O3/Cu core-shell nanostructures. *Applied Surface Science*, *261*, 166–173. doi:10.1016/j.apsusc.2012.07.134

Beygi, H., Vafaeenezhad, H., & Sajjadi, S. A. (2012). Modeling the electroless nickel deposition on aluminum nanoparticles. *Applied Surface Science*, *258*(19), 7744–7750. Advance online publication. doi:10.1016/j.apsusc.2012.04.132

Bhatia, R., Sidhu, H. S., & Sidhu, B. S. (2015). High Temperature Behavior of Cr3C2-NiCr Coatings in the Actual Coal-Fired Boiler Environment. *Metallurgical and Materials Transactions E, 2*(1), 1011–1022.

Bhattacharyya, M., Kumar, A. N., & Kapuria, S. (2008). Synthesis and characterization of Al/SiC and Ni/Al2O3 functionally graded materials. *Materials Science and Engineering A*, *487*(1–2), 524–535. doi:10.1016/j.msea.2007.10.040

Bhavsar, A. N., & Patel, V. A. (2016). Influence of process parameters of TIG welding process on mechanical properties of SS304L welded joint. *International Research Journal of Engineering and Technology,* 977-984.

Boily, J. F., & Seward, T. M. (2005). Palladium (II) chloride complexation: Spectrophotometric investigation in aqueous solutions from 5 to 125° C and theoretical insight into Pd-Cl and Pd-OH2 interactions. *Geochimica et Cosmochimica Acta*, *69*(15), 3773–3789. doi:10.1016/j.gca.2005.03.015

Boing, D., Oliveira, A. J. D., & Schroeter, R. B. (2020). Evaluation of wear mechanisms of PVD and CVD coatings deposited on cemented carbide substrates applied to hard turning. *International Journal of Advanced Manufacturing Technology*, *106*(11), 5441–5451. doi:10.100700170-020-05000-x

Bolelli, G., Bellucci, D., Cannillo, V., Lusvarghi, L., Sola, A., Stiegler, N., & De Nardo, L. (2014). Suspension thermal spraying of hydroxyapatite: Microstructure and in vitro behaviour. *Materials Science and Engineering C*, *34*, 287–303. doi:10.1016/j.msec.2013.09.017 PMID:24268261

Bolelli, G., Lusvarghi, L., & Giovanardi, R. (2008). A Comparison between the Corrosion Resistances of Some HVOF-Sprayed Metal Alloy Coatings. *Surface and Coatings Technology*, *202*(19), 4793–4809. doi:10.1016/j.surfcoat.2008.04.056

Borca-Tasciuc, T., & Chen, G. (2004). Experimental Techniques for Thin-Film Thermal Conductivity Characterization. In T. M. Tritt (Ed.), *Thermal Conductivity: Theory, Properties and Applications* (pp. 205–237). Kluwer Academic. doi:10.1007/0-387-26017-X_9

Brachmann, E., Seifert, M., Neumann, N., Alshwawreh, N., Uhlemann, M., Menzel, S., Acker, J., Herold, S., Hoffmann, V., & Gemming, T. (2019). Electroless-Deposited Platinum Antennas for Wireless Surface Acoustic Wave Sensors. *Materials (Basel)*, *12*(7), 1002. doi:10.3390/ma12071002

Brackx, E., Pages, S., Dugne, O., & Podor, R. (2015). Recent analytical developments for powder characterization. *The European Physical Journal. Special Topics*, *224*(9), 1787–1804. doi:10.1140/epjst/e2015-02499-2

Brogan, J. A. (2000). Thermal-spraying of polymers and polymer blends. *MRS Bulletin*, *25*(7), 48–53. doi:10.1557/mrs2000.124

Brunner, H., & Picalek, J. (2012). Patent No. EP3026143A1. Retrieved from https://patents.google.com/patent/US20170335462A1/en

Bulloch, J., & Callagy, A. (1999). An in situ wear-corrosion study on a series of protective coatings in large induced draft fans. *Wear*, *233-235*, 284–292. doi:10.1016/S0043-1648(99)00229-X

Burlacov, I., Jirkovský, J., Kavan, L., Ballhorn, R., & Heimann, R. B. (2007). Cold gas dynamic spraying (CGDS) of TiO2 (anatase) powders onto poly (sulfone) substrates: Microstructural characterisation and photocatalytic efficiency. *Journal of Photochemistry and Photobiology A Chemistry*, *187*(2-3), 285–292. doi:10.1016/j.jphotochem.2006.10.023

Bush, T. B., Khalkhali, Z., Champagne, V., Schmidt, D. P., & Rothstein, J. P. (2017). Optimization of cold spray deposition of high-density polyethylene powders. *Journal of Thermal Spray Technology*, *26*(7), 1548–1564. doi:10.100711666-017-0627-5

Buytoz, S., Ulutan, M., Islak, S., Kurt, B., & Celik, O. N. (2013). Microstructural and wear characteristics of high velocity oxygen fuel (HVOF) sprayed NiCrBSi–SiC composite coating on SAE 1030 steel. *Arabian Journal for Science and Engineering*, *38*(6), 1481–1491. doi:10.100713369-013-0536-y

Cabezon, C. G., Hernandez, C. G., Mendez, M. L. R., Herranz, G., & Pedrosa, F. M. (2020). Role of carbon and nitrogen in the improvement of corrosion resistance of new powder metallurgy Co-Cr-Mo alloys. *Corrosion Reviews*. Advance online publication. doi:10.1515/corrrev-2019-0109

Cabrales, L., & Abidi, N. (2012). Microwave plasma induced grafting of oleic acid on cotton fabric surfaces. *Applied Surface Science*, *258*(10), 4636–4641. doi:10.1016/j.apsusc.2011.12.130

Cabrera, G., Torres, F., Caicedo, J. C., Aperador, W., Amaya, C., & Prieto, P. (2012). Improvement of electrochemical surface properties in steel substrates using a nanostructured CrN/AlN multilayer coating. *Journal of Materials Engineering and Performance*, *21*(1), 128–136. doi:10.100711665-010-9798-7

Cadavid, E., Parra-Velásquez, C., & Vargas, F. (2016). Estudio de llamas oxiacetilénicas usadas en la proyección térmica. *Revista Colombiana de Materiales*, (9), 15-26.

Cai, B., Tan, Y., He, L., Tan, H., & Gao, L. (2013). Tribological properties of TiC particles reinforced Ni-based alloy composite coatings. *Transactions of Nonferrous Metals Society of China*, *23*(6), 1681–1688. doi:10.1016/S1003-6326(13)62648-5

Cameron, D. (2014). Introduction to Films and Coatings: Technology and Recent Development. In Comprehensive Materials Processing (Volume 4.01, pp. 1-2). Elsevier.

Camp, M., Carboneras, M., Lopez, M. D., Torres, B., Rodrigo, P., Otero, E., & Rams, J. (2009). Corrosion Resistance of Thermally Sprayed Al and Al/SiC Coatings on Mg. *Surface and Coatings Technology*, *203*(20-21), 3224–3230. doi:10.1016/j.surfcoat.2009.03.057

Camurlu, H. E. (2019). Al matrix FGM having 5, 15 and 25% B 4 C p reinforced layers. In Machines. Technologies. Materials (Vol. 13). Scientific Technical Union of Mechanical Engineering "Industry 4.0."

Canakci, A., Varol, T., Özkaya, S., & Erdemir, F. (2014). Microstructure and Properties of Al-B4C Functionally Graded Materials Produced by Powder Metallurgy Method. *Universal Journal of Materials Science*, *2*(5), 90–95. doi:10.13189/ujms.2014.020502

Cantarero, A. (2015). Raman scattering applied to materials science. *Procedia Mater. Sci.*, *9*, 113–122. doi:10.1016/j.mspro.2015.04.014

Cao, X. Q., Vassen, R., & Stoever, D. (2004). Ceramic Materials for Thermal Barrier Coatings. *Journal of the European Ceramic Society*, *24*(1), 1–10. doi:10.1016/S0955-2219(03)00129-8

Cao, X. Q., Vassen, R., & Stöver, D. (2004). Ceramic materials for thermal barrier coatings. *JournaloftheEuropeanCeramicSociety*, *24*(1), 1–10.

Cao, X. Q., Vassen, R., Tietz, F., & Stoever, D. (2006). New double-ceramic-layer thermal barrier coatings based on zirconia-rare earth composite oxides. *Journal of the European Ceramic Society*, *26*(3), 247–251. doi:10.1016/j.jeurceramsoc.2004.11.007

Cao, Y., Wang, Q., Ning, X., & Liu, Y. (2019). Evaluation of the Phase Stability, and Mechanical and Thermal Properties of $Ba(Sr_{1/3}Ta_{2/3})O_3$ as a Potential Ceramic Material for Thermal Barrier Coatings. *Ceramics International*, *45*(10), 12989–12993. doi:10.1016/j.ceramint.2019.03.227

Capus, J. (2013). PM light alloys gaining applications in automotive sector. *Metal Powder Report*, *68*(6), 12–15. doi:10.1016/S0026-0657(13)70210-8

Carlsson, J-O., & Martin, P. M. (2010). *Chemical Vapor Deposition*. Elsevier. DOI: . doi:10.1016/B978-0-8155-2031-3.00007-7

Cavaliere, P. (2017). *Cold-Spray Coatings: Recent Trends and Future perspectives*. Springer.

Celik, S., Albayrak, A. T., Akyuz, S., & Ozel, A. E. (2020). The Role of Multivariant Analysis on the Interpretation of FTIR and Raman Spectra. In Design of Experiments for Chemical, Pharmaceutical, Food, and Industrial Applications (pp. 104-129). IGI Global. doi:10.4018/978-1-7998-1518-1.ch005

Çengel, Y. A., & Ghajar, A. J. (2015). *Heat and Mass Transfer: Fundamentals & Applications* (5th ed.). McGraw-Hill.

Champagne, V. (2007). *The cold spray Materials Deposition Process: Fundamentals and Application*. Woodhead Publishing Ltd.

Champagne, V. K. (2008). The repair of magnesium rotorcraft components by cold spraying. *Journal of Failure Analysis and Prevention*, *8*(2), 164–175. doi:10.100711668-008-9116-y

Champagne, V. K., & Helfritch, D. J. (2013). A demonstration of the antimicrobial effectiveness of various copper surfaces. *Journal of Biological Engineering*, *7*(8), 1–7. doi:10.1186/1754-1611-7-8 PMID:23537176

Champagne, V. K., Leyman, P. F., & Helfritch, D. (2008). Magnesium repair by cold spraying. *Plating and Surface Finishing*, *95*, 34.

Chang, C., Lei, Z., Li, Y., & Wang, Z. (2016). Electrochemical Research of a Stable Electroless Silver Bath. *Journal of the Electrochemical Society*, *163*(3), 121–125. doi:10.1149/2.1101603jes

Charbonnier, M., Goepfert, Y., Romand, M., & Leonard, D. (2004). Electroless plating of glass and silicon substrates through surface pretreatments involving plasma-polymerization and grafting processes. *The Journal of Adhesion*, *80*(12), 1103–1130. doi:10.1080/00218460490884132

Chawla, N., Fillari, G., & Narasimham, K. S. (1999). Recent advances in ferrous powder metallurgy. In Powder Materials: Current Research and Industrial Practices(as held at the 1999 TMS Fall Meeting) (pp. 247-255). Academic Press.

Chebiam, R. V., Dubin, V. M., & Simka, H. S. (2004). Patent No. US20040096592A1. Retrieved from https://patents.google.com/patent/US20040096592A1/en

Chen, F., Jia, M., She, Y., Wu, Y., Shen, Q., & Zhang, L. (2020). Mechanical behavior of AlN/Mo functionally graded materials with various compositional structures. *Journal of Alloys and Compounds*, *816*, 152512. doi:10.1016/j.jallcom.2019.152512

Cheng, J., Gan, X., Chen, S., Lai, Y., Xiong, H., & Zhou, K. (2019). Properties and microstructure of copper/nickel-iron-coated graphite composites prepared by electroless plating and spark plasma sintering. *Powder Technology*, *343*, 705–713. doi:10.1016/j.powtec.2018.11.057

Chen, J., Ren, S., He, X., & Qu, X. (2017). Properties and microstructure of nickel-coated graphite flakes/copper composites fabricated by spark plasma sintering. *Carbon*, *121*, 25–34. doi:10.1016/j.carbon.2017.05.082

Chen, J., Xie, Z., Zeng, W., & Wu, W. (2017). Toughening mechanisms of ZTA ceramics at cryogenic temperature (77 K). *Ceramics International*, *43*(5), 3970–3974. doi:10.1016/j.ceramint.2016.11.072

Chen, J., Zhao, G., Matsuda, K., & Zou, Y. (2019b). Microstructure evolution and corrosion resistance of Ni–Cu–P amorphous coating during crystallization process. *Applied Surface Science*, *484*, 835–844. doi:10.1016/j.apsusc.2019.04.142

Chen, J., Zhao, G., Zhang, Y., Duan, S., Matsuda, K., & Zou, Y. (2019a). Metastable phase evolution and nanoindentation behavior of amorphous Ni–Cu–P coating during heat treatment process. *Journal of Alloys and Compounds*, *805*, 597–608. doi:10.1016/j.jallcom.2019.07.068

Chen, K., Liu, Y., Fu, Y., Huang, J., Feng, X., Wang, J., & Li, H. (2020). Loading of Zn/ZnO particles in the precursor feedstock affects the characteristics of liquid plasma sprayed nano-ZnO coatings for photocatalytic applications. *Nanotechnology*, *31*(18), 185301. doi:10.1088/1361-6528/ab6cda PMID:31945757

Chen, L., Jiang, Y., Chong, X., & Feng, J. (2018). Synthesis and Thermophysical Properties of $RETa_3O_9$ (RE = Ce, Nd, Sm, Eu, Gd, Dy, Er) as Promising Thermal Barrier Coatings. *Journal of the American Ceramic Society*, *101*(3), 1266–1278. doi:10.1111/jace.15268

Chen, M. L., Qu, Y. Y., Yang, L., & Gao, H. (2008). Structures and antifouling properties of low surface energy non-toxic antifouling coatings modified by nano-SiO_2 powder. *Science in China. Series B, Chemistry*, *51*(9), 848–852. doi:10.100711426-008-0069-5

Chen, P. Y., Wong, H. C., Hu, S. F., Huang, C. Y., & Liu, R. S. (2011). Functional electroless gold Ohmic contacts in light emitting diodes. *Applied Physics Letters*, *99*(6), 063511. doi:10.1063/1.3624831

Chen, P., Zhou, Z., Jiang, W., Luo, W., Yang, J., Zhu, J., & Fan, Y. (2018). Enhancing the thermoelectric performance of filled skutterudite nanocomposites in a wide temperature range via electroless silver plating. *Scripta Materialia*, *146*, 136–141. doi:10.1016/j.scriptamat.2017.11.019

Chen, S., Brown, L., Levendorf, M., Cai, W., Ju, S. Y., Edgeworth, J., Li, X., Magnuson, C. W., Velamakanni, A., Piner, R. D., Kang, J., Park, J., & Ruoff, R. S. (2011). Oxidation Resistance of Graphene-Coated Cu and Cu/Ni Alloy. *ACS Nano*, *5*(2), 1321–1327. doi:10.1021/nn103028d PMID:21275384

Chen, S., Zhou, X., Song, W., Sun, J., Zhang, H., Jiang, J., Deng, L., Dong, S., & Cao, X. (2019). Mg_2SiO_4 as a Novel Thermal Barrier Coating Material for Gas Turbine Applications. *Journal of the European Ceramic Society*, *39*(7), 2397–2408. doi:10.1016/j.jeurceramsoc.2019.02.016

Chen, W. G., Duan, X. M., Dong, T., Gao, L. N., Chen, M. Z., & Ye, H. (2008). Study on Steam Treatment Process for Steel. *Surface Technology*, *37*(6), 57–60.

Chen, Y., Cao, M., Xu, Q., & Zhu, J. (2003). Electroless nickel plating on silicon carbide nanoparticles. *Surface and Coatings Technology*, *172*(1), 90–94. doi:10.1016/S0257-8972(03)00320-7

Chevalier, J., & Gremillard, L. (2009). Ceramics for medical applications: A picture for the next 20 years. *Journal of the European Ceramic Society*, *29*(7), 1245–1255. doi:10.1016/j.jeurceramsoc.2008.08.025

Chin, B., Ding, P., Sun, B., Chiang, T., Angelo, D., Hashim, I., & Chen, F. (1998). Barrier and seed layers for damascene copper metallization. *Solid State Technology*, *41*(7), 141–145.

Chinglenthoiba, C., Ramkumar, K., Shanmugaraja, T., & Sharma, S. (2017). Study on Nanotechnology, Nanocoating and Nanomaterial. *IJCAM,* 17-25.

Choi, E. B., & Lee, J. H. (2019). Dewetting behavior of Ag in Ag-coated Cu particle with thick Ag shell. *Applied Surface Science*, *480*, 839–845. doi:10.1016/j.apsusc.2019.02.221

Choudhuri, A., Mohanty, P. S., & Karthikeyan, J. (2009). *Bioceramic composite coatings by cold spraying technology*. Addit. Manuf. Process.

Choudhury, B. S., Sen, R. S., Oraon, B., & Majumdar, G. (2009). Statistical study of nickel and phosphorus contents in electroless Ni–P coatings. *Surface Engineering*, *25*(5), 410–414. doi:10.1179/174329407X239081

Choy, K. (2003). Chemical vapour deposition of coatings. *Progress in Materials Science*, *48*(2), 57–170. doi:10.1016/S0079-6425(01)00009-3

Choy, K. L. (2003). Chemical vapor deposition of coatings. *Progress in Materials Science*, *48*(2), 57–170. doi:10.1016/S0079-6425(01)00009-3

Christy, A. A., Ozaki, Y., & Gregoriou, V. G. (2001). *Modern Fourier transform infrared spectroscopy*. Elsevier.

Chu, C. C. (2002). *Textile-based biomaterials for surgical applications*. CRC.

Chu, K., Liu, Z., Jia, C., Chen, H., Liang, X., Gao, W., Tian, W., & Guo, H. (2010). Thermal conductivity of SPS consolidated Cu/diamond composites with Cr-coated diamond particles. *Journal of Alloys and Compounds*, *490*(1-2), 453–458. doi:10.1016/j.jallcom.2009.10.040

Chun, D. M., Kim, M. H., Lee, J. C., & Ahn, S. H. (2008). TiO2 coating on metal and polymer substrates by nano-particle deposition system (NPDS). *CIRP Annals*, *57*(1), 551–554. doi:10.1016/j.cirp.2008.03.111

Cinca, N., Vilardell, A. M., Dosta, S., Concustell, A., Garcia Cano, I., Guilemany, J. M., & Peiró, F. (2016). A new alternative for obtaining nanocrystalline bioactive coatings: Study of hydroxyapatite deposition mechanisms by cold gas spraying. *Journal of the American Ceramic Society*, *99*(4), 1420–1428. doi:10.1111/jace.14076

Ciurczak, E. W., & Igne, B. (2015). *Pharmaceutical and medical applications of near-infrared spectroscopy*. CRC Press.

Clarke, D. R., & Levi, C. G. (2003). Materials design for the next generation thermal barrier coatings. *Annual Review of Materials Research*, *33*(1), 383–417.

Clarke, D. R., Oechsner, M., & Padture, N. P. (2012). Thermal-Barrier Coatings for More Efficient Gas-Turbine Engines. *MRS Bulletin*, *37*(10), 891–898. doi:10.1557/mrs.2012.232

Clarke, D. R., & Phillpot, S. R. (2005). Thermal Barrier Coating Materials. *Materials Today*, *8*(6), 22–29. doi:10.1016/S1369-7021(05)70934-2

CoatingsP. V. D. (n.d.). https://www.pvd-coatings.co.uk/applications/

Company, A. G. A. (n.d.). *There is no better one*. Retrieved from: https://www.linde-gas.ee

Conradi, M., Kocijan, A., Kek-Merl, D., Zorko, M., & Verpoest, I. (2014). Mechanical and anticorrosion properties of nanosilica-filled epoxy-resin composite coatings. *Applied Surface Science*, *292*, 423–437. doi:10.1016/j.apsusc.2013.11.155

Corti, C. W., & Holliday, R. J. (2004). Commercial aspects of gold applications: From materials science to chemical science. *Gold Bulletin*, *37*(1-2), 20–26. doi:10.1007/BF03215513

Creighton, J. R., & Ho, P. (2001). *Introduction to chemical vapor deposition (CVD)*. ASM International.

Ctibor, P., Lechnerova, R., & Benes, V. (2006). Quantitative analysis of pores of two types in a plasma-sprayed coating. *Materials Characterization*, *56*(4-5), 297–304. doi:10.1016/j.matchar.2005.11.016

Cui, R., Han, Y., Zhu, Z., Chen, B., Ding, Y., Zhang, Q., Wang, Q., Ma, G., Pei, F., & Ye, Z. (2019). Investigation of the structure and properties of electrodeposited Cu/graphene composite coatings for the electrical contact materials of an ultrahigh voltage circuit breaker. *Journal of Alloys and Compounds*, *777*, 1159–1167. doi:10.1016/j.jallcom.2018.11.096

Cui, X., Hutt, D. A., & Conway, P. P. (2012). Evolution of microstructure and electrical conductivity of electroless copper deposits on a glass substrate. *Thin Solid Films*, *520*(19), 6095–6099. doi:10.1016/j.tsf.2012.05.068

Dangsheng, X. (2002). The Effect of Rare Earth Compounds on the Friction and Wear Properties of PM Ni-based Alloy. *PM Technology-China, 20*(82), 67-70.

Danninger, H., & Gierl, C. (2008). New alloying systems for ferrous powder metallurgy precision parts. *Science of Sintering*, *40*(1), 33–46. doi:10.2298/SOS0801033D

Daoush, W. M. (2008). Processing and characterization of CNT/Cu nanocomposites by powder technology. *Powder Metallurgy and Metal Ceramics*, *47*(9–10), 531–537. doi:10.1007/s11106-008-9055-x

Das, I., & De, G. (2015). Zirconia based superhydrophobic coatings on cotton fabrics exhibiting excellent durability for versatile use. *Scientific Reports*, *5*(1), 18503. doi:10.1038rep18503 PMID:26678754

Das, S., Weerasiri, L. D., & Yang, W. (2017). Influence of surface tension on bubble nucleation, formation and onset of sliding. *Colloids and Surfaces. A, Physicochemical and Engineering Aspects*, *516*, 23–31. doi:10.1016/j.colsurfa.2016.12.010

Davis, J. R. (2004). *Handbook of Thermal Spray Technology*. ASM International.

Davis, J. R. (Ed.). (2001). *Surface engineering for corrosion and wear resistance*. ASM international.

Davis, J. R. (Ed.). (2004). *Handbook of thermal spray technology*. ASM International.

Davis, R., El-Shafei, A., & Hauser, P. (2011). Use of atmospheric pressure plasma to confer durable water repellent functionality and antimicrobial functionality on cotton/polyester blend. *Surface and Coatings Technology*, *205*(20), 4791–4797. doi:10.1016/j.surfcoat.2011.04.035

Dcleng. (n.d.). Retrieved from: https://www.dcleng.com.au/index.php?f=flame-spray-ceramic

Deckert, C. A. (1995). Electroless copper plating a review: Part II. *Plating and Surface Finishing*, *82*(3), 58–64.

Deepa, J. P., Rajan, T. P. D., Pavithran, C., & Pai, B. C. (2012). Formation of Electroless Ni-B Coating on Boron Carbide Particles for Composite Processing. *Materials Science Forum*, *710*, 359–364. www.scientific.net/MSF.710.359 doi:10.4028/www.scientific.net/MSF.710.359

DeForce, Eden, Potter, Champagne, Leyman, & Helfritch. (2007). Application of aluminium coatings for the corrosion protection of magnesium by cold spraying. *TRI Serv. Corros.*, 1-16.

Delichere, P., Goff, A. H., & Joiret, S. (1988). Study of thin corrosion films by in situ Raman spectroscopy combined with direct observation of nuclear reactions. *Surface and Interface Analysis*, *12*(7), 419–423. doi:10.1002ia.740120710

Deloro Stellite Guns. (n.d.). Retrieved from: https://www.stellite.com/us/en/products/stellite-equipment/hvof-jet-kote/jet-kote-III.html

Denka spherical aluminal Denka Company Limited. (n.d.). Retrieved from: https://www.denka.co.jp/eng/product/detail_00040/

Dennis, J. K., & Such'Nickel, T. E. (1993). Chromium Plating (3rd ed.). Woodhead Publishing Ltd.

Dewan, P. (2009). Nanotech enabled anti-graffiti clear coating system. *Bombay Technologists*, *59*, 79–83.

Dhomne, S., & Mahalle, A. M. (2019). Thermal Barrier Coating Materials for SI Engine. *Journal of Materials Research and Technology*, *8*(1), 1532–1537. doi:10.1016/j.jmrt.2018.08.002

Dikici, B., Tekmen, C., Gavgali, M., & Cocen, U. (2011). The effect of electroless Ni coating of SiC particles on the corrosion behavior of A356 Based Squeeze Cast Composite. Strojniski Vestnik/Journal of Mechanical Engineering, 57(1), 11–20. doi:10.5545v-jme.2010.111

Dincer, M. (2014). *Sinter hardening studies of ASTALOY 85 Mo powder with varying copper contents* (Master's thesis). Middle East Technical University.

Ding, D., Pan, Z., Cuiuri, D., & Li, H. (2015). Wire-feed additive manufacturing of metal components: Technologies, developments and future interests. *International Journal of Advanced Manufacturing Technology*, *81*(1-4), 465–481. doi:10.100700170-015-7077-3

Dini, J. W., Beat, T. G., Cowden, W. C., Ryan, L. E., & Hewitt, W. B. (1992, June). Plating under reduced pressure. In Proceedings of the 79th AESF Annual Technical Conference. Atlanta: American Electroplaters and Surface Finishers Society.

Djokić, S. S. (2005). Electroless Deposition of Metals and Alloys. In Modern Aspects of Electrochemistry (pp. 51–133). doi:10.1007/0-306-47604-5_2

Dobrzanski, L. A., Brytan, Z., Grande, M. A., & Rosso, M. (2007). Properties of duplex stainless steels made by powder metallurgy. *Archives of Materials Science and Engineering*, *28*(4), 217–223.

Dong, H., Wang, D., Pei, Y., Li, H., Li, P., & Ma, W. (2013). Optimization and Thermal Cycling Behavior of $La_2Ce_2O_7$ Thermal Barrier Coatings. *Ceramics International, 39*(2), 1863–1870. doi:10.1016/j.ceramint.2012.08.034

Donghai, Y., Chengyong, W., & Fenglin, Z. (2007). Development of Cutting Tool Coating Material. *Tool Engineering, 41*(6), 25–32.

Dooley, R., & Wiertel, E. (2009). A survey of erosion and corrosion resistant materials being used on boiler tubes in waste to energy boilers. *Proceedings of the 17the Annual North American Waste-to-Energy Conference, NAWTEC17.* 10.1115/NAWTEC17-2334

Dorfman, M. R. (2018). Thermal spray coatings. In Handbook of Environmental Degradation Of Materials: Third Edition (pp. 469–488). doi:10.1016/B978-0-323-52472-8.00023-X

Dosta, S., Couto, M., & Guilemany, J. M. (2013). Cold spray deposition of a WC-25Co cermet onto Al7075-T6 and carbon steel substrates. *Acta Materialia, 61*(2), 643–652. doi:10.1016/j.actamat.2012.10.011

Duan, W., Xie, A., Shen, Y., Wang, X., Wang, F., Zhang, Y., & Li, J. (2011). Fabrication of Superhydrophobic Cotton Fabrics with UV Protection Based on CeO_2 Particles. *Industrial & Engineering Chemistry Research, 50*(8), 4441–4445. doi:10.1021/ie101924v

Duan, Z., & Heberlein, J. (2002). Arc instabilities in a plasma spray torch. *Journal of Thermal Spray Technology, 11*(1), 44–51. doi:10.1361/105996302770348961

Dubbert, W., Schwirn, K., Völker, D., & Apel, P. (2014). *Use of nanomaterials in coatings.* Federal Environment Agency.

Durul, E. N. A., & Nurbas, M. (2010). Corrosion Behavior of Different Thermal Spray Coatings and Hard Chromium Electroplating on A286 Super Alloy. *Advanced Materials Research, 154-155*, 226–229. doi:10.4028/www.scientific.net/AMR.154-155.226

Dykhuizen & Smith. (1998). Gas dynamic principle of cold spray. *Thermal Spray Technology*, 205-212.

Dykhuizen, R. C., Smith, M. F., Glimore, D. L., Neiser, R. A., Jiang, X., & Samapth, S. (1999). Impact of high velocity cold spray particles. *Journal of Thermal Spray Technology, 8*(4), 559–564. doi:10.1361/105996399770350250

Eddings, E. G. (2005). *Mechanical Engineers' Handbook.* John Wiley & Sons, Inc.

Edris, H., Mc-Cartney, D. G., & Sturgeon, A. J. (1997). Micro-structural Characterization of HVOF Sprayed Coatings of Inconel 625. *Journal of Materials, 32*, 863–868.

Edwin, A., Anand, V., & Prasanna, K. (2017). Sustainable development through functionally graded materials: An overview. *Rasayan Journal of Chemistry, 10*(1), 149–152. doi:10.7324/RJC.2017.1011578

Eklund, J., Phother, J., Sadeghi, E., Joshi, S., & Liske, J. (2019). High-Temperature Corrosion of HVAF-Sprayed Ni-Based Coatings for Boiler Applications. *Oxidation of Metals, 91*(5), 729–747. doi:10.100711085-019-09906-0

Eklund, J., Sadeghimeresht, E., Simon, J. P., Liske, J., Markocsan, N., & Joshi, S. (2018). Effect of Water Vapor on the Oxidation Behavior of HVAF-Sprayed NiCr and NiCrAlY Coatings. *Mater. Corros., 69*(10), 1431–1440. doi:10.1002/maco.201710005

El Mahallawy, N., Bakkar, A., Shoeib, M., Palkowski, H., & Neubert, V. (2008). Electroless Ni–P coating of different magnesium alloys. *Surface and Coatings Technology, 202*(21), 5151–5157. doi:10.1016/j.surfcoat.2008.05.037

El-Eskandarany, M. S. (2015). *Mechanical alloying: nanotechnology, materials science and powder metallurgy* (Vol. 2). Elsevier.

El-Galy, I. M., Saleh, B. I., & Ahmed, M. H. (2019). Functionally graded materials classifications and development trends from industrial point of view. SN Applied Sciences, 1(11), 1–23. doi:10.1007/s42452-019-1413-4

Eliaz, N., Shemesh, G., & Latanision, R. M. (2002). Hot Corrosion in Gas Turbine Components. *Engineering Failure Analysis*, *9*(1), 31–43. doi:10.1016/S1350-6307(00)00035-2

El-Shafei, A., El-Bisi, M. K., Zaghloul, S., & Refai, R. (2017). Herbal Textile Finishes-Natural Antibacterial Finishes for Cotton Fabric. *Egyptian Journal of Chemistry*, *60*(2), 161–180.

Elsharkawi, M., & Esawi, A. M. K. (2018). Development of an Electroless Plating Process for Multi-wall Carbon Nanotubes (MWCNTS) to Improve Their Dispersion and Wettability in Molten Aluminum. In Minerals, Metals and Materials Series (pp. 29–39). doi:10.1007/978-3-319-72853-7_3

El-Zairy, E. M. R., & Morgan, N. N. (2015). Functionalization of Polyester Fabric with UV-Protection and Antibacterial Property by Means of Atmospheric Glow Discharge. *International Journal of Textile Science*, *4*(5), 113–117.

Engström, U., Lindberg, C., & Tengzelius, J. (1992). Powders and processes for high performance PM steels. *Powder Metallurgy*, *35*(1), 67–73. doi:10.1179/pom.1992.35.1.67

Erdemir, F., Canakci, A., & Varol, T. (2015). Microstructural characterization and mechanical properties of functionally graded Al2024/SiC composites prepared by powder metallurgy techniques [English Edition]. *Transactions of Nonferrous Metals Society of China*, *25*(11), 3569–3577. doi:10.1016/S1003-6326(15)63996-6

Eutectic Castolin Eutalloy, S. F. (n.d.). *Eutalloy - Eutalloy Spray and Fusing ®*. Retrieved from: https://www.castolin.com/sites/default/files/Eutalloy-powder-spray-fusing

Faizan-Ur-Rab, M., Zahiri, S. H., Masood, S. H., Phan, T. D., Jahedi, M., & Nagaraja, R. (2016). Application of a holistic 3D model to estimate state of cold spray titanium particles. *Materials & Design*, *89*, 1227–1241. doi:10.1016/j.matdes.2015.10.075

Fan, G., Xu, R., Tan, Z., Zhang, D., & Li, Z. (2014). Development of Flake Powder Metallurgy in Fabricating Metal Matrix Composites: A Review. *Acta Metallurgica Sinica. English Letters*, *27*(5), 806–815. doi:10.1007/s40195-014-0148-x

Fan, L., Wang, Q., Yang, P., Chen, H., Hong, H., Zhang, W., & Ren, J. (2018). Preparation of nickel coating on ZTA particles by electroless plating. *Ceramics International*, *44*(10), 11013–11021. doi:10.1016/j.ceramint.2018.03.055

Fantozzi, D., Matikainen, V., Uusitalo, M., Koivuluoto, H., & Vuoristo, P. (2017). Chlorine induced high temperature corrosion of Inconel 625 sprayed coatings deposited with different thermal spray techniques. *Surface and Coatings Technology*, *318*, 233–243. doi:10.1016/j.surfcoat.2016.12.086

Fauchais, P. (2004). Understanding Plasma Spraying. *Journal of Physics. D, Applied Physics*, *37*(9), 86–108. doi:10.1088/0022-3727/37/9/R02

Fauchais, P. L., Heberlein, J. V. R., & Boulos, M. I. (2014). Industrial Applications of Thermal Spraying Technology. In *Thermal Spray Fundamentals* (pp. 1401–1566). Springer. doi:10.1007/978-0-387-68991-3_18

Fauchais, P. L., Heberlein, J. V., & Boulos, M. I. (2014). *Thermal spray fundamentals: from powder to part*. Springer Science & Business Media. doi:10.1007/978-0-387-68991-3

Fauchais, P., Coudert, J. F., & Vardelle, M. (1997). Transient Phenomena in Plasma Torches and in Plasma Sprayed Coating Generation. *Journal de Physique. IV*, *7*(C4), 187–198. doi:10.1051/jp4:1997414

Fernandes, K. S., Alvarenga, E. D. A., Brandão, P. R. G., & Lins, V. D. F. C. (2011). Infrared-spectroscopy analysis of zinc phosphate and nickel and manganese modified zinc phosphate coatings on electrogalvanized steel. *REM. Revista Escola de Minas*, *64*(1), 45–49. doi:10.1590/S0370-44672011000100005

Ferrari, A. M., Pini, M., Neri, P., & Bondioli, F. (2015). Nano-TiO_2 Coatings for Limestone: Which Sustainability for Cultural Heritage. *Coatings*, *5*(3), 232–245. doi:10.3390/coatings5030232

Ferreira, S. C., Rocha, L. A., Ariza, E., Sequeira, P. D., Watanabe, Y., & Fernandes, J. C. S. (2011). Corrosion behaviour of Al/Al3Ti and Al/Al3Zr functionally graded materials produced by centrifugal solid-particle method: Influence of the intermetallics volume fraction. *Corrosion Science*, *53*(6), 2058–2065. doi:10.1016/j.corsci.2011.02.010

Ferrer, M. Y., Vargas, F., Vera López, E., Moreno Téllez, C. M., & Peña Rodríguez, G. (2017). Evaluación de la morfología, resistencia al desgaste ya la corrosión de recubrimientos de ZrO2-Al2O3/Ni depositados sobre sustratos de acero al carbono mediante proyección térmica por llama. *Ingeniare. Revista Chilena de Ingeniería*, *25*(4), 721–732.

Flame Spray Guns. (n.d.). Retrieved from: https://www.fst.nl/systems/thermal-spray-guns-and-torches/egun-hvof-spray-gun/

Fotovvati, B., Namdari, N., & Dehghanghadikolaei, A. (2019). On Coating Techniques for Surface Protection: A Review. *Journal of Manufacturing and Materials Processing*, *3*(1), 28. doi:10.3390/jmmp3010028

Franco, D., Vargas, F., & López Gómez, M. E. (2016). Study of thermal shock resistance of flame sprayed coatings manufactured to protect molds used in glass containers industry. *Revista Chilena de Ingeniería*, *24*(2), 239–248. doi:10.4067/S0718-33052016000200007

Friedrich, C., Gadow, R., & Schirmer, T. (2001). Lanthanum hexaluminate a new material for atmospheric plasma spraying of advanced thermal barrier coatings. *Journal of Thermal Spray Technology*, *10*(4), 592–598. doi:10.1361/105996301770349105

Fukuda, M. (2017). 22 - Advanced USC technology development in Japan. *Materials for Ultra-Supercritical and Advanced Ultra-Supercritical Power Plants*, 733-754.

Galedari, S. A., Mahdavi, A., Azarmi, F., Huang, Y., & McDonald, A. (2019). A comprehensive review of corrosion resistance of thermally-sprayed and thermally-diffused protective coatings on steel structures. *J. Therm. Spray Tech.*, *28*(4), 645–677. doi:10.100711666-019-00855-3

Ganesan, S., Padmanabhan, S., Hemanandh, J., & Venkatesan, S. P. (2019). Influence of substrate temperature on coated engine piston head using multi-response optimisation techniques. *International Journal of Ambient Energy*, 1-8.

Gangotri, D. L., & Chaware, A. D. (2004). *Paint India*. Academic Press.

Ganvir, A. (2016). *Microstructure and Thermal Conductivity of Liquid Feedstock Plasma Sprayed Thermal Barrier Coatings* (Doctoral dissertation). University West.

Gao, J., Wu, Y., Liu, L., & Hu, W. (2005). Crystallization behavior of nanometer-sized Al2O3 composite coatings prepared by electroless deposition. *Materials Letters*, *59*(2–3), 391–394. doi:10.1016/j.matlet.2004.09.032

Garcia, B., Lamzoudi, A., Pillier, F., Le, H. N. T., & Deslouis, C. (2002). Oxide/Polypyrrole Composite Films for Corrosion Protection of Iron. *Journal of the Electrochemical Society*, *149*(12), B560. doi:10.1149/1.1517581

Garcia, M., Martinez, H. V., & Ortiz, A. (2010). Compoforging of Al-Si metal matrix composites reinforced with beta-SiC: An alternative technique. *Open Industrial & Manufacturing Engineering Journal*, *3*, 1–6.

Garg, A. K., & De Jonghe, L. C. (1993). Metal-coated colloidal particles. *Journal of Materials Science*, *28*(13), 3427–3432. doi:10.1007/BF01159817

Gärtner, F., Stoltenhoff, T., Schmidt, T., & Kreye, H. (2006). The cold spray process and its potential for industrial applications. *Journal of Thermal Spray Technology*, *15*(2), 223–232. doi:10.1361/105996306X108110

Gashti, M. P., Pakdel, E., & Alimohammadi, F. (2016). *Nanotechnology-based coating techniques for smart textiles*. Academic Press.

Gasik, M. M. (2010). Functionally Graded Materials: Bulk processing techniques. *International Journal of Materials & Product Technology*, *39*(1/2), 20. Advance online publication. doi:10.1504/IJMPT.2010.034257

Gebert, A., & Bouaifi, B. (2006). Surface Protection by Means of Build-up Welding. *Modern Surface Technology*, 263-296.

Ge, J., Turunen, M. P. K., & Kivilahti, J. K. (2003). Surface modification of a liquid-crystalline polymer for copper metallization. *Journal of Polymer Science. Part B, Polymer Physics*, *41*(6), 623–636. doi:10.1002/polb.10407

Ghane, M., & Ghorbani, E. (2016). Investigation into the UV-protection of woven fabrics composed of metallic weft yarns. *AUTEX Research Journal*, *16*(3), 154–159. doi:10.1515/aut-2015-0021

Ghelichi, R., & Guaglians, M. (2009). Coating by the cold spraying process. Frat Ed Integrita Strut Ed, 3, 30-44.

Ghosh Chaudhuri, R., & Paria, S. (2012). Core/Shell Nanoparticles: Classes, Properties, Synthesis Mechanisms, Characterization, and Applications. *Chemical Reviews*, *112*(4), 2373–2433. doi:10.1021/cr100449n

Ghosh, S. (2019). Electroless copper deposition: A critical review. *Thin Solid Films*, *669*, 641–658. doi:10.1016/j.tsf.2018.11.016

Gildersleeve, E., Viswanathan, V., & Sampath, S. (2019). Molten Silicate Interactions with Plasma Sprayed Thermal Barrier Coatings: Role of Materials and Microstructure. *Journal of the European Ceramic Society*, *39*(6), 2122–2131. doi:10.1016/j.jeurceramsoc.2019.01.023

Gill, A. S., & Kumar, S. (2019). Surface characteristics investigation of tool steel machined by powder metallurgy tool in EDA. *Materials Research Express*, *6*(12), 1265b1.

Gil, L., & Staia, M. H. (1999). Microstructure and properties of HVOF thermal sprayed NiWCrBSi coatings. *Surface and Coatings Technology*, *120–121*, 423–429. doi:10.1016/S0257-8972(99)00494-6

Gil, L., & Staia, M. H. (2002). Influence of HVOF parameters on the corrosion resistance of NiWCrBSi coatings. *Thin Solid Films*, *420–421*, 446–454. doi:10.1016/S0040-6090(02)00815-5

Gläsel, H. J., Bauer, F., Ernst, H., Findeisen, M., Hartmann, E., Langguth, H., Mehnert, R., & Schubert, R. (2000). Preparation of scratch and abrasion resistant polymeric nanocomposites by monomer grafting onto nanoparticles, Characterization of radiation-cured polymeric nanocomposites. *Macromolecular Chemistry and Physics*, *201*(18), 2765–2770. doi:10.1002/1521-3935(20001201)201:18<2765::AID-MACP2765>3.0.CO;2-9

Glassman, I., Yetter, R. A., & Glumac, N. G. (2014). *Combustion*. Academic press.

Goff, A. H., Joiret, S., & Abourazzouk, D. (1998). Raman investigation of crevice corrosion in nickel-chromium dental alloys containing beryllium. *Electrochimica Acta*, *43*(1-2), 53–62. doi:10.1016/S0013-4686(97)00234-X

Goffredo, G. B., & Munafò, P. (2015). Preservation of Historical Stone Surfaces by TiO_2 Nanocoatings. *Coatings*, *5*(2), 222–231. doi:10.3390/coatings5020222

Gokarneshan, N., Gopalakrishnan, P. P., & Jeyanthi, B. (2012). Influence of Nanofinishes on the Antimicrobial Properties of Fabrics. *ISRN Nanomaterials*, *2012*, 193836. Advance online publication. doi:10.5402/2012/193836

Gok, M. G., & Goller, G. (2016). Production and characterization of GZ/CYSZ alternative thermal barrier coatings with multilayered and functionally graded designs. *Journal of the European Ceramic Society*, *36*(7), 1755–1764. doi:10.1016/j.jeurceramsoc.2016.01.036

Goldschmidt, A., & Streitberger, H. J. (2003). *BASF handbook on basics of coating technology*. Primedia.

Gonzalez, R., Ashrafizadeh, H., Lopera, A., Mertiny, P., & McDonald, A. (2016). A Review of Thermal Spray Metallization of Polymer-Based Structures. *Journal of Thermal Spray Technology*, *25*(5), 897–919. doi:10.1007/s11666-016-0415-7

González, R., Garcia, M. A., Penuelas, I., Cadenas, M., del Rocío Fernández, M., Battez, A. H., & Felgueroso, D. (2007). Microstructural study of NiCrBSi coatings obtained by different processes. *Wear*, *263*(1-6), 619–624. doi:10.1016/j.wear.2007.01.094

Goodman, P. (2002). Current and future uses of gold in electronics. *Gold Bulletin*, *35*(1), 21–26. doi:10.1007/BF03214833

Grasso, S., Sakka, Y., & Maizza, G. (2009). Electric current activated/assisted sintering (ECAS): a review of patents 1906–2008. Science and Technology of Advanced Materials, 10(5), 053001. doi:10.1088/1468-6996/10/5/053001

Green, T. A., Liew, M. J., & Roy, S. (2003). Electrodeposition of gold from a thiosulfate-sulfite bath for microelectronic applications. *Journal of the Electrochemical Society*, *150*(3), C104–C110. doi:10.1149/1.1541006

Grewal, H. S., Singh, H., & Agrawal, A. (2013). Microstructural and mechanical characterization of thermal sprayed nickel–alumina composite coatings. *Surface and Coatings Technology*, *216*, 78–92. doi:10.1016/j.surfcoat.2012.11.029

Grimanelis, D., & Eyre, T. S. (2007). Wear characteristics of a diffusion bonded sintered steel with short term surface treatments. *Wear*, *262*(1-2), 93–103. doi:10.1016/j.wear.2006.04.006

Grujicic, M., Saylor, J., Beasley, D., DeRosset, W., & Helfritch, D. (2003). Computational analysis of the interfacial bonding between feed-powder particles and the substrate in the cold-gas dynamic spray process. *Applied Surface Science*, *219*(3-4), 211–227. doi:10.1016/S0169-4332(03)00643-3

Grujicic, M., Zaho, C. L., DeRosset, W. S., & Helfritch, D. (2004). Adiabatic shear instability based mechanism for particles substrate bonding in the cold-gas dynamic-spray process. *Materials & Design*, *25*, 68–81.

Guetta, S., Berger, M., Borit, F., Guipont, V., Jeandin, M., Bousite, M., Ichikawa, Y., Sakaguchi, K., & Ogawa, K. (2009). Influence of Particle Velocity on Adhesion of Cold-Sprayed Splats. *Journal of Thermal Spray Technology*, *18*(3), 331–342. doi:10.100711666-009-9327-0

Guilemany, J. M., Fernández, J., Delgado, J., Benedetti, A. V., & Climent, F. (2002). Effects of Thickness Coating on the Electrochemical Behaviour of Thermal Spray Cr_3C_2-NiCr Coatings. *Surface and Coatings Technology*, *153*(2-3), 107–113. doi:10.1016/S0257-8972(01)01679-6

Güler, O., Varol, T., Alver, Ü., & Çanakçı, A. (2019). The effect of flake-like morphology on the coating properties of silver coated copper particles fabricated by electroless plating. *Journal of Alloys and Compounds*, *782*, 679–688. doi:10.1016/j.jallcom.2018.12.229

Guns Oerlikon MetcoH. V. O. F. (n.d.). Retrieved from: https://www.oerlikon.com/metco/en/products-services/coating-equipment/thermal-spray/spray-guns/spray-guns-hvof/hvof-diamond-jet/

GunsE. (n.d.). Retrieved from: https://www.castolin.com/product/hvof-castojet-cjk5

GunsG. T. V. Online. (n.d.). Available: https://www.gtv-mbh.com/_old/gtv-mbh-englisch/www.gtv-mbh.de/cms/front_content5d9e.html?idcat=84

GunsT. (n.d.). Retrieved from: http://thermico.de/EN/products/engineering/spray-gun-technology.html

GunsU. (n.d.). Retrieved from: http://www.uniquecoat.com/m2-hvaf-spray-gun.html

Guo, L., Guo, H., Ma, G., Abbas, M., & Gong, S. (2012). Ruddlesden-Popper Structured $BaLa_2Ti_3O_{10}$, a Highly Anisotropic Material for Thermal Barrier Coatings. *Ceramics International*, *38*(5), 4345–4352. doi:10.1016/j.ceramint.2012.02.017

Guo, N., Luan, B., He, F., Li, Z., & Liu, Q. (2014). Influence of flake thickness on the shape and distribution of Al2O3 particles in Al matrix composites fabricated by flake powder metallurgy. *Scripta Materialia*, *78–79*, 1–4. doi:10.1016/j.scriptamat.2013.12.019

Guo, R. H., Jiang, S. Q., Yuen, C. W. M., Ng, M. C. F., & Zheng, G. H. (2012). Influence of nickel ions for electroless Ni-P plating on polyester fabric. *Journal of Coatings Technology and Research*, *9*(2), 171–176. Advance online publication. doi:10.1007/s11998-009-9227-8

Guo, R. H., Jiang, S. X., Yuen, C. W. M., Ng, M. C. F., & Lan, J. W. (2013). Optimization of electroless nickel plating on polyester fabric. *Fibers and Polymers*. Advance online publication. doi:10.100712221-013-0459-y

Guo, W., Zhang, S., Zhu, N., Luo, D., & Wu, P. (2019). Recovery of high purity secondary silver from waste Ag/Cu electrical contacts. *Process Safety and Environmental Protection*, *127*, 197–205. doi:10.1016/j.psep.2019.05.027

Gupta, B. (2017). Few Studies on Biomedical Applications of Functionally Graded Material. Barkha Gupta International Journal of Engineering Technology Science and Research, 4(3), 39–43. Retrieved from www.ijetsr.com

Gupta, R., Tripathy, A., & Chaubey, A. K. (2020). Development of CNT Reinforced Al Matrix FGM for Advanced Applications in Lightweight and High Strength Components. In Applications of Robotics in Industry Using Advanced Mechanisms (pp. 350–356). doi:10.1007/978-3-030-30271-9_32

Gupta, A., & Talha, M. (2015, November 1). Recent development in modeling and analysis of functionally graded materials and structures. *Progress in Aerospace Sciences*, *79*, 1–14. doi:10.1016/j.paerosci.2015.07.001

Gutzeit, G., Talmey, P., & Lee, W. G. (1958). Chemical nickel plating processes and baths therefor. *US Patent, 2*.

Hadj Mostefa, A., Merdaci, S., & Mahmoudi, N. (2018). An Overview of Functionally Graded Materials FGM. In Proceedings of the Third International Symposium on Materials and Sustainable Development (pp. 267–278). doi:10.1007/978-3-319-89707-3_30

Hai, H. T., Ahn, J. G., Kim, D. J., Lee, J. R., Chung, H. S., & Kim, C. O. (2006). Developing process for coating copper particles with silver by electroless plating method. *Surface and Coatings Technology*, *201*(6), 3788–3792. doi:10.1016/j.surfcoat.2006.03.025

Hajdu, J., & Zabrocky, S. (2000). The future of electroless nickel. *Metal Finishing*, *98*(5), 42–46. doi:10.1016/S0026-0576(00)81730-4

Hajizamani, M., Alizadeh, A., & Ehsani, N. (2012). Deposition of a Ni3P nano-scale layer on B4C nanoparticles by simple electroless plating in an acidic bath. Applied Nanoscience (Switzerland), 2(4), 417–421. doi:10.1007/s13204-011-0055-7

Hanlon, J. F., Kelsey, R. J., & Forcinio, H. (1998). *Handbook of Package Engineering*. Technomic. doi:10.1201/b18045

Han, X., Liang, Z., Wang, W., Chen, J., Xue, C., & Zhao, H. (2015). Characterization and synthesis of ZTA nanopowders and ceramics by rotating packed bed (RPB). *Ceramics International*, *41*(3), 3568–3573. doi:10.1016/j.ceramint.2014.10.184

Hanyaloglu, S. C., Aksakal, B., & McColm, I. J. (2001). Reactive sintering of electroless nickel-plated aluminum powders. *Materials Characterization*, *47*(1), 9–16. doi:10.1016/S1044-5803(01)00146-2

Haoliang, T., Guo, J., Erbao, L., Fuyuan, W., Changliang, W., Mengqiu, G., Junguo, G., & Yongjing, C. (2020). Preparation and Performance of Thermal Barrier Coatings made of BNw-containing modified Nd_2O_3-doped Yttria-stabilized Zirconia. *Ceramics International*, *46*(1), 500–507. doi:10.1016/j.ceramint.2019.08.289

HardKarb 12 Co Thermal Spray Flexicord I Saint-Gobain Coating Solutions. (n.d.). Retrieved from: https://www.coatingsolutions.saint-gobain.com/materials/flexicord/carbide-series/hardkarb-12-co

Harsha, S., Dwivedi, D., & Agarwal, A. (2008). Influence of CrC addition in NiCrSiB flame sprayed coatings on microstructure, microhardness and wear behaviour. *International Journal of Advanced Manufacturing Technology*, *38*(1-2), 93–101. doi:10.100700170-007-1072-2

Hashemi, A., Krasheninnikov, A. V., Puska, M., & Komsa, H. P. (2019). Efficient method for calculating Raman spectra of solids with impurities and alloys and its application to two-dimensional transition metal dichalcogenides. *Physical Review Materials*, *3*(2), 023806. doi:10.1103/PhysRevMaterials.3.023806

Hassani-Gangaraja, Veysset, Champagne, Nelson, & Schuh. (2019). Adiabatic shear instability is not necessary for adhesion in cold spray. *Scripta Materialia, 162*, 515-51.

Hayez, V., Guillaume, J., Hubin, A., & Terryn, H. (2004). Micro-Raman spectroscopy for the study of corrosion products on copper alloys: Setting up of a reference database and studying works of art. *Journal of Raman Spectroscopy: JRS*, *35*(8-9), 732–738. doi:10.1002/jrs.1194

Hazledine, F. (2019). Oxy-Acetylene Welding. *Proceedings - Institution of Mechanical Engineers*, *96*(1), 99–189. doi:10.1243/PIME_PROC_1919_096_010_02

Heinonen, S., Huttunen-Saarivirta, E., Nikkanen, J. P., Raulio, M., Priha, O., Laakso, J., Storgards, E., & Levänen, E. (2014). Antibacterial properties and chemical stability of superhydrophobic silver-containing surface produced by sol-gel route. *Colloids and Surfaces. A, Physicochemical and Engineering Aspects*, *453*, 149–161. doi:10.1016/j.colsurfa.2014.04.037

He, L., Tan, Y., Wang, X., Xu, T., & Hong, X. (2014). Microstructure and wear properties of Al_2O_3–CeO_2/Ni-base alloy composite coatings on aluminum alloys by plasma spray. *Applied Surface Science*, *314*, 760–767. doi:10.1016/j.apsusc.2014.07.047

Henao, J., Cruz-bautista, M., Hincapie-Bedoya, J., Ortega-Bautista, B., Corona-Castuera, J., Giraldo-Betancur, A. L., & Poblano-Salas, C. A. (2018). HVOF hydroxyapatite/titania-graded coatings: Microstructural, mechanical, and in vitro characterization. *Journal of Thermal Spray Technology*, *27*(8), 1302–1321. doi:10.100711666-018-0811-2

Henao, J., & Sharma, M. M. (2018). Characterization, Deposition Mechanisms, and Modeling of Metallic Glass Powders for Cold Spray. In *Cold-Spray Coatings* (pp. 251–272). Springer. doi:10.1007/978-3-319-67183-3_8

Herd, S., Wood, R. J. K., Wharton, J. A., & Higgs, C. F. III. (2018). Explicit fracture modelling of cemented tungsten carbide (WC-Co) at the mesoscale. *Materials Science and Engineering A*, *712*, 521–530. doi:10.1016/j.msea.2017.11.109

Her, E. K., Ko, T. J., Shin, B., Roh, H., Dai, W., Seong, W. K., Kim, H. Y., Lee, K. R., Oh, K. H., & Moon, M. W. (2013). Superhydrophobic Transparent Surface of Nanostructured Poly(Methyl Methacrylate) Enhanced by a Hydrolysis Reaction. *Plasma Processes and Polymers*, *10*(5), 481–488. doi:10.1002/ppap.201200131

Herman, H., Sampath, S., & McCune, R. (2000). Thermal spray: Current status and future trends. *MRS Bulletin*, *25*(7), 17–25. doi:10.1557/mrs2000.119

Herzog, R., Warnken, N., Steinbach, I., Hallstedt, B., Walter, C., Müller, J., Hajas, D., Münstermann, E., Schneider, J. M., Nickel, R., Parkot, D., Bobzin, K., Lugscheider, E., Bednarz, P., Trunova, O., & Singheiser, L. (2006). Integrated Approach for the Development of Advanced, Coated Gas Turbine Blades. *Advanced Engineering Materials*, *8*(6), 535–562. doi:10.1002/adem.200500277

Heuer, S., Lienig, T., Mohr, A., Weber, T., Pintsuk, G., Coenen, J. W., Gormann, F., Theisen, W., & Linsmeier, C. (2019). Ultra-fast sintered functionally graded Fe/W composites for the first wall of future fusion reactors. *Composites. Part B, Engineering*, *164*, 205–214. doi:10.1016/j.compositesb.2018.11.078

Holmberg, K., & Matthews, A. (2009). *Coatings tribology: properties, mechanisms, techniques and applications in surface engineering*. Elsevier.

Honda, K., Atake, T., & Saito, Y. (1992). Infrared spectra of oxide films on SUS430 stainless steel formed at temperatures between 473 and 1073 K. *Journal of Materials Research*, *7*(1), 80–84. doi:10.1557/JMR.1992.0080

Hossain, K. M. Z., Patel, U., Kennedy, A. R., Macri-Pellizzeri, L., Sottile, V., Grant, D. M., & Ahmed, I. (2018). Porous calcium phosphate glass microspheres for orthobiologic applications. *Acta Biomaterialia*, *72*, 396–406. doi:10.1016/j.actbio.2018.03.040 PMID:29604438

Huang, F. L., Wei, Q. F., Xu, W. Z., & Li, Q. (2007). Preparation and characterizations of PTFE gradient nanostructure on silk fabric. *Surface Review and Letters*, *14*(4), 547–551. doi:10.1142/S0218625X07009827

Huang, R., Sone, M., Ma, W., & Fukanuma, H. (2015). The effect of heat treatment on the mechanical properties of cold sprayed coatings. *Surface & Coating Technology*, *261*, 278–288.

Hu, B., & Tian, G. (2012). Morphological instability of γ'phase in nickel-based powder metallurgy superalloys. *Chin Shu Hsueh Pao*, *48*(3), 257–263. doi:10.3724/SP.J.1037.2011.00731

Hussain, McCartney, Shipway, & Zhang. (2009). Bonding Mechanism in cold Spraying: The Contribution of Metallurgical and Mechanical components. *Journal of Thermal Spray Technology, 18*(3), 364-379.

Hussain, T., Dudziak, T., Simms, N. J., & Nicholls, J. R. (2013). Fireside Corrosion Behavior of HVOF and Plasma-Sprayed Coatings in Advanced Coal/Biomass Co-fired Power Plants. *J. Therm. Spray Technol.*, *22*(5), 797–807. doi:10.100711666-013-9887-x

Hussain, T., McCartney, D. G., Shipway, P. H., & Zhang, D. (2009). Bonding mechanisms in cold spraying: The contributions of metallurgical and mechanical components. *Journal of Thermal Spray Technology*, *18*(3), 364–379. doi:10.100711666-009-9298-1

Hussain, T., McCartney, P., & Shipway, P. H. (2011). Shipway, "Impact phenomenon in cold spraying of titanium on various ferrous alloys. *Surface and Coatings Technology*, *205*(21-22), 5021–5027. doi:10.1016/j.surfcoat.2011.05.003

HVAF. High Velocity Air-Fuel. (n.d.). Retrieved from: http://www.hvaf.com/index.html

Hwang, Y. S., & Rapp, R. A. (1989). Thermo-Chemistry and Solubility of Oxides in Sodium Sulfate-Vanadate Solutions. *Corrosion*, *45*(11), 933–937. doi:10.5006/1.3585003

Irissou, E., Legoux, J.-G., Ryabinin, A. N., Jodoin, B., & Moreau, C. (2008). Review on cold spray process and technology: Part I- Intellectual Property. *J. Therm. Spray Technol.*, *17*(4), 495–516. doi:10.100711666-008-9203-3

Islam, M. R., Yamada, M., Abrosimov, N. V., Kiyama, M., & Tatsumi, M. (2004). Raman scattering characterization of residual strain and alloy composition in bulk Si1-xGex crystal. *The European Physical Journal Applied Physics*, *27*(1-3), 325–328. doi:10.1051/epjap:2004124

Islam, M., & Shehbaz, T. (2011). Effect of synthesis conditions and post-deposition treatments on composition and structural morphology of medium-phosphorus electroless Ni–P films. *Surface and Coatings Technology*, *205*(19), 4397–4400. doi:10.1016/j.surfcoat.2011.03.042

Ivanovaa, N. A., & Philipchenkob, A. B. (2012). Superhydrophobic chitosan-based coatings for textile processing. *Applied Surface Science*, *263*, 783–787. doi:10.1016/j.apsusc.2012.09.173

Izatt, R. M., Eatough, D., & Christensen, J. J. (1967). A study of Pd 2+(aq) hydrolysis. Hydrolysis constants and the standard potential for the Pd, Pd 2+ couple. *Journal of the Chemical Society A: Inorganic, Physical, Theoretical*, 1301-1304.

Jacob, L. J., & Deigner, H. P. (2018). Nanoparticles and Nanosized Structures in Diagnostics and Therapy. In Precision Medicine (pp. 229-252). Academic Press. doi:10.1016/B978-0-12-805364-5.00010-X

Jadhav, A. D., Padture, N. P., Jordan, E. H., Gell, M., Miranzo, P., & Fuller, E. R. Jr. (2006). Low-Thermal-Conductivity Plasma-Sprayed Thermal Barrier Coatings with Engineered Microstructures. *Acta Materialia*, *54*(12), 3343–3349. doi:10.1016/j.actamat.2006.03.024

Jafari, M., Enayati, M. H., Salehi, M., Nahvi, S. M., & Park, C. G. (2013). Microstructural and mechanical characterizations of a novel HVOF-sprayed WC-Co coating deposited from electroless Ni-P coated WC-12Co powders. *Materials Science and Engineering A*, *578*, 46–53. doi:10.1016/j.msea.2013.04.064

Jain, A., Kodas, T. T., Jairath, R., & Hampden-Smith, M. J. (1993). Selective and blanket copper chemical vapor deposition for ultra-large-scale integration. *Journal of Vacuum Science & Technology. B, Microelectronics and Nanometer Structures: Processing, Measurement, and Phenomena: An Official Journal of the American Vacuum Society*, *11*(6), 2107–2113.

James, W. B., & West, G. T. (2002). *Ferrous powder metallurgy materials. In powder technologies and applications* (Vol. 7). ASM International.

Jasinski, R., & Iob, A. (1988). FTIR measurements of iron oxides on low alloy steel. *Journal of the Electrochemical Society*, *135*(3), 551–556. doi:10.1149/1.2095656

Jha, D. K., Kant, T., & Singh, R. K. (2013). A critical review of recent research on functionally graded plates. *Composite Structures*, *96*, 833–849. doi:10.1016/j.compstruct.2012.09.001

Jia, Q., Li, D., Zhang, Z., Zhang, N., & Zhao, W. (2019). Comparison of oxidation resistance of Al–Ti and Al–Ni intermetallic formed in situ by thermal spraying. Materials Research Express, 6(9), 096408. doi:10.1088/2053-1591/ab2cfc

Jiang, D., Wang, Y., Wang, S., Liu, R., & Han, J. (2019). Thermal Conductivity of Air Plasma Sprayed Yttrium Heavily-doped Lanthanum Zirconate Thermal Barrier Coatings. *Ceramics International*, *45*(3), 3199–3206. doi:10.1016/j.ceramint.2018.10.222

Jiang, L., Li, Z., Fan, G., Cao, L., & Zhang, D. (2012a). Strong and ductile carbon nanotube/aluminum bulk nanolaminated composites with two-dimensional alignment of carbon nanotubes. *Scripta Materialia*, *66*(6), 331–334. doi:10.1016/j.scriptamat.2011.11.023

Jiang, L., Li, Z., Fan, G., Cao, L., & Zhang, D. (2012b). The use of flake powder metallurgy to produce carbon nanotube (CNT)/aluminum composites with a homogenous CNT distribution. *Carbon*, *50*(5), 1993–1998. doi:10.1016/j.carbon.2011.12.057

Jiang, L., Li, Z., Fan, G., & Zhang, D. (2011). A flake powder metallurgy approach to Al2O3/Al biomimetic nanolaminated composites with enhanced ductility. *Scripta Materialia*, *65*(5), 412–415. doi:10.1016/j.scriptamat.2011.05.022

Jin, G., Takeuchi, M., Honda, S., Nishikawa, T., & Awaji, H. (2005). Properties of multilayered mullite/Mo functionally graded materials fabricated by powder metallurgy processing. *Materials Chemistry and Physics*, *89*(2–3), 238–243. doi:10.1016/j.matchemphys.2004.03.031

Jnido, G., Ohms, G., & Viöl, W. (2020). One-Step Deposition of Polyester/TiO2 Coatings by Atmospheric Pressure Plasma Jet on Wood Surfaces for UV and Moisture Protection. *Coatings*, *10*(2), 184. doi:10.3390/coatings10020184

Jodion, B., Richer, P., Berube, G., Ajdelszatajn, L., & Yanouzi, M. (2007). Pulsed gas dynamic spraying process analysis, development and selected coating examples. *Surface Coating Technology*, *201*, 7544–7551.

Jordan, E. H., Gell, M., Kumar, R., & Jiang, C. (2020). *U.S. Patent Application No. 16/584,820*. Washington, DC: US Patent Office.

Joshi, M., Bhattacharyya, A., & Ali, S. W. (2008) Characterization techniques for nanotechnology applications in textiles, *Indian Journal of Fibre and Textile Research, 33*(3), 304-317.

Joshi, M., & Adak, B. (2018). *Advances in Nanotechnology Based Functional, Smart and Intelligent Textiles: A Review Comprehensive*. Elsevier. doi:10.1016/B978-0-12-803581-8.10471-0

Joshi, M., & Adak, B. (2018). *Nanotechnology-based textiles: A solution for the emerging automotive sector*. Walter de Gruyter GmbH.

Joshi, M., & Bhattacharyya, A. (2011). Nanotechnology-a new route to high-performance functional textiles. *Textile Progress*, *43*(3), 155–233. doi:10.1080/00405167.2011.570027

Joshi, M., Bhattacharyya, A., Agarwal, N., & Parmar, S. (2012). Nanostructured coatings for super hydrophobic textiles. *Bulletin of Materials Science*, *35*(6), 933–938. doi:10.100712034-012-0391-6

Kahraman, N., & Gulenc, B. (2002). Abrasive wear behaviour of powder flame sprayed coatings on steel substrates. *Materials & Design*, *23*(8), 721–725. doi:10.1016/S0261-3069(02)00075-4

Kairet, T. (2007). *A contribution to the study of cold gas dynamic spraying of copper: Influence of the powder characteristics on the mechanical properties of the coating* (Doctoral thesis). University Libre deBruxelles, Belgium.

Kai, X. Z., Li, Z. Q., Fan, G. L., Guo, Q., Xiong, D. B., Zhang, W. L., Su, Y. S., Lu, W. J., Moon, W. J., & Zhang, D. (2013). Enhanced strength and ductility in particulate-reinforced aluminum matrix composites fabricated by flake powder metallurgy. *Materials Science and Engineering A*, *587*, 46–53. doi:10.1016/j.msea.2013.08.042

Kai, X., Li, Z., Fan, G., Guo, Q., Tan, Z., Zhang, W., & Zhang, D. (2013). Strong and ductile particulate reinforced ultrafine-grained metallic composites fabricated by flake powder metallurgy. *Scripta Materialia*, *68*(8), 555–558. doi:10.1016/j.scriptamat.2012.11.024

Kalita, G., Muhammed, E., Ayhan, E., Sharma, S., & Shinde, M. (2014). Low temperature deposited graphene by surface wave plasma CVD as effective oxidation resistive barrier. *Corrosion Science*, *78*, 183–187. doi:10.1016/j.corsci.2013.09.013

Kamal, S., Sharma, K. V., Pedapati, S. R., & Mamat, O. (2017). Thermal spray coatings for hot corrosion resistance, engineering applications of nanotechnology, Topics in Mining. *Metallurgy and Materials Engineering,* 235–268.

Kamnis, S., & Gu, S. (2006). 3-D modelling of kerosene-fuelled HVOF thermal spray gun. *Chemical Engineering Science*, *61*(16), 5427–5439. doi:10.1016/j.ces.2006.04.005

Kang, C., Lu, H., Yuan, S., Hong, D., Yan, K., & Liang, B. (2012). Superhydrophilicity/superhydrophobicity of nickel micro-arrays fabricated by electroless deposition on an etched porous aluminum template. *Chemical Engineering Journal*, *203*, 1–8. doi:10.1016/j.cej.2012.06.128

Karter, M. J. (2014). *Fire loss in the United States during 2013*. National Fire Protection Association.

Karthikeyan, J. (2004). Cold spraying Technology: International Status and USA Efforts. Academic Press.

Karthikeyan, J. (2007). *Development of oxidation resistant coatings on GRCop-84 substrates by cold spraying processs*. NASA/CR.

Karthikeyan, J., Berndt, C. C., Tikkanen, J., Reddy, S., & Herman, H. (1997). Plasma spray synthesis of nanomaterial powders and deposits. *Materials Science and Engineering A*, *238*(2), 275–286. doi:10.1016/S0921-5093(96)10568-2

Karuppasamy, S., Sivan, V., Natarajan, S., Babu, S. K., Duraiselvam, M., & Dhanuskodi, R. (2019). Establishment of Wear Resistant HVOF Coatings for 50CrMo4 Chromium Molybdenum Alloy Steel as an Alternative for Hard Chrome Plating. *Journal of The Institution of Engineers (India): Series C*, *100*(4), 635-645.

Kathirvelu, S., D'souza, L., & Dhurai, B. (2009). UV protection finishing of textiles using ZnO nanoparticles. *Indian Journal of Fibre and Textile Research*, *34*(3), 267–273.

Kaur, M., Singh, H., & Prakash, S. (2008). A survey of the literature on the use of high velocity oxy-fuel spray technology for high temperature corrosion and erosion-corrosion resistant coatings. *Anti-Corrosion Methods and Materials*, *55*(2), 86–96. doi:10.1108/00035590810859467

Kaur, M., Singh, H., & Prakash, S. (2009). High-temperature corrosion studies of HVOF-sprayed Cr_3C_2-NiCr coating on SAE-347H boiler steel. *Thermal Spray Technology*, *18*(4), 619–632. doi:10.100711666-009-9371-9

Kaur, M., Singh, H., & Prakash, S. (2012). High-temperature behavior of a high-velocity oxy-fuel sprayed Cr_3C_2-NiCr coating. *Metallurgical and Materials Transactions. A, Physical Metallurgy and Materials Science*, *43*(8), 2979–2993. doi:10.100711661-012-1118-4

Kaushal, G., Singh, H., & Prakash, S. (2011a). Comparative high temperature analysis of HVOF sprayed and detonation gun sprayed Ni–20Cr coating in laboratory and actual boiler environment. *Oxidation of Metals*, *76*(3-4), 169–191. doi:10.100711085-011-9245-7

Kaushal, G., Singh, H., & Prakash, S. (2011b). High temperature corrosion behaviour of HVOF-sprayed Ni-20Cr coating on boiler steel in molten salt environment at 900°C. *International Journal of Surface Science and Engineering*, *5*(5/6), 415–433. doi:10.1504/IJSURFSE.2011.044388

Kawahra, Y. (2016). An Overview on Corrosion-Resistant Coating Technologies in Biomass/Waste-to-Energy Plants in Recent Decades. *Coatings*, *6*(3), 1–24.

Kawasaki, A., & Watanabe, R. (1997). Evaluation of thermomechanical performance for thermal barrier type of sintered functionally graded materials. *Composites. Part B, Engineering*, *28*(1–2), 29–35. doi:10.1016/S1359-8368(96)00017-0

Kennedy, D., Xue, Y., & Mihaylova, E. (2005). *Current and future applications of surface engineering*. Academic Press.

Keppler, H., Dubrovinsky, L. S., Narygina, O., & Kantor, I. (2008). Optical Absorption and Radiative Thermal Conductivity of Silicate Perovskite to 125 Gigapascals. *Science*, *322*(5907), 1529–1532. doi:10.1126cience.1164609 PMID:19056982

Kermiko HVAF Guns. (n.d.). Retrieved from: https://kermetico.com/

Kessel, H., Hennicke, J., Kirchner, R., & Kessel, T. (2010). Rapid sintering of novel materials by FAST/SPS—Further development to the point of an industrial production process with high cost efficiency. FCT Systeme GmbH. Retrieved from www.fct-systeme.de

Keuler, J. N., Lorenzen, L., Sanderson, R. N., & Linkov, V. (1997). Optimizing Palladium Conversion in Electroless Palladium Plating of Alumina Membranes. *Plating and Surface Finishing*, *84*(8), 34–40.

Keyvani, A., Saremi, M., & Sohi, M. H. (2011). Oxidation resistance of YSZ-alumina composites compared to normal YSZ TBC coatings at 1100°C. *Journal of Alloys and Compounds*, *509*(33), 8370–8377. doi:10.1016/j.jallcom.2011.05.029

Khanna, A. S. (2008). Nanotechnology in High Performance Paint Coatings. *Asian J. Exp. Sci.*, *21*(2), 25–32.

Khanna, A. S., & Jha, K. S. (1998). Degradation of Materials under Hot Corrosion Conditions. *Transactions of the Indian Institute of Metals*, *51*(5), 279–290.

Kilicarslan, A., Toptan, F., & Kerti, I. (2012). Electroless nickel-phosphorus coating on boron carbide particles. *Materials Letters*, *76*, 11–14. doi:10.1016/j.matlet.2012.02.037

Kim, B., Kim, S., & Kim, H. (2018). Effects of Alloying Elements (Cr, Mn) on Corrosion Properties of the High-Strength Steel in 3.5% NaCl Solution. *Advances in Materials Science and Engineering*, *2018*, 1–13. doi:10.1155/2018/7638274

Kim, B., Koncar, V., & Dufour, C. (2006). Polyaniline-coated PET conductive yarns: Study of electrical, mechanical, and electro-mechanical properties. *Journal of Applied Polymer Science*, *101*(3), 1252–1256. doi:10.1002/app.22799

Kim, H. J., Hwang, S. Y., Lee, C. H., & Juvanon, P. (2003). Assessment of wear performance of flame sprayed and fused Ni-based coatings. *Surface and Coatings Technology*, *172*(2-3), 262–269. doi:10.1016/S0257-8972(03)00348-7

Kim, J. H., Kim, J. D., Park, H. M., Lee, S. B., Yi, J. W., & Jung, S. (2013). A homogeneous Ni-P coating with a unique scalelike structure deposited on VGCNFs by an electroless deposition method. *Applied Surface Science*, *265*, 527–531. doi:10.1016/j.apsusc.2012.11.039

Kim, Y. K. (2010). *Nanotechnology-based advanced coatings and functional finishes for textiles*. Woodhead. doi:10.1533/9781845697785.2.251

Klar, E., & Samal, P. K. (2007). *Powder metallurgy stainless steels: processing, microstructures, and properties*. ASM international.

Klinkov, S., Kosarev, V., & Rein, M. (2005). Cold spray deposition: Significance of particle impact phenomena. *Aerospace Sci. Technology*, *9*, 582–591.

Knight, R., & Smith, R. W. (1998). *Thermal Spray Forming of Materials, Powder Metal Technologies and Applications* (Vol. 7). ASM Handbook. ASM International.

Knuuttila, J., Ahmaniemi, S., & Mantyla, T. (1999). Wet abrasion and slurry erosion resistance of thermally sprayed oxide coatings. *Wear*, *232*(2), 207–212. doi:10.1016/S0043-1648(99)00147-7

Kobayashi, Y., Salgueiriño-Maceira, V., & Liz-Marzán, L. M. (1630–1633). Salgueiriñ O-Maceira, V., & Liz-Marzá, L. M. (2001). Deposition of Silver Nanoparticles on Silica Spheres by Pretreatment Steps in Electroless Plating. *Chemistry of Materials*, *13*(5), 1630–1633. Advance online publication. doi:10.1021/cm001240g

Kodama, S. H. U. N. I. C. H. I., Horiuchi, M. A. S. A. O., Kuroda, K., & Kunii, T. (1989,April).Ultra-blacknickel-phosphorus alloy optical absorber. In *6th IEEE Conference Record., Instrumentation and Measurement Technology Conference* (pp.495-497) .IEEE. 10.1109/IMTC.1989.36909

Koivluoto. (2010). *Microstructural characteristics and corrosion properties of cold sprayed coating*. Tampere University of Technology.

Koivuluoto, H., Coleman, A., Murray, K., Kearns, M., & Vuoristo, P. (2012). High pressure cold sprayed (HPCS) and low pressure cold sprayed (LPCS) coatings prepared from OFHC Cu feedstock: Overview from powder characteristics to coating properties. *Journal of Thermal Spray Technology*, *21*(5), 1065–1075. doi:10.100711666-012-9790-x

Koivuluoto, H., & Vuoristo, P. (2009). Effect of ceramic particles on properties of cold sprayed Ni–20Cr + Al_2O_3 coatings. *J. Therm. Spray Technol.*, *18*(4), 555–562. doi:10.100711666-009-9345-y

Kolawole, M. Y., Aweda, J. O., Iqbal, F., Ali, A., & Abdulkareem, S. (2019). Mechanical Properties of Powder Metallurgy Processed Biodegradable Zn-Based Alloy for Biomedical Application. *International Journal of Materials and Metallurgical Engineering*, *13*(12), 558–563.

Kondo, K., Akolkar, R. N., Barkey, D. P., & Yokoi, M. (Eds.). (2014). *Copper electrodeposition for nanofabrication of electronics devices* (Vol. 14). Springer. doi:10.1007/978-1-4614-9176-7

Kong, F. Z., Zhang, X. B., Xiong, W. Q., Liu, F., Huang, W. Z., Sun, Y. L., Tu, J. P., & Chen, X. W. (2002). Continuous Ni-layer on multiwall carbon nanotubes by an electroless plating method. *Surface and Coatings Technology*, *155*(1), 33–36. doi:10.1016/S0257-8972(02)00032-4

Konnov, A. Y., Demchenko, M. V., Sisanbaev, A. V., Nafikov, M. Z., & Farkhshatov, M. N. (2019). Estimation of the Corrosion Resistance of the Metal Coatings Formed by Powder Metallurgy. *Russian Metallurgy (Metally)*, *2019*(12), 1285–1288. doi:10.1134/S0036029519120097

Konyashin, I. Y., & Chukalovskaya, T. V. (1997). A technique for measurement of porosity in protective coatings. *Surface and Coatings Technology*, *88*(1-3), 5–11. doi:10.1016/S0257-8972(95)02758-0

Koropchak, J. A., & Roychowdhury, S. B. (1990). Evidence for aerosol ionic redistribution within aerosols produced by chrome electroplating. *Environmental Science & Technology*, *24*(12), 1861–1863. doi:10.1021/es00082a012

Kozlov, A. S., Palanisamy, T., & Narasimhan, D. (2002). *U.S. Patent No. 6,387,542*. Washington, DC: U.S. Patent and Trademark Office.

Krishna, M. E., & Patowari, P. K. (2013). Parametric optimisation of electric discharge coating process with powder metallurgy tools using Taguchi analysis. *Surface Engineering*, *29*(9), 703–711. doi:10.1179/1743294413Y.0000000207

Krishnan, K. H., John, S., Srinivasan, K. N., Praveen, J., Ganesan, M., & Kavimani, P. M. (2006). An overall aspect of electroless Ni-P depositions—A review article. *Metallurgical and Materials Transactions. A, Physical Metallurgy and Materials Science*, *37*(6), 1917–1926. doi:10.100711661-006-0134-7

Kucuk, A., Berndt, C. C., Senturk, U., Lima, R. S., & Lima, C. R. C. (2000). Influence of plasma spray parameters on mechanical properties of yttria stabilized zirconia coatings. I: Four point bend test. *Materials Science and Engineering A*, *284*(1-2), 29–40. doi:10.1016/S0921-5093(00)00799-1

Kumar Chauhan, P., & Khan, S. (2019). Microstructural examination of aluminium-copper functionally graded material developed by powder metallurgy route. Materials Today: Proceedings. doi:10.1016/j.matpr.2019.10.007

Kumar, A., Singh, K., & Pandey, O. P. (2011). Sintering behavior of nanostructured WC-Co composite. *Ceramics International*, *37*(4), 1415–1422. doi:10.1016/j.ceramint.2011.01.001

Kumar, S. (2015). *Clean Hydrogen Production Methods* (1st ed.). Springer International Publishing., doi:10.1007/978-3-319-14087-2.

Kumar, S., Handa, A., & Kumar, M. (2018). Combating hot corrosion of boiler tubes- A study. *Engineering Failure Analysis*, *94*, 379–395. doi:10.1016/j.engfailanal.2018.08.004

Kumar, S., Kumar, M., & Jindal, N. (2020). *Overview of cold spray coatings applications and comparisons: a critical review. World Journal of Engineering.*

Kundu, S., Das, S. K., & Sahoo, P. (2014). Properties of electroless nickel at elevated temperature-a review. *Procedia Engineering*, *97*(0), 1698–1706. doi:10.1016/j.proeng.2014.12.321

Kunimine, T., Sato, H., Miura-Fujiwara, E., & Watanabe, Y. (2016). New Processing Routes for Functionally Graded Materials and Structures through Combinations of Powder Metallurgy and Casting. In F. Ebrahimi (Ed.), Advances in Functionally Graded Materials and Structures. doi:10.5772/62393

Kuroda, S., Kawakita, J., Watanabe, M., & Katanoda, H. (2008). Warm spraying- A novel coating process based on high-velocity impact of solid particles. *Sci. Technol. Adv. Mater., 9*(033002), 1-17. DOI: 1088/1468-6996/9/3/033002

Kutnahorsky, M. R. (2009). *Electroless copper deposition: A sustainable approach* (Doctoral dissertation).

Kutz, M. (Ed.). (2011). *Applied plastics engineering handbook: processing and materials*. William Andrew.

Kwon, K. C., & Kim, S. Y. (2014). Extended thermal stability in metal-chloride doped graphene using graphene over layers. *Chemical Engineering Journal*, *244*, 355–363. doi:10.1016/j.cej.2014.01.100

Laima, L., Xiaoyue, T., Xiaoyu, D., Zelong, L., Guangnan, L., Xiang, Z., Xiaoyong, Z., & Yucheng, W. (2016). Fabrication of W/Cu Graded Heat-sink Materials by Electroless Plating and Powder Metallurgy. *Rare Metal Materials and Engineering*, *45*(8), 1983–1987. doi:10.1016/S1875-5372(16)30158-8

Lajevardi, S. A., Shahrabi, T., & Szpunar, J. A. (2017). Tribological Properties of Functionally Graded Ni-Al 2 O 3 Nanocomposite Coating. *Journal of the Electrochemical Society*, *164*(6), D275–D281. doi:10.1149/2.0731706jes

Landillon, V., Cassan, D., Morel, M. H., & Cuq, B. (2008). Flowability, cohesive, and granulation properties of wheat powders. *Journal of Food Engineering*, *86*(2), 178–193. doi:10.1016/j.jfoodeng.2007.09.022

Lara, J. P. H., Labra, M. P., Hernández, F. R. B., Serrano, J. A. R., Dávila, E. O. A., Thangarasu, P., & Ramirez, A. H. (2017). Structural Evolution and Electrical Properties of $BaTiO_3$ Doped with Gd^{3+}. *Materials Research*, *20*(2), 538–542. doi:10.1590/1980-5373-mr-2016-0606

Lawley, A., & Murphy, T. F. (2003). Metallography of powder metallurgy materials. *Materials Characterization*, *51*(5), 315–327. doi:10.1016/j.matchar.2004.01.006

Le Bras, M., Wilkie, C. A., & Bourbigot, S. (2005). *Fire retardancy of polymers: new applications of mineral fillers*. Academic Press.

Lee, P. W., Trudel, Y., German, R. M., Ferguson, B. L., Eisen, W. B., Mover, K., . . . Davidson, G. M. (1998). ASM handbook. Powder Metal Technologies and Applications, 7, 1146.

Lee, H., Jung, S., Lee, S., You, Y., & Ko, K. H. (2005). Correlation between Al2O3 particles and interface of Al- Al2O3 coatings by cold spray. *Applied Surface Science*, *252*(5), 1891–1898. doi:10.1016/j.apsusc.2005.03.148

Lee, H., Yu, Y., Lee, Y., Hong, Y., & Ko, K. (2004). Cold Spray of SiC and Al2O3 with soft metal incorporation: A Technical Contribution. *Journal of Thermal Spray Technology*, *13*(2), 184–189. doi:10.1361/10599630419355

Legoux, J.-G., Irissou, E., Desauluniers, S., Bobyn, J., Harvey, B., Wong, W., Gagnon, E., & Yue, S. (2010). *Characterization and performance evaluation of a helium recovery system designed for cold spraying. NRC Publ Arch.* NP Arc.

Leivo, E., Wilenius, T., Kinos, T., Vuoristo, P., & Mäntylä, T. (2004). Properties of thermally sprayed fluoropolymer PVDF, ECTFE, PFA and FEP coatings. *Progress in Organic Coatings*, *49*(1), 69–73. doi:10.1016/j.porgcoat.2003.08.011

Leon, C. A., & Drew, R. A. L. (2000). Preparation of nickel-coated powders as precursors to reinforce MMCs. *Journal of Materials Science*, *35*(19), 4763–4768. doi:10.1023/A:1004860326071

León, C., & Drew, R. A. (2002). The influence of nickel coating on the wettability of aluminum on ceramics. *Composites. Part A, Applied Science and Manufacturing*, *33*(10), 1429–1432. doi:10.1016/S1359-835X(02)00161-6

León-Patiño, C. A., Aguilar-Reyes, E. A., Bedolla-Becerril, E., Bedolla-Jacuinde, A., & Méndez-Díaz, S. (2013). Dry sliding wear of gradient Al-Ni/SiC composites. *Wear*, *301*(1–2), 688–694. doi:10.1016/j.wear.2012.11.052

Liang, P., Dong, S., Zeng, J., Yuan, J., Jiang, J., Deng, L., Zhou, X. & Cao, X. (2019). $La_2Hf_2O_7$ Ceramics as Potential Top-Coat Materials for Thermal/Environmental Barrier Coatings. *Ceramics International*, *45*(17), 22432-22436. doi:10.1016/j.ceramint.2019.07.235

Li, C. J., Yang, H., & Li, H. (1999). Effect of gas conditions on HVOF flame and properties of WC-Co coatings. *Materials and Manufacturing Processes*, *14*(3), 383–395. doi:10.1080/10426919908914834

Li, C.-J., Li, W.-Y., & Liao, H. (2006). Examination of the critical velocity for deposition of particles in cold spraying. *Journal of Therm. Spray Technol.*, *15*(2), 212–222. doi:10.1361/105996306X108093

Li, C.-J., Li, W.-Y., Wang, Y.-Y., Yang, G.-J., & Fukanuma, H. (2005). A theoretical model for prediction of deposition efficiency in cold spraying. *Thin Solid Films*, *489*(1-2), 79–85. doi:10.1016/j.tsf.2005.05.002

Li, G. J., Huang, X. X., & Guo, J. K. (2001). Fabrication of Ni-coated Al2O3 powders by the heterogeneous precipitation method. *Materials Research Bulletin*, *36*(7-8), 1307–1315. doi:10.1016/S0025-5408(01)00628-6

Li, H., Ng, B. S., Khor, K. A., Cheang, P., & Clyne, T. W. (2004). Raman spectroscopy determination of phases within thermal sprayed hydroxyapatite splats and subsequent in vitro dissolution examination. *Acta Materialia*, *52*(2), 445–453. doi:10.1016/j.actamat.2003.09.046

Lim, H. S., Park, S. H., Koo, S. H., Kwark, Y.-J., Thomas, E. L., Jeong, Y., & Cho, J. H. (2010). Super amphiphilic Janus Fabric. *Langmuir*, *26*(24), 19159–19162. doi:10.1021/la103829c PMID:21073162

Link, R. (1998). Modern Pre-alloyed PM Steels and Advanced Process Technology. *PM World Congress Microstructure*, 596-606.

Lin, W. H., & Chang, H. F. (1998). Effect of chelating agents on the structure of electroless copper coating on alumina powder. *Surface and Coatings Technology*, *107*(1), 48–54. Advance online publication. doi:10.1016/S0257-8972(98)00553-2

Lin, Y.-S., & Chiu, S.-S. (2004). Effects of oxidation and particle shape on critical volume fractions of silver-coated copper powders in conductive adhesives for microelectronic applications. *Polymer Engineering and Science*, *44*(11), 2075–2082. doi:10.1002/pen.20212

Lin, Y.-S., & Chiu, S.-S. (2008). Electrical Properties of Copper-Filled Electrically Conductive Adhesives and Pressure-Dependent Conduction Behavior of Copper Particles. *Journal of Adhesion Science and Technology*, *22*(14), 1673–1697. doi:10.1163/156856108X320537

Li, S., & Jinjin, D. (2007). Improvement of hydrophobic properties of silk and cotton by hexafluoropropene plasma treatment. *Applied Surface Science*, *253*(11), 5051–5055. doi:10.1016/j.apsusc.2006.11.027

Litzenburger, A. (2000). Criteria for, and examples of optimal choice of flame-retardants. *Polymers & Polymer Composites*, *8*, 581–592.

Liu, B., Liu, Y., Zhu, C., Xiang, H., Chen, H., Sun, L., Gao, Y., & Zhou, Y. (2019). Advances on Strategies for Searching for Next Generation Thermal Barrier Coating Materials. *Journal of Materials Science and Technology*, *35*(5), 833–851. doi:10.1016/j.jmst.2018.11.016

Liu, G., Yang, L., Wang, L., Wang, S., Chongyang, L., & Wang, J. (2010). Corrosion behavior of electroless deposited Ni–Cu–P coating in flue gas condensate. *Surface and Coatings Technology*, *204*(21-22), 3382–3386. doi:10.1016/j.surfcoat.2010.03.056

Liu, Q., Huang, S., & He, A. (2019). Composite Ceramics Thermal Barrier Coatings of Yttria Stabilized Zirconia for Aero-Engines. *Journal of Materials Science and Technology*, *35*(12), 2814–2823. doi:10.1016/j.jmst.2019.08.003

Liu, S. L., & Zheng, X. P. (2009). Microstructure and Properties of AC-HVAF Sprayed Ni60/WC Composite Coating. *Journal of Alloys and Compounds*, *480*(2), 254–258. doi:10.1016/j.jallcom.2009.02.028

Liu, X. J., Sun, W. C., Dong, Y. R., Ma, M., Zhang, Y. G., Yang, F., & Ruan, Y. Y. (2019). Microstructure and corrosion resistance of Ni-P gradient coating/stannate conversion film on magnesium alloy. *Journal of Materials Research*, *34*(6), 1064–1072. doi:10.1557/jmr.2019.21

Liu, X., Song, X., Wang, H., Liu, X., Tang, F., & Lu, H. (2018). Complexions in WC-Co cemented carbides. *Acta Materialia*, *149*, 164–178. doi:10.1016/j.actamat.2018.02.018

Liu, Y. B., Lim, S. C., Lu, L., & Lai, M. O. (1994). Recent development in the fabrication of metal matrix-particulate composites using powder metallurgy techniques. *Journal of Materials Science*, *29*(8), 1999–2007. doi:10.1007/BF01154673

Liu, Y., Tang, J., Wang, R., Lu, H., Li, L., Kong, Y., Qia, K., & Xin, J. H. (2007). Artificial lotus leaf structures from assembling carbon nanotubes and their applications in hydrophobic textiles. *Journal of Materials Chemistry*, *17*(11), 1071–1078. doi:10.1039/B613914K

Liu, Y., Wang, X., Qia, K., & Xin, J. H. (2008). Functionalization of cotton with carbon nanotubes. *Journal of Materials Chemistry*, *18*(29), 3454–3460. doi:10.1039/b801849a

Liu, Y., & Zhao, Q. (2004). Study of electroless Ni–Cu–P coatings and their anti-corrosion properties. *Applied Surface Science*, *228*(1-4), 57–62. doi:10.1016/j.apsusc.2003.12.031

Li, W. Y., Liao, H., Li, C. J., Bang, H. S., & Coddet, C. (2007). Numerical simulation of deformation behavior of Al particles impacting on Al substrate and effect of surface oxide films on interfacial bonding in cold spraying. *Applied Surface Science*, *253*(11), 5084–5091. doi:10.1016/j.apsusc.2006.11.020

Li, W.-Y., & Li, C.-J. (2005). Optimal design of a novel cold spraying gun nozzle at a limited space. *Journal of Therm Spray Technol.*, *14*(3), 391–396. doi:10.1361/105996305X59404

Li, W.-Y., Zhang, C., Guo, X. P., Zhang, G., Liao, H. L., Li, C.-J., & Coddet, C. (2008). Effect of standoff distance on coating deposition characteristics. *Materials & Design*, *29*(2), 297–304. doi:10.1016/j.matdes.2007.02.005

Li, X., Yang, K., Wang, D., Deng, C., & Zhou, Y. (2020). Property of TiO2-15MgAl2O4 Electrical-Heating Coating Prepared by Atmospheric Plasma Spraying and Hydrogen Heat Treatment. *Coatings*, *10*(2), 177. doi:10.3390/coatings10020177

Li, Y. (2015). Process of electroless silver coated Al-Mg alloy powder. Xiyou Jinshu Cailiao Yu Gongcheng. *Rare Metal Materials and Engineering*, *44*(11), 2633–2636. doi:10.1016/S1875-5372(16)60008-5

Li, Y., Liu, R., Zhang, J., Luo, G., Shen, Q., & Zhang, L. (2019). Fabrication and microstructure of W-Cu composites prepared from Ag-coated Cu powders by electroless plating. *Surface and Coatings Technology*, *361*, 302–307. doi:10.1016/j.surfcoat.2019.01.030

Li, Y., Wang, R., Qi, F., & Wang, C. (2008). Preparation, characterization and microwave absorption properties of electroless Ni-Co-P-coated SiC powder. *Applied Surface Science*, *254*(15), 4708–4715. doi:10.1016/j.apsusc.2008.01.076

Li, Y., Xie, Y., huang, L., Liu, X., & Zheng, X. (2012). Effect of physical vapor deposited Al2O3 film on TGO growth in YSZ/CoNiCrAlY coatings. *Ceramics International*, *38*(6), 5113–5121. doi:10.1016/j.ceramint.2012.03.014

López-Honorato, E., Dessoliers, M., Shapiro, I. P., Wang, X., & Xiao, P. (2012). Improvements to the sintering of yttria-stabilized zirconia by the addition of Ni. *Ceramics International*, *38*(8), 6777–6782. doi:10.1016/j.ceramint.2012.05.073

Loto, C. A. (2016). Electroless Nickel Plating – A Review. Silicon, 8(2), 177–186. doi:10.1007/s12633-015-9367-7

Lu, G., & Zangari, G. (2002). Corrosion resistance of ternary Ni- P based alloys in sulfuric acid solutions. *Electrochimica Acta*, *47*(18), 2969–2979. doi:10.1016/S0013-4686(02)00198-6

Lugscheider, E., Herbst, C., & Znao, D. (1998). Parameter Studies on High- Velocity Oxy-Fuel Spraying of MCrAlY Coatings. *Surface and Coatings Technology*, *108-109*, 16–23. doi:10.1016/S0257-8972(98)00630-6

Luo, H., Leitch, M., Zeng, H., & Luo, J. L. (2018). Characterization of microstructure and properties of electroless duplex Ni-WP/Ni-P nano-ZrO2 composite coating. *Materials Today Physics, 4*, 36-42.

Luo, H., Sui, Y., Qi, J., Meng, Q., Wei, F., & He, Y. (2017). Copper matrix composites enhanced by silver/reduced graphene oxide hybrids. *Materials Letters*, *196*, 354–357. doi:10.1016/j.matlet.2017.03.084

Luo, L., Yu, J., Luo, J., & Li, J. (2010). Preparation and characterization of Ni-coated Cr3C2 powder by room temperature ultrasonic-assisted electroless plating. *Ceramics International*, *36*(6), 1989–1992. doi:10.1016/j.ceramint.2010.03.003

Luo, Q. G., Dai, Y., Liu, S. L., Yu, K., Shen, Q., & Zhang, L. M. (2014). Fabrication and Properties of W-Cu Functionally Graded Material by Tape-Casting. *Key Engineering Materials*, *616*, 66–71. doi:10.4028/www.scientific.net/KEM.616.66

Lynam, C., Moulton, S. E., & Wallace, G. G. (2007). Carbon-Nanotube Biofibers. *Advanced Materials*, *19*(9), 1244–1248. doi:10.1002/adma.200601986

Lyphout, C., Bjorklund, S., Karlsson, M., Runte, M., Reisel, G., & Boccaccio, P. (2014). Screening Design of Supersonic Air Fuel Processing for Hard Metal Coatings. *J. Therm. Spray Technol.*, *23*(8), 1323–1332. doi:10.100711666-014-0139-5

Maatta, A., Kanerva, U., & Vuoristo, P. (2011). Structure and tribological characteristics of HVOF coatings sprayed from powder blends of Cr_3C_2–25NiCr and NiCrBSi alloy. *J. Therm.Spray Technol.*, *20*(1-2), 366–371. doi:10.100711666-010-9579-8

Maev, R. G., & Leshchynsky, V. (2009). *Introduction to low pressure gas dynamic spray: physics and technology*. John Wiley & Sons.

Maev, R., & Leshchynsky, V. (2006). Air gas dynamic spraying of powder mixtures: Theory and Applications. *Journal of Thermal Spray Technology*, *15*(2), 198–205. doi:10.1361/105996306X108048

Mahade, S., Narayan, K., Govindarajan, S., Björklund, S., Curry, N., & Joshi, S. (2019). Exploiting suspension plasma spraying to deposit wear-resistant carbide coatings. *Materials (Basel)*, *12*(15), 2344. doi:10.3390/ma12152344 PMID:31344804

Mahamood, R. M., Akinlabi, E. T., Shukla, M., & Pityana, S. (2012). Functionally graded material: An overview. In S. I. Ao, L. Gelman, D.W.L Hukins, A. Hunter, A. M. Korsunsky (Ed.), Proceedings of the World Congress on Engineering (pp. 1593-1597). London, UK: Imperial College London.

Makhlouf, A. S. H. (2011). *Current and advanced coating technologies for industrial applications*. Woodhead., doi:10.1533/9780857094902.1.3

Makino, Y., Mizuuchi, K., Agari, Y., & Ito, M. (2016). Characterization of Interfaces in Diamond/Metal Composites by Raman Scattering Spectroscopy. *Journal of the Japan Society of Powder and Powder Metallurgy*, *63*(7), 479–483. doi:10.2497/jjspm.63.479

Mallory, G. O., & Hajdu, J. B. (1990). Electroless plating: fundamentals and applications (W. Andrew, Ed.). American Electroplaters and Surface Finishers Society.

Mallory, G. O., & Hajdu, J. B. (Eds.). (1990). *Electroless plating: fundamentals and applications*. Cambridge University Press.

Marder, A. R. (2000). The metallurgy of zinc-coated steel. *Progress in Materials Science*, *45*(3), 191–271. doi:10.1016/S0079-6425(98)00006-1

Marie,O.J.(1966). *U.S.PatentNo.3,282,810*. Washington, DC: U.S. Patent and Trademark Office.

Marrocco, T., McCartney, D. G., & Sturgeon, A. J. (2006). Production of titanium deposits by cold-gas dynamic spray: Numerical modeling and experimental characterization. *Journal of Thermal Spray Technology*, *15*(2), 263–272. doi:10.1361/105996306X108219

Martin, A., Rodriguez, J., Fernandez, J. E., & Vijande, R. (2001). Sliding wear behaviour of plasma sprayed WC–NiCrBSi coatings at different temperatures. *Wear, 251*(1-12), 1017–1022.

Martina, F., Mehnen, J., Williams, S. W., Colegrove, P., & Wang, F. (2012). Investigation of the benefits of plasma deposition for the additive layer manufacture of Ti–6Al–4V. *Journal of Materials Processing Technology*, *212*(6), 1377–1386. doi:10.1016/j.jmatprotec.2012.02.002

Ma, S., Xu, E., Zhu, Z., Liu, Q., Yu, S., Liu, J., Zhong, H., & Jiang, Y. (2018). Mechanical and wear performances of aluminum/sintered-carbon composites produced by pressure infiltration for pantograph sliders. *Powder Technology*, *326*, 54–61. doi:10.1016/j.powtec.2017.12.027

Matejicek, J., & Sampath, S. (2001). Intrinsic residual stresses in single splats produced by thermal spray processes. *Acta Materialia*, *49*(11), 1993–1999. doi:10.1016/S1359-6454(01)00099-4

Mathiazhagan, A., & Joseph, R. (2011). Nanotechnology-A New Prospective in Organic Coating –Review. *International Journal of Chemical Engineering and Applications*, *2*(4), 225–237. doi:10.7763/IJCEA.2011.V2.108

Matikainen, V., Koivuluoto, H., & Vuoristo, P. (2020). A study of Cr3C2-based HVOF-and HVAF-sprayed coatings: Abrasion, dry particle erosion and cavitation erosion resistance. *Wear*, *446-447*, 203188. doi:10.1016/j.wear.2020.203188

Mauer, G., Hospach, A., & Vaßen, R. (2013). Process development and coating characteristics of plasma spray-PVD. *Surface and Coatings Technology*, *220*, 219–224. doi:10.1016/j.surfcoat.2012.08.067

Mavrogenis, A. F., Dimitriou, R., Parvizi, J., & Babis, G. C. (2009). Biology of implant osseointegration. *Journal of Musculoskeletal & Neuronal Interactions*, *9*(2), 61–71. PMID:19516081

Ma, W., Jarligo, M. O., Mack, D. E., Malzbender, J., Vaßen, R., & Stöver, D. (2008). New Generation Perovskite Thermal Barrier Coating Materials. *Journal of Thermal Spray Technology*, *17*(5-6), 831–837. doi:10.100711666-008-9239-4

Ma, W., Mack, D., Malzbender, J., Vaben, R., & Stover, D. (2008). Yb2O3 and Gd2O3 doped strontium zirconate for thermal barrier coatings. *Journal of the European Ceramic Society*, *28*(16), 3071–3081. doi:10.1016/j.jeurceramsoc.2008.05.013

Mazaheri, H., & Allahkaram, S. R. (2012). Deposition, characterization and electrochemical evaluation of Ni-P-nano diamond composite coatings. *Applied Surface Science*, *258*(10), 4574–4580. doi:10.1016/j.apsusc.2012.01.031

Meier, G. H. (1989). A review of advances in high-temperature corrosion. *Materials Science and Engineering A*, *120-121*(1), 1–11. doi:10.1016/0921-5093(89)90712-0

Melero, H. C., Sakai, R. T., Vignatti, C. A., Benedetti, A. V., Fernández, J., Guilemany, J. M., & Suegama, P. H. (2018). Corrosion resistance evaluation of HVOF produced hydroxyapatite and TiO2-hydroxyapatite coatings in Hanks' solution. *Materials Research*, *21*(2). Advance online publication. doi:10.1590/1980-5373-mr-2017-0210

Meng, M., Leech, A., & Le, H. (2019). Mechanical properties and tribological behaviour of electroless Ni–P–Cu coatings on corrosion-resistant alloys under ultrahigh contact stress with sprayed nanoparticles. *Tribology International*, *139*, 59–66. doi:10.1016/j.triboint.2019.06.031

Meng, X., Zhang, J., Zhao, J., Liang, Y., & Zhang, Y. (2011). Influence of gas temperature on microstructure and properties of cold spray 304SS coating. *Journal of Materials Science and Technology*, *27*(9), 809–815. doi:10.1016/S1005-0302(11)60147-3

Meng, Y., Zhang, J., Duan, C., Chen, C., Feng, X., & Shen, Y. (2015). Microstructures and properties of W–Cu functionally graded composite coatings on copper substrate via high-energy mechanical alloying method. *Advanced Powder Technology*, *26*(2), 392–400. doi:10.1016/j.apt.2014.11.009

Metallisation Spray Guns. (n.d.b). Retrieved from: https://www.metallisation.com/product-category/hvof/

Metallizing Company Guns. (n.d.a). Retrieved from: https://www.mecpl.com/products/coating-equipments/hvof-spray-coatings.php

Metco, O. (n.d.). *Plasma Guns for Thermal Spray*. Retrieved from: https://www.oerlikon.com/metco/en/products-services/coating-equipment/thermal-spray/spray-guns/coating-equipment-plasma/

Metzger, W., Schmidt, H., Ott, R., & Pappe, G. (1973). *U.S. Patent No. 3,753,667*. Washington, DC: U.S. Patent and Trademark Office.

Mikli, V., Kaerdi, H., Kulu, P., & Besterci, M. (2001). Characterization of powder particle morphology. Proceedings of the Estonian Academy of Sciences: Engineering, 22–34.

Milanti, A., Matikainen, V., Koivuluoto, H., Bolelli, G., Lusvarghi, L., & Vuoristo, P. (2015). Effect of Spraying Parameters on the Microstructural and Corrosion Properties of HVAF-Sprayed Fe-Cr-Ni-B-C Coatings. *Surface and Coatings Technology*, *277*, 81–90. doi:10.1016/j.surfcoat.2015.07.018

Miller A.R. (2009). *History of Thermal Barrier Coatings for Gas Turbine Engines: Emphasizing NASA's Role From 1942 to 1990*. NASA/TM-2009-215459.

Milovanović, N., Dordević, B., Tatić, U., Sedmak, S., & Štrbački, S. (2017). Low-temperature corrosion damage and repair of boiler bottom panel tubes. *Structural Integrity and Life*, *17*(2), 125–131.

Mirazimi, J., Abachi, P., & Purazrang, K. (2016). Spark plasma sintering of ultrafine YSZ reinforced Cu matrix functionally graded composite. *Acta Metallurgica Sinica. English Letters*, *29*(12), 1169–1176. doi:10.1007/s40195-016-0512-0

Mirzaali, M. J., Bobbert, F. S. L., Li, Y., & Zadpoor, A. A. (2019). Additive Manufacturing of Metals Using Powder Bed-Based Technologies. In *Additive Manufacturing* (pp. 94–127). CRC Press. doi:10.1201/9780429466236-4

Mishra, S. K. (2002). *Advanced Techniques for Surface Engineering of Industrial Materials*. http://eprints.nmlindia.org/4215/1/128-140-S_K_Mishra.PDF

Mishra, A., Mohapatra, H. S., & Butola, B. S. (2019). Imparting protection against UV radiations In situ coatings of Titanium dioxide on textiles. *International Letters of Chemistry. Physics and Astronomy*, *82*, 14–20.

Mishra, R. S., Tiwari, D., & Singh, K. K. (2013). Nanocoating: A novel drift in nanotechnology. *International Journal of Engineering Innovation & Research*, *2*(2), 128–131.

Mitchell, R. J., Rae, C. M. F., & Tin, S. (2005). Grain boundary transformations during isothermal exposure of powder metallurgy nickel base superalloys for turbine disc applications. *Materials Science and Technology*, *21*(1), 125–132. doi:10.1179/174328405X16261

Miyajima, M., Kitamura, K., & Matsumoto, K. (2017). Characterization of Tribochemical Reactions on Steel Surfaces. *Nippon Steel Sumitomo Met Tech Rep*, *114*, 101–107.

Mizuhashi, S., Cordonier, C. E. J., Honma, H., & Takai, O. (2015). Stabilization of an Electroless Platinum Plating Bath Using S-Bearing Additives. *Journal of the Electrochemical Society*, *162*(10), D497–D502. doi:10.1149/2.0081510jes

Moazzenchi, B., & Montazer, M. (2019). Click electroless plating of nickel nanoparticles on polyester fabric: Electrical conductivity, magnetic and EMI shielding properties. *Colloids and Surfaces. A, Physicochemical and Engineering Aspects*, *571*, 110–124. Advance online publication. doi:10.1016/j.colsurfa.2019.03.065

Mocarski, S., Hall, D. W., Chernenkoff, R. A., Yeager, D. A., & McHugh, C. O. (1996). Master alloys to obtain premixed hardenable powder metallurgy steels. *Powder Metallurgy*, *39*(2), 130–137. doi:10.1179/pom.1996.39.2.130

Mohan, P. (2014). *Study the effects of welding parameters on TIG welding of Aluminium plate* (Doctoral dissertation).

Monticelli, C., Frignani, A., & Zucchi, F. (2004). Investigation on the Corrosion Process of Carbon Steel Coated by HVOF WC/Co Cermets in Neutral Solution. *Corrosion Science*, *46*(5), 1225–1237. doi:10.1016/j.corsci.2003.09.013

Moridi, A., Hassani-Gangaraj, S. M., Guagliano, M., & Dao, M. (2014). Cold spray coating: Review of material systems and future perspectives. *Surface Engineering*, *30*(6), 369–395. doi:10.1179/1743294414Y.0000000270

Moskal, G., Swadźba, L., Hetmańczyk, M., Witala, B., Mendala, B., Mendala, J., & Sosnowy, P. (2012). Characterization of the Microstructure and Thermal Properties of $Nd_2Zr_2O_7$ and $Nd_2Zr_2O_7$/YSZ Thermal Barrier Coatings. *Journal of the European Ceramic Society*, *32*(9), 2035–2042. doi:10.1016/j.jeurceramsoc.2011.12.004

Mostaghimi, J., Pershin, L., & Yugeswaran, S. (2017). Heat Transfer in DC AND RF Plasma Torches. In F. Kulacki (Ed.), *Handbook of Thermal Science and Engineering*. Springer. doi:10.1007/978-3-319-32003-8_27-1

Moustafa, S. F., Abdel-Hamid, Z., Baheig, O. G., & Hussien, A. (2011). Synthesis of WC hard materials using coated powders. *Advanced Powder Technology*, *22*(5), 596–601. doi:10.1016/j.apt.2010.08.011

Moustafa, S. F., El-Badry, S. A., Sanad, A. M., & Kieback, B. (2002). Friction and wear of copper–graphite composites made with Cu-coated and uncoated graphite powders. *Wear*, *253*(7-8), 699–710. doi:10.1016/S0043-1648(02)00038-8

Muehlbrger, E. (2007). *Method and apparatus for low pressure cold spraying*. US Patent 6759085.

Muench, F., Eils, A., Toimil-Molares, M. E., Hossain, U. H., Radetinac, A., Stegmann, C., Kunz, U., Lauterbach, S., Kleebe, H.-J., & Ensinger, W. (2014). Polymer activation by reducing agent absorption as a flexible tool for the creation of metal films and nanostructures by electroless plating. *Surface and Coatings Technology*, *242*, 100–108. doi:10.1016/j.surfcoat.2014.01.024

Müler, K., Bugnicourt, E., Latorre, M., Jorda, M., Sanz, Y. E., Lagaron, J. M., & … . (2017). Review on the Processing and Properties of Polymer Nanocomposites and Nanocoatings and Their Applications in the Packaging, Automotive and Solar Energy Fields. *Nanomaterials (Basel, Switzerland)*, *7*(74), 74. doi:10.3390/nano7040074 PMID:28362331

Müller, E., Drašar, Č., Schilz, J., & Kaysser, W. A. (2003). Functionally graded materials for sensor and energy applications. *Materials Science and Engineering A*, *362*(1–2), 17–39. doi:10.1016/S0921-5093(03)00581-1

Mu, Z., Geng, H.-R., Li, M.-M., Nie, G.-L., & Leng, J.-F. (2013). Effects of Y2O3 on the property of copper based contact materials. *Composites. Part B, Engineering*, *52*, 51–55. doi:10.1016/j.compositesb.2013.02.036

Nagano, M., Nagashima, S., Maeda, H., & Kato, A. (1999). Sintering behaviour of Al_2TiO_5 base ceramics and their thermal properties. *Ceramics International*, *25*(8), 681–687. doi:10.1016/S0272-8842(98)00083-2

Naga, S. M., Awaad, M., Bondioli, F., Fino, P., & Hassan, A. M. (2017). Thermal diffusivity of ZTA composites with different YSZ quantity. *Journal of Alloys and Compounds*, *695*, 1859–1862. doi:10.1016/j.jallcom.2016.11.019

Nakahara, S. (1977). Microporosity induced by nucleation and growth processes in crystalline and non-crystalline films. *Thin Solid Films*, *45*(3), 421–432. doi:10.1016/0040-6090(77)90229-2

Narasimhan, K. S., & Amuda, M. O. H. (2017). Powder Characterization. In Reference Module in Materials Science and Materials Engineering. doi:10.1016/B978-0-12-803581-8.10141-9

Narasimhan, K. S. (1996). Recent advances in ferrous powder metallurgy. *Advanced Performance Materials*, *3*(1), 7–27. doi:10.1007/BF00136857

Narasimhan, K. S. (2001). Sintering of powder mixtures and the growth of ferrous powder metallurgy. *Materials Chemistry and Physics*, *67*(1-3), 56–65. doi:10.1016/S0254-0584(00)00420-X

Narayanan, T. S., & Seshadri, S. K. (2004). Formation and characterization of borohydride reduced electroless nickel deposits. *Journal of Alloys and Compounds*, *365*(1-2), 197–205. doi:10.1016/S0925-8388(03)00680-7

Narcus, H. (1947). Practical copper reduction on nonconductors. *Metal Finishing*, *45*(9), 64–67.

Nassef, A., El-Garaihy, W., & El-Hadek, M. (2017). Characteristics of Cold and Hot Pressed Iron Aluminum Powder Metallurgical Alloys. *Metals*, *7*(5), 170. doi:10.3390/met7050170

Nastic, A., Merati, A., Bielawski, M., Bolduc, M., Fakolujo, O., & Nganbe, M. (2015). Instrumented and Vickers indentation for the characterization of stiffness, hardness and toughness of zirconia toughened Al2O3 and SiC armor. *Journal of Materials Science and Technology*, *31*(8), 773–783. doi:10.1016/j.jmst.2015.06.005

Natesan, K. (1976). Corrosion-Erosion behavior of materials in a coal-gasification environment. *Corrosion*, *32*(9), 364–370. doi:10.5006/0010-9312-32.9.364

Nathan, A., Ahnood, A., Cole, M. T., Lee, S., Suzuki, Y., Hiralal, P., ... Haque, S. (2012). Flexible electronics: the next ubiquitous platform. *Proceedings of the IEEE, 100*, 1486-1517.

Nautiyal, H., & Tyagi, R. (2020). High temperature erosive wear behavior of high velocity oxy-fuel sprayed Cr3C225 (Ni20Cr) coating on (AISI-316) austenitic steel. *Journal of Tribology*, *142*(7), 1–16. doi:10.1115/1.4046290

Nemani, S. K., Annavarapu, R. K., Mohammadian, B., Raiyan, A., Heil, J., Haque, M. A., … Sojoudi, H. (2018). Surface Modification of Polymers: Methods and Applications. Adv. Mater. Interfaces, 1-26.

Nemat-Alla, M. M., Ata, M. H., Bayoumi, M. R., & Khair-Eldeen, W. (2011). Powder Metallurgical Fabrication and Microstructural Investigations of Aluminum/Steel Functionally Graded Material. *Materials Sciences and Applications*, *02*(12), 1708–1718. doi:10.4236/msa.2011.212228

Nguyen, S. T., Nakayama, T., Suematsu, H., Suzuki, T., Takeda, M., & Niihara, K. (2016). Low Thermal Conductivity $Y_2Ti_2O_7$ as a Candidate Material for Thermal/Environmental Barrier Coatings. *Ceramics International*, *42*(9), 11314–11323. doi:10.1016/j.ceramint.2016.04.052

Nguyen-Tri, P., Nguyen, T. A., Carriere, P., & Xuan, C. N. (2018). Nanocomposite Coatings: Preparation, Characterization, Properties, and Applications. *International Journal of Corrosion*, *2018*, 1–19. doi:10.1155/2018/4749501

Nienow, A. W., Harnby, N., & Edwards, M. F. (1992). Introduction to mixing problems. In Mixing in the Process Industries (pp. 1–24). doi:10.1016/B978-075063760-2/50022-6

Ning, X.-J., Kim, J.-H., Kim, H.-J., & Lee, C. (2009). Characteristics and heat treatment of cold-sprayed Al–Sn binary alloy coatings. *Applied Surface Science*, *255*(7), 3933–3939. doi:10.1016/j.apsusc.2008.10.074

Noori, H., Mousavian, R. T., Khosroshahi, R. A., Brabazon, D., & Damadi, S. R. (2016). Effect of SiC particle morphology on Co–P electroless coating characteristics. *Surface Engineering*, *32*(5), 391–396. doi:10.1179/1743294415Y.0000000035

Nouicer, E., Benlahreche, F., & Nouicer, A. (2017). Spectroscopic Characterization of Titanium Alloy Surface in a Biological Medium. *Acta Physica Polonica A*, *131*(1), 28–31. doi:10.12693/APhysPolA.131.28

Nouri, S., Sahmani, S., Asayesh, M., & Aghdam, M. M. (2019). Study on the oxidation resistance of γ -TiAl intermetallic alloy coated via different diffusion coating processes. Materials Research Express, 6(10), 106522. doi:10.1088/2053-1591/ab36d5

O'Brien, P. (2001). Encyclopedia of Materials: Science and Technology. Academic Press.

Odhiambo, J. G., Lei, W. G., Zhao, Y. T., & Lei, C. L. (2019). Porosity and its significance in plasma sprayed coatings. *Coatings*, *9*(460), 1–24. doi:10.3390/coatings9070460

Oguocha, I. N. A., Taheri, R., Yannacopoulos, S., Uju, W. A., Sammynaiken, R., Wettig, S., & Hu, Y. F. (2010). Temperature effects on the chemical composition of nickel–phosphorus alloy thin films. *Thin Solid Films*, *518*(8), 2045–2049. doi:10.1016/j.tsf.2009.09.007

Oh, K. W., Hong, K. H., & Kim, S. H. (1999). Electrically conductive textiles by in situ polymerization of aniline. *Journal of Applied Polymer Science*, *74*(8), 2094–2101. doi:10.1002/(SICI)1097-4628(19991121)74:8<2094::AID-APP26>3.0.CO;2-9

Ohkubo, N., Miyakusu, K., Uematsu, Y., & Kimura, H. (1994). Effect of alloying elements on the mechanical properties of the stable austenitic stainless steel. *ISIJ International*, *34*(9), 764–772. doi:10.2355/isijinternational.34.764

Oita, M., Matsuoka, M., & Iwakura, C. (1997). Deposition rate and morphology of electroless copper film from solutions containing 2, 2¢-dipyridyl. *Electrochimica Acta*, *42*(9), 1435–1440. doi:10.1016/S0013-4686(96)00367-2

Oksa, M., Auerkari, P., Salonen, J., & Varis, T. (2014). Nickel-based HVOF coatings promoting high temperature corrosion resistance of biomass-fired power plant boilers. *Fuel Processing Technology*, *125*, 236–245. doi:10.1016/j.fuproc.2014.04.006

Oksa, M., Tuurna, S., & Varis, T. (2013). Increased lifetime for biomass and waste to energy power plant boilers with HVOF coatings: High temperature corrosion testing under chlorine-containing molten salt. *J. Therm Spray Technol.*, *22*(5), 783–796. doi:10.100711666-013-9928-5

Oluwafunmilade, A. K., Abidine, Z., Ando, Y., Noda, Y., & Iino, M. (2020). Deposition of Porous Photocatalytic TiO 2 Film for Dye-Sensitized Solar Cells by Low-Powered Atmospheric Plasma Spray Equipment. In *Renewable Energy and Sustainable Buildings* (pp. 945–953). Springer. doi:10.1007/978-3-030-18488-9_79

Oluwalowo, A., Nguyen, N., Zhang, S., Park, J. G., & Liang, R. (2019). Electrical and thermal conductivity improvement of carbon nanotube and silver composites. *Carbon*, *146*, 224–231. doi:10.1016/j.carbon.2019.01.073

Ordonez, A. A., & Prieto, M. (2012). Technical and methodological aspects of fourier transform infrared spectroscopy in food microbiology research. In *Fourier Transform Infrared Spectroscopy in Food Microbiology* (pp. 1–18). Springer. doi:10.1007/978-1-4614-3813-7_1

Oukach, S., Hamdi, H., El Ganaoui, M., & Pateyron, B. (2020). Protective Plasma Sprayed Coating forThermo-Sensitive Substrates. In *MATEC Web of Conferences* (*Vol. 307*, p. 01039). EDP Sciences. 10.1051/matecconf/202030701039

Ozaki, T., Zhang, Y., Komaki, M., & Nishimura, C. (2003). Preparation of palladium-coated V and V?15Ni membranes for hydrogen purification by electroless plating technique. *International Journal of Hydrogen Energy*, *28*(3), 297–302. doi:10.1016/S0360-3199(02)00065-4

Padmavathi, K. R., & Ramakrishnan, R. (2014). Tribological Behaviour of Aluminium Hybrid Metal Matrix Composite. *Procedia Engineering*, *97*, 660–667. doi:10.1016/j.proeng.2014.12.295

Padture, N. P., Gell, M., & Jordan, E. H. (2002). Thermal barrier coatings for gas turbine engine applications. *Science*, *296*(5566), 280–284. doi:10.1126cience.1068609 PMID:11951028

Palaniappa, M., Babu, G. V., & Balasubramanian, K. (2007). Electroless nickel–phosphorus plating on graphite powder. *Materials Science and Engineering A*, *471*(1-2), 165–168. doi:10.1016/j.msea.2007.03.004

Palaniappa, M., & Seshadri, S. K. (2008). Friction and wear behavior of electroless Ni–P and Ni–W–P alloy coatings. *Wear*, *265*(5-6), 735–740. doi:10.1016/j.wear.2008.01.002

Palathai, T., Tharajak, J., & Sombatsompop, N. (2008). Hardness, adhesion index and microstructure of PEEK coating on Al or Fe substrate by LVOF flame spray. *Materials Science and Engineering A*, *485*(1-2), 66–73. doi:10.1016/j.msea.2007.07.049

Pancrecious, J. K., Ulaeto, S. B., Ramya, R., Rajan, T. P. D., & Pai, B. C. (2018). Metallic composite coatings by electroless technique–a critical review. *International Materials Reviews*, *63*(8), 488–512. doi:10.1080/09506608.2018.1506692

Pandey, V. K., Patel, B. P., & Guruprasad, S. (2016). Mechanical properties of Al/Al 2 O 3 and Al/B 4 C composites. *Advanced Materials Research*, *5*(4), 263–277. doi:10.12989/amr.2016.5.4.263

Pang, J. N., Jiang, S. W., Lin, H., & Wang, Z. Q. (2016). Significance of sensitization process in electroless deposition of Ni on nanosized Al2O3 powders. *Ceramics International*, *42*(3), 4491–4497. doi:10.1016/j.ceramint.2015.11.137

Pan, Z., Ding, D., Wu, B., Cuiuri, D., Li, H., & Norrish, J. (2018). Arc welding processes for additive manufacturing: a review. In *Transactions on Intelligent Welding Manufacturing* (pp. 3–24). Springer. doi:10.1007/978-981-10-5355-9_1

Papyrin, A., Kosarev, V., Klinkov, S., Alkimov, A., & Fomin, V. (2007). Cold spray technology. Elsevier.

Papyrin, A., Kosarev, V., Klinkov, S., Alkhimov, A., & Fomin, V. M. (2006). *Cold spray technology*. Elsevier.

Park, J. H., Kim, Y. H., Baek, H. J., & Cho, S. M. (2019). A study on process development of super-TIG welding for 9% nickel steel with Alloy 625. *Journal of Manufacturing Processes*, *40*(February), 140–148. doi:10.1016/j.jmapro.2019.03.017

Park, J., Lee, S., Kang, S., Jeon, J., Lee, S. H., Kim, H. K., & Choi, H. (2015). Complex effects of alloy composition and porosity on the phase transformations and mechanical properties of powder metallurgy steels. *Powder Technology*, *284*, 459–466. doi:10.1016/j.powtec.2015.07.029

Park, S., Kim, J., & Park, C. H. (2015). Superhydrophobic Textiles: Review of Theoretical Definitions, Fabrication and Functional Evaluation. *Journal of Engineered Fibers and Fabrics*, *10*(4), 1–18. doi:10.1177/155892501501000401

Pasandideh-Fard, M., Pershin, V., Chandra, S., & Mostaghimi, J. (2002). Splat shapes in a thermal spray coating process: Simulations and experiments. *Journal of Thermal Spray Technology*, *11*(2), 206–217. doi:10.1361/105996302770348862

Patil, S. R., & Waghmare, C. A. (2013). Optimization of MIG welding parameters for improving the strength of welded joints. *Int. J. Adv. Engg. Res. Studies, 14*, 16.

Pattison, Celotto, Khan, & O'Neill. (2008). Standoff distance and bow shock phenomena in the Cold Spray process. *Surface and Coatings Technology, 202*(8), 1443-1454.

Paul, T. (1956). *U.S. Patent No. 2,766,138*. Washington, DC: U.S. Patent and Trademark Office.

Paunovic, M. (1968). Electrochemical aspects of electroless deposition of metals. *Plating (East Orange, N.J.), 55*, 1161–1167.

Pauw, L. (1958). A method of measuring specific resistivity and Hall effect of discs of arbitrary shape. *Philips Research Reports, 13*(2), 1–9.

Pawlowski, L. (2008). *The science and engineering of thermal spray coatings* (2nd ed.). John Wiley & Sons. doi:10.1002/9780470754085

Pázmán, J., Mádai, V., Tóth, J., & Gácsi, Z. (2010). Investigation of the electroless nickel plated SiC particles in metal matrix composites. *Powder Metallurgy Progress, 10*(2), 102–109.

Peng, Y., Yang, C., Chen, K., Popuri, S. R., Lee, C.-H., & Tang, B.-S. (2012). Study on synthesis of ultrafine Cu–Ag core–shell powders with high electrical conductivity. *Applied Surface Science, 263*, 38–44. doi:10.1016/j.apsusc.2012.08.066

Perez Andrade, L. (2020). *Additive Manufacturing by Cold Spray of Nickel-based Superalloys Inconel 718* (Doctoral dissertation). Cinvestav, Queretaro, México.

Petrovicova, E., & Schadler, L. S. (2002). Thermal spraying of polymers. *International Materials Reviews, 47*(4), 169–190. doi:10.1179/095066002225006566

Pettit, F. (2011). Hot corrosion of metals and alloys. *Oxidation of Metals, 76*(1-2), 1–21. doi:10.100711085-011-9254-6

Physical Vapor Deposition, P. V. D. (n.d.). http://www.sigmaaldrich

Picalek, J., Holger, B., Akhtar, S., Bejan, I., & Pinnau, J. (2016). Patent No. 15/520,558. Washington, DC: US Patent Office.

Pinzón, A. V., Urrego, K. J., González-Hernández, A., Rincon Ortiz, M., & Vargas, F. (2018). Corrosion protection of carbon steel by alumina-titania ceramic coatings used for industrial applications. *Ceramics International, 44*(17), 21765–21773. doi:10.1016/j.ceramint.2018.08.273

Poerschke, D. L., Jackson, R. W., & Levi, C. G. (2017). Silicate Deposit Degradation of Engineered Coatings in Gas Turbines: Progress toward Models and Materials Solutions. *Annual Review of Materials Research, 47*(1), 297–330. doi:10.1146/annurev-matsci-010917-105000

Poirier, T., & Quercia, G. (2015). Formation of hollow vitreous and semi-crystalline microspheres in slag flame spraying. *Ceramics International, 41*(1), 369–377. doi:10.1016/j.ceramint.2014.08.080

Pollock, T. M., Lipkin, D. M., & Hemker, K. J. (2012). Multifunctional Coating Interlayers for Thermal-Barrier Systems. *MRS Bulletin, 37*(10), 923–931. doi:10.1557/mrs.2012.238

Pompanon, F., Laporte, J., Fouvry, S., & Alquier, O. (2019). Normal force and displacement amplitude influences on silver-plated electrical contacts subjected to fretting wear: A basic friction energy – contact compliance formulation. *Wear, 426–427*, 652–661. doi:10.1016/j.wear.2018.12.010

Prashar, G., & Vasudev, H. (2020, February). *Hot corrosion behavior of super alloys*. Paper presented at the International conference on material processing and charcterziation. GLA University, Mathura.

Pratihar, S. K., Sharma, A. D., Basu, R. N., & Maiti, H. S. (2004). Preparation of nickel coated YSZ powder for application as an anode for solid oxide fuel cells. *Journal of Power Sources, 129*(2), 138–142. doi:10.1016/j.jpowsour.2003.11.023

Praxair HVOF Guns. (n.d.). Retrieved from: https://www.praxairsurfacetechnologies.com/en/components-materials-and-equipment/coating-equipment/thermal-spray-coating-systems/high-velocity-oxygen-fuel-hvof

Prince, R. M. R., Selvakumar, N., Arulkirubakaran, D., Sreedharan, C. E. S., Thulasiram, R., & Kumar, R. M. (2020). Surface structural features and wear analysis of a multilayer Ti6Al4V-B_4C thin film coated AISI 1040 steel. *Materials Research Express*, *7*, 1–17.

Proner, A. (1999). Revêtements par projection thermique : Conclusion, perspectives d'avenir | Techniques de l'Ingénieur. *Techniques de l'ingénieur.* Retrieved from: https://www.techniques-ingenieur.fr/base-documentaire/materiaux-th11/traitements-de-surface-des-metaux-par-voie-seche-et-en-milieu-fondu-42360210/revetements-par-projection-thermique-m1645/conclusion-perspectives-d-avenir-m1645niv10008.html

Purohit, P., & Vagge, S. T. (2016). Evaluation of alumina incorporated combined ceramic layer thermal barrier coating. *Surface and Coatings Technology*, *307*, 871–878. doi:10.1016/j.surfcoat.2016.10.022

Radka, J., Ctibor, P., & Benes, V. (2004). Analysis of porous structure in plasma-sprayed coating. *Image Analysis & Stereology*, *23*, 45–52.

Rahimipour, M. R., & Sobhani, M. (2013). Evaluation of centrifugal casting process parameters for in situ fabricated functionally gradient Fe-TiC composite. *Metallurgical and Materials Transactions. B, Process Metallurgy and Materials Processing Science*, *44*(5), 1120–1123. doi:10.1007/s11663-013-9903-z

Rahmawan, Y., Xu, L. B., & Yang, S. (2013). Self-assembly of nanostructures towards transparent, superhydrophobic surfaces. *Journal of Materials Chemistry. A, Materials for Energy and Sustainability*, *1*(9), 2955–2969. doi:10.1039/C2TA00288D

Ramachandran, C. S., Balasubramanian, V., Ananthapadmanabhan, P. V., & Viswabaskaran, V. (2012). Influence of the intermixed interfacial layers on the thermal cycling behavior of atmospheric plasma sprayed lanthanum zirconate based coatings. *Ceramics International*, *38*(5), 4081–4096. doi:10.1016/j.ceramint.2012.01.066

Ramalho, A., & Miranda, J. C. (2005). Friction and wear of electrolessNiP and NiP+ PTFE coatings. *Wear*, *259*(7-12), 828–834. doi:10.1016/j.wear.2005.02.052

Ramaseshan, R., Seshadri, S., & Nair, N. (2001). Electroless nickel-phosphorus coating on Ti and Al elemental powders. *Scripta Materialia*, *45*(2), 183–189. doi:10.1016/S1359-6462(01)01013-2

Ramesh, M. R., Prakash, S., Nath, S. K., Sapra, P. K., & Venkataraman, B. (2010). Solid particle erosion of HVOF sprayed WC–Co/NiCrFeSiB coatings. *Wear*, *269*(3-4), 197–205. doi:10.1016/j.wear.2010.03.019

Rana, A., Singh, B., & Sharma, R. (2018). Raman Study of Ga Doped Ge-S Glassy Alloy. *International Journal of Chemical and Physical Sciences*, *7*, 638–641.

Rana, R. S., Purohit, R., & Das, S. (2012). Reviews on the influences of alloying elements on the microstructure and mechanical properties of aluminum alloys and aluminum alloy composites. *Int. J. Scientific Res. Publ.*, *2*(6), 1–7.

Rapp, R. A. (1986). Chemistry and electrochemistry of the hot corrosion of metals. *Corrosion*, *42*(10), 568–577. doi:10.5006/1.3583026

Rath, S., Hsieh, M. L., Etchegoin, P., & Stradling, R. A. (2003). Alloy effects on the Raman spectra of $Si_{1-x}Ge_x$ and calibration protocols for alloy compositions based on polarization measurements. *Semiconductor Science and Technology*, *18*(6), 566–575. doi:10.1088/0268-1242/18/6/330

Redjdal, O., Zaid, B., Tabti, M. S., Henda, K., & Lacaze, P. C. (2013). Characterization of thermal flame sprayed coatings prepared from FeCr mechanically milled powder. *Journal of Materials Processing Technology*, *213*(5), 779–790. doi:10.1016/j.jmatprotec.2012.11.018

Reid, J., Mayer, S., Broadhent, E., Klawuhn, E., & Ashtiani, K. (2000). Factors influencing damascene feature fill using copper PVD and electroplating. *Solid State Technology*, *43*(7), 86–86.

Reinshagen, J. H., & Witsberger, J. C. (1991). Processing and properties of 400 series PM stainless steels. *Metal Powder Report*, *46*(5), 38–44. doi:10.1016/0026-0657(91)91757-W

Reis, M. A. L. D., Neto, N. M. B., Sousa, M. E. S. D., Araujo, P. T., Simões, S., Vieira, M. F., ... Assuncao, P. D. (2018). Raman spectroscopy fingerprint of stainless steel-MWCNTs nanocomposite processed by ball-milling. *AIP Advances*, *8*(1), 015323. doi:10.1063/1.5018745

Ren, C., He, Y. D., & Wang, D. R. (2011). Cyclic oxidation behavior and thermal barrier effect of YSZ-(Al2O3/YAG) double-layer TBCs prepared by the composite sol–gel method. *Surface and Coatings Technology*, *206*(6), 1461–1468. doi:10.1016/j.surfcoat.2011.09.025

Ren, K., Wang, Q., Shao, G., Zhao, X., & Wang, Y. (2020). Multicomponent High-Entropy Zirconates with Comprehensive Properties for Advanced Thermal Barrier Coating. *Scripta Materialia*, *178*, 382–386. doi:10.1016/j.scriptamat.2019.12.006

Ren, X., Tian, Z., Zhang, J., & Wang, J. (2019). Equiatomic Quaternary $(Y_{1/4}Ho_{1/4}Er_{1/4}Yb_{1/4})_2SiO_5$ Silicate: A Perspective Multifunctional Thermal and Environmental Barrier Coating Material. *Scripta Materialia*, *168*, 47–50. doi:10.1016/j.scriptamat.2019.04.018

Richer, P., Yandouzi, M., Beauvais, L., & Jodoin, B. (2010). Oxidation behaviour of CoNiCrAlY bond coats produced by plasma, HVOF and cold gas dynamic spraying. *Surface and Coatings Technology*, *204*(24), 3962–3974. doi:10.1016/j.surfcoat.2010.03.043

Riva, A., Algaba, I. M., & Pepio, M. (2006). Action of a finishing product in the improvement of the ultraviolet protection provided by cotton fabrics. Modelisation of the effect. *Cellulose (London, England)*, *13*(6), 697–704. doi:10.100710570-006-9085-9

Rodrıguez, J., Martın, A., Fernández, R., & Fernández, J. E. (2003). An experimental study of the wear performance of NiCrBSi thermal spray coatings. *Wear*, *255*(7-12), 950–955. doi:10.1016/S0043-1648(03)00162-5

Rodriguez, M., Staia, M., Gil, L., Arenas, F., & Scagni, A. (2000). Effect of Heat Treatment on Properties of Nickel Hard Surface Alloy Deposited by HVOF. *Surface Engineering*, *16*(5), 415–420. doi:10.1179/026708400101517404

Rodriguez-Suarez, T., Bartolomé, J. F., & Moya, J. S. (2012). Mechanical and tribological properties of ceramic/metal composites: A review of phenomena spanning from the nanometer to the micrometer length scale. *Journal of the European Ceramic Society*, *32*(15), 3887–3898. doi:10.1016/j.jeurceramsoc.2012.06.026

Roe, B., & Zhang, X. (2009). Durable Hydrophobic Textile Fabric Finishing Using Silica Nanoparticles and Mixed Silanes. *Textile Research Journal*, *79*(12), 1115–1122. doi:10.1177/0040517508100184

Romanov, D., Moskovskii, S., Konovalov, S., Sosnin, K., Gromov, V., & Ivanov, Y. (2019). Improvement of copper alloy properties in electro-explosive spraying of ZnO-Ag coatings resistant to electrical erosion. *Journal of Materials Research and Technology*, *8*(6), 5515–5523. doi:10.1016/j.jmrt.2019.09.019

Ropp, R. C., & Libowitz, G. G. (1978). The nature of the alumina-rich phase in the System La_2O_3-$A1_2O_3$. *Journal of the American Ceramic Society*, *61*(11-12), 473–475. doi:10.1111/j.1151-2916.1978.tb16119.x

Rout, T. K., Jha, G., Singh, A. K., Bandyopadhyay, N., & Mohanty, O. N. (2003). Development of conducting polyaniline coating: A novel approach to superior corrosion resistance. *Surface and Coatings Technology*, *167*(1), 16–24. doi:10.1016/S0257-8972(02)00862-9

Roy, S., & Sahoo, P. (2012). Corrosion study of electroless Ni-PW coatings using electrochemical impedance spectroscopy. *Portugaliae Electrochimica Acta*, *30*(3), 203–220. doi:10.4152/pea.201203203

Roy, S., & Sahoo, P. (2013). Tribological performance optimization of electroless Ni-PW coating using weighted principal component analysis. *Tribology in Industry*, *35*(4), 297–307.

Roy, S., & Sahoo, P. (2014). Parametric optimization of corrosion and wear of electroless Ni-P-Cu coating using grey relational coefficient coupled with weighted principal component analysis. *International Journal of Mechanical and Materials Engineering*, *9*(1), 10. doi:10.118640712-014-0010-y

Ruigang, W., Wei, P., Jian, C., Mengning, J., & Minghao, F. (2002). Fabrication and characterization of machinable Si3N4/h-BN functionally graded materials. *Materials Research Bulletin*, *37*(7), 1269–1277. doi:10.1016/S0025-5408(02)00773-0

S. (2009). Electroless deposition of nickel and copper on titanium substrates: Characterization and application. *Journal of Alloys and Compounds, 472*(1-2), 595-601.

Saaedi, J., Coyle, T. W., Arabi, H., Mirdamadi, S., & Mostaghimi, J. (2010). Effects of HVOF Process Parameters on the Properties of Ni-Cr Coatings. *J. Therm. Spray Technol.*, *19*(3), 521–530. doi:10.100711666-009-9464-5

Sadeghi, E., Markocsan, N., & Joshi, S. (2019). Advances in Corrosion-Resistant Thermal Spray Coatings for Renewable Energy Power Plants. Part I: Effect of Composition and Microstructure. *J Therm Spray Tech.*, *28*(8), 1749–1788. doi:10.100711666-019-00938-1

Sadeghimeresht, E. (2018). *Ni-Based Coatings for High Temperature Corrosion Protection.* (Doctoral dissertation). University West, Department of Engineering Science, Division of Subtractive and Additive Manufacturing.

Sadeghimeresht, E., Hooshyar, H., Markocsan, N., Joshi, S., & Nylen, P. (2016a). Oxidation Behavior of HVAF-Sprayed NiCoCrAlY Coating in H_2-H_2O Environment. *Oxidation of Metals*, *86*(3-4), 299–314. doi:10.100711085-016-9637-9

Sadeghimeresht, E., Markocsan, N., & Nylen, P. (2016b). A Comparative Study on Ni-Based Coatings Prepared by HVAF, HVOF, and APS Methods for Corrosion Protection Applications. *J. Therm.Spray Technol.*, *25*(8), 1604–1616. doi:10.100711666-016-0474-9

Sadeghimeresht, E., Markocsan, N., Nylen, P., & Bjorklund, S. (2016c). Corrosion Performance of Bi-Layer Ni/Cr_2C_3-NiCr HVAF Thermal Spray Coating. *Applied Surface Science*, *369*, 470–481. doi:10.1016/j.apsusc.2016.02.002

Sadeghimeresht, E., Reddy, L., Hussain, T., Huhtakangas, M., Markocsan, N., & Joshi, S. (2018a). Influence of KCl and HCl on HighTemperature Corrosion of HVAF-Sprayed NiCrAlY and NiCrMo Coatings. *Materials & Design*, *148*, 17–29. doi:10.1016/j.matdes.2018.03.048

Sadeghimeresht, E., Reddy, L., Hussain, T., Markocsan, N., & Joshi, S. (2018b). Chlorine-Induced High Temperature Corrosion of HVAF Sprayed Ni-Based Alumina and Chromia Forming Coatings. *Corrosion Science*, *132*, 170–184. doi:10.1016/j.corsci.2017.12.033

Sadeghimeresht, E., Zhang, P., Chen, S., Li, X. H., Markocsan, N., & Joshi, S. (2019). Effects of Surface Finish on the Initial Oxidation of HVAF-Sprayed NiCoCrAlY Coatings. *Surface and Coatings Technology*, *364*, 43–56. doi:10.1016/j.surfcoat.2019.02.068

Sadeq, B. R., Sahib, B. S., & Alher, M. (2020). Study of the Effects of Welding Process on the Microstructure of Worn Carbon Steel Shaft. *IOP Conference Series. Materials Science and Engineering*, *671*(1), 012161. Advance online publication. doi:10.1088/1757-899X/671/1/012161

Safranek, W. H., Dahle, F. B., & Faust, C. L. (1948). Electroforming—A Literature Review and a NewApplication. *Plating (East Orange, N.J.)*, *35*, 39–49.

Sahoo, P., & Das, S. K. (2011). Tribology of electroless nickel coatings–a review. *Materials & Design*, *32*(4), 1760–1775. doi:10.1016/j.matdes.2010.11.013

Sahoo, P., & Roy, S. (2017). Tribological behavior of electroless Ni-P, Ni-PW and Ni-P-Cu coatings: A comparison. *International Journal of Surface Engineering and Interdisciplinary Materials Science*, *5*(1), 1–15. doi:10.4018/IJSEIMS.2017010101

Saint-Gobain Coating Solutions. (n.d.). *Al_2O_3 and TiO_2, TI-TEX Ceramic Supra series-ref : 98210 4547 000*. Retrieved from: https://www.coatingsolutions.saint-gobain.com/sites/imdf.coatingsolutions.com/files/

Saint-Gobain Coating Solutions. (n.d.). *Coatings Al_2O_3, TI-ELITE Ceramic Supra series-ref : 98210 4447 000 Typical composition (%)-T TIP*. Retrieved from: https://www.coatingsolutions.saint-gobain.com/sites/imdf.coatingsolutions.com/files/

Šalak, A., & Selecká, M. (2012). *Manganese in powder metallurgy steels*. Springer Science & Business Media. doi:10.1007/978-1-907343-75-9

Šalak, A., Selecká, M., & Danninger, H. (2005). *Machinability of powder metallurgy steels*. Cambridge International Science Publishing.

Salgado, L., Filho, F. A., Neto, R. L., Rossi, J. L., Colosio, M. A., & Filho, C. H. (2001). P/M high-speed steels for valve seat insert application. *SAE Transactions*, 261–264. doi:10.4271/2001-01-0395

Samal, C. P., Parihar, J. S., & Chaira, D. (2013). The effect of milling and sintering techniques on mechanical properties of Cu-graphite metal matrix composite prepared by powder metallurgy route. *Journal of Alloys and Compounds*, *569*, 95–101. doi:10.1016/j.jallcom.2013.03.122

Samal, P., & Newkirk, J. (2015). ASM handbook: Vol. 7. *Powder metallurgy*. ASM International. doi:10.31399/asm.hb.v07.9781627081757

Sancaktar, E., & Bai, L. (2011). Electrically Conductive Epoxy Adhesives. *Polymers*, *3*(1), 427–466. doi:10.3390/polym3010427

Santacruz, G., Shigueaki Takimi, A., Vannucchi de Camargo, F., Pérez Bergmann, C., & Fragassa, C. (2019). Comparative study of jet slurry erosion of martensitic stainless steel with tungsten carbide HVOF coating. *Metals*, *9*(5), 600. doi:10.3390/met9050600

Santangelo, M., Silwal, B., & Purdy, A. (2016). *Vibration assisted robotic hot-wire gas tungsten arc welding (GTAW) for additive manufacturing of large metallic parts*. Solid Freeform Fabrication.

Sarkar, A. K. (2004). An evaluation of UV protection imparted by cotton fabrics dyed with natural colorants. *BMC Dermatology*, *4*(15), 1–8. doi:10.1186/1471-5945-4-15 PMID:15509304

Sauvet, G., Dupond, S., Kazmierski, K., & Chojnowski, J. (2003). Biocidal polymers active by contact. V. Synthesis of polysiloxanes with biocidal activity. *Journal of Applied Polymer Science*, *75*(8), 1005–1012. doi:10.1002/(SICI)1097-4628(20000222)75:8<1005::AID-APP5>3.0.CO;2-W

Saxena, V., Rani, R. U., & Sharma, A. K. (2006). Studies on ultra high solar absorber black electroless nickel coatings on aluminum alloys for space application. *Surface and Coatings Technology*, *201*(3-4), 855–862.

Schade, C. T., Schaberl, J. W., & Lawley, A. (2008). Stainless Steel AISI Grades for PM Applications. *International Journal of Powder Metallurgy*, *44*(3), 57–67.

Schartel, B. (2010). Phosphorus-based Flame Retardancy Mechanisms-Old Hat or a Starting Point for Future Development. *Materials (Basel)*, *3*(10), 4710–4745. doi:10.3390/ma3104710 PMID:28883349

Schmidt, T., Gartner, F., Assadi, H., & Kreye, H. (2006). Devlopement of a generalized parameter window for cold spray deposition. *Acta Materialia*, *54*(3), 729–742. doi:10.1016/j.actamat.2005.10.005

Schwarz, A., Cardoen, J., Westbroek, P., Van Langenhove, L., Bruneel, E., Van Driessche, I., & Hakuzimana, J. (2010). Steps Towards a Textile-Based Transistor: Development of the Gate and Insulating Layer. *Textile Research Journal*, *80*(16), 1738–1746. doi:10.1177/0040517510365948

Schwarz, A., Hakuzimana, J., Kaczynska, A., Banaszczyk, J., Westbroek, P., McAdams, E., ... Tseles, D. (2010). Gold coated para-aramid yarns through electroless deposition. *Surface and Coatings Technology*, *204*(9-10), 1412–1418. doi:10.1016/j.surfcoat.2009.09.038

Science Direct. (n.d.). Retrieved from: https://www.sciencedirect.com/topics/engineering/spray-process

Seeböck, R., Esrom, H., Charbonnier, M., Romand, M., & Kogelschatz, U. (2001). Surface modification of polyimide using dielectric barrier discharge treatment. *Surface and Coatings Technology*, *142*, 455–459.

Selvakumar, N., & Barshilia, H. C. (2012). Review of physical vapor deposited (PVD) spectrally selective coatings for mid- and high-temperature solar thermal applications. *Solar Energy Materials and Solar Cells*, *98*, 1–23. doi:10.1016/j.solmat.2011.10.028

Selvam, M. (2010). Electroless silver deposition on ABS plastic using Co (II) as reducing agent. *Transactions of the IMF*, *88*(4), 198-203. 10.1179/002029610X12694310988964

Sevillano, F., Poza, P., Munez, C. J., Vezzu, S., Rech, S., & Trentin, A. (2013). Cold-Sprayed Ni-Al_2O_3 Coatings for Applications in Power Generation Industry. *Journal of Thermal Spray Technology, 22,* 772-782.

Shacham-Diamand, Y., Dubin, V., & Angyal, M. (1995). Electroless copper deposition for ULSI. *Thin Solid Films*, *262*(1-2), 93–103. doi:10.1016/0040-6090(95)05836-2

Shacham-Diamand, Y., Inberg, A., Sverdlov, Y., Bogush, V., Croitoru, N., Moscovich, H., & Freeman, A. (2003). Electroless processes for micro- and nanoelectronics. *Electrochimica Acta*, *48*(20–22), 2987–2996. doi:10.1016/S0013-4686(03)00364-5

Shanmugam, K., Thirumalaikumarasamy, D., & Balasubramanian, V. (2014). Establishing empirical relationships to predict porosity level and corrosion rate of atmospheric plasma-sprayed alumina coatings on AZ31B magnesium alloy. *J. Magnes. Alloy*, *2*(2), 140–153. doi:10.1016/j.jma.2014.05.002

Shanmugasundaram, D., & Chandramouli, R. (2009). Tensile and impact behaviour of sinter-forged Cr, Ni and Mo alloyed powder metallurgy steels. *Materials & Design*, *30*(9), 3444–3449. doi:10.1016/j.matdes.2009.03.020

Sharma, R., Agarwala, R. C., & Agarwala, V. (2006). Development of copper coatings on ceramic powder by electroless technique. *Applied Surface Science*, *252*(24), 8487–8493. doi:10.1016/j.apsusc.2005.11.085

Sharma, S. B., Agarwala, R. C., Agarwala, V., & Ray, S. (2002). Dry sliding wear and friction behavior of Ni-P-ZrO2-Al2O3 composite electroless coatings on aluminum. *Materials and Manufacturing Processes*, *17*(5), 637–649. doi:10.1081/AMP-120016088

Sheban, M., Abu-Dalo, M., Ababneh, A., & Andreescu, S. (2007). Effect of benzotriazole derivatives on the corrosion of steel in simulated concrete pore solutions. *Anti-Corrosion Methods and Materials*, *54*(3), 135–147. doi:10.1108/00035590710748605

Shen, G. X., Chen, Y. C., & Lin, C. J. (2005). Corrosion protection of 316 L stainless steel by a TiO_2 nanoparticle coating prepared by sol-gel method. *Thin Solid Films*, *489*(1-2), 130–136. doi:10.1016/j.tsf.2005.05.016

Shih, S., Zhang, Y., & Li, X. (1989). Sub-melting point hot corrosion of alloys and coatings. *Journal of Materials Science and Engineering: A, 120-121*(1), 277–282.

Shi, J. M., Zhang, L. X., Liu, H., Sun, Z., & Feng, J. C. (2019). Reliable brazing of SiBCN ceramic and TC4 alloy using AgCuTi filler with the assist of laser melting deposited FGM layers. *Journal of Materials Science*, *54*(4), 2766–2778. doi:10.1007/s10853-018-3025-6

Shijitha, T., Baiju, K. V., Shukla, S., Patil, K., & Warrier, K. G. K. (2009). Novelelectrolessprocessforcoppercoatingofflyashusingtitania/ultraviolet-radiation/metalcatalyst-system. *Applied Surface Science*, *255*(13-14), 6696–6704. doi:10.1016/j.apsusc.2009.02.066

Shin, J., Kim, H., Song, K. H., & Choe, J. (2015). Synthesis of Silver-coated Copper Particles with Thermal Oxidation Stability for a Solar Cell Conductive Paste. *Chemistry Letters*, *44*(9), 1223–1225. doi:10.1246/cl.150424

Shkodkin, Oklyuev, & Buzdygar. (2010). *Peculiarities of Gas Dynamic Spray Applications in Russia.* Academic Press.

Shourgeshty, M., Aliofkhazraei, M., & Alipour, M. M. (2016). *Introduction to High-Temperature Coatings, High Temperature Corrosion.* IntechOpen. Available from: https://www.intechopen.com/books/high-temperature-corrosion/introduction-to-high-temperature-coatings

Shukla, S., & Koelling, K. W. (2009). Classical nucleation theory applied to homogeneous bubble nucleation in the continuous microcellular foaming of the polystyrene– CO2 system. *Industrial & Engineering Chemistry Research*, *48*(16), 7603–7615. doi:10.1021/ie8011243

Sidky, P. S., & Hocking, M. G. (1999). Review of inorganic coatings and coating processes for reducing wear and corrosion. *British Corrosion Journal*, *34*(2), 171–183. doi:10.1179/000705999101500815

Siegmann, S., & Abert, C. (2013). 100 years of thermal spray: About the inventor Max Ulrich Schoop. *Surface and Coatings Technology*, *220*, 3–13. doi:10.1016/j.surfcoat.2012.10.034

Sierra, M. T. H., Sánchez, M. G. B., Báez, J. E., Camacho, L. D. A., Miranda, J. S. G., & Moreno, K. J. (2019). Improvement Effect of Green Lubricants on the Tribological and Mechanical Performance of 4140 Steel. *Applied Sciences (Basel, Switzerland)*, *9*(22), 4896. doi:10.3390/app9224896

Singh, Sidhu, Kalsi, & Karthikeyan. (2013). Development of Cold Spray from innovation to emerging future coating technology. *Journal of Brazilian Soc Mech*, 231-245.

Singh, H., & Prakash, S. (2009). *Plasma Spray Coatings for Superalloys: Characterization and HighTemperature Oxidation Behavior.* VDM Verlag.

Singh, H., Puri, D., & Prakash, S. (2005). High Temperature Oxidation Behaviour of Plasma Sprayed NiCrAlY Coatings on Ni-Based Superalloys in Air. *Transactions of the Indian Institute of Metals*, *59*, 215–227.

Singh, H., Sidhu, B. S., Puri, D., & Prakash, S. (2007). Use of plasma spray technology for deposition of high temperature oxidation/corrosion resistant coatings–a review. *Materials and Corrosion*, *58*(2), 92–102. doi:10.1002/maco.200603985

Singh, H., Sidhu, T. S., & Kalsi, S. (2012). *Cold spray technology: Future of coating deposition processes*. Gurunaka Dev University.

Skapin, S., Kolar, D., & Suvorov, D. (1993). X-ray diffraction and microstructural investigation of the Al_2O_3-La_2O_3-TiO_2 system. *Journal of the American Ceramic Society*, *76*(9), 2359–2362. doi:10.1111/j.1151-2916.1993.tb07777.x

Skoglund, P., Litstrom, O., & Flodin, A. (2013). Improvement of Powder Metallurgy Gears for Engines and Transmissions. *SAE International*, *26*, 1–8. doi:10.4271/2013-32-9102

Slotwinski, J. A., Garboczi, E. J., Stutzman, P. E., Ferraris, C. F., Watson, S. S., & Peltz, M. A. (2014). Characterization of Metal Powders Used for Additive Manufacturing. Journal of Research of the National Institute of Standards and Technology, 119(), 460. doi:10.6028/jres.119.018

Smith, B. C. (2011). *Fundamentals of Fourier transform infrared spectroscopy*. CRC Press., doi:10.1201/b10777

Sobolev, V., Guilemany, J., & Nutting, J. (2004). *High velocity oxy-fuel spraying theory, structure-property Relationships and Application*. Maney Publishing.

So, K. P., Lee, I. H., Duong, D. L., Kim, T. H., Lim, S. C., An, K. H., & Lee, Y. H. (2011). Improving the wettability of aluminum on carbon nanotubes. *Acta Materialia*, *59*(9), 3313–3320. doi:10.1016/j.actamat.2011.01.061

Šolic, S., Schauperl, Z., & Tropša, V. (2019). Wear behaviour of TiAlN PVD coating on deep cryogenically treated HSS substrate. *Kovove Mater*, *57*, 429–434.

Somiya, S. (2013). *Handbook of Advanced Ceramics: Materials* (2nd ed.). Applications, Processing, and Properties., doi:10.1016/C2010-0-66261-4

Souza, A. D., & Neville, A. (2003). Linking Electrochemical Corrosion Behaviour and Corrosion Mechanisms of Thermal Spray Cermet Coatings (WC-CrNi and WC/CrC-CoCr). *Materials Science and Engineering A*, *352*(1-2), 202–211. doi:10.1016/S0921-5093(02)00888-2

Sova, Grigoriev, Okunkova, & Smurov. (2013). Potential of cold gas dynamic spray as additive manufacturing technology. *International Journal of Advance Manufacturing Technology, 69*, 2269-2278.

Spinel Supra Thermal Spray Flexicord. (n.d.). *Ceramic Series | Saint-Gobain Coating Solutions*. Available: https://www.coatingsolutions.saint-gobain.com/materials/flexicord/ceramic-supra-series/spinel-supra

Spray, T. (n.d.). Thermal Spray Materials Guide, Metco, Sulzer. *Oerlikon metco*. Retrieved from: https://www.oerlikon.com/ecomaXL/files

Sreedhar, G., Alam, M. D. M., & Raja, V. S. (2009). Hot corrosion behavior of plasma sprayed YSZ/Al2O3 dispersed NiCrAlY coatings on Inconel-718 superalloy. *Surface and Coatings Technology*, *204*(3), 291–299. doi:10.1016/j.surfcoat.2009.07.026

Steinfeldt, M., Gleich, A. V., Petschow, U., Haum, R., Chudoba, T., & Haubold, S. (2004) Nachhaltigkeitseffekte durch Herstellung und Anwendung nano-technologischer Produkte. Schriftenreihe des IÖW 17704.

Steyer, P., Millet, J.-P., Anderbouhr, S., & Jacquot, P. (2001). Effect of Ni Interlayer on Corrosion Behaviour of Crn Pvd Coating Deposited onto Plastic Mould Steels. *Surface Engineering*, *17*(4), 1743–2944. doi:10.1179/026708401101517962

Stoltenhoff, T., Kreye, H., & Ritcher, H. (2001). An Analysis of the Cold Spray Process and its Coatings. *Journal of Thermal Spray Technology*, *11*(4), 542–550.

Stoltenhoff, T., & Ritcher, H. J. (2002). An analysis of the cold spray process and its coatings. *Journal of Thermal Spray Technology*, *11*(4), 542–550. doi:10.1361/105996302770348682

Strangman, T. E. (1985). Thermal barrier coatings for turbine airfoils. *Thin Solid Films*, *127*(1-2), 93–106. doi:10.1016/0040-6090(85)90215-9

Stringer, J. (1987). High temperature corrosion of super alloys. *Materials Science and Technology*, *3*(7), 482–493. doi:10.1080/02670836.1987.11782259

Stuart, B. H. (2004). *Infrared spectroscopy: Fundamentals and applications*. John Wiley & Sons., doi:10.1002/0470011149

Su, S. H., Conroy, S., Lin, T. L., Sheu, M. S., & Loh, I. H. (2004). Surface coatings. Encyclopedia of Biomaterials and Biomedical Engineering, 2, 1412-1431.

Sudagar, J., Tamilarasan, R., Sanjith, U., Rajendran, R., & Kumar, R. (2017). Electroless Deposition of Nanolayered Metallic Coatings. In Nanoscaled Films and Layers. doi:10.5772/intechopen.68220

Sudagar, J., Lian, J., & Sha, W. (2013). Electroless nickel, alloy, composite and nano coatings–A critical review. *Journal of Alloys and Compounds*, *571*, 183–204. doi:10.1016/j.jallcom.2013.03.107

Sułowski, M., & Matusiewicz, P. (2013). Comparison of mechanical properties of Mn-Cr-Mo steels sintered in different conditions. *Diffusion and Defect Data, Solid State Data. Part B, Solid State Phenomena*, *197*, 33–40. doi:10.4028/www.scientific.net/SSP.197.33

Sun, C., Zhou, X., & Xie, C. (2019). Effect of processing conditions on Al-based amorphous/nanocrystalline coating by cold-spraying. *Surface and Coatings Technology*, *362*, 97–104. doi:10.1016/j.surfcoat.2019.01.096

Sundaram, R. M., Sekiguchi, A., Sekiya, M., Yamada, T., & Hata, K. (2018). Copper/carbon nanotube composites: Research trends and outlook. *Royal Society Open Science*, *5*(11), 180814. doi:10.1098/rsos.180814

Sundararajan, G., Sen, D., & Sivakumar, G. (2005). The tribological behaviour of detonation sprayed coatings: The importance of coating process parameters. *Wear*, *258*(1-4), 377–391. doi:10.1016/j.wear.2004.03.022

Sun, G., Xu, X., Bickett, J. R., & Williams, J. F. (2001). Durable and Regenerable Antibacterial Finishing of Fabrics with a New Hydantoin Derivative. *Industrial & Engineering Chemistry Research*, *40*(4), 1016–1021. doi:10.1021/ie000657t

Supra Thermal Spray Flexicords, A. Z. Ceramic Series | Saint-Gobain Coating Solutions. (n.d.). Retrieved from: https://www.coatingsolutions.saint-gobain.com/materials/flexicord/ceramic-supra-series/az-supra

Surya, M. S., & Prasanthi, G. (2018). Tribological Behaviour of Aluminum Silicon Carbide Functionally Graded Material. Tribology in Industry, 40(2), 247–253. doi:10.24874/ti.2018.40.02.08

Sutcliffe, C. J., Papworth, A., Gallagher, C., Fox, P., Morgan, R., & O'Neill, W. (2003). Cold gas dynamic manufacturing a new approach to near-net shape metal component fabrication. *Mater.Res.Soc. Proc.*, *758*, 73–84.

Święch, D., Paluszkiewicz, C., Piergies, N., Pięta, E., Borkowska, U. L., & Kwiatek, W. (2018). Identification of Corrosion Products on Fe and Cu Metals using Spectroscopic Methods. *Acta Physica Polonica A*, *133*(2), 286–288. doi:10.12693/APhysPolA.133.286

Tailor, S., Modi, A., & Modi, S. C. (2019). Thermally Sprayed Thin Copper Coatings by W-HVOF. *Journal of Thermal Spray Technology*, *28*(1-2), 273–282. doi:10.100711666-018-0770-7

Tanasković, D., Dordevic, B., Gajin, M., Arandelović, M., & Gostović, N. (2018, January). Damages of burner pipes due to the working conditions and its repair welding. *Procedia Structural Integrity*, *13*, 404–409. doi:10.1016/j.prostr.2018.12.067

Tanasković, D., Đorđević, B., Sedmak, S., & Mihajlo, A. (2018). The effect of exploitation conditions on the damage of a roller reducer toothed shaft and its repair. *Machine Design*, *10*(4), 157–162. doi:0.24867/MD.10.2018.4.157-162

Tanasković, D., Đorđević, B., Tatić, U., Sedmak, A., & Opačić, M. (2017). Repair Welding of Gear Shafts of Service Rollers at the Železara Smederevo. *8th International Scientific Conference IRMES: Machine Elements and Systems in Energy Sector Development of Power Production Systems,* 365–370.

Tao, Y., Xiong, T., Sun, C., Kong, L., Cui, X., Li, T., & Song, G. L. (2010). Microstructure and Corrosion Performance of a Cold Sprayed Aluminium Coating on AZ91D Magnesium Alloy. *Corrosion Science*, *52*(10), 3191–3197. doi:10.1016/j.corsci.2010.05.023

Tarasi, F., Medraj, M., Dolatabadi, A., Oberste-Berghaus, J., & Moreau, C. (2011). High-Temperature Performance of Alumina-Zirconia Composite Coatings containing Amorphous Phases. *Advanced Functional Materials*, *21*(21), 4143–4151. doi:10.1002/adfm.201100471

Tavares, J., Swanson, E. J., & Coulombe, S. (2008). Plasma Synthesis of Coated Metal Nanoparticles with Surface Properties Tailored for Dispersion. *Plasma Processes and Polymers*, *5*(8), 759–769. doi:10.1002/ppap.200800074

Temitope & Jen. (2019). A comparative review on cold gas dynamic spraying processes and technologies. *Manufacturing Review*.

Tengzelius, J., & Grinder, O. (2008). Powder Metallurgy in Denmark, Finland, and Sweden. *International Journal of Powder Metallurgy*, *50*(2), 57–66.

Textor, T., Schollmeyer, H., Bahners, T., & Schollmeyer, E. (2004). Modification of low energy polymer surfaces by immobilization of fluorinated carboxylates with zirconium-based coupling agents. *Journal of Applied Polymer Science*, *94*(2), 789–795. doi:10.1002/app.20685

Thomas, F. A. P., Stoks, J. A. W., & Buegman, A. (2001). *US Patent 6291054*. Washington, DC: US Patent Office.

Thorpe, M. L., & Richter, H. J. (1992). A pragmatic analysis and comparison of HVOF processes. *Journal of Thermal Spray Technology*, *1*(2), 161–170. doi:10.1007/BF02659017

Thorvaldsson, A., Edvinsson, P., Glantz, A., Rodriguez, K., Walkenstrom, P., & Gatenholm, P. (2012). Superhydrophobic behaviour of plasma modified electro spun cellulose nanofiber-coated microfibers. *Cellulose (London, England)*, *19*(5), 1743–1748. doi:10.100710570-012-9751-z

Tian, Y., Zhang, S. T., Ma, J. H., & Xie, J. G. (2006). Development and application of abradable sealing coating. *Nonferrous Metals*, *1*, 96–99.

Tian, B., Liu, P., Song, K., Li, Y., Liu, Y., Ren, F., & Su, J. (2006). Microstructure and properties at elevated temperature of a nano-Al2O3 particles dispersion-strengthened copper base composite. *Materials Science and Engineering A*, *435–436*, 705–710. doi:10.1016/j.msea.2006.07.129

Tien, S. K., & Duh, J. G. (2004). Thermal reliability of electroless Ni–P–W coating during the aging treatment. *Thin Solid Films*, *469*, 268–273. doi:10.1016/j.tsf.2004.08.179

Tien, S. K., Duh, J. G., & Chen, Y. I. (2004). Structure, thermal stability and mechanical properties of electroless Ni–P–W alloy coatings during cycle test. *Surface and Coatings Technology*, *177*, 532–536. doi:10.1016/j.surfcoat.2003.08.032

Tien, S. K., Duh, J. G., & Chen, Y. I. (2004). The influence of thermal treatment on the microstructure and hardness in electroless Ni–P–W deposit. *Thin Solid Films*, *469*, 333–338. doi:10.1016/j.tsf.2004.08.146

Tobar, M. J., Alvarez, C., Amado, J. M., Rodriguez, G., & Yanez, A. (2006). Morphology and characterization of laser clad composite NiCrBSi–WC coatings on stainless steel. *Surface and Coatings Technology*, *200*(22-23), 6313–6317. doi:10.1016/j.surfcoat.2005.11.093

Toben, M. P., Martin, J. L., Ohta, Y., Takizawa, Y., & Enomoto, H. (2002). Patent No. US6383269B1.

Tocalo Company. (n.d.). Retrieved from: https://www.tocalo.co.jp/english/technical/arc.html

Tong, J., Zhang, J., Wang, Y., Min, F., Wang, X., Zhang, H., & Ma, J. (2019). Preparation of Co-plated WC powders by a non-precious-Co-activation triggered electroless plating strategy. *Advanced Powder Technology*, *30*(10), 2311–2319. doi:10.1016/j.apt.2019.07.012

Toozandehjani, M. (2017). *Synthesis And Characterization of Carbon Nanotubes and Nano Alumina-Reinforced Aluminum Matrix Nanocomposites*. Academic Press.

Trademark Odekerken, J. M. (1972). *U.S. Patent No. 3,644,183*. Washington, DC: U.S. Patent and Office.

Travessa, D. N., Silva, M. J., & Cardoso, K. R. (2017). Niobium Carbide-Reinforced Al Matrix Composites Produced by High-Energy Ball Milling. *Metallurgical and Materials Transactions. B, Process Metallurgy and Materials Processing Science*, *48*(3), 1754–1762. doi:10.1007/s11663-017-0959-z

Tripathy, A., Gupta, R., Sarangi, S. K., & Chaubey, A. K. (2020). Design and Fabrication of Aluminium/Alumina Ultrafine Composite and Functionally Graded Material Using Powder Metallurgy Route. In M. Shunmugam (Ed.), Advances in Unconventional Machining and Composites. doi:10.1007/978-981-32-9471-4_62

Tsai, T. K., Wu, S. S., Liu, W. L., Hsieh, S. H., & Chen, W. J. (2007). ElectrolessCoWP as a diffusion barrier between electroless copper and silicon. *Journal of Electronic Materials*, *36*(11), 1408–1414. doi:10.100711664-007-0223-4

Tsai, W. T., Chen, H. P., & Hsien, W. Y. (2002). A review of uses, environmental hazards and recovery/recycle technologies of perfluorocarbons (PFCs) emissions from the semiconductor manufacturing processes. *Journal of Loss Prevention in the Process Industries*, *15*(2), 65–75. doi:10.1016/S0950-4230(01)00067-5

Tsai, Y. Y., Wu, F. B., Chen, Y. I., Peng, P. J., Duh, J. G., & Tsai, S. Y. (2001). Thermal stability and mechanical properties of Ni–W–P electroless deposits. *Surface and Coatings Technology*, *146*, 502–507. doi:10.1016/S0257-8972(01)01462-1

Tsui, Y., Doyle, C., & Clyne, T. W. (1998). Plasma sprayed hydroxyapatite coatings on titanium substrates Part 1: Mechanical properties and residual stress levels. *Biomaterials*, *199*(22), 2015–2029. doi:10.1016/S0142-9612(98)00103-3 PMID:9870753

Tu, T. P., Liu, M. S., & Mao, Z. Y. (1997). Erosion resistance of Ni–WC self-fluxing alloy coating at high temperature. *Wear*, *209*(1-2), 43–48. doi:10.1016/S0043-1648(96)07457-1

Twardowski, A., Makowski, P., Malachowski, A., Hrynyk, R., Pietrowski, P., & Tyczkowski, J. (2012). Plasma treatment of thermoactive membrane textiles for superhydrophobicity. *Journal of Materials Science*, *18*(2), 163–166.

TWI-global. (n.d.). *What are the most common applications of flame spraying?* Retrieved from: https://www.twi-global.com/technical-knowledge/faqs/faq-what-are-the-most-common-applications-of-flame-spraying. (Accessed March 9th 2020).

Übeyli, M., Balci, E., Sarikan, B., Öztaş, M. K., Camuşcu, N., Yildirim, R. O., & Keleş, Ö. (2014). The ballistic performance of SiC-AA7075 functionally graded composite produced by powder metallurgy. *Materials & Design*, *56*, 31–36. doi:10.1016/j.matdes.2013.10.092

Uddin, A. J. (2010). Coatings for technical textile yarns. In *Technical Textile Yarns* (pp. 140–184). Woodhead Publishing. doi:10.1533/9781845699475.1.140

Ugla, A. A., Khaudair, H. J., & Almusawi, A. R. (2019). Metal Inert Gas Welding-Based-Shaped Metal Deposition in Additive Layered Manufacturing: A Review. *International Journal of Mechanical and Materials Engineering*, *13*(3), 244–257.

United States Department of Labor. (n.d.). Retrieved from: https://www.osha.gov/SLTC/noisehearingconservation/

Upadhyay, R. K., & Kumaraswamidhas, L. A. (2014). Surface modification by multilayered W/W2N coating. *Surface Engineering*, *30*(7), 475–482. doi:10.1179/1743294414Y.0000000260

Uusitalo, M. A., Vuoristo, P. M. J., & Mantyla, T. A. (2002). High Temperature Corrosion of Coatings and Boiler Steels in Reducing Chlorine-Containing Atmosphere. *Surface and Coatings Technology*, *161*(2-3), 275–285. doi:10.1016/S0257-8972(02)00472-3

Uysal, M., Karslioğlu, R., Alp, A., & Akbulut, H. (2013). The preparation of core-shell Al2O3/Ni composite powders by electroless plating. *Ceramics International*, *39*(5), 5485–5493. doi:10.1016/j.ceramint.2012.12.060

Vaben, R., Cao, X., Tietz, F., Basu, D., & Stover, D. (2000). Zirconates as new materials for thermal barrier coatings. *Journal of the American Ceramic Society*, *83*, 2023–2028.

Valova, E., Dille, J., Armyanov, S., Georgieva, J., Tatchev, D., Marinov, M., Delplancke, J.-L., Steenhaut, O., & Hubin, A. (2005). Interface between electroless amorphous Ni–Cu–P coatings and Al substrate. *Surface and Coatings Technology*, *190*(2-3), 336–344. doi:10.1016/j.surfcoat.2004.03.050

Van Der Biest, O., Joos, E., Vleugels, J., & Baufeld, B. (2006). Electrophoretic deposition of zirconia layers for thermal barrier coatings. *Journal of Materials Science*, *41*(24), 8086–8092. doi:10.1007/s10853-006-0635-1

van Lente, H., & van Til, J. I. (2008). Articulation of sustainability in the emerging field of nanocoatings. *Journal of Cleaner Production*, *16*(8-9), 967–976. doi:10.1016/j.jclepro.2007.04.020

Van Steenkiste, T., Smith, J., & Teets, R. (2002). Aluminium Coatings via Kinetic Spray with Relatively Large particle powders. *Surface Coating Technology*, *154*, 237–252.

Vardelle, A., Moreau, C., Akedo, J., Ashrafizadeh, H., Berndt, C. C., Berghaus, J. O., Boulos, M., Brogan, J., Bourtsalas, A. C., Dolatabadi, A., Dorfman, M., Eden, T. J., Fauchais, P., Fisher, G., Gaertnet, F., Gindrat, M., Henne, R., Hyland, M., Irissou, E., ... Vuoristo, P. (2016). The 2016 Thermal Spray Roadmap. *Journal of Thermal Spray Technology*, *25*(8), 1376–1440. doi:10.100711666-016-0473-x

Vardelle, A., Moreau, C., Themelis, N. J., & Chazelas, C. (2015). A perspective on plasma spray technology. *Plasma Chemistry and Plasma Processing*, *35*(3), 491–509. doi:10.100711090-014-9600-y

Vargas, F., Ageorges, H., Fournier, P., Fauchais, P., & López, M. E. (2010). Mechanical and tribological performance of Al2O3-TiO2 coatings elaborated by flame and plasma spraying. *Surface and Coatings Technology*, *205*(4), 1132–1136. doi:10.1016/j.surfcoat.2010.07.061

Vasiljevic, J., Gorjanc, M., Tomšic, B., Orel, B., Jerman, I., Mozetic, M., Vesel, A., & Simoncic, B. (2013). The surface modification of cellulose fibres to create super-hydrophobic, oleophobic and self-cleaning properties. *Cellulose (London, England)*, *20*(1), 277–289. doi:10.100710570-012-9812-3

Vaskova, H. (2011). A powerful tool for material identification: Raman spectroscopy. *Int. J. Math. Model. Methods Appl. Sci.*, *5*, 1205–1212.

Vaßen, R., Bakan, E., Mack, D., Schwartz-Lückge, S., Sebold, D., Sohn, Y. J., Zhou, D., & Guillon, O. (2020). Performance of YSZ and $Gd_2Zr_2O_7$/YSZ Double Layer Thermal Barrier Coatings in Burner Rig Tests. *Journal of the European Ceramic Society*, *40*(2), 480–490. doi:10.1016/j.jeurceramsoc.2019.10.021

Vassen, R., Cao, X., Tietz, F., Basu, D., & Stöver, D. (2000). Zirconates as New Materials for Thermal Barrier Coatings. *Journal of the American Ceramic Society*, *83*(8), 2023–2028. doi:10.1111/j.1151-2916.2000.tb01506.x

Vaßen, R., Jarligo, M. O., Steinke, T., Mack, D. E., & Stöver, D. (2010). Overview on advanced thermal barrier coatings. *Surface and Coatings Technology*, *205*(4), 938–942. doi:10.1016/j.surfcoat.2010.08.151

Vaßen, R., Kagawa, Y., Subramanian, R., Zombo, P., & Zhu, D. (2012). Testing and Evaluation of Thermal-Barrier Coatings. *MRS Bulletin*, *37*(10), 911–916. doi:10.1557/mrs.2012.235

Vasudev, H., Thakur, L., Bansal, A., Singh, H., & Zafar, S. (2019). High temperature oxidation and erosion behaviour of HVOF sprayed Inconel-718 coating. *Surface and Coatings Technology*, *362*, 366–380. doi:10.1016/j.surfcoat.2019.02.012

Verstak, A., & Baranovski, V. (2003). *Activated Combustion HVAF Coatings for Protection Against Wear and High Temperature Corrosion.* Paper presented at the International Thermal Spray Conference and Exposition, Orlando, FL.

Verstak, A., & Baranovski, V. (2006). AC-HVAF sprayed tungsten carbide: properties and applications. *Thermal Spray 2006: Building on 100 Years of Success*, 15-18.

Vihodceva, S., & Kukle, S. (2013). Improvement of UV Protection Properties of the Textile from Natural Fibres by the Sol-gel Method. *IOP Conf. Series: Materials Science and Engineering*, *49*, 1-4. 10.1088/1757-899X/49/1/012022

Vilardell, A. M., Cinca, N., Dosta, S., Cano, I. G., & Guilemany, J. M. (2019). Feasibility of using low pressure cold gas spray for the spraying of thick ceramic hydroxyapatite coatings. *International Journal of Applied Ceramic Technology*, *16*(1), 221–229. doi:10.1111/ijac.13088

Villafuerte, J. (2010). Current and future applications of cold spraying technology. *Metal Finishing*, *108*(1), 37–39. doi:10.1016/S0026-0576(10)80005-4

Villafuerte, J. (Ed.). (2015). *Modern cold spray: materials, process, and applications*. Springer. doi:10.1007/978-3-319-16772-5

Viscusi, A., Astarita, A., Gatta, R. D., & Rubino, F. (2019). A perspective review on the bonding mechanisms in cold gas dynamic spray. *Surface Engineering*, *35*(9), 743–771. doi:10.1080/02670844.2018.1551768

Wang, G., Jong, Y.W., Balakumar, S., Seah, C.H., & Hara, T. (2004). Nanotechnology and Microelectronics: Materials, Processing, Measurement, and Phenomena. *Journal of Vacuum Science & Technology B*.

Wang, D., Dong, S., Zeng, J., Liang, P., Liao, H., Wang, Y., & Cao, X. (2020). Influence of doping Mg2+ or Ti4+ captions on the microstructures, thermal radiation and thermal cycling behavior of plasma-sprayed Gd2Zr2O7 coatings. *Ceramics International*, *46*(9), 13054–13065. doi:10.1016/j.ceramint.2020.02.076

Wang, D., Zhang, Z. B., Li, Y. M., & Xu, C. H. (2014). Highly Transparent and Durable Superhydrophobic Hybrid Nanoporous Coatings Fabricated from Polysiloxane. *ACS Applied Materials & Interfaces*, *6*(13), 10014–10021. doi:10.1021/am405884x PMID:24955659

Wang, F., Arai, S., & Endo, M. (2005). The preparation of multi-walled carbon nanotubes with a Ni-P coating by an electroless deposition process. *Carbon*, *43*(8), 1716–1721. doi:10.1016/j.carbon.2005.02.015

Wang, H., Jia, J., Song, H., Hu, X., Sun, H., & Yang, D. (2011). The preparation of Cu-coated Al2O3 composite powders by electroless plating. *Ceramics International*, *37*(7), 2181–2184. doi:10.1016/j.ceramint.2011.03.013

Wang, H., Xia, W., & Jin, Y. (1996). A study on abrasive resistance of Ni-based coatings with a WC hard phase. *Wear*, *195*(1-2), 47–52. doi:10.1016/0043-1648(95)06762-0

Wang, J., Chong, X., Zhou, R., & Feng, J. (2017). Microstructure and Thermal Properties of $RETaO_4$ (RE = Nd, Eu, Gd, Dy, Er, Yb, Lu) as Promising Thermal Barrier Coating Materials. *Scripta Materialia*, *126*, 24–28. doi:10.1016/j.scriptamat.2016.08.019

Wang, J., Li, L., & Tao, W. (2016). Crack initiation and propagation behavior of WC particles reinforced Fe-based metal matrix composite produced by laser melting deposition. *Optics & Laser Technology*, *82*, 170–182. doi:10.1016/j.optlastec.2016.03.008

Wang, L., Wang, Y., Sun, X. G., He, J. Q., Pan, Z. Y., Zhou, Y., & Wu, P. L. (2011). Influence of pores on the thermal insulation behavior of thermal barrier coatings prepared by atmospheric plasma spray. *Materials & Design*, *32*(1), 36–47. doi:10.1016/j.matdes.2010.06.040

Wang, M. C. (1994). Properties of high density powder forged iron based alloy. *Powder Metallurgy*, *37*(3), 201–205. doi:10.1179/pom.1994.37.3.201

Wang, W. F. (2005). Effect of alloying elements and processing factors on the microstructure and hardness of sintered and induction-hardened Fe–C–Cu alloys. *Materials Science and Engineering A*, *402*(1-2), 92–97. doi:10.1016/j.msea.2005.04.016

Wang, X., Xiang, H., Sun, X., Liu, J., Hou, F., & Zhou, Y. (2014). Thermal Properties of a Prospective Thermal Barrier Material: $Yb_3Al_5O_{12}$. *Journal of Materials Research*, *29*(22), 2673–2681. doi:10.1557/jmr.2014.319

Wang, Y., Gao, Y., Li, Y., Li, M., Sun, L., Zhai, W., & Li, K. (2020). Research on synergistic lubrication effect of silver modified Cu–Ni-graphite composite. *Wear*, *444–445*, 203140. doi:10.1016/j.wear.2019.203140

Wang, Y., Gao, Y., Li, Y., Zhang, C., Sun, L., & Zhai, W. (2017). Research on nickel modified graphite/Cu composites interface. *Surface and Coatings Technology*, *328*, 70–79. doi:10.1016/j.surfcoat.2017.08.036

Wang, Y., Normand, B., Mary, N., Yu, M., & Liao, H. (2014). Microstructure and Corrosion Behavior of Cold Sprayed SiC_p/Al 5056 Composite Coatings. *Surface and Coatings Technology*, *251*, 264–275. doi:10.1016/j.surfcoat.2014.04.036

Warke, V. S., Sisson, R. D. Jr, & Makhlouf, M. M. (2009). FEA model for predicting the response of powder metallurgy steel components to heat treatment. *Materials Science and Engineering A*, *518*(1-2), 7–15. doi:10.1016/j.msea.2009.04.003

Wei, Q. L., Guo, H. B., Gong, S. K., & Xu, H. B. (2008). Novel microstructure of EBPVD double ceramic layered thermal barrier coatings. *Thin Solid Films*, *516*(16), 5736–5739. doi:10.1016/j.tsf.2007.07.032

Wei, X., Roper, K., Roper, D. K., & Martin, R. E. (2014). Tin Sensitization for Electroless Plating Review Fundamental chromatography View project Optoplasmonic Heating with Gold Nanoparticle Arrays View project Tin Sensitization for Electroless Plating Review. *Journal of the Electrochemical Society*, *161*(5), 235–242. doi:10.1149/2.047405jes

Wen, G., Guo, Z. X., & Davies, C. K. L. (2000). Microstructural characterisation of electroless-nickel coatings on zirconia powder. *Scripta Materialia*, *43*(4), 307–311. doi:10.1016/S1359-6462(00)00409-7

Wessel, J. K. (2004). Handbook of Advanced Materials. In J. K. Wessel (Ed.), Handbook of Advanced Materials. doi:10.1002/0471465186

Westbroek, P., Schwarz, A., Hakuzimana, J., Gasana, E., Van Langenhove, L., & Tseles, D. (2007). Corrosion resistant textiles electrodes obtained through electroless coating of fibers with gold. In eRA-2 Conference, Athens, GA.

Whang, T. J., Hsieh, M. T., & Kao, Y. C. (2010). Studies of single-step electrode position of CuInSe2 thin films with sodium citrate as a complexing agent. *Applied Surface Science*, *257*(5), 1457–1462. doi:10.1016/j.apsusc.2010.08.072

Windsor, B. (2005, March). *Hygienic Coatings & Surfaces*. PRA Coatings Technology Centre.

Witz, G., Shklover, V., Steurer, W., Bachegowda, S., & Bossmann, H.-P. (2007). Phase Evolution in Yttria-Stabilized Zirconia Thermal Barrier Coatings Studied by Rietveld Refinement of X-Ray Powder Diffraction Patterns. *Journal of the American Ceramic Society*, *90*(9), 2935–2940. doi:10.1111/j.1551-2916.2007.01785.x

Wojcik, H., Lehninger, D., Neumann, V., & Bartha, J. W. (2011, September). *Characterization of barrier and seed layer integrity for copper interconnects. In 2011 Semiconductor Conference Dresden*. IEEE.

Wu, C., Niu, M., Bao, S., Sun, Y., & Zhang, X. (2020). Friction and Wear Properties of Ni3Si Alloy with Ti Addition at High Temperatures. *Materials (Basel)*, *13*(4), 982. doi:10.3390/ma13040982 PMID:32098307

Wu, F. B., Tien, S. K., Chen, W. Y., & Duh, J. G. (2004). Microstructure evaluation and strengthening mechanism of Ni–P–W alloy coatings. *Surface and Coatings Technology*, *177*, 312–316. doi:10.1016/j.surfcoat.2003.09.010

Wu, F. B., Tien, S. K., Duh, J. G., & Wang, J. H. (2003). Surface characteristics of electroless and sputtered Ni–P–W alloy coatings. *Surface and Coatings Technology*, *166*(1), 60–66. doi:10.1016/S0257-8972(02)00725-9

Wu, L., Zhang, J., Lia, B., & Wang, A. (2013). Mimic nature, beyond nature: Facile synthesis of durable superhydrophobic textiles using organosilanes. *Journal of Materials Chemistry. B, Materials for Biology and Medicine*, *1*(37), 4756–4763. doi:10.1039/c3tb20934b PMID:32261159

Wu, P., Hu, M. Y., Chong, X. Y., & Feng, J. (2018). The Glass-like Thermal Conductivity in ZrO_2-Dy_3TaO_7 Ceramic for promising Thermal Barrier Coating Application. *Applied Physics Letters*, *112*(13), 131903. doi:10.1063/1.5022610

Wurtz, A. (1844). Sur LHydrure De. *Ann. Chim. Phys*, *11*, 250–251.

Xie, X., Guo, H., & Gong, S. (2010). Mechanical properties of $LaTi_2Al_9O_{19}$ and thermal cycling behaviours of plasma-sprayed $LaTi_2Al_9O_{19}$/YSZ thermal barrier coatings. *Journal of Thermal Spray Technology*, *19*(6), 1179–1185. doi:10.100711666-010-9529-5

Xie, X., Guo, H., Gong, S., & Xu, H. (2011). Lanthanum-titanium-aluminum oxide: A novel thermal barrier coating material for applications at 1300°C. *Journal of the European Ceramic Society*, *31*(9), 1677–1683. doi:10.1016/j.jeurceramsoc.2011.03.036

Xie, Y., Yin, S., Chen, C., Planche, M.-P., Liao, H., & Lupoi, R. (2016). New insights into the coating/substrate interfacial bonding mechanism in cold spray. *Scripta Materialia*, *125*(December), 1–4. doi:10.1016/j.scriptamat.2016.07.024

Xi, J. Q., Schubert, M. F., Kim, J. K., Schubert, E. F., Chen, M., Lin, S. Y., Liu, W., & Smart, J. A. (2007). Optical thin-film materials with low refractive index for broadband elimination of Fresnel reflection. *Nature Photonics*, *1*(3), 176–179. doi:10.1038/nphoton.2007.26

Xing, Y., Yang, X., & Dai, J. (2007). Antimicrobial finishing of cotton textile based on water glass by sol-gel method. *Journal of Sol-Gel Science and Technology*, *43*(2), 187–192. doi:10.100710971-007-1575-1

Xu, B., & Cai, Z. (2008). Fabrication of a superhydrophobic ZnO nanorod array film on cotton fabrics via a wet chemical route and hydrophobic modification. *Applied Surface Science*, *254*(18), 5899–5904. doi:10.1016/j.apsusc.2008.03.160

Xu, C., Du, L., Yang, B., & Zhang, W. (2011). Study on Salt Spray Corrosion of Ni-Graphite Abradable Coating with 80Ni20Al and 96NiCr-4Al as Bonding Layers. *Surface and Coatings Technology*, *205*(17-18), 4154–4161. doi:10.1016/j.surfcoat.2011.03.007

Xue, C.-H., Chen, J., Yin, W., Jia, S.-T., & Ma, J.-Z. (2012). Superhydrophobic conductive textiles with antibacterial property by coating fibers with silver nanoparticles. *Applied Surface Science*, *258*(7), 2468–2472. doi:10.1016/j.apsusc.2011.10.074

Xue, C.-H., Jia, S.-T., Chen, H.-Z., & Wang, M. (2008). Superhydrophobic cotton fabrics prepared by sol-gel coating of TiO_2 and surface hydrophobization. *Science and Technology of Advanced Materials*, *9*(3), 1–5. doi:10.1088/1468-6996/9/3/035001 PMID:27877998

Xue, C.-H., Jia, S.-T., Zhang, J., & Tian, L.-Q. (2009). Superhydrophobic surfaces on cotton textiles by complex coating of silica nanoparticles and hydrophobization. *Thin Solid Films*, *517*(16), 4593–4598. doi:10.1016/j.tsf.2009.03.185

Xu, L., & He, J. (2012). Antifogging and Antireflection Coatings Fabricated by Integrating Solid and Mesoporous Silica Nanoparticles without Any Post-Treatments. *ACS Applied Materials & Interfaces*, *4*(6), 3293–3299. doi:10.1021/am300658e PMID:22663342

Xu, Q. F., Wang, J. N., & Sanderson, K. D. (2010). Organic–Inorganic Composite Nanocoatings with Super hydrophobicity, Good Transparency, and Thermal Stability. *ACS Nano*, *4*(4), 2201–2209. doi:10.1021/nn901581j PMID:20302323

Xu, R., Zhou, K., & Hu, M. (2009). Preparation of core-shell Cu-Ag bimetallic powder via electroless coating. Journal of Wuhan University of Technology-Mater. *Science Editor*, *24*(4), 637–639. doi:10.100711595-009-4637-0

Xu, T., & Xie, C. S. (2003). Tetrapod-like nano-particle ZnO/acrylic resin composite and its multi-function property. *Progress in Organic Coatings*, *46*(4), 297–301. doi:10.1016/S0300-9440(03)00016-X

Xu, X., Cui, Z. D., Zhu, S. L., Liang, Y. Q., & Yang, X. J. (2014). Preparation of nickel-coated graphite by electroless plating under mechanical or ultrasonic agitation. *Surface and Coatings Technology*, *240*, 425–431. doi:10.1016/j.surfcoat.2013.12.070

Xu, X., Luo, X., Zhuang, H., Li, W., & Zhang, B. (2003). Electroless silver coating on fine copper powder and its effects on oxidation resistance. *Materials Letters*, *57*(24–25), 3987–3991. doi:10.1016/S0167-577X(03)00252-0

Xu, Y., & Hutchings, I. M. (2006). Cold spray deposition of thermoplastic powder. *Surface and Coatings Technology*, *201*(6), 3044–3050. doi:10.1016/j.surfcoat.2006.06.016

Xu, Z., He, Z., Song, Y., Fu, X., Rommel, M., Luo, X., Hartmaier, A., Zhang, J., & Fang, F. (2018). Topic review: Application of Raman spectroscopy characterization in micro/nano-machining. *Micromachines*, *9*(7), 361. doi:10.3390/mi9070361 PMID:30424294

Xu, Z., Hodgson, M. A., & Cao, P. (2015). A comparative study of powder metallurgical (PM) and wrought Fe–Mn–Si alloys. *Materials Science and Engineering A*, *630*, 116–124. doi:10.1016/j.msea.2015.02.021

Yadav, R., Anubhav, V., Mathur, P., & Sheikh, J. (2019). Antibacterial, UV protective and antioxidant linen obtained by natural dyeing with henna. *Cellulose Chemistry and Technology*, *53*(3-4), 357–362. doi:10.35812/CelluloseChemTechnol.2019.53.36

Yang, L., Newton, J., & Fieselmann, B. (1989). Raman Spectroscopy of a. SiGe: H Alloys. *MRS Online Proceedings Library Archive, 149*.

Yang, H., Zhu, S., & Pan, N. (2004). Studying the mechanisms of titanium dioxide as ultraviolet-blocking additive for films and fabrics by an improved scheme. *Journal of Applied Polymer Science*, *92*(5), 3201–3210. doi:10.1002/app.20327

Yang, J., Han, Y., Shahid, M., Pan, W., Zhao, M., Wu, W., & Wan, C. (2018). A Promising Material for Thermal Barrier Coating: Pyrochlore-related Compound Sm_2FeTaO_7. *Scripta Materialia*, *149*, 49–52. doi:10.1016/j.scriptamat.2018.02.005

Yang, L., Ran, L., & Yi, M. (2011). Carbon fiber knitted fabric reinforced copper composite for sliding contact material. *Materials & Design*, *32*(4), 2365–2369. doi:10.1016/j.matdes.2010.11.063

Yang, R., Liu, S., Chen, J., Cui, H., Liu, M., Zhu, F., Yang, Y., Xie, M., Sun, X., & Li, X. (2019). Porous Y2O3 fiber-reinforced silver composite exhibiting enhanced mechanical and electrical properties. *Ceramics International*, *45*(2), 1881–1886. doi:10.1016/j.ceramint.2018.10.079

Yang, W., Thordarson, P., Gooding, J. J., Ringer, S. P., & Braet, F. (2007). Carbon nanotubes for biological and biomedical applications. *Nanotechnology*, *18*(41), 412001. doi:10.1088/0957-4484/18/41/412001

Yazhinia, K. B., & Prabu, H. G. (2015). Study on flame-retardant and UV-protection properties of cotton fabric functionalized with ppy–ZnO–CNT nanocomposite. *RSC Advances*, *5*(61), 49062–49069. doi:10.1039/C5RA07487H

Yeh, J. M., & Chin, C. P. (2003). Structure and properties of poly (*o*-methoxyaniline)-clay nanocomposite materials. *Journal of Applied Polymer Science*, *88*(4), 1072–1078. doi:10.1002/app.11829

Yekane, Deshmukh, & Suryavanshi. (n.d.). Corrosion Protection of Light Metal Using Cold Spray Technique. *Journal of Computer Science*, 72-78.

Yin, B., Zhou, H. D., Chen, J. M., & Yan, F. Y. (2011). Microstructure and high temperature sliding wear behavior of laser remelted NiCrBSi/35(WC–12Ni) composite coating. *Surface Engineering*, *27*(6), 458–463. doi:10.1179/026708410X12687356948517

Yin, H. M., Sun, L. Z., & Paulino, G. H. (2004). Micromechanics-based elastic model for functionally graded materials with particle interactions. *Acta Materialia*, *52*(12), 3535–3543. doi:10.1016/j.actamat.2004.04.007

Yin, S., Cavaliere, P., Aldwell, B., Jenkins, R., Liao, H., Li, W., & Lupoi, R. (2018). Cold spray additive manufacturing and repair: Fundamentals and applications. *Additive Manufacturing*, *21*, 628–650. doi:10.1016/j.addma.2018.04.017

Yin, S., Xie, Y., Suo, X., Liao, H., & Wang, X. (2015). Interfacial bonding features of Ni coating on Al substrate with different surface pretreatments in cold spraying. *Materials Letters*, *138*, 143–147. doi:10.1016/j.matlet.2014.10.016

Yoon, S., Kim, H., & Lee, C. (2007). Fabrication of automotive heat exchange using kinetic spraying process. Surface Coating Technology, 201, 9524-9932.

Young, M. A. (1965). Health hazards of electroplating. *Journal of Occupational and Environmental Medicine*, *7*(7), 348–352. doi:10.1097/00043764-196507000-00008 PMID:14321464

Yousaf, S., Alhnan, M. A., Abdallah, A., Abdallah, B., Khan, I., & Ahmed, W. (2015). *Nanocoatings in medicine: antiquity and modern times.* Elsevier. doi:10.1016/B978-0-323-28990-0.00016-6

Yuan, J., Song, W., Zhang, H., Zhou, X., Dong, S., Jiang, J., Deng, L., & Cao, X. (2019). $TaZr_{2.75}O_8$ Ceramics as a Potential Thermal Barrier Coating Material for High-Temperature Applications. *Materials Letters*, *247*, 82–85. doi:10.1016/j.matlet.2019.03.102

Yuan, Y., Gan, X., Lai, Y., Zhao, Q., & Zhou, K. (2019). Microstructure and properties of graphite/copper composites fabricated with Cu-Ni double-layer coated graphite powders. *Composite Interfaces*, •••, 1–15. doi:10.1080/09276440.2019.1644051

Yu, C. J., Eifert, H. H., Banhart, J., & Baumeister, J. (1998). Metal foaming by a powder metallurgy method: Production, properties and applications. *Materials Research Innovations*, *2*(3), 181–188. doi:10.1007100190050082

Yu, H. S., Luo, S. F., & Wang, Y. R. (2001). A comparative study on the crystallization behavior of electroless Ni- P and Ni- Cu- P deposits. *Surface and Coatings Technology*, *148*(2-3), 143–148. doi:10.1016/S0257-8972(01)01345-7

Yu, H. S., Sun, X., Luo, S. F., Wang, Y. R., & Wu, Z. Q. (2002). Multifractal spectra of atomic force microscope images of amorphous electroless Ni–Cu–P alloy. *Applied Surface Science*, *191*(1-4), 123–127. doi:10.1016/S0169-4332(02)00170-8

Yu, S., Guo, Z., & Liu, W. (2015). Biomimetic transparent and superhydrophobic coatings: From nature and beyond nature. *Chemical Communications*, *51*(10), 1775–1794. doi:10.1039/C4CC06868H PMID:25406877

Yusong, P., Qianqian, S., Chengling, P., & Jing, W. (2013). Prediction of mechanical properties of multilayer gradient hydroxyapatite reinforced poly(vinyl alcohol) gel biomaterial. *Journal of Biomedical Materials Research. Part B, Applied Biomaterials*, *101*(5), 729–735. doi:10.1002/jbm.b.32875

Yu, Y., Lindqvist, B., & Tengzelius, J. (2004). Höganäs Chair–an industry/university cooperation. *Powder Metallurgy*, *47*(4), 304–308. doi:10.1179/pom.2004.47.4.304

Yu, Z., Lv, L., Ma, Y., Di, H., & He, Y. (2016). Covalent modifications of graphene oxide by metronidazole for reinforced anti-corrosion properties of epoxy coatings. *RSC Advances*, *6*(22), 18217–18226. doi:10.1039/C5RA23595B

Zhang, C., Kong, L., Li, H., & Wu, Y. (2019). Electroless Cu plating on the microstructure and hardness of Al–Cu powder metallurgy alloy. Materials Research Express, 6(8), 086577. doi:10.1088/2053-1591/ab1e1a

Zhang, B. (2016). Green Electroless Plating. In B. Zhang (Ed.), *Amorphous and Nano Alloys Electroless Depositions* (pp. 383–409). Elsevier., doi:10.1016/B978-0-12-802685-4.00007-8.

Zhang, B., Wenyuan, W., Shaohua, Y., Shiwei, L., Yao, L., Xue, B., & Ganfeng, T. (2010). Process optimization of electroless copper plating and its influence on electrochemical properties of AB5-type hydrogen storage alloy. *Journal of Rare Earths*, *28*(6), 922–926.

Zhang, C., Chan, K. C., Wu, Y., & Liu, L. (2012). Pitting Initiation in Fe- Based Amorphous Coatings. *Acta Materialia*, *60*(10), 4152–4159. doi:10.1016/j.actamat.2012.04.005

Zhang, C., Chen, F., Huang, Z., Jia, M., Chen, G., Ye, Y., Lin, Y., Liu, W., Chen, B., Shen, Q., Zhang, L., & Lavernia, E. J. (2019). Additive manufacturing of functionally graded materials: A review. *Materials Science and Engineering A*, *764*, 138209. doi:10.1016/j.msea.2019.138209

Zhang, N., Khan, T., Guo, H., Shi, S., Zhong, W., & Zhang, W. (2019). Functionally Graded Materials: An Overview of Stability, Buckling, and Free Vibration Analysis. *Advances in Materials Science and Engineering*, •••, 1–18. doi:10.1155/2019/1354150

Zhang, N., Liu, S., & Xu, Y.-J. (2012). Recent progress on metal core@semiconductor shell nanocomposites as a promising type of photocatalyst. *Nanoscale*, *4*(7), 2227. doi:10.1039/c2nr00009a

Zhang, S. D., Wu, J., Qi, W. B., & Wang, J. Q. (2016). Effect of Porosity Defects on the Long-Term Corrosion Behaviour of Fe-Based Amorphous Alloy Coated Mild Steel. *Corrosion Science*, *110*, 57–70. doi:10.1016/j.corsci.2016.04.021

Zhang, X., Zhang, Y., Tian, B., An, J., Zhao, Z., Volinsky, A. A., Liu, Y., & Song, K. (2019). Arc erosion behavior of the Al2O3-Cu/(W, Cr) electrical contacts. *Composites. Part B, Engineering*, *160*, 110–118. doi:10.1016/j.compositesb.2018.10.040

Zhao, H., Feng, Y., Zhou, Z., Qian, G., Zhang, J., Huang, X., & Zhang, X. (2020). Effect of electrical current density, apparent contact pressure, and sliding velocity on the electrical sliding wear behavior of Cu–Ti3AlC2 composites. *Wear*, *444–445*, 203156. doi:10.1016/j.wear.2019.203156

Zhao, M., Pan, W., Wan, C., Qu, Z., Li, Z., & Yang, J. (2017). Defect Engineering in Development of Low Thermal Conductivity Materials: A Review. *Journal of the European Ceramic Society*, *37*(1), 1–13. doi:10.1016/j.jeurceramsoc.2016.07.036

Zhao, Q., Wang, C., Liu, Y., & Wang, S. (2007). Bacterial adhesion on the metal-polymer composite coatings. *International Journal of Adhesion and Adhesives*, *27*(2), 85–91. doi:10.1016/j.ijadhadh.2006.01.001

Zhao, Y., Min, X., Fang, M., Wu, X., Liu, Y., & Huang, Z. (2019). Dependence of Crystal Structure on Mechanical and Thermophysical Properties of Magnetoplumbite-type $LnMgAl_{11}O_{19}$ Ceramics with Substitution of Ln^{3+} Ions. *International Journal of Applied Ceramic Technology*, *16*(4), 1596–1605. doi:10.1111/ijac.13216

Zhao, Y., Tang, Y., Wang, X., & Lin, T. (2010). Superhydrophobic cotton fabric fabricated by electrostatic assembly of silica nanoparticles and its remarkable buoyancy. *Applied Surface Science*, *256*(22), 6736–6742. doi:10.1016/j.apsusc.2010.04.082

Zheng, X., Tan, J., Zhang, Q., Wang, M., & Meng, L. (2017). Effect of laser surface texturing depth on the adhesion of electroless plated nickel coating on alumina. *Surface and Coatings Technology*, *311*, 151–156. doi:10.1016/j.surfcoat.2017.01.002

Zhou, Liu, Li, Zhu, & Xiong. (2019). *Microstructure and properties of Fe-Cr-Ni alloy coating on T10steel by laser cladding*. Material Research Express.

Zhou, C.-E., Kan, C., Matinlinna, J. P., & Tsoi, J. K. (2017). Regenerable Antibacterial Cotton Fabric by Plasma Treatment with Dimethylhydantoin: Antibacterial Activity against S. aureus. *Coatings*, *7*(11), 1–14. doi:10.3390/coatings7010011

Zhou, J., Walleser, J. K., Meacham, B. E., & Branagan, D. J. (2010). Novel in Situ transformable coating for elevated-temperature applications. *J. Therm. Spray Technol.*, *19*(5), 950–957. doi:10.100711666-010-9498-8

Zhou, P., Cai, W., Yang, Y., Li, X., Zhang, T., & Wang, F. (2019). Effect of ultrasonic agitation during the activation process on the microstructure and corrosion resistance of electroless Ni-WP coatings on AZ91D magnesium alloy. *Surface and Coatings Technology*, *374*, 103–115. doi:10.1016/j.surfcoat.2019.05.080

Zhou, X., Xu, Z., Fan, X., Zhao, S., Cao, X., & He, L. (2014). $Y_4Al_2O_9$ Ceramics as a Novel Thermal Barrier Coating Material for High-Temperature Applications. *Materials Letters*, *134*, 146–148. doi:10.1016/j.matlet.2014.07.027

Zhou, Y., Gan, G., Ge, Z., Song, P., & Feng, J. (2020). Microstructure and Thermophysical Properties of CeO_2-doped $SmTaO_4$ Ceramics for Thermal Barrier Coatings. *Journal of Materials Research*, *35*(3), 242–251. doi:10.1557/jmr.2020.15

Zhou, Z., Wang, L., Wang, F., & Liu, Y. (2009). Formation and Corrosion Behavior of Fe-Based Amorphous Metallic Coatings Prepared by Detonation Gun Spraying. *Transactions of Nonferrous Metals Society of China*, *19*(3), 634–638. doi:10.1016/S1003-6326(10)60123-9

Zhu, J., Hsu, C. M., Yu, Z. F., Fan, S. H., & Cui, Y. (2010). Nanodome Solar Cells with Efficient Light Management and Self-Cleaning. *Nano Letters*, *10*(6), 1979–1984. doi:10.1021/nl9034237 PMID:19891462

Zimmermann, J., Reifler, F. A., Fortunato, G., Gerhardt, L.-C., & Seeger, S. (2008). A Simple, One-Step Approach to Durable and Robust Superhydrophobic Textiles. *Advanced Functional Materials*, *18*(22), 3662–3669. doi:10.1002/adfm.200800755

Zou, G., Cao, M., Lin, H., Jin, H., Kang, Y., & Chen, Y. (2006). Nickel layer deposition on SiC nanoparticles by simple electroless plating and its dielectric behaviors. *Powder Technology*, *168*(2), 84–88. doi:10.1016/j.powtec.2006.07.002

About the Contributors

Supriyo Roy graduated in Mechanical Engineering from The Institution of Engineers, India in 2010. He obtained his Master of Mechanical Engineering and Ph.D. degree from Jadavpur University, Kolkata, India. He is Associate professor in Mechanical Engineering Department of Haldia Institute of Technology, Haldia, India. His research domain is Tribology of Electroless Nickel Coatings. He has so far to his credit 19 publications in international journals and conference proceedings and also co author of four book chapters.

* * *

Sevim Akyuz is a professor of physics, currently working at the Physics Department of Istanbul Kultur University (Istanbul-Turkey). Her research interest is on experimental and theoretical molecular spectroscopy and mainly focused on the study of a relationship between the structure and spectral properties of macromolecules using combined theoretical and experimental methods. She has 182 scientific papers as SCI record and total number of citations according to Web of Knowledge is 1335. Her H-index according to Web of Knowledge is 22.

Partha Pratim Bag, is presently working as Assistant Professor in SRM University Sikkim, Department of Chemistry, Sikkim, India. He obtained his Ph.D. from IISER Kolkata in 2013 under the supervision of Professor C. Malla Reddy. Successively, he completed two Postdoctoral Studies from China (2014-2016 and 2016-2018). He was Assistant Professor in Department of basic Science and Humanities (Chemistry), Dumka Engineering College, Jharkhand from 2018-2020. During his 10 years research tenure, he published several top lined journals including research articles and book chapters. He is also the Editorial Board member of international journal like American Journal of Applied Chemistry, International Journal of Creative Research Thoughts, Journal of Emerging Technologies and Innovative Research etc. He is life member and Fellow of Indian Chemical Society, life member of Life member of Material Research Society of India (MRSI). He won several national/international awards; International Young Scientist (2015), Best Young Scientist (2019), Best Professor Award (2020) etc. His research interest is Crystal Engineering, Solid State Chemistry and Supramolecular energetic materials.

Nuray Bekoz Ullen is currently working in the Department of Metallurgical and Materials Engineering, Istanbul University-Cerrahpaşa (IUC), Turkey. She obtained her BSc, MSc and PhD degree at Department of Metallurgical and Materials Engineering from Istanbul University, Turkey in 2003, 2006 and 2011, respectively. Her main research interest includes powder metallurgy, metal foams, coating technology, surface treatments, material characterizations, machinability and welding. In recent years, she has focused on improving the surface properties of dense porous materials. She has collaborated actively with researchers in several other disciplines of mechanical, chemistry and physics. She has taught on engineering materials and deformation mechanisms, highly porous material production techniques and characterization, failure of materials. She had role as a consultant in master theses in the fields of material characterization. She has published several research articles on material production, characterization and processing.

Bárbara Bermúdez-Reyes: Research Professor at the Center for Research and Innovation in Aeronautical Engineering of the Faculty of Mechanical and Electrical Engineering of the Autonomous University of Nuevo Leon. PhD in Sciences of Metallurgy and Materials Science from the Michoacana University of San Nicolas de Hidalgo. She is General Secretary of the University Space Engineering Consortium-Mexico Chapter for the period 2018-2010. She is member of Space Science and Technology Network. She has collaborated in projects with Mexican Air Force and has evaluated projects of Mexican Naval Research Institute. She is part of organizing committee of National Cansat Contest from 2015. Her research lines are: Design and processing of aerospace ceramic-ceramic and metal-ceramic composites materials and protective coatings for satellite structures.

Sefa Celik is an Associate Professor at the Physics Department of Istanbul University (Turkey). His research is focused on the elucidation of the structure of biomacromolecules using combined experimental (molecular spectroscopy) and theoretical methods (ab initio, molecular dynamics), particularly on the molecular docking studies for shedding light on new drug design. He has 23 scientific papers as SCI record.

Mintu Maan Dutta was born in Nazira, Sibsagar, Assam, India in May 10, 1992. He obtained his B.Sc. degree from Cotton College in 2012 and M.Sc (Physical Chemistry) degree in 2014, from Gauhati University, Assam, India. He obtained his Ph.D. degree in April 2020 from Department of Chemistry, Gauhati University. His PhD. thesis entitled "Surface functionalized magnetic nanoparticles as catalyst for organic synthesis". His current research interest includes heterogeneous catalysis, organic transformations. He has six publications in reputed journals that includes Dalton Transactions, Catalysis Communications, Chemistry Select, etc.

Astrid L. Giraldo is Physical Engineer and Master in Science-Physics graduated from the Universidad Nacional de Colombia, campus Manizales. Doctor in Science- Materials science graduated from Cinvestav Campus Querétaro. Currently, she works as CONACYT commissioned to the Cinvestav unit Querétaro. Co-founder of the Mexican Laboratory of Thermal Spray - CENAPROT. Her research area is focused on multifunctional materials; especially in bioceramics: synthesis of raw material such hydroxyapatite from natural sources and bioglasses, hydroxyapatite-based composite materials to coating manufacture using different thermal spray technologies and their characterization. Additionally, she is working on the microencapsulation processes development for controlled drug release.

Mridusmita Goswami is working as a guest faculty in the Department of Chemistry, North Gauhati College, Guwahati, Assam, India. She had obtained her B. Sc. Degree in Chemistry (2009) and M. Sc.

Degree in Physical Chemistry (2011) from Gauhati University. She was awarded the PhD degree in the year 2018 from Gauhati University. Her research area includes adsorption, catalysis, nanotechnology.

Onur Güler was born in 1989 in Nazilli/Aydin, Turkey. He graduated from his primary, secondary and high school education in Nazilli. He started to have undergraduate education in the department of Metallurgical and Materials Engineering in the Karadeniz Technical University in 2008, Trabzon, Turkey. He graduated from the department in 2012. After he started to get master's degree, he worked on the production of metal matrix composites by powder metallurgy for his master thesis in the Karadeniz Technical University and graduated from master degree with his thesis titled 'The Production of ZA27 /Graphite/Al2O3 Hybrid Nanocomposites and Investigation of Their Wear Properties' in 2015. He started to study on PhD studies in same department in 2016, Karadeniz Technical University, Trabzon, Turkey. He continues his PhD studies on the production of electroless deposited metallic powders and bulk materials produced from layered metallic and ceramic powders produced by electroless deposition. Besides, he works as research assistant in the department of Metallurgical and Materials Engineering, Karadeniz Technical University, Trabzon, Turkey.

John Henao was born in Colombia. He studied Materials Engineering at the University of Antioquia in Medellín, Colombia. Afterwards, he obtained his master degree at the University of Limoges, France. He completed his Ph.D. in Engineering and Advanced Technologies at the Universitat de Barcelona, Spain; where he also worked at the Centre of Thermal Spray. During his PhD, he traveled to the faculty of engineering Enzo Ferrari at Modena, Italy, to work in the simulation of cold sprayed particles. With his PhD completed, he got a posdoctoral position at the National University of Mexico (UNAM). He now is research scientist at the National Council of Science and Technology of Mexico (CONACYT) and works at CIATEQ A.C. in the field of Thermal Spray and Biomaterials. His research work concerns mainly different aspects of thermal spraying and application of various materials, in particular, for biomedical applications, with current research that is focused on HVOF, Flame Spray, Plasma Spray and Cold Spray.

Magdaline Muigai is a graduate assistant and MSc student at the Dedan Kimathi University of Technology, Nyeri. She has a BSc in mechanical engineering from the same institution.

Fredrick Mwema is a lecturer and postdoctoral fellow at the Dedan Kimathi University of Technology and the University of Johannesburg at the departments of mechanical engineering respectively. Currently, Dr FM Mwema is the Head of Department (HOD) of the mechanical engineering department at the Dedan Kimathi University of Technology, Nyeri, Kenya. Additionally, he is the director at the Centre for nanomaterials and nanoscience research (CNSR) at Dedan Kimathi University of Technology, Kenya. He has BSc and MSc degrees in mechanical engineering from the Jomo Kenyatta University of Agriculture (2011 and 2015 respectively) and a PhD degree from the University of Johannesburg (2019). He has interests in materials and characterization, advanced manufacturing, finite element modeling, surface engineering, coating, and thin-film technologies. He has published more than 60 articles in journals, book chapters, and conference proceedings. He has written two book monographs on fused deposition modelling and finite element simulation of severe plastic deformation processes and is currently working on two more books. He has over 7 years of teaching and training of undergraduate mechanical engineering students and one year of masters teaching. Currently, Dr Mwema mentors several young students and researchers with postgraduate supervisions of 4 masters students and 2 PhD candidates.

Japheth Obiko is currently enrolled as a PhD student in the School of Chemical and Metallurgical Engineering, University of the Witwatersrand-Johannesburg, South Africa since 2017. He received an MSc in Mechanical Engineering from Jomo Kenyatta University of Agriculture and Technology in 2015 and BSc (Mechanical Engineering) from Jomo Kenyatta University of Agriculture and Technology, Kenya in 2012. J. O is a registered graduate engineer with the Engineers Board of Kenya. Currently, J. O. is Tutorial Fellow in the Department of Mining, Materials and Petroleum Engineering at Jomo Kenyatta University of Agriculture and Technology. J. O teaches courses in solid and structural mechanics of materials, vibration mechanics, manufacturing process and failure analysis.

Aysen E. Ozel is a Professor at the Physics Department of Istanbul University (Turkey). Her research is focused on the molecular spectroscopy and on the molecular modeling. She has 34 scientific papers as SCI record.

Carlos Poblano-Salas is a metallurgical engineer graduated from Universidad Nacional Autonoma de Mexico where he also obtained a MSc degree in metallurgy. He obtained a PhD degree in materials science at the University of Sheffield, UK. Currently, he is director of the Advanced Materials Department at CIATEQ A.C. His research lines include the fabrication of different metallic, ceramic,and composite materials by additive manufacturing and thermal spraying; he also works in powder metallurgy and corrosion science projects.

Gaurav Prashar is currently working as Assistant Professor and HOD in the Mechanical Engineering Department at Rayat-Bahra institute of Engineering and nano-technology, Hoshiarpur- India. He has obtained his M.Tech. (Production Engineering) from Guru Nanak Dev Engineering College, Ludhiana-India. He had total 12 years of experience in teaching. His research areas are thermal spray coatings for wear, and corrosion. He has authored 5 research publications in various International Journals and Conferences of repute.The research outcomes have been published in reputed journals (SCI/Scopus) such as Engineering failure analysis, Materials today proceedings, CRC Press (Taylor and Francis), and IGI global. Moreover, he is a dedicated reviewer of reputed journal such as IGI Global journal.

Pritee Purohit is currently working as Assistant Professor in the Department of Mechanical Engineering at Army Institute of Technology Pune. She has completed her PhD in 2017 from Government College of Engineering Pune, affiliated to Savitribai Phule Pune University Pune. She has completed her UG in Mechanical engineering and PG in (Mechanical - Heat Power) in 2004 and 2006 respectively. She is having 15 years of teaching experience. Her fields of research are heat transfer, fluid dynamics, high temperature corrosion and thermal barrier coatings. She has 25 plus national and International conference and Journal Publications. She is having one SCI publication out of 7 International Journal Publications. She has filed 2 patents in the field of Mechanical Engineering.

Sudesna Roy has over 10 years of post PhD experience in research and teaching. During her tenure in KIIT she has taught undergraduate as well and graduate classes in manufacturing processes, materials science, fatigue, fracture and materials engineering. She was the recipient of the DST women scientist award in 2015 and has also received an external grant from DST, SERB in 2017. She has published several papers in reputed journals and presented in many national and international conferences. Here research interest is currently in the area of coatings, i.e thin film deposition, cold spraying and thermal spraying. She has worked extensively in the area of magnetron sputtering, transparent oxide ceramics, optical materials, rare earth oxides, thermal spraying and alloy solidification modeling. She has completely her doctoral degree from Northeastern University, Boston, Massachusetts and post-doctoral experience at Boston University, Boston, Massachusetts, USA for a tenure of 3 yrs. Her educational background comprises of a B.Tech degree in Ceramic Engineering from N.I.T. Rourkela and M.Tech degree in Metallurgy from I.I.T. Kharagpur.

Gourisankar Roymahapatra, FICS, FIC, is working as Associate Professor of Chemnistry at School of Applied Science and Humanities in Haldia Institute of Technology, West Bengal, India. He completed Master(s) in chemistry on both Physical Chemistry (2003) and Organic Chemistry (2005) from DAV PG College, CSJM University, Kanpur. He was awarded Ph.D. (2014) from Jadavpur University, Kolkata. His early carrier strated as Senior Chemist at Mitsubishi Chemicals (2003-11). Then he joined as Lecturer in Chemistry at Global Institute of Science and Technology (2011-15), Assistant Professor of Chemistry at Haldia Institute Technology (2015-16). He got several prestigious awards, like 'Bharat Gaurav Award' (2018) from India International Friendship Society, New Delhi, 'Distinguished Young Scientist Award' (2014) from World Science Congress, Kolkata, India, 'Rasayan Ratna' (2017) from SONAR TORI, Purba Medinipur, West Bengal, India, 'Per-Excellent' award in NCQC-2007 and 'Gold' award in CCQC-2010 from QCFI, India. He is the Fellow of the Indian Chemical Society and Institution of Chemists (India). He holds the respected Vice-president position of Haldia Vigyan Parishad, a premier scince society of Bengal. His research articles published on several top lined international journals including book chapters. He is also the Editorial Board member and Reviewer of several international journals. His current research interests are - designing of new molecules; synthesis and theoretical studies (Hydrogen Storage Materials, Super Alkali, Catalysts, Antibiotics, Anti Carcinogecic and Anti-Viral drug design).

Bhavana Shrigadi has completed her UG and PG from College of Engineering Pune in Metallurgy. Currently working in Cummins India Ltd. Her area of research is Corrosion and coating.

Oscar Sotelo-Mazon is currently professor at the Universidad Autonoma del Estado de Morelos (FCQeI), Mexico, and is member of the Mexican system of researchers (SNI) level 1. He obtained his M.SC. and Ph.D. in Materials Science from the Universidad Autonoma del Estado de Morelos in 2011 and 2015, respectively. He carried out his first postdoctoral studies at the Universidad Nacional Autonoma de Mexico unit Instituto de Ciencias Fisicas in 2018. After that, He carried out his second postdoctoral studies at the CIATEQ A.C. unit Queretaro in 2019. He has been working on the field of corrosion at high temperature, corrosion inhibitors, thermal spray, and biomaterials.

Varol T. was born in Trabzon (1980). He completed his primary, secondary and high school education in Trabzon. Temel Varol started the Mechanical Engineering education in the Atatürk University, Erzurum. He worked on the production of metal matrix composites by powder metallurgy during master student in the Karadeniz Technical University and presented master thesis (The Production of AA2024 Matrix-B4C Particle Reinforced Composites by Powder Metallurgy Method and Investigation of Their Properties) at 2012. His first article's name is "The effect of mechanical alloying on Al2O3distribution and properties of Al2O3particle reinforced Al-MMCs". He worked on the fabrication and characterization of nanoparticle reinforced electrical contact materials during PhD student and completed PhD thesis in the Karadeniz Technical University (January 2016). The current studies of Varol are related to powder metallurgy, mechanical alloying, composites/nanocomposites, functionally graded materials and biomaterials.

Shashikant Vagge has completed his PhD from IIT Bombay in 2008. He has completed his UG and PG from College of Engineering Pune in Metallurgy. Teaching subject at UG & PG/ PhD level Corrosion & surface protection, Corrosion Engineering, Surface Engineering, Surface Processing of Materials, High temperature corrosion. Gideed M. Tech. & B. Tech. Dissertations, projects are generally related with different coatings and study of their corrosion behavior.

Fabio Vargas, born in 1973 in Bucaramanga-Colombia, is a Metallurgical Engineer and Master in Metallurgical Engineering from the Industrial University of Santander, Doctor in Materials Engineering from the University of Antioquia and Doctor in Science of Ceramic Processes and Surface Treatment from the Limoges University. During his first 6 years of professional experience he carried out research projects for the oil industry in Colombia, in 2003 he started his career as a professor at the University of Antioquia, where in 2019 he reached the rank of full professor. He coauthored more than 30 papers and 2 book chapter about thermal spraying coatings, as well as, a patent and two industrial secrets, granted in the name of the University of Antioquia and ECOPETROL-Colombian oil company.

Rafael Vargas-Bernal received a Bachelor degree in Communications and Electronics Engineering from the University of Guanajuato in 1995, and the degrees of Master of Science and Doctorate in Science with Specialty in Electronics from the National Institute of Astrophysics, Optics and Electronics (INAOE) in 1997 and 2000, respectively. Since January 2002, he has been a professor-researcher at the Higher Technological Institute of Irapuato (ITESI), Mexico and, particularly since 2006, he has worked in the Department of Materials Engineering where he has established himself as a researcher. He has authored 15 articles in journals, 26 chapters in books and about 100 conference articles. He is a member of the National System of Researchers (SNI-Mexico). He regularly serves as a reviewer of scientific articles in RSC Advances and Royal Society Open Science as well as Standards in Semiconductor Equipment and Materials International (SEMI). His research interests include two-dimensional materials, nanomaterials, space materials, composite materials, gas sensors and biosensors.

Hitesh Vasudev is currently working as Associate Professor at Lovely Professional University, Phagwara-India. He has received Ph.D. in Mechanical Engineering from Guru Nanak Dev Engineering College, Ludhiana-India. His area of research is thermal spray coatings especially for the development of new materials used for high temperature erosion and oxidation resistance, and microwave processing of materials. He has contributed extensively in thermal spray coatings in repute journals which include Surface coatings and Technology and Engineering Failure analysis in various publications such as Elsevier, Taylor&Francis, Springer, IGI Global and InTech Open. Moreover, he is a dedicated reviewer of reputed journals such as Ceramics International, Journal of Material Engineering Performance, Material Research Express, Engineering Research Express and IGI Global journals etc. He has authored more than 25 International publications in various International Journals and conferences. He has published 10 book chapters in various publications. He has a teaching experience of more than eight years. He has also published one patent in thermal spraying. He has also got "Research Excellence" award in 2019 from Lovely Professional University-Phagwara-India. He has an experience of coordinating conferences.

Index

www.ingramcontent.com/pod-product-compliance
Lightning Source LLC
LaVergne TN
LVHW060416300926
849162LV00006B/1444